Bioinspired Structures and Design

Master simple to advanced biomaterials and structures with this essential text. Featuring topics ranging from bionanoengineered materials to bioinspired structures for spacecraft and bioinspired robots, and covering issues such as motility, sensing, control, and morphology, this highly illustrated text walks the reader through key scientific and practical engineering principles, discussing properties, applications, and design. Presenting case studies for the design of materials and structures at the nano, micro, meso, and macro scales, and written by some of the leading experts on the subject, this is the ideal introduction to this emerging field for students in engineering and science as well as researchers.

Wole Soboyejo currently serves as the Senior Vice President and Provost at Worcester Polytechnic Institute, Massachusetts.

Leo Daniel is a Professor of Aerospace Engineering and Provost of the College of Engineering and Technology at Kwara State University in Nigeria.

Bioinspired Structures and Design

WOLE SOBOYEJO
Worcester Polytechnic Institute

LEO DANIEL
Kwara State University

CAMBRIDGE
UNIVERSITY PRESS

CAMBRIDGE
UNIVERSITY PRESS

University Printing House, Cambridge CB2 8BS, United Kingdom

One Liberty Plaza, 20th Floor, New York, NY 10006, USA

477 Williamstown Road, Port Melbourne, VIC 3207, Australia

314–321, 3rd Floor, Plot 3, Splendor Forum, Jasola District Centre, New Delhi – 110025, India

79 Anson Road, #06–04/06, Singapore 079906

Cambridge University Press is part of the University of Cambridge.

It furthers the University's mission by disseminating knowledge in the pursuit of
education, learning, and research at the highest international levels of excellence.

www.cambridge.org
Information on this title: www.cambridge.org/9781107015586
DOI: 10.1017/9781139058995

First published 2020

Printed in the United Kingdom by TJ International Ltd, Padstow Cornwall

A catalogue record for this publication is available from the British Library.

Library of Congress Cataloging-in-Publication Data
Names: Soboyejo, W. O., editor. | Daniel, Leo, editor.
Title: Bioinspired structures and design / [edited by] Wole Soboyejo, Worcester Polytechnic Institute, Massachusetts, Leo Daniel, Kwara State University, Nigeria.
Description: Cambridge, United Kingdom ; New York, NY, USA : Cambridge University Press, 2020. | Includes bibliographical references and index.
Identifiers: LCCN 2019053726 (print) | LCCN 2019053727 (ebook) | ISBN 9781107015586 (hardback) | ISBN 9781139058995 (epub)
Subjects: LCSH: Bionics. | Biomimicry. | Biomimetic materials.
Classification: LCC TA164.2 .B56 2020 (print) | LCC TA164.2 (ebook) | DDC 660.6–dc23
LC record available at https://lccn.loc.gov/2019053726
LC ebook record available at https://lccn.loc.gov/2019053727

ISBN 978-1-107-01558-6 Hardback

Contents

Contributors

Iwnetim I. Abate
Department of Material Science and Engineering, Stanford University, Stanford, CA, USA

Nicolas Anuku
Princeton Institute for Science and Technology of Materials (PRISM), Princeton University, Princeton, NJ, USA
Department of Chemistry and Chemical Technology, Bronx Community College, Bronx, NY, USA

Sina Askarinejad
Department of Engineering, University of Cambridge, UK

Bo Cheng
Department of Mechanical Engineering, The Pennsylvania State University, University Park, PA, USA

Steven van Dessel
Department of Civil and Environmental Engineering – Architectural Engineering Program, Worcester Polytechnic Institute, Worcester, MA, USA

Jing Du
Department of Mechanical Engineering, Pennsylvania State University, State College, PA, USA

Frank E. Fish
Department of Biology, West Chester University, West Chester, PA, USA

Daniel Floryan
Department of Mechanical and Aerospace Engineering, Princeton University, Princeton, NJ, USA

Sergio Granados-Focil
Department of Chemistry, Clark University, Worcester, MA, USA

Yasamin Aliashrafi Jodat
NeuroBionics and Neuro-Electric Medicine Laboratory, Department of Mechanical Engineering, Stevens Institute of Technology, Hoboken, NJ, USA

Kiavash Kiaee
NeuroBionics and Neuro-Electric Medicine Laboratory, Department of Mechanical Engineering, Stevens Institute of Technology, Hoboken, NJ, USA

Karen A. Malatesta
Princeton Institute for Science and Technology of Materials (PRISM), and Department of Mechanical and Aerospace Engineering, Princeton University, Princeton, NJ, USA

Manu Sebastian Mannoor
NeuroBionics and Neuro-Electric Medicine Laboratory, Department of Mechanical Engineering, Stevens Institute of Technology, Hoboken, NJ, USA

Xinrui Niu
Department of Mechanical and Biomedical Engineering, City University of Hong Kong, Kowloon, Hong Kong, China

John David Obayemi
Department of Mechanical Engineering, and Department of Biomedical Engineering, Worcester Polytechnic Institute, Worcester, MA, USA

Olushola S. Odusanya
Biotechnology and Genetic Engineering Advanced Laboratory, Sheda Science and Technology Complex (SHESTCO), Abuja, Nigeria

Nima Rahbar
Department of Civil and Environmental Engineering, and Department of Mechanical Engineering, Worcester Polytechnic Institute, Worcester, MA, USA

Robert O. Ritchie
Materials Sciences Division, Lawrence Berkeley National Laboratory, and Department of Materials Science & Engineering, University of California, Berkeley, CA, USA

Ali A. Salifu
Department of Mechanical Engineering, and Department of Biomedical Engineering, Worcester Polytechnic Institute, Worcester, MA, USA

Wanliang Shan
Department of Mechanical Engineering, University of Nevada, Reno, NV, USA

Yantao Shen
Department of Electrical and Biomedical Engineering, University of Nevada, Reno, NV, USA

Alexander J. Smits
Department of Mechanical and Aerospace Engineering, Princeton University, Princeton, NJ, USA

Wole Soboyejo
Department of Mechanical Engineering, and Department of Biomedical Engineering, Worcester Polytechnic Institute, Worcester, MA, USA
Department of Mechanical and Aerospace Engineering, Princeton University, Princeton, NJ, USA

Birgitt Boschitsch Stogin
Department of Mechanical Engineering, Materials Research Institute, The Pennsylvania State University, University Park, PA, USA

Loza F. Tadesse
Department of Bioengineering, Stanford University, Stanford, CA, USA

Ting Tan
Department of Civil and Environmental Engineering, The University of Vermont, Burlington, VT, USA

Mingjiang Tao
Department of Civil and Environmental Engineering – Civil Engineering Program, Worcester Polytechnic Institute, Worcester, MA, USA

Tyler Van Buren
Department of Mechanical and Aerospace Engineering, Princeton University, Princeton, NJ, USA
Department of Mechanical Engineering, University of Delaware, Newark, DE, USA

Lin Wang
Department of Materials Science and Engineering, Materials Research Institute, The Pennsylvania State University, University Park, PA, USA

Tak-Sing Wong
Department of Mechanical Engineering, Materials Research Institute, Department of Biomedical Engineering, The Pennsylvania State University, University Park, PA, USA

Elizabeth A. Zimmermann
Materials Sciences Division, Lawrence Berkeley National Laboratory, and Department of Materials Science & Engineering, University of California, Berkeley, CA, USA

Preface

Since the pioneering work of Leonardo da Vinci, the subject of bioinspired design has received considerable attention. The efforts to develop new concepts for bioinspired design and biomimetics have also broadened considerably in recent years, with contributions from biologists, chemists, physicists, materials scientists, and engineers from different disciplines. These efforts have resulted in insights for the design of more robust materials, airplanes, robots, marine vehicles, nonstick paints, nanostructures for the targeting and treatment of disease, and biomedical materials and structures.

This book presents a materials and mechanics perspective of bioinspired design. It is divided into three parts, with individual chapters that are written by some of the leading researchers in the field. Part I presents an introduction to bioinspired design and aerospace bioinspired design. This is followed by Part II, in which multifunctional approaches to bioinspired design and biomimetics are presented. Finally, the book concludes with Part III, which draws inspiration from natural phenomena such as the swimming of fish. In each of these sections, a combination of scientific observations and models is used to provide insights for the design of multifunctional materials.

We would like to thank the authors of the chapters (Robert O. Ritchie, Alexander Lex Smits, Ali A. Salifu, John David Obayemi, Elizabeth A. Zimmerman, Sina Askarinejad, Nima Rahbar, Ting Tan, Xinrui Niu, Jing Du, Manu Sebastian Mannoor, Yasamin Aliashrafi Jodat, Kiavash Kiaee, Iwnetim I. Abate, Loza F. Tadesse, Nicolas Anuku, Olushola S. Odusanya, Karen A. Malatesta, Bo Cheng, Tak-Sing Wong, Lin Wang, Birgitt Boschitsch Stogin, Yantao Shen, Wanliang Shan, Frank E. Fish, Steven Van Dessel, Mingjiang Tao, Sergio Granados-Focil, Daniel Floryan, and Tyler Van Buren) for their efforts in preparing and revising the chapters. We are also grateful to Cambridge University Press (Lisa Bonvissuto, Julia Ford, and Steven Elliot) for their patience during the years that it took to develop the text. This book would not have been possible without the review support of Alex Pottinger, Donna Zuidema, and Kim Hollan at the Worcester Polytechnic Institute. We thank them for their careful work in coordinating the communications with the different authors and the publishers. Finally, we would like to thank our families for their support and tolerance of the many hours that we spent on the preparation of this text.

We hope that the book will serve as a useful text to researchers, as well as those engaged in senior-level undergraduate courses or graduate courses in science and engineering departments. We also hope that the chapters in the book will be useful to researchers and engineers/designers who are interested in emerging new ideas in the field of bioinspired design.

Part I

Materials

1 Bioinspired and Biomimetic Design of Multilayered and Multiscale Structures

Ali A. Salifu, Manu Sebastian Mannoor, and Wole Soboyejo*

1.1 Introduction

In recent years, there has been significant interest in the fields of bioinspired design and biomimetics [1]. Bioinspired design involves the use of scientific and engineering principles in the design of engineering components and structures that are inspired by biological systems. In contrast, biomimetics involves the design of engineering components and structures that copy biological systems. Hence, airfoils and aircraft wings are examples of bioinspired design that are inspired by bird flight but guided by the principles of lift and drag from aerodynamics. In contrast, the early idea of an airplane with flapping wings is an example of biomimetics, which is based on the simple idea of copying nature without thinking carefully about the underlying scientific principles that enable such natural systems to function in the way that they do.

Although archaeological digs reveal numerous examples of relics that were clearly inspired by nature in the remains of ancient civilizations [1,2], the work of Leonardo da Vinci [3] is perhaps one of the earliest recorded studies of flight that was inspired by careful observations of birds. Leonardo da Vinci produced a codex on flight in 1505–1506, which was entitled "Codex on the Flight of Birds" [3]. He was particularly fascinated by the possibility of human mechanical flight, inspired by his observations of birds in flight [3]. Leonardo produced about 500 sketches of flying machines [3]. He also made some observations and identified some concepts that would find their place in the development of airplanes in the twentieth century when the Wright brothers [4] and Alberto Santos-Dumont [5] developed the first aircraft.

More recently, in the twentieth and twenty-first centuries, careful experimental studies of nature have provided the basis for the design and development of future components and systems. These include the pioneering work of Currey [6,7] who initiated some of the early studies of shells that enabled subsequent efforts on bioinspired design and biomimetics in the twentieth and twenty-first centuries. More recently, Chen and coworkers [8], Wegst et al. [9], Beese et al. [10], Rahbar et al. [11], Martini et al. [12], Huang et al. [13], Li et al. [14], Wang et al. [15], Dooley et al. [16], Niu et al. [17],

* The authors are grateful to Worcester Polytechnic Institute for financial support of this review. Appreciation is also extended to the National Institutes of Health (Grant No. P01DE10956) and the National Science Foundation (Grant No. 0231418) for financial support. Finally, the authors would like to thank Cambridge University Press for the encouragement and support of this text on bioinspired design.

Wang et al. [18], Farmer et al. [19], Tan et al. [20], and Du et al. [21] have developed bioinspired design and biomimetic concepts for the design of robust multifunctional structures. These include models for the design of layered structures that could enable the development of robust materials with remarkable mechanical properties.

Similarly, recent advances in our understanding of fluid mechanics and controls have enabled the development of naval structures that are inspired by careful studies of fish locomotion [22], the hydrodynamics of the propulsion of jellyfish [23] and marine mammals [24], experimental studies of fish locomotion by Moored et al. [22] and Melli et al. [25], and models of insect locomotion by Ayali et al. [26]. These are all areas in which recent work has provided new insights for bioinspired design. They are also areas that could find future applications in multifunctional robotics.

Beyond these potential applications, significant progress has been made in the development of bioinspired and biomimetic structures for applications in medicine [27], dentistry [28], batteries [29,30], bionic structures [31,32], micro- and nanostructures [33], and molecular devices [34]. These include the design of bioinspired dental multilayers that are inspired by the structure of natural teeth (Figures 1.1–1.3) with improved crack-growth resistance by virtue of their functionally graded architectures [11,13,17]; the development of soft robotic hands (Figure 1.4) [35]; the development of bionic leaves for solar energy harvesting (Figure 1.5) [31]; and the development of metal-free flow batteries that utilize quinones to move electrons between electrodes, inspired by the electron transfer mediating roles of quinone molecules during metabolism in plants and animals (Figure 1.6) [30].

The underlying concepts associated with these examples will be examined in this chapter along with their implications for bioinspired design and biomimetics. For simplicity, the chapter is divided into four sections. Following the introduction (Section 1.1), the bioinspired design of layered structures will be examined in

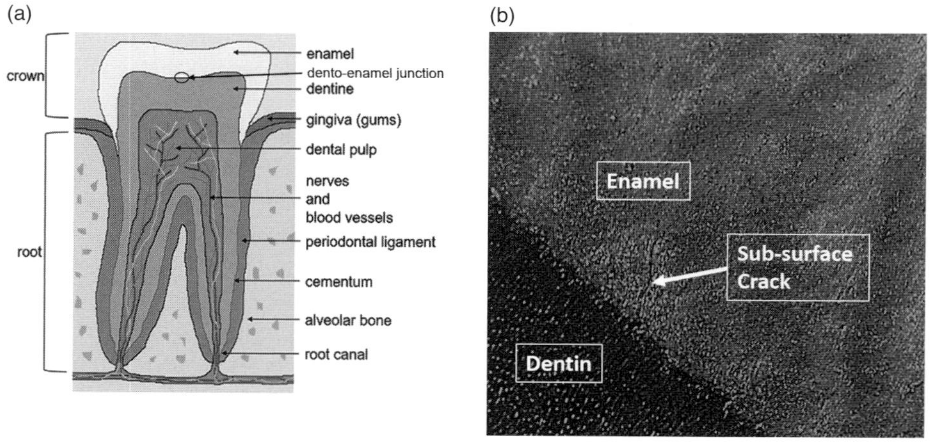

Figure 1.1 Structure of human tooth: (a) schematic and (b) enamel damage (shown by the white arrow) at the dento-enamel junction. (Courtesy of Ivory Kilpatrick, Princeton University.)

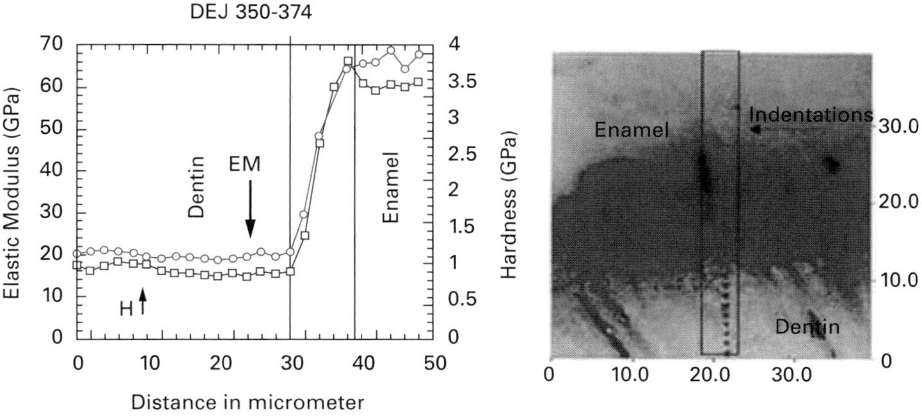

Figure 1.2 Variations in the Young's moduli across the dento-enamel junction. (Reprinted from Marshall et al. [36] with permission from John Wiley and Sons, Inc.)

Section 1.2. This will be followed by a review of the bioinspired design of multifunctional materials in Section 1.3. These include tissue-engineered materials, bionic materials, and materials for energy harvesting. We then conclude with a summary of insights for the bioinspired design and the biomimetic design of materials in Section 1.4.

1.2 Bioinspired Design of Layered Structures

Several materials in nature have layered hierarchical structures that can provide insights into the management of stresses in layered composite materials. These include bamboo [20,37], the dento-enamel junction [13,17,20,21], and turtle/tortoise shells [38,39], in which layered structures have evolved to reduce or manage the stresses that might lead to fracture during exposure to loading under different conditions. Hence, these materials will be explored in this section as potential sources of bioinspiration for the design of robust layered structures.

1.2.1 Bamboo

The layered hierarchical structure of bamboo (Figure 1.7) has evolved due to the wind loading of its cantilevered structure [20,37]. This induces the lowest stress at the tip of the bamboo, intermediate stress in the middle, and the highest stress at the bottom [20]. Consequently, the volume fraction of cellulose and hemicellulose fibers in the lignin matrix has evolved to be highest in the regions with the highest stress, intermediate in the regions with intermediate stress, and lowest in the regions subjected to the lowest stress (Figure 1.7).

Hence, the functionally graded structure of bamboo provides us with insights for the design of composites in which variations in the fiber volume fractions can be used to manage or support stresses. One can, therefore, envisage composites in which the fiber volume fractions are varied to manage or reduce the effective stresses under a range of

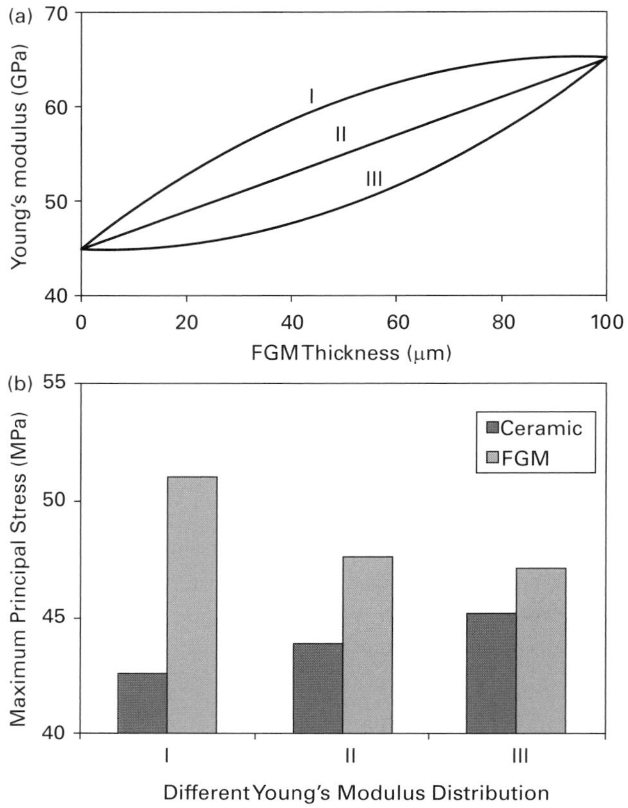

Figure 1.3 The effects of different distributions of Young's moduli of the functionally graded material (FGM) layer on the top ceramic layer in a bioinspired ceramic/FGM/composite dental multilayer structure: (a) the distribution of Young's moduli in the FGM layer and (b) the maximum tensile stresses on the top ceramic and FGM layers corresponding to the different Young's moduli distributions. (Reprinted from Huang et al. [13], copyright 2007, with permission from Elsevier.)

wide loading conditions. Askarinejad et al. [37] and Tan et al. [20] have also shown that the hierarchical and layered structure of bamboo can promote significant improvements in the resistance-curve behavior of Moso culm bamboo in the crack-arrestor orientation that gives rise to significant toughening by crack bridging (Figure 1.8).

1.2.2 Dental Multilayers

In the case of the functionally graded structure of dental multilayers (Figure 1.2), analytical and finite element models of contact-induced deformation [13] have revealed that the approximate linear gradients in Young's moduli across the dento-enamel junction (Figure 1.2) result in lower principal stresses (Figure 1.9) in the top ceramic layers. These serve as a model for natural teeth (Figures 1.1 and 1.2). However, in the

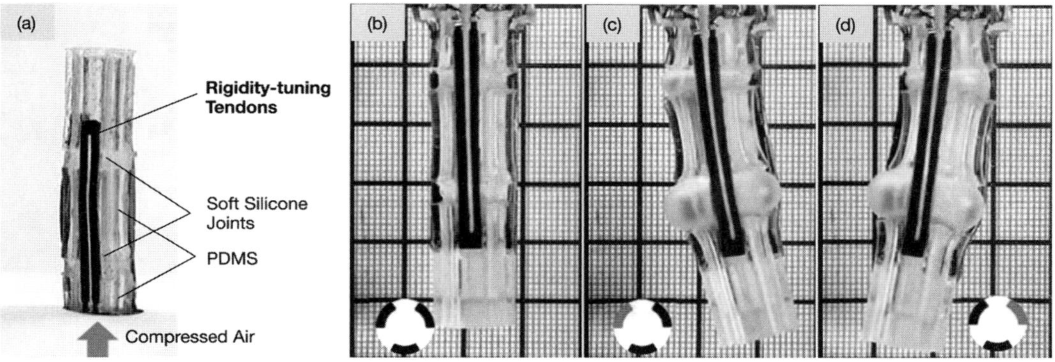

Figure 1.4 Bioinspired soft robotic gloves. (Reprinted from Shan et al. [35] © IOP Publishing. Reproduced with permission. All rights reserved.)

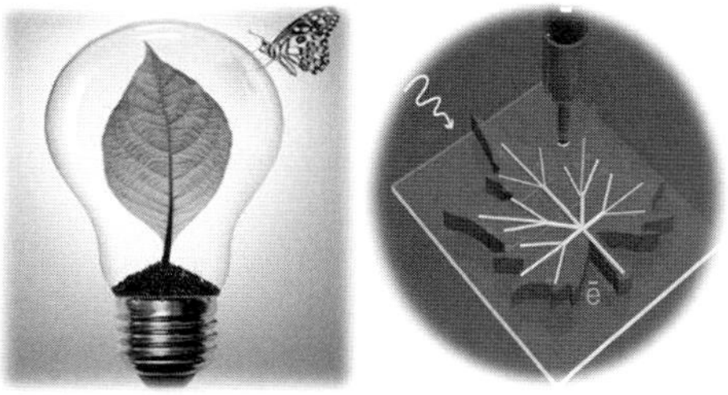

Figure 1.5 Bionic leaves for solar energy harvesting. (Adapted from Mannoor [32].)

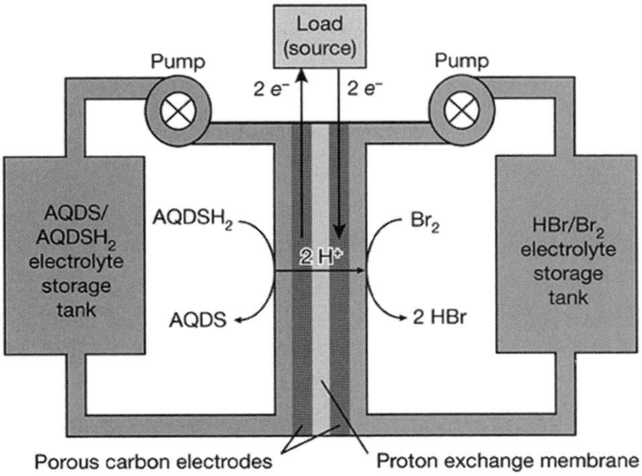

Figure 1.6 Schematic of a bioinspired flow battery utilizing 9,10-anthraquinone-2,7-disulphonic acid (AQDS), a rhubarb-like quinone compound, as an electrode. (Reprinted with permission from Springer Nature from Huskinson et al. [30].)

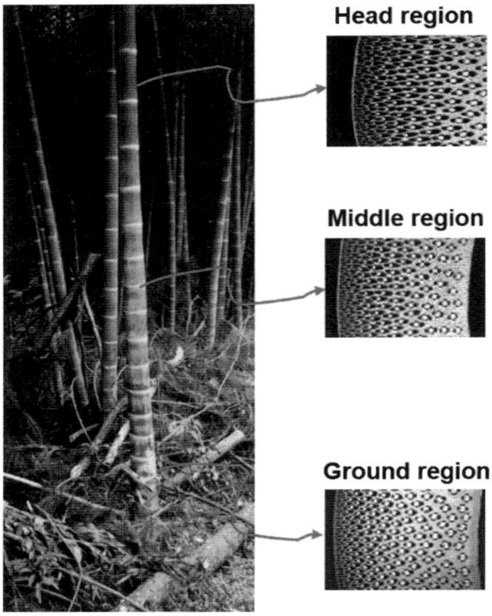

Figure 1.7 Layered functionally graded structure of Moso culm bamboo. (Adapted from Ghavami et al. [40].)

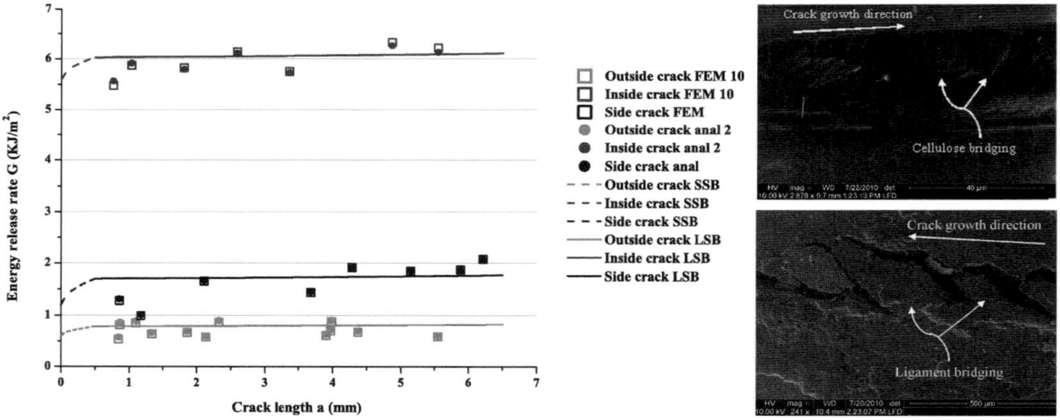

Figure 1.8 Resistance-curve behavior of bamboo along different crack growth planes. (Reprinted from Tan et al. [20], copyright 2011, with permission from Elsevier.)

case of typical ungraded interfaces in dental ceramic crowns, the sharp changes in Young's moduli (from one layer to the other) result in high-stress concentrations that give rise to high principal stresses in the top ceramic layers (Figure 1.9) [13]. These stresses are much reduced when almost linear or sigmoidal functionally graded nanocomposites are used to mimic the variations in Young's moduli associated with natural teeth (Figure 1.9).

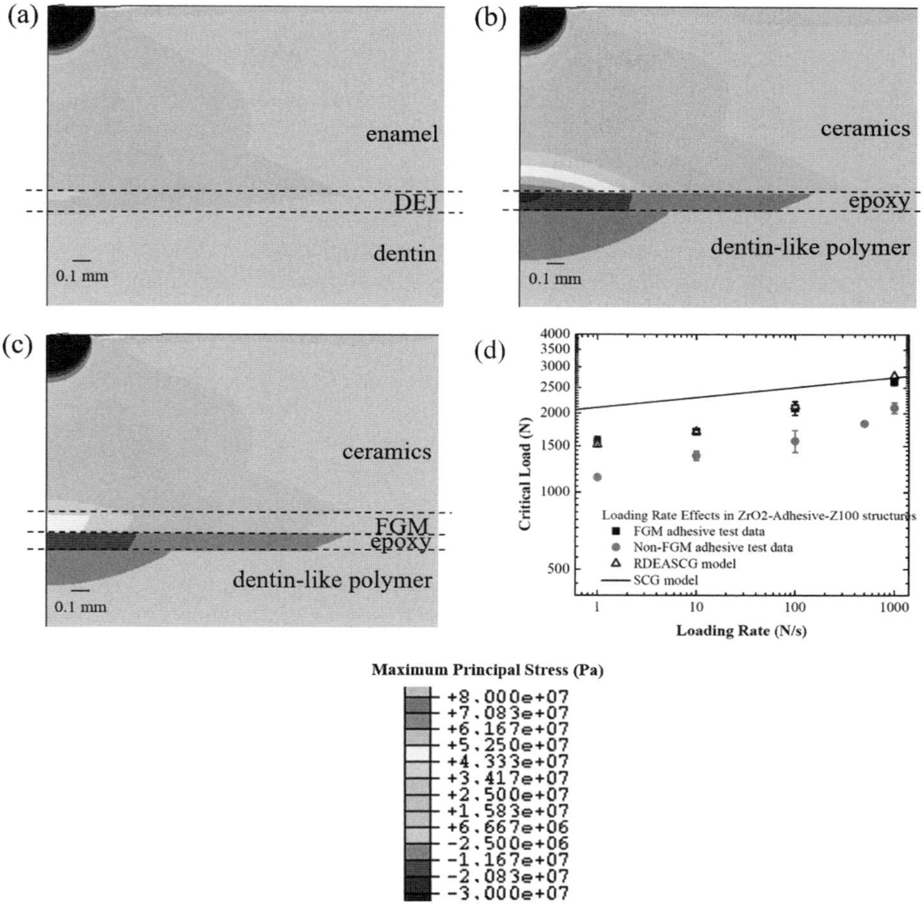

Figure 1.9 Principal stresses in finite element models of: (a) natural tooth, (b) typical crown multilayer, (c) synthetic functionally graded crown structure inspired by the functionally graded structure of natural teeth, and (d) the effect of loading rate on the critical loads of zirconia/FGM/substrate and zrconia/non-FGM/substrate dental multilayers. (Parts (a), (b), and (c) reprinted from Huang et al. [13], copyright 2007, with permission from Elsevier.)

The critical loads for cracking l_R, and the time to rupture t_R, have also been estimated from mechanism-based models. The models incorporate the loading-rate dependent mechanical properties of the individual layers into slow crack-growth fracture mechanics models of cracking in the top ceramic layers. This gives

$$D = \frac{K_{Ic}^{N}\left({c_o}^{1-\frac{N}{2}} - {c_f}^{1-\frac{N}{2}}\right)}{\left(\frac{N}{2}-1\right)\nu_o \beta^N}, \tag{1.1}$$

where

$$\int_0^{t_R} \sigma(t)dt = D \tag{1.2}$$

and D is a parameter dependent on material geometry and properties, and independent of load and time; K_{I_C} is the fracture toughness; c_o is the initial radial crack size in the ceramic subsurface; c_f is the final radial crack size; N is the crack velocity exponent; v_o is the crack velocity parameter; β is a crack geometry coefficient; t is the time, and $\sigma(t)$ is a time-dependent expression of the stress state at the center of the subsurface of the top ceramic layer.

For monotonic loading at a loading rate of $\dot{P}$, with a predicted time to rupture of t_R, this gives a predicted critical load of

$$P_C = \dot{P} t_R, \tag{1.3}$$

where P_C is the critical load.

Plots of the loading-rate dependence of the critical loads are presented in Figure 1.9d. These show that the trends in the loading-rate dependence of P_C are only consistent with the experimental data when the layer-dependent mechanical properties are incorporated into the models. However, in the absence of this, the predicted pop-in loads are much less than the experimental measurements on the bioinspired functionally graded multilayers. Hence, the bioinspired dental multilayers increase the predicted and measured values of the critical loads that are related to subcritical cracking under lower top layer stresses associated with bioinspired functionally graded structures.

1.2.3 Turtle/Tortoise Shells

In this section, it is interesting to explore the layered structure of the turtle/tortoise shell (Figure 1.10) and how this contributes to the management of contact stresses in shell structures. This has been studied by Damiens et al. [38] and Owoseni et al. [39]. The studies of Owoseni et al. [39] on *Kinixys erosa* tortoise reveal a layered structure that consists of an outer layer of cortical bone, a middle layer of trabecular bone, and an inner layer of cortical bone (Figure 1.10).

Because such a structure can absorb energy due to external contact of the turtle/tortoise shell (Figure 1.10), it can reduce the potential effects of contact stresses that can occur due to contact with potential predators. The layered structure of the shell is, therefore, a likely outcome of evolution to promote the survival of turtles/tortoises. It also has implications for the design of protective helmets for soldiers and on contact sports in which bioinspired porous structures can be designed for protection from impact loads (Figure 1.10b–d).

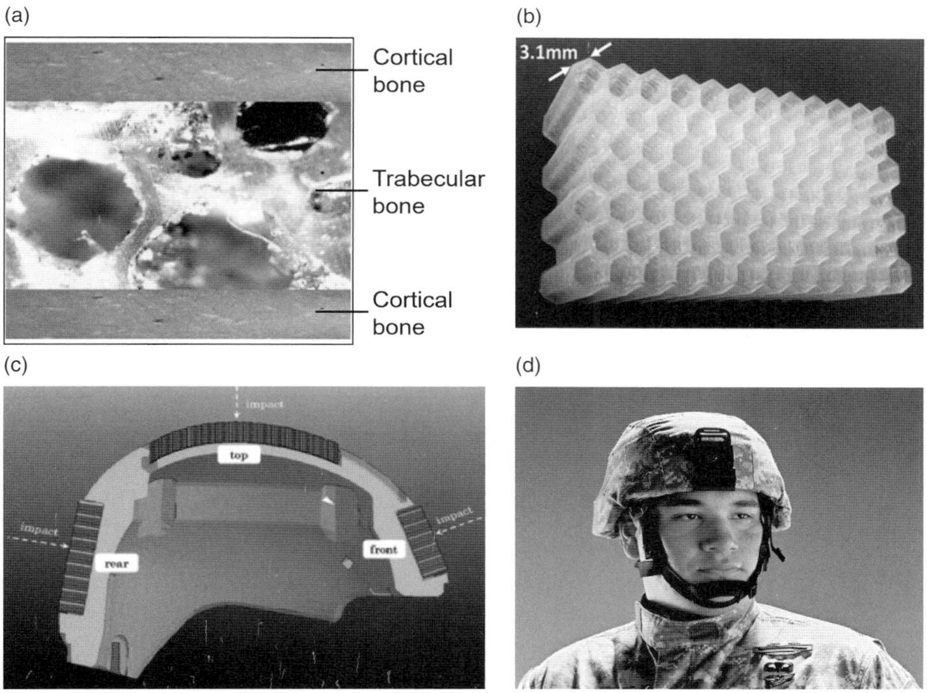

Figure 1.10 The structure and bioinspired applications of turtle/tortoise shell: (a) layered structure of *Kinixys erosa* tortoise bone (reprinted from Owoseni et al. [39] ©Trans Tech Publications); (b) plastic energy–absorbing porous structure (reprinted from Bates et al. [41], copyright 2016, with permission from Elsevier); (c) energy–absorbing inner lining structure of a protective helmet (reprinted from Caserta et al. [52], copyright 2011, with permission from Elsevier); and (d) a soldier wearing an energy-absorbing helmet (U.S. Army Photo/Public Domain).

1.3 Bioinspired and Biomimetic Design of Tissue-Engineered and Nano- and Microscale Structures

This section presents some emerging frontiers in biomimetics and bioinspired design of tissue-engineered structures and multiscale devices at the nano-, micro-, and macroscales. These include tissue engineering for organ reconstruction or repair; bionic devices for radio-frequency hearing and energy harvesting [31,32]; and nano- and microparticles for targeting and treating diseases such as cancer [42–46]. In each of these cases, selected examples are presented to show how bioinspired solutions can provide new opportunities for the design of functional materials and structures.

1.3.1 Tissue/Regenerative Engineering

The advent of tissue engineering (from the latter part of the twentieth century) offers new opportunities for the repair and regeneration of tissues and organs, using methods

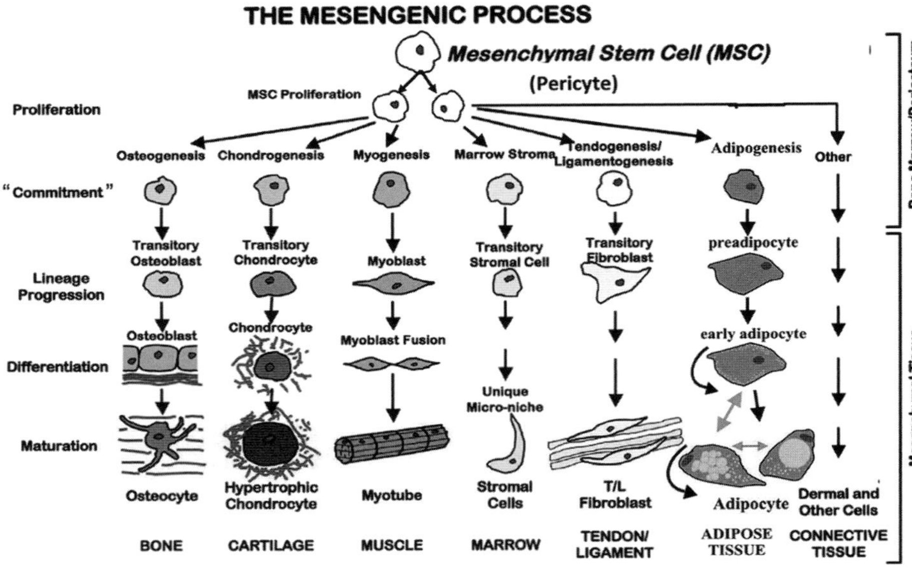

Figure 1.11 The mesengenic process: mesenchymal stem cells and cells derived from them can be used to engineer different tissues of the human body. (Reprinted from DiMarino et al. [51] under creative commons attribution license [CC BY 3.0].)

that are inspired by our understanding of the conditions (pressure, temperature, pH, etc.) that are required for the spreading, proliferation, differentiation, and mineralization processes. In this way, a combination of life science and materials science techniques can be used to form different types of tissues and organs from stem cells (Figure 1.11) and fully differentiated cells.

However, although significant progress has been made in the development of simple bioengineered tissues for the repair of skin burns, bone, and simple tissues/organs [47–49], there is still a need to develop improved methods for the fabrication of next-generation bioengineered tissues with integrated multiscale neurons and vascular structures that can transport signals and materials through multifunctional tissue and organs. This has inspired Gershlak and coworkers [50] to use the veins in spinach leaves as vascular structures in the design of novel natural scaffolds for next-generation tissue/regenerative engineering (Figure 1.12). Such structures could offer a new generation of native plant/vein structures that could enable the development of future tissue-engineered structures. However, further work is needed to test these structures under in vitro and in vivo conditions. These are some of the challenges for future work.

Bionic structures also offer important opportunities for the development of novel systems with "superhuman" characteristics [31,32]. For example, radio-frequency (RF) hearing can be engineered in tissue-engineered human ears that have integrated coils/sensors built into the ear structure. These can be engineered by chondrocytes (cartilage-forming cells) growing on scaffolds. Such systems have been developed that can sense

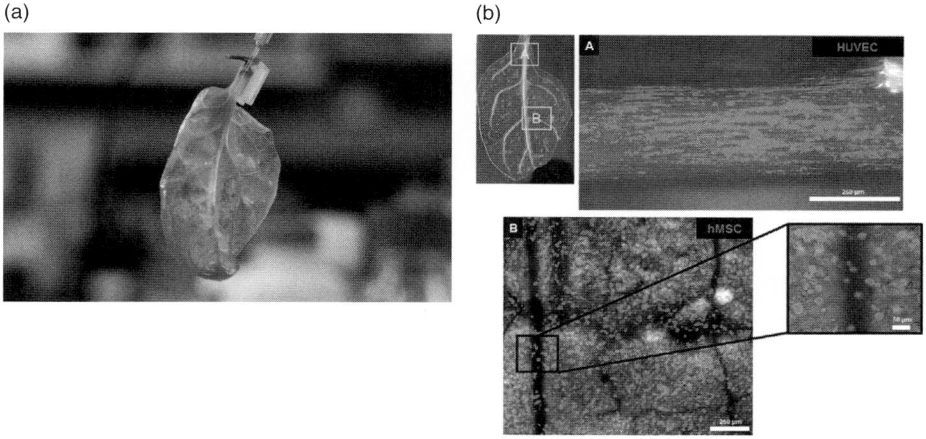

Figure 1.12 Decellularized spinach leaves as vascularized scaffolds: (a) simulation of blood flow through the veins using ponceau red stain (photo by Matthew Burgos, Worcester Polytechnic Institute); and (b) human umbilical vein endothelial cells (HUVECs) growing within the vein and human mesenchymal stem cells (hMSCs) growing on the surface of the spinach leaf scaffold (reprinted from Gershlak et al. [50], copyright 2017, with permission from Elsevier).

audio signals in the RF regime (Figure 1.13). These structures could enable future soldiers to detect RF signals that would otherwise be undetected (Figure 1.13a–d).

Another example of bionic bioinspired design is in the development of energy-harvesting (from incident light) leaf structures (Figure 1.14). This has inspired significant efforts in the development of structures that mimic photosynthetic pathways for energy harvesting [32]. Three-dimensional (3D) bionic leaf structures can be fabricated via the 3D printing of thylakoid structures extracted from leaves. Such structures have been printed into multilayered structures for charge extraction and storage (Figures 1.5 and 1.14).

1.3.2 Nano- and Microparticles for Disease Detection and Treatment

Inspired by the engulfing of viruses by biological cells (Figure 1.15a), ligand-conjugated magnetic or gold nanoparticles have been developed for the specific detection and treatment of diseases such as cancer [42–46]. These rely on the use of molecular recognition units (MRUs), which are mostly peptides and specific antibodies, that attach specifically to receptors that are overexpressed on the surfaces of cancer cells [44,45]. The MRUs increase the adhesive interactions between the MRU-conjugated nanoparticles and cell membranes [45]. They also enhance the uptake of the nanoparticles via receptor-mediated endocytosis (Figure 1.15a–d).

Once within the cells, the presence of the nanoparticles can be used to enhance magnetic resonance imaging (MRI) (Figure 1.15e) or the localized delivery of drugs to the cells. The nanoparticles are then exocytosed from the cells after spending a number of days within the cells. In this way, nature's example of receptor-mediated endocytosis

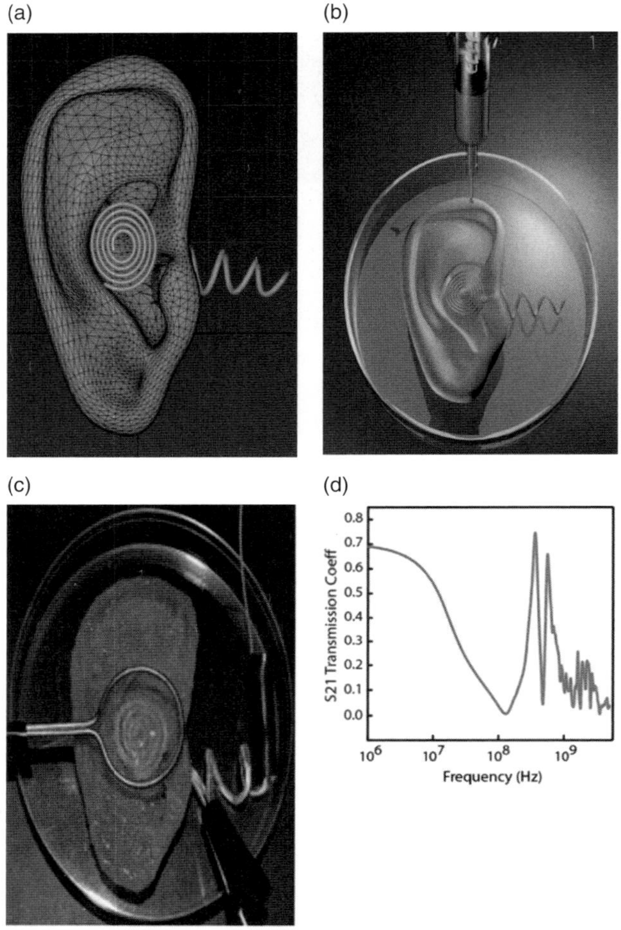

Figure 1.13 Bionic tissue engineering for RF hearing: (a) CAD design of bionic ear, (b) schematic of 3D printing of bionic ear, (c) 3D printed bionic ear connected to electrodes for testing, and (d) plot of the transmission coefficient at different test frequencies. (Reprinted with permission from Mannoor et al. [31], copyright 2013, American Chemical Society.)

can be used to provide the inspiration for the specific detection and treatment of diseases such as cancer [42–46].

Before concluding, it is important to note that recent advances in robotics, additive manufacturing, wireless communications, and the advent of cheaper sensors offer several opportunities for the development of the next generation of bioinspired robotic structures for multiple functions. They also offer the opportunity for the development of scalable manufacturing processes that could enable the future development of multiscale bioinspired structures for multiple applications. These are clearly some of the prospects for the future development of multifunctional bioinspired and biomimetic structures.

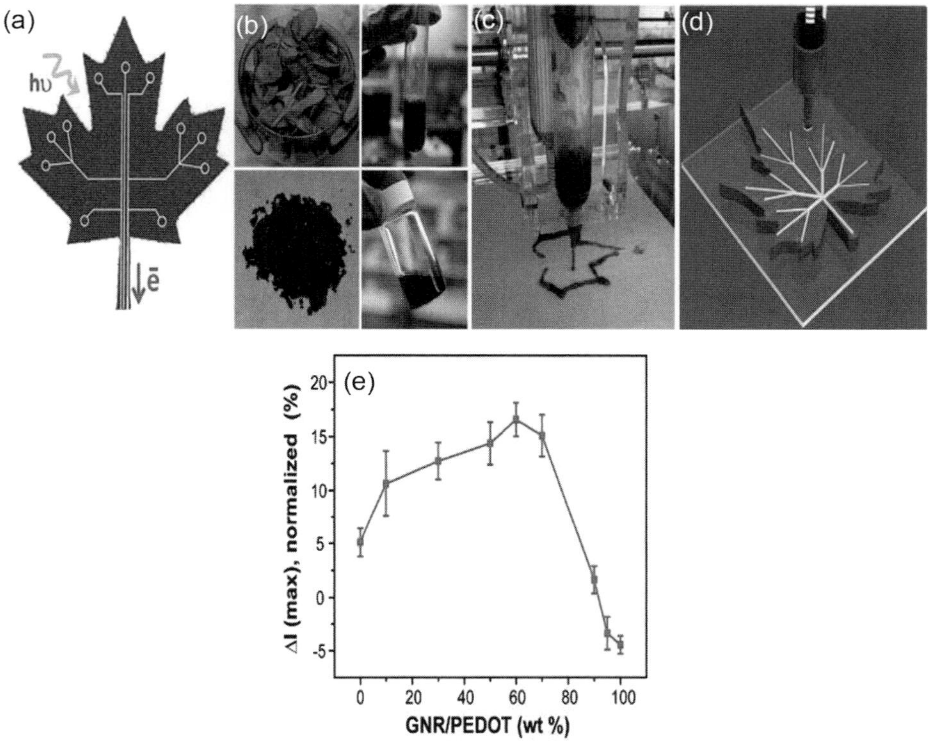

Figure 1.14 Printing of thylakoid structures for charge extraction and storage: (a) schematic of bionic leaf, (b) thylakoid extraction from spinach leaves, (c) 3D printing of extracted thylakoids mixed with a conductive matrix consisting of graphene nanoribbons and PEDOT:PSS (GNR/PEDOT), (d) schematic of the 3D printed leaf structure showing the vasculature for electrodes and water input, and (e) plot of normalized current versus the GNR/PEDOT composition of the printed leaf structures. (Adapted from Mannoor [32].)

1.4 Summary and Concluding Remarks

This chapter presents an introduction to bioinspired and biomimetic structures for different applications. These include both multilayered structures and multifunctional structures that are inspired by nature. In the case of multilayered structures, the structures and properties of natural teeth, bamboo, and turtle/tortoise shells are used as sources of inspiration for the design of layered structures with improved deformation and fracture resistance. In the case of multifunctional structures, examples of bioinspired nano-, micro- and macroscale structures are presented for the development of tissue-engineered structures and bionic structures. The tissue-engineered structures show the potential for evolution from biomedical implants to a paradigm of tissue engineering and regenerative medicine. They can be integrated with the concept of

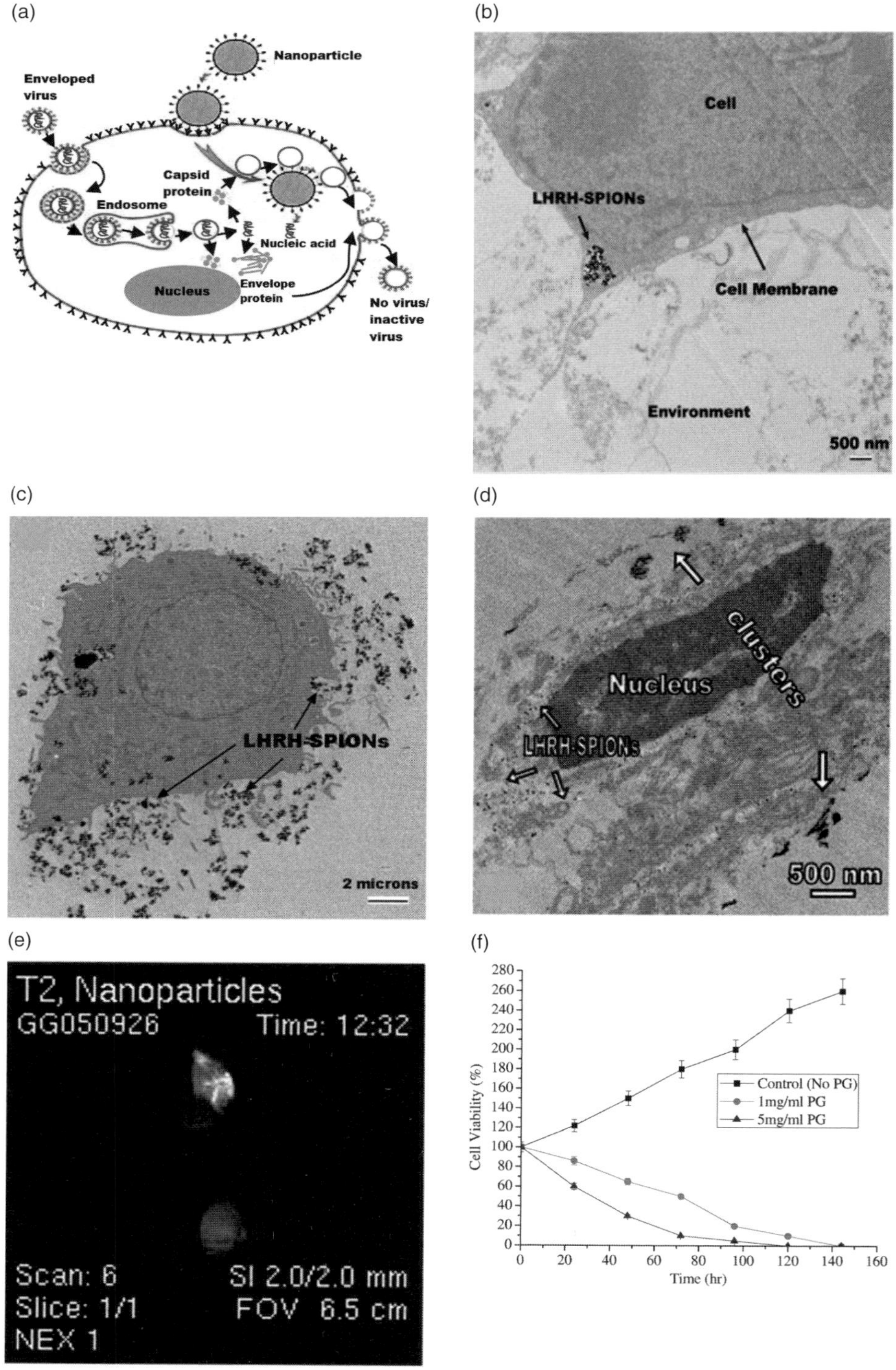
(a)
Nanoparticle
Enveloped virus
Capsid protein
Endosome
Nucleic acid
Nucleus
Envelope protein
No virus/ inactive virus
(b)
Cell
LHRH-SPIONs
Cell Membrane
Environment
500 nm
(c)
LHRH-SPIONs
2 microns
(d)
clusters
Nucleus
LHRH-SPIONs
500 nm
(e)
T2, Nanoparticles
GG050926
Time: 12:32
Scan: 6
SI 2.0/2.0 mm
Slice: 1/1
FOV 6.5 cm
NEX 1
(f)
Cell Viability (%)
Time (hr)
Control (No PG)
1mg/ml PG
5mg/ml PG
0
20
40
60
80
100
120
140
160
180
200
220
240
260
280
0
20
40
60
80
100
120
140
160

bionic structures. Finally, the bioinspired development of nanoparticles can provide the basis for the development of nanoparticles for the specific detection and treatment of diseases such as cancer.

References

1. Meyers, M. A., & Chen, P. -Y. (2014). *Biological materials science: Biological materials, bioinspired materials, and biomaterials*. Cambridge: Cambridge University Press.
2. Ratner, B. D. (2013). A history of biomaterials. In B. D. Ratner, A. S. Hoffman, F. J. Schoen, & J. E. Lemons (Eds.), *Biomaterials science: An introduction to materials in medicine* (3rd ed.). Amsterdam: Academic Press.
3. Jakab, P. L. (2013). *Leonardo da Vinci's codex on the flight of birds*. Washington, DC: Smithsonian National Air and Space Museum.
4. Padfield, G. D., & Lawrence, B. (2003). The birth of flight control: An engineering analysis of the Wright Brothers' 1902 glider. *The Aeronautical Journal*, 107, 697–718.
5. de Barros, H. L. (2006). *Santos-Dumont and the invention of the airplane*. Rio de Janeiro: Brazilian Ministry of Science and Technology and Brazilian Center for Research in Physics.
6. Currey, J. D., & Taylor, J. D. (1974). The mechanical behaviour of some molluscan hard tissues. *Journal of Zoology*, 173, 395–406.
7. Currey, J. D. (1976). Further studies on the mechanical properties of mollusc shell material. *Journal of Zoology*, 180, 445–453.
8. Chen, P. Y., Joanna, M. K., & Meyers, M. A. (2012). Biological materials: Functional adaptations and bioinspired designs. *Progress in Materials Science*, 57, 1492–1704.
9. Wegst, U. G. K., Bai, H., Saiz, E., Tomsia, A. P., & Ritchie, R. O. (2015). Bioinspired structural materials. *Nature Materials*, 14, 23–36.
10. Beese, A. M., Sarkar, S., Nair, A., et al. (2013). Bio-inspired carbon nanotube-polymer composite yarns with hydrogen bond-mediated lateral interactions. *ACS Nano*, 7, 3434–3446.
11. Rahbar, N., & Soboyejo, W. O. (2011). Design of functionally graded dental multilayers. *Fatigue and Fracture of Engineering Materials and Structures*, 34, 887–897.
12. Martini, R., & Barthelat, F. (2016). Stretch-and-release fabrication, testing, and optimization of a flexible ceramic armor inspired from fish scales. *Bioinspiration and Biomimetics*, 11, 066001. doi:10.1088/1748-3190/11/6/066001.
13. Huang, M., Rahbar, N., Wang, R., Thompson, R. D., & Soboyejo, W. O. (2007). Bioinspired design of dental multilayers. *Materials Science and Engineering: A*, 464, 315–320.

Figure 1.15 Nanoparticles (NPs) for disease detection and treatment: (a) antagonistic viruses in diseased cells controlled with NPs, (b) specific entry of magnetic NPs into cancer cells, (c) magnetic NPs entering a cancer cell, (d) magnetic NPs within breast cancer cell, (e) T2 MRI images of breast tumor (top: with NPs, bottom: without NPs), and (f) breast cancer cell death at different durations and anti-cancer drug concentrations. (Part (a) courtesy of Jingjie Hu, Princeton University; (b), (c), and (e) reprinted from Meng et al. [44], copyright 2009; (d) reprinted from Zhou et al. [42], copyright 2006; and (f) reprinted from Dozie-Nwachukwu et al. [46], copyright 2017, all with permission from Elsevier.)

14. Li, L., & Ortiz, C. (2015). A natural 3D interconnected laminated composite with enhanced damage resistance. *Advanced Functional Materials*, 25, 3463–3471.
15. Wang, L. F., & Boyce, M. C. (2010). Bioinspired structural material exhibiting post-yield lateral expansion and volumetric energy dissipation during tension. *Advanced Functional Materials*, 20, 3025–3030.
16. Dooley, C., & Taylor, D. (2017). Self-healing materials: What can nature teach us? *Fatigue and Fracture of Engineering Materials and Structures*, 40, 655–669.
17. Niu, X., Rahbar, N., Farias, S., & Soboyejo, W. (2009). Bio-inspired design of dental multilayers: Experiments and model. *Journal of the Mechanical Behavior of Biomedical Materials*, 2, 596–602.
18. Wang, R. Z., Suo, Z., Evans, A. G., Yao, N., & Aksay, I. A. (2001). Deformation mechanisms in nacre. *Journal of Materials Research*, 16, 2485–2493.
19. Farmer, B. L., Holmes, D. M., Vandeperre, L. J., Stearn, R. J., & Clegg, W. J. (2002). The growth of bamboo-structured carbon tubes using a copper catalyst. *MRS Proceedings*, 740, I3.8. doi: 10.1557/PROC-740-I3.8.
20. Tan, T., Rahbar, N., Allameh, S. M., et al. (2011). Mechanical properties of functionally graded hierarchical bamboo structures. *Acta Biomaterialia*, 7, 3796–3803.
21. Du, J., Niu, X., Rahbar, N., & Soboyejo, W. (2013). Bio-inspired dental multilayers: Effects of layer architecture on the contact-induced deformation. *Acta Biomaterialia*, 9, 5273–5279.
22. Moored, K. W., Dewey, P. A., Leftwich, M. C., Bart-Smith, H., & Smits, A. J. (2011). Bioinspired propulsion mechanisms based on manta ray locomotion. *The Marine Technology Society Journal*, 45, 110–118.
23. Gemmell, B. J., Colin, S. P., Costello, J. H., & Dabiri, J. O. (2015). Suction-based propulsion as a basis for efficient animal swimming. *Nature Communications*, 6, 8790. doi: 10.1038/ncomms9790.
24. Fish, F. E., Howle, L. E., & Murray, M. M. (2008). Hydrodynamic flow control in marine mammals. *Integrative and Comparative Biology*, 48, 788–800.
25. Melli, J., & Rowley, C. W. (2010). Models and control of fish-like locomotion. *Experimental Mechanics*, 50, 1355–1360.
26. Ayali, A., Borgmann, A., Büschges, A., Couzin-Fuchs, E., Daun-Gruhn, S., & Holmes, P. (2015). The comparative investigation of the stick insect and cockroach models in the study of insect locomotion. *Current Opinion in Insect Science*, 12, 1–10.
27. Su, H., Shang, W., Li, G., Patel, N., & Fischer G. S. (2017). An MRI-guided telesurgery system using a Fabry-Perot interferometry force sensor and a pneumatic haptic device. *Annals of Biomedical Engineering*, 45, 1917–1928.
28. Zhang, Y., Chai, H., & Lawn, B. R. (2010). Graded structures for all-ceramic restorations. *Journal of Dental Research*, 89, 417–421.
29. Abate, I., & Tadesse, L. (2018). Bioinspired design for energy storage devices. In L. Daniel & W. O. Soboyejo (Eds.), *Bioinspired design*, Cambridge: Cambridge University Press.
30. Huskinson, B., Marshak, M. P., Suh, C., et al. (2014). A metal-free organic-inorganic aqueous flow battery. *Nature*, 505, 195–198.
31. Mannoor, M. S., Jiang, Z., James, T., et al. (2013). 3D printed bionic ears. *Nano Letters*, 13, 2634–2639.
32. Mannoor, M. S. (2014). *Bionic nanosystems*. [PhD thesis.] Princeton: Princeton University.
33. James, T. (2014). *Nanoscale patterning and 3D assembly for biomedical applications*. [PhD thesis.] Baltimore: Johns Hopkins University.
34. Štacko, P., Kistemaker, J. C. M., van Leeuwen, T., Chang, M. -C., Otten, E., & Feringa, B. L. (2017). Locked synchronous rotor motion in a molecular motor. *Science*, 356, 964–968.

35. Shan, W., Diller, S., Tutcuoglu, A., & Majidi, C. (2015). Rigidity-tuning conductive elastomer. *Smart Materials and Structures*, 24, 065001. doi: 10.1088/0964-1726/24/6/065001.
36. Marshall, G. W. Jr., Balooch, M., Gallagher, R. R., Gansky, S. A., & Marshall, S. J. (2001) Mechanical properties of the dentinoenamel junction: AFM studies of nanohardness, elastic modulus, and fracture. *Journal of Biomedical Materials Research*, 54, 87–95.
37. Askarinejad, S., Kotowski, P., Youssefian, S., & Rahbar, N. (2016). Fracture and mixed-mode resistance curve behavior of bamboo. *Mechanics Research Communications*, 78, 79–85.
38. Damiens, R., Rhee, H., Hwang, Y., et al. (2012). Compressive behavior of a turtle's shell: Experiment, modeling, and simulation. *Journal of the Mechanical Behavior of Biomedical Materials*, 6, 106–112.
39. Owoseni, T. A., Olukole, S. G., Gadu, A. I., Malik, I. A., & Soboyejo, W. O. (2016). Bioinspired design. *Advanced Materials Research*, 1132, 252–266.
40. Ghavami, K., Allameh, S. M., Sánchez, M. L., & Soboyejo, W. O. (2003). Multiscale study of bamboo *Phyllostachys edulis*. First Inter American Conference on Non-Conventional Materials and Technologies in the Eco-Construction and Infrastructure-IAC-NOCMAT.
41. Bates, S. R. G., Farrow, I. R., & Trask, R. S. (2016). 3D printed polyurethane honeycombs for repeated tailored energy absorption. *Materials and Design*, 112, 172–183.
42. Zhou, J., Leuschner, C., Kumar, C., Hormes, J. F., & Soboyejo, W. O. (2006). Sub-cellular accumulation of magnetic nanoparticles in breast tumors and metastases. *Biomaterials*, 27, 2001–2008.
43. Obayemi, J. D., Danyuo, Y., Dozie-Nwachukwu, S., et al. (2016). PLGA-based microparticles loaded with bacterial-synthesized prodigiosin for anticancer drug release: Effects of particle size on drug release kinetics and cell viability. *Materials Science and Engineering: C*, 66, 51–65.
44. Meng, J., Fan, J., Galiana, G., et al. (2009). LHRH-functionalized superparamagnetic iron oxide nanoparticles for breast cancer targeting and contrast enhancement in MRI. *Materials Science and Engineering: C*, 29, 1467–1479.
45. Meng, J., Paetzell, E., Bogorad, A., & Soboyejo, W. O. (2010). Adhesion between peptides/antibodies and breast cancer cells. *Journal of Applied Physics*, 107, 114301. doi: 10.1063/1.3430940.
46. Dozie-Nwachukwu, S. O., Danyuo, Y., Obayemi, J. D., Odusanya, O. S., Malatesta, K., & Soboyejo, W. O. (2017). Extraction and encapsulation of prodigiosin in chitosan microspheres for targeted drug delivery. *Materials Science and Engineering C*, 71, 268–278.
47. Chandrasekaran, A. R., Venugopal, J., Sundarrajan, S., & Ramakrishna, S. (2011). Fabrication of a nanofibrous scaffold with improved bioactivity for culture of human dermal fibroblasts for skin regeneration. *Biomedical Materials*, 6, 015001. doi:10.1088/1748-6041/6/1/015001.
48. Eweida, A. M., Nabawi, A. S., Abouarab, M., et al. (2014). Enhancing mandibular bone regeneration and perfusion via axial vascularization of scaffolds. *Clinical Oral Investigations*, 18, 1671–1678.
49. Lo, K. W., Jiang, T., Gagnon, K. A., Nelson, C., & Laurencin, C. T. (2014). Small-molecule based musculoskeletal regenerative engineering. *Trends in Biotechnology*, 32, 74–81.
50. Gershlak, J. R., Hernandez, S., Fontana, G., et al. (2017). Crossing kingdoms: Using decellularized plants as perfusable tissue engineering scaffolds. *Biomaterials*, 125, 13–22.
51. DiMarino, A. M., Caplan, A. I., & Bonfield, T. L. (2013). Mesenchymal stem cells in tissue repair. *Frontiers in Immunology*, 4, 201. doi: 10.3389/fimmu.2013.00201.
52. Caserta, G. D., Iannucci, L., & Galvanetto, U. (2011). Shock absorption performance of a motorbike helmet with honeycomb reinforced liner. *Composite Structures*, 93, 2748–2759.

2 Human Cortical Bone as a Structural Material

Hierarchical Design and Biological Degradation

Elizabeth A. Zimmermann and Robert O. Ritchie

2.1 Introduction

Nature has developed a wide range of materials with specific properties matched to function by combining minerals and organic polymers into hierarchical structures spanning multiple length-scales. For instance, some materials, such as antler, mimic bone structure with a lower mineralization to provide toughness [1,2], whereas many fish scales have graded material properties from the hard, penetration-resistant outer layer to the adaptive lamellae in the collagen fibril subsurface [3,4]. Indeed, biological systems represent an inexhaustible source of inspiration to materials scientists by offering potential solutions for the development of new generations of structural and functional materials [5]. Nature's key role here is in the complex hierarchical assembly of the structural architectures [6]. The concept of multiscale hierarchical structures, where the microstructure at each level is tailored to local needs, allows the adaptation and optimization of the material form and structure at each level of hierarchy to meet specific functions. Indeed, the complexity and symbiosis of structural biological materials has generated enormous interest of late, primarily because these composite biological systems exhibit mechanical properties that are invariably far superior to those of their individual constituents [7].

Of the many natural composites found in biological systems, one of the most intriguing and sophisticated materials is bone. Like other biological materials, bone is primarily comprised of collagen molecules, water, and mineral nanoparticles that hierarchically assemble to form an extremely tough, lightweight, adaptive, and multifunctional composite material (Figure 2.1) [8]. While bone can be simply viewed as a protective and supportive framework for the body, it actually has a much more active role than ordinary man-made load-bearing structures. This dynamic organ is constantly remodeling and changing shape to adapt to the daily forces placed upon it. Thus, the structure of bone, like all natural materials, has specifically evolved to possess the mechanical properties demanded by its (primarily mechanical) function [9]. A prime example of this adaptive property is the long-term evolution of different bones in the body to their primary function. For instance, long bones, such as the femur or the tibia, resist bending and buckling; short bones, e.g., vertebra, resist compression; and plate-like bones, such as the

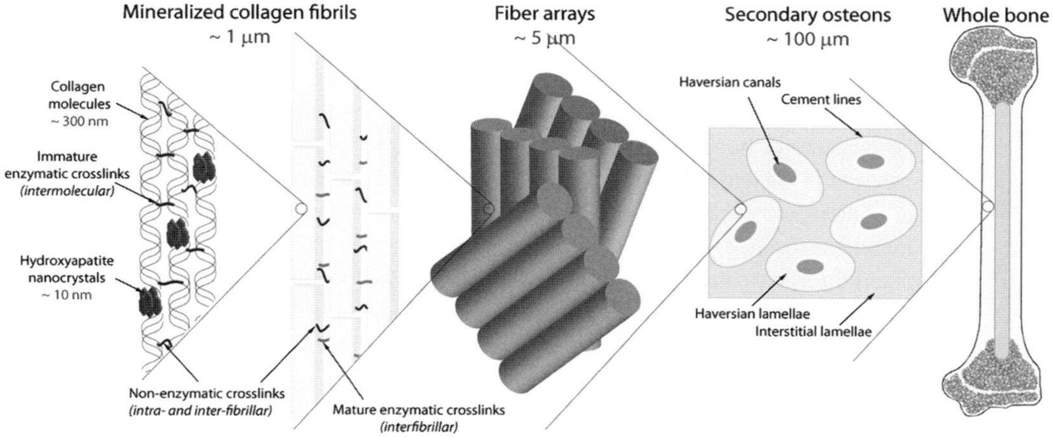

Figure 2.1 *Hierarchical structure of human cortical bone.* Bone and other natural materials are able to generate their unique combination of strength and toughness from their hierarchical structure spanning nano- to macroscale features. The dense cortical bone is mainly found within the diaphysis of long bones and the shell surrounding the porous trabecular bone. At the microstructural level, the cortical bone consists of cylindrical features called osteons, which are roughly 250 μm wide. At the center of the osteon is a vascular channel termed the Haversian canal, which is concentrically surrounded by lamellae. At the outer bounds of the osteon is an interface of material called the cement line with a high content of mineral. The lamellae consist of arrays of collagen fibers, which are assembled from arrays of collagen fibrils. The fibril is a composite of collagen and mineral. Essentially, an array of collagen molecules is stabilized with cross-links with mineral nanoplatelets embedded in between the heads and tails of the collagen molecules. (Adapted from Zimmermann et al. [38].)

skull, protect vital organs. The diversity of structures within this family reflects the fine-tuning or adaptation of the structure to its function [9].

Bone's remarkable ability to grow and remodel (resorption and replacement of old material) occurs first via specialized cells that permanently remove (i.e., osteoclast bone cells) and then deposit (i.e., osteoblast bone cells) tissue material [10]. Through this remodeling process, bone adapts its structure to changing external conditions, which are thought, in the case of mechanical stimuli, to be signaled by damaged material, i.e., microcracks, which blocks signaling between osteocytes [11]. While remodeling continually renews and improves bone in healthy conditions, imbalances in the remodeling process (e.g., more bone resorption than bone formation or quicker bone deposition) brought upon by aging and disease influence the form and integrity of the newly deposited material throughout its hierarchical structure. Thus, with aging, disease, and abnormal environments, the primary factor governing bone's mechanical integrity is *bone quality*, which encompasses the characteristics of the bone-matrix nano/microstructure that can influence mechanical properties such as stiffness (resistance to elastic deformation), strength (resistance to plastic deformation), and toughness (resistance to fracture) [12,13].

Through the lessons learned from biological materials to date, it is clear that sustaining bone's natural hierarchical, multidimensional character is the key to attaining an optimal combination of strength and toughness, which surpasses the ability of the basic mineral and matrix components alone. As such, an understanding of the physics-based mechanisms of bone fracture and how they relate to its hierarchical, multidimensional structure [8] is necessary to discern the origins of bone's multiscale fracture resistance in relation to these structural length-scales, which have each been adapted to generate fracture resistance. Traditionally, toughness has been thought of as the ability of a material to dissipate deformation energy without propagation of a crack. However, in materials such as bone that allow stable crack propagation, fracture is best understood as a "conflict" [14] – the result of a mutual competition of *intrinsic* damage mechanisms ahead of a crack tip that promote cracking and *extrinsic* shielding mechanisms mainly behind the tip that impede it [15,16]. Intrinsic toughening mechanisms increase the microstructural resistance to crack initiation, as exemplified by the role of plasticity ahead of the crack tip in metals. Extrinsic toughening involves microstructural mechanisms that act primarily behind a growing crack tip to inhibit further crack growth by effectively reducing the crack-driving force actually experienced at the crack tip, as illustrated by crack-tip shielding mechanisms, such as crack bridging [17].

This review describes the mechanistic role of each level of bone's multiscale architecture in developing its fracture resistance as a natural structural material. Specifically, we show how the smaller structural length-scales in bone (typically below 1 μm) develop strength and resist crack initiation through intrinsic toughening mechanisms that promote "plasticity" (or, more correctly, inelasticity). Furthermore, we describe how corresponding features in the microstructure (typically above ~1 μm) develop extrinsic toughness by interacting with, and shielding, growing cracks. By the same token, we give examples of how aging, vitamin D deficiency, and high strain rates can change this multiple length-scale structure, causing degradation in these toughening mechanisms. From the lessons learned in studying healthy and diseased bone, not only can we gain a better insight into the role of disease, but we can also establish how perturbations in the bone-matrix nano/microstructure can affect its damage tolerance. In many respects, bone, with its hierarchical architecture and fracture resistance, provides a prime example of the natural design of structural materials.

2.2 Basic Concepts

2.2.1 Hierarchical Structure of Human Cortical Bone

The innate hierarchical structure of human cortical bone, spanning some seven architectural levels from the individual collagen molecules and mineral platelets to the whole bone [8], is the basis for its unique mechanical properties. As in most biological materials, each level has distinct features and is built upon the organized assembly of the level below it. Here, strength is derived from the smaller length-scales, while the toughness is achieved additionally from phenomena acting at much larger length-scales [15].

At its smallest length-scales, bone is a composite of collagen molecules and hydroxyapatite nanoparticles. The collagen molecules (roughly 150 nm long) form an array, where the adjacent collagen molecules are staggered by 67 nm (Figure 2.1). The array is stabilized by the presence of cross-links produced enzymatically (DHLNL and HLNL[1]) and occurring at four specific points along the collagen molecule [18,19]. This array of collagen molecules becomes mineralized with hydroxyapatite mineral platelets roughly $50 \times 50 \times 2$ nm in size. The mineral platelets first nucleate in between the heads and tails of the collagen molecules [20,21], with the *c*-axis of their hexagonal crystal structure aligned parallel to the long axis of the collagen [22]. In bone, further mineralization occurs outside of the gap zones and on the surface of the fibril [23–25]. Overall, this composite structure of collagen molecules embedded with mineral platelets is termed a *mineralized collagen fibril* and is the basis of many biological mineralized tissues, from tendons and skin to hard mineralized tissues, such as bone, teeth, antler, and fish scales [9].

As bone matures, additional cross-links form within and between the mineralized collagen fibrils. The immature intermolecular enzymatic cross-links between the collagen molecules are converted to stable nonreducible cross-links (i.e., pyridinoline and pyrrole) between neighboring fibrils [18,26]. Again, the enzymatic cross-links (both immature and mature) can only occur at four specific points along the collagen molecule, which limits their numbers but allows the molecules and fibrils to register like pieces in a puzzle. Another type of cross-link, which specifically plays a role in aging and diabetes, are nonenzymatic cross-links known as advanced glycation end-products (AGEs). AGEs occur via a glucose-mediated reaction to form both intra- and interfibrillar links along the collagen backbone [18]. As they are not limited to a certain region of the backbone, their magnitude can markedly increase with age. Indeed, the enzymatic cross-link profile is known to stabilize around 10–15 years of age, whereas AGEs can increase up to five-fold with age [26–29].

The mineralized collagen fibrils hierarchically assemble into fibers (~1 μm diameter) and lamellae (~5 μm thick sheets) to form osteons, which are the most prominent motif on the microstructural scale. The osteons are cylindrical features roughly 200 μm in diameter and consist of lamellae concentrically surrounding a central vascular channel (~90 μm in diameter) called the Haversian canal [30]. The osteons have a prominent border called the cement line, which separates individual osteons from the surrounding tissue. The cement line, which is roughly 5 μm thick, plays a key role in generating toughness in bone and is thought to be either collagen-deficient or to have a higher mineralization than the surrounding tissue [31].

2.2.2 Toughening Mechanisms in Cortical Bone

The primary design concerns for structural materials are their mechanical properties. Here, we will focus on the development of fracture resistance, which is a function of strength and toughness (damage tolerance). Strength implies resistance to permanent ("plastic") deformation, while toughness is the resistance to fracture or crack growth.

[1] DHLNL and HLNL are, respectively, dehydrodihydroxynorleucine and dehydrohydroxylysinonorleucine.

Thus the properties of strength and toughness are typically mutually exclusive, as developing strength implies the suppression of plasticity, while the development of toughness implies the promotion of plasticity [14].

Biological materials are a prime example of how these two conditions of strength and toughness can be expressed by assembling a structure across multiple dimensions with basic building blocks characterized by distinctly different mechanical properties. In bone, the natural assembly of stiff hydroxyapatite mineral crystals (with an elastic modulus, E ~150 GPa [32]) and viscoelastic collagen molecules (E ~ 5 GPa [33]) into a multilevel structure creates a material with a versatile combination of strength and toughness; such damage tolerance has evolved to meet the body's basic structural needs and allows cortical bone to resist fracture through multiscale toughening mechanisms that develop as it is loaded.

To understand the role that each structural hierarchy plays in the deformation and fracture of bone, the salient toughening mechanisms can be classified into intrinsic mechanisms relating primarily to the resistance to crack formation and extrinsic mechanisms relating solely to the resistance to crack growth (Figure 2.2) [16,34]. The intrinsic fracture resistance originates from small (submicron) length-scales, such as at the scale of the mineralized collagen fibrils, where bone generates "plasticity" to promote toughness. Extrinsic fracture resistance develops at larger length-scales, such as at the scale of the osteonal structures, whose larger dimensions can markedly influence the crack path. Thus, healthy human bone resists fracture through a combination of intrinsic and extrinsic toughening mechanisms acting *in concert* at multiple length-scales throughout the structure.

2.2.2.1 Intrinsic Toughening Mechanisms

The intrinsic resistance to fracture is the inherent response of the material to deformation and describes the material's ability to plastically deform prior to failure. While the intrinsic fracture resistance is measured in terms of the *crack-initiation toughness* or the *strength* at millimeter length-scales, materials actually develop this intrinsic fracture resistance at their smallest length-scales. In bone and other biological materials, significant contributions to the intrinsic fracture resistance are developed at the scale of the mineralized collagen fibril. Indeed, fibrillar-level deformation is key to attaining bone's mechanical properties. Here, we overview the various mechanisms responsible for the intrinsic toughness of bone.

Intrafibrillar toughening mechanisms: The collagen–mineral composite structure of the fibril absorbs energy and deforms during mechanical loading. Previous experimental and molecular dynamics studies on unmineralized and mineralized collagen fibrils indicate that during elastic (recoverable) deformation, fibrils deform through stretching of the collagen–mineral composite [35–38]. Experimentally, this has been observed in small-angle x-ray scattering experiments, where the fibrillar strain linearly increases with tissue strain (i.e., applied strain) in the elastic region, implying stretching of the fibril. Molecular dynamics studies have shown that in unmineralized fibril arrays, elastic deformation occurs through molecular stretching and uncoiling as well as stretching of the gap regions, where the mineral would reside [39,40], while mineralization of the collagen fibril limits the deformation in the gap region and increases the stiffness of the fibril [39]. Cross-linking between the collagen molecules can also influence the

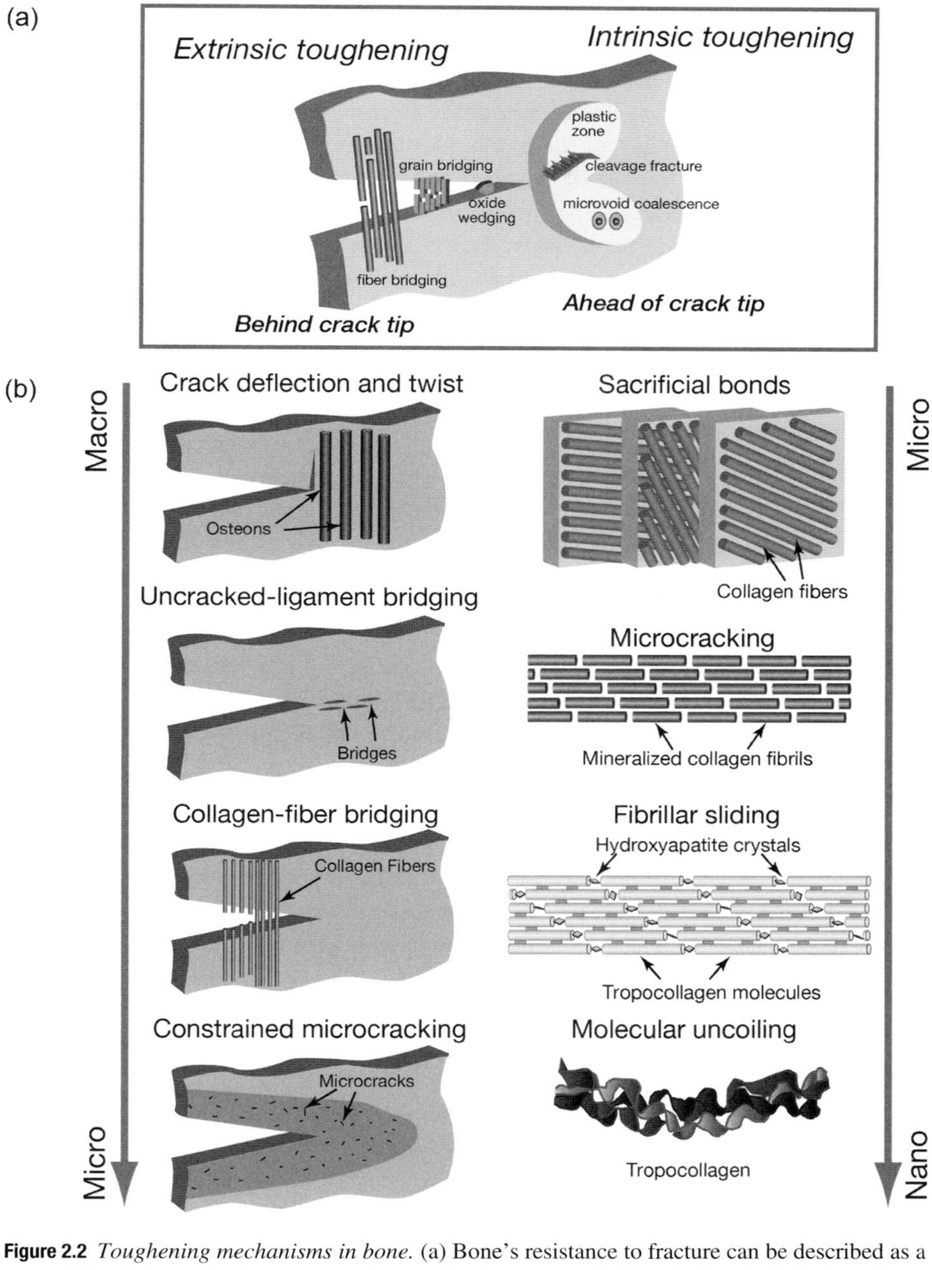

Figure 2.2 *Toughening mechanisms in bone.* (a) Bone's resistance to fracture can be described as a competition between intrinsic toughening mechanisms promoted largely by plasticity to resist microstructural damage and extrinsic crack-tip shielding mechanisms that specifically resist crack propagation. (b) The intrinsic mechanisms mainly contribute to toughness at small length-scales through inelastic deformation within the collagen–mineral composite, sliding between fibrils, and sacrificial bonding. Additionally, larger structural features, such as the osteonal structures and collagen fibers, are able to interact with a growing crack and shield the crack tip through mechanisms of crack bridging or the deflection and/or twisting of the crack path. (Adapted from Launey et al. [15].)

deformation mechanisms, by further promoting fibrillar stretching [41,42]. Thus, the deformation mechanism in bone during elastic stretching is most likely dominated by stretching of the collagen–mineral composite structure.

Interfibrillar sliding: Strength tests provide ample evidence of inelastic deformation in bone. For a long time, this plasticity was thought to be associated directly with collagen deformation. However, recent studies on unmineralized collagen fibrils show that the hierarchical structure is actually more viscous and able to generate more plasticity than the collagen molecules alone [37,43]. Similarly, in mineralized fibrils from bone samples, small-angle x-ray scattering (SAXS) studies have shown that fibrillar sliding is an important mechanism for the generation of plasticity. In SAXS experiments, the strain in the fibril linearly increases with the tissue strain in the elastic region, as discussed previously. Then, as plasticity sets in, the strain in the fibril reaches a constant value even though the whole sample is continuing to deform [36,38,44]. Thus, when plastic deformation begins, the fibrils are no longer stretching, but are instead slipping past one another while being held at a constant strain. Changes in the cross-linking profile, specifically cross-linking between fibrils, may play a key role in resisting the sliding process by constraining the allowable deformation [38].

Sacrificial bonding: Here, bonding between the hierarchical length-scales (i.e., collagen molecules, fibrils, fibers, lamellae) plays a role in generating inelastic deformation. In between fibrils, fibers, or lamellae, a "glue" exists in the form of molecules, noncollagenous proteins, or collagen fibrils [45–47], which links the hierarchical length-scales. These bonds between the length-scales are hypothesized to first stretch to resist deformation within the proteinous "glue" that separates different hierarchical layers and interfaces [48]. With further deformation, these bonds can break and reform to absorb more energy (mechanism of "sacrificial" bonding), as seen in the osteopontin layer that surrounds the collagen fibrils [45]. Overall, sacrificial bonding is an important mechanism in the generation of inelastic deformation and intrinsic toughness in bone and is a prime example of how hierarchical structures are able to transfer deformation to higher length-scales to improve toughness [38].

2.2.2.2 Extrinsic Toughening Mechanisms

Extrinsic toughening mechanisms promote toughness by "shielding" a growing crack from the applied stresses and strains; these mechanisms act primarily in the crack wake and are invariably a function of the path of the crack in relation to the structural features of the material (Figure 2.2). The extrinsic toughness can be directly quantified using techniques derived from fracture mechanics, specifically in terms of the crack-driving force, most commonly the linear-elastic stress-intensity factor, K, or the nonlinear-elastic J-integral [49]. As these mechanisms operate principally behind the crack tip, the extrinsic toughness can increase as the crack grows (bone is a good example of this as it exhibits stable crack extension, as in Figure 2.3); consequently, the measurement of a single-value toughness, such as the plane-strain fracture toughness K_{Ic},[2] does not

[2] The fracture toughness, K_{Ic}, is defined as the critical stress intensity at fracture instability; if measured correctly, it characterizes the crack-initiation toughness.

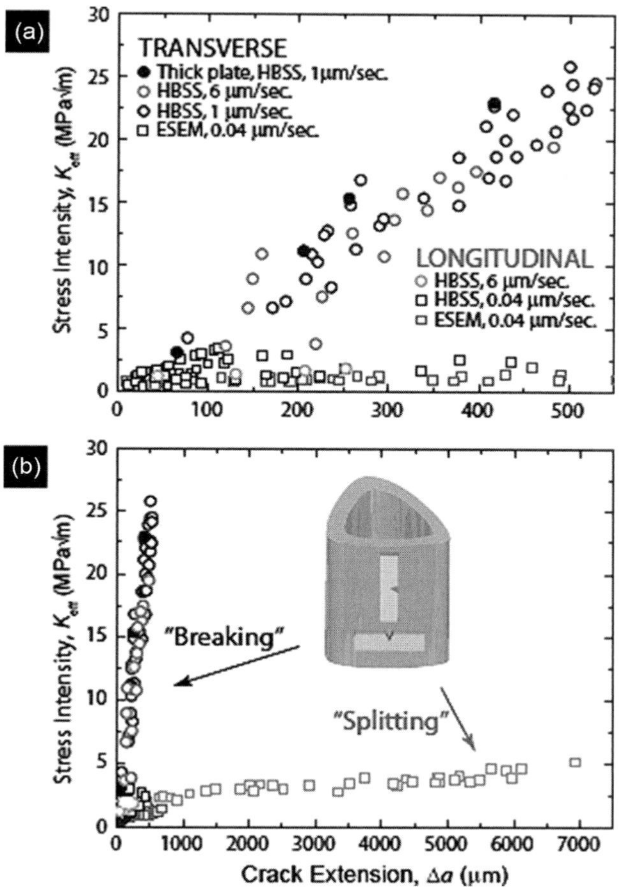

Figure 2.3 *Crack-growth resistance behavior in bone.* The toughness of human cortical bone increases as a crack extends due to the creation of extrinisc toughening mechanisms. As such, a single value measurement of the toughness cannot completely characterize the material behavior. Instead, the stress intensity can be measured as a function of crack extension from the notch, which is called a crack-growth resistance curve or R-curve. Example R-curves are shown for (a) short and (b) long crack extensions for the longitudinal (splitting) and transverse (breaking) directions in humerus bone. The R-curve behavior comes from the hierarchical structure's interaction with the crack through extrinsic toughening mechanisms, mainly at large length-scales that are able to shield the crack tip from the full stress intensity. Clearly, cracks oriented in the transverse orientation are able to generate more toughness during crack extension than the longitudinal orientation, due to the potency of such active shielding mechanisms. (Adapted from Koester et al. [56].)

faithfully characterize the material behavior. Therefore, the toughness is best measured, e.g., in terms of K or J, as a function of crack extension, a, which is termed the crack-growth resistance or R-curve (Figure 2.3). The slope of the R-curve quantifies the degree to which the extrinsic toughening mechanisms are active, i.e., it characterizes the *crack-growth toughness*. In materials with a rising R-curve, mechanisms within the microstructure become activated as the crack extends. These mechanisms, such as the deflection of a crack or the presence of bridging "ligaments" spanning the crack surfaces (Figure 2.2), shield the crack tip from the full stress intensity, thereby requiring a higher applied driving force to cause further crack extension. Because these mechanisms shield the crack or influence the path of the growing crack, they are most effective on larger length-scales, generally on the order of 1 to 100 μm.

Constrained microcracking: Microcracking is a common way for bone to absorb energy during deformation, with microcrack densities of 0.11 cracks/mm^2 common in bone [50]. Microcracks occur along the many interfaces in bone, but primarily along the outer osteonal boundaries. Because these cement lines have a higher mineralization in relation to the surrounding bone matrix, microcracks preferentially form along, or close to, this brittle interface as well as in the interstitial bone [51,52].

There has been debate in the literature regarding the direct role of microcracking on toughness. Constrained microcracking has been noted as a toughening mechanism in ceramics [53]. The basic idea is that microcracks will preferentially form in the highly stressed region near the tip of a growing (macro) crack (a microcracking zone). For these microcracks to form, they must open or dilate; however, the corresponding increase in volume is constrained by less microcracked (and hence, less dilated) regions surrounding this zone, which results in a compressive stress acting on the crack wake thereby reducing the stress intensity at the crack tip [53]. Whereas calculations have shown that the direct contribution of microcracking via this mechanism to the extrinsic toughness in bone is minimal [54], bone's ability to microcrack specifically along the brittle interfaces acts in a more important way by facilitating other extrinsic mechanisms, such as crack deflection and uncracked ligament bridging, as described next.

Crack bridging: Crack bridging is a common phenomenon in many monolithic and composite ceramic materials; in bone it acts to shield the crack tip from the full stress intensity thereby increasing the material's extrinsic toughness. The basic idea is that as the crack extends, some material remains intact, or uncracked, behind the crack tip; as this material spans across the mating crack surfaces it carries load that would otherwise be used to further propagate the crack (Figure 2.4) [55].

In cortical bone, bridging mainly takes the form of uncracked matrix or collagen fibril bridges. The latter are often observed spanning the crack wake in many biological materials; however, the degree to which collagen fibrils contribute to the extrinsic toughness is thought to be relatively small [54]. Another form of bridging in bone occurs when larger regions of the matrix span the crack wake, which is referred to as uncracked ligament bridging (Figure 2.4). In this scenario, microcracks form ahead of the growing crack, resulting in "mother" and "daughter" cracks. The matrix in between the microcracks, usually on the scale of 10s of μm, spans the crack wake to shield the crack tip from the full stress intensity. This process is the primary toughening mechanism for fracture in the longitudinal (splitting) orientation, where the crack grows parallel to the long axis of the bone. In this orientation, microcracks easily form along the weak interfaces (i.e., the lamellar interfaces or hypermineralized

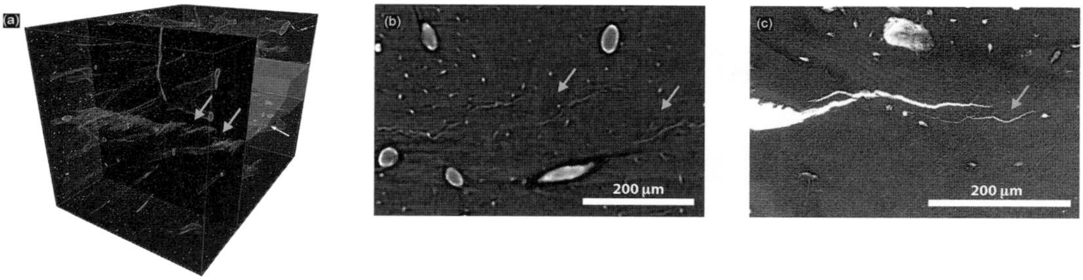

Figure 2.4 *Crack-bridging promotes extrinsic toughness.* The R-curve behavior of human cortical bone originates from extrinsic toughening mechanisms that shield the crack tip. In the longitudinal orientation, especially, crack bridges play an important role. These bridges are intact material spanning the crack wake. (a) 3D synchrotron computed microtomography images of a fracture toughness sample after testing show how the crack (gray shaded area) grew from the notch (small arrow) and in relation to the Haverisan canals (gray lines). In the crack wake, the larger arrows indicate the intact material spanning the crack wake. (b) The same crack bridges are also apparent when looking at individual slices from tomography that are in a plane perpendicular to the crack growth direction and (c) in the scanning electron microscope while monitoring crack extension during mechanical testing. (Adapted from Zimmermann et al. [38].)

cement lines), while leaving intact matrix between the main growing crack and the microcracks initiated ahead of it [46].

Crack deflection and twist: Crack deflection is an effective toughening mechanism when it deviates the crack path from the plane of maximum mechanical driving force; likewise, crack twist is a similar out-of-plane deviation through the thickness. Both of these mechanisms can have a marked effect on toughness, increasing it extrinsically by a factor of two or more (Figure 2.3). Cracks in most materials tend to follow a primary path governed by the maximum mechanical driving force, specifically, the path of maximum tensile stress where the strain-energy release rate, G, is at a maximum or where the mode II stress intensity is zero, i.e., $K_{\mathrm{II}} = 0$. When the crack deviates from this path, the crack tip experiences a lower *local* driving force; as a result, the toughness is effectively increased as higher applied stresses are required to sustain further crack extension [56,57].

In cortical bone, crack deflection/twist most commonly occurs when growing cracks encounter osteons (Figure 2.5); this is particularly prevalent when bone is loaded in the transverse orientation where the crack path is nominally orthogonal to the orientation of the osteons. The interfaces of the osteons, the hypermineralized cement lines, provide a prime location for cracks to markedly deflect, resulting in the rough fracture surfaces that are generally characteristic of bone fractures [56]. Indeed, multiaxial loading studies have shown that bone's toughness will be lowest when the direction of the maximum mechanical driving force (where $K_{\mathrm{II}} = 0$) is in the same direction as the path of weakest microstructural resistance (along the osteonal direction). For this reason, bone is significantly tougher in the transverse ("breaking") orientation than the longitudinal ("splitting") orientation [57,58].

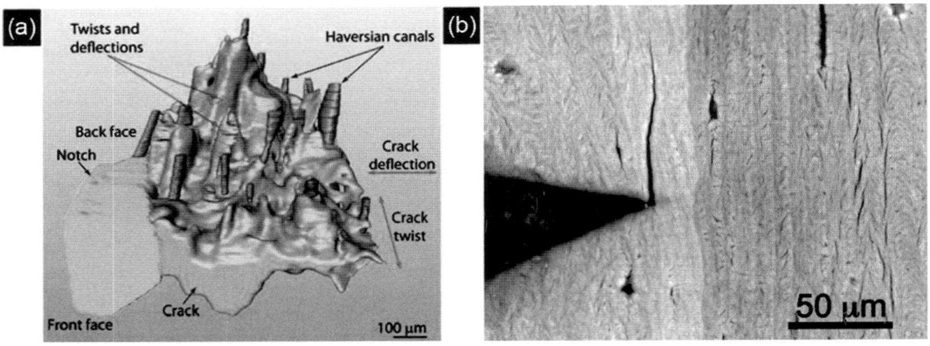

Figure 2.5 *Crack deflection and twist promote extrinsic toughness.* Another important extrinsic toughening mechanism in human cortical bone is crack deflection and twisting. When the growing crack is oriented perpendicular to the osteons and Haversian canals in human cortical bone, crack deflection/bridging is particularly important. Here, the aligned interfaces within the osteon (i.e., the cement lines and lamellae) provide an interface for the crack to deflect upon as it grows. Crack deflection increases the toughness of bone because the local stress intensity at the crack tip requires a higher driving force for further crack extension. (a) 3D synchrotron computed microtomography of a fracture toughness sample after testing illustrates how a crack (medium gray) can grow from a notch (lighter gray) in a wavy character with deflections and twists. (b) Crack deflections are also visible when using scanning electron microscopy during crack extension. (Adapted from Launey et al. [15].)

In summary, bone's toughness is highly dependent on the ability of its hierarchical structure to develop intrinsic mechanisms at small (submicron) length-scales to promote inelastic deformation as well as extrinsic mechanisms at larger length-scales to shield growing cracks (Figure 2.2). What is particularly interesting about these multiscale toughening mechanisms is that there is a coupling between the length-scales. If "plasticity" via fibrillar sliding at small scales is inhibited by, for example, excessive collagen cross-linking, the bone can still dissipate energy by microcracking (which can be considered as a form of inelasticity); this in turn promotes crack deflection and uncracked-ligament bridging, which is the basis of extrinsic toughening at higher length-scales. In this way, healthy human cortical bone is able to sustain a level of microdamage without catastrophic failure, which may signal the remodeling process to repair the damaged tissue. Further studies, however, are required to fully understand exactly how this multidimensional structure resists fracture at nanometer length-scales, particularly in light of very recent evidence of the presence of "dilatational bands," i.e., tiny ellipsoidal voids that form between the mineral aggregates and osteocalcin and osteopontin proteins surrounding the collagen fibrils [59]. We believe that the coupling between the divergent structural length-scales and how this affects the individual intrinsic and extrinsic toughening mechanisms represents the key to deciphering the fracture resistance of bone and may provide insight into understanding the specific mechanisms of how bone diseases may alter this structure and lead to increased fracture risk.

2.3 Prior Work

2.3.1 Age-Related Deterioration in Toughness

An ongoing clinical issue regarding the elderly is cortical bone's age-related deterioration in mechanical properties. Indeed, the main debate as to the cause of the aging-related increase in fracture risk is whether the increased fracture risk stems from a loss in bone *quantity* or a deterioration in bone *quality* [60,61]. Bone quantity can more readily be assessed in clinical settings via x-ray microtomography scans, where the 3D bone architecture and mineral content can be spatially viewed, or through dual emission x-ray absorption, which is the clinical standard for the diagnosis of osteoporosis despite its drawbacks. While the loss in bone mass (or bone-mineral density) with age is a reality, the landmark study by Hui et al. [61] found that bone mass was not the sole predictor of fracture risk; other characteristics of the bone tissue play a role in determining its resistance to fracture. These characteristics describe the bone's quality and include the composition and distribution of mineral and collagen, the characteristics and structural integrity of each hierarchical length-scale (i.e., osteon size and distribution), as well as the amount of microdamage present [12]. Numerous investigations of aging-related changes in bone quality at multiple length-scales corroborate the loss of fracture resistance due to a multiscale deterioration in structure.

Aging-related loss in intrinsic toughness: At the smallest length-scales in bone, i.e., at the scale of the collagen molecules, aging can induce marked changes in the cross-linking profile. While the amount of enzymatic cross-links in bone stabilizes at a chronological age of ~10 to 15 years [26,28], the occurrence of nonenzymatic cross-links progressively increases with age [27–29]. These latter AGEs occur at both intra- and interfibrillar regions [18]. AGEs form due to a glucose-mediated reaction between amino acids in the collagen and have been strongly associated with increased fracture risk [62].

In a recent study, Zimmermann et al. [38] investigated how aging-related changes at small length-scales affect deformation at the collagen fibril length-scale. The mechanical properties at small length-scales of young and aged sample groups were investigated with synchrotron radiation SAXS and wide-angle x-ray diffraction (WAXD). In the SAXS/WAXD experiments, the x-rays are scattered by the periodic nanoscale structure of the collagen fibril, allowing the strain within the fibril and the mineral to be measured when x-ray diffraction patterns are collected during a mechanical tensile test. The aged fibrils were found to strain less than their young counterparts at the same applied strain (i.e., tissue strain (Figure 2.6a). The restricted strain in the fibril with aging correlates with a three-fold greater amount of AGEs cross-links in the aged bone than the young bone (Figure 2.6b) [38]).

Aging-related loss in extrinsic toughness: The extrinsic toughness of bone is primarily realized at the scale of the osteons. As described earlier, the deflection of cracks as they impinge on the cement lines and the formation of uncracked ligament bridges, due to the interaction of the main growth crack with microcracks initiated ahead, leads to rising R-curve toughness behavior due to the shielding effect of these mechanisms

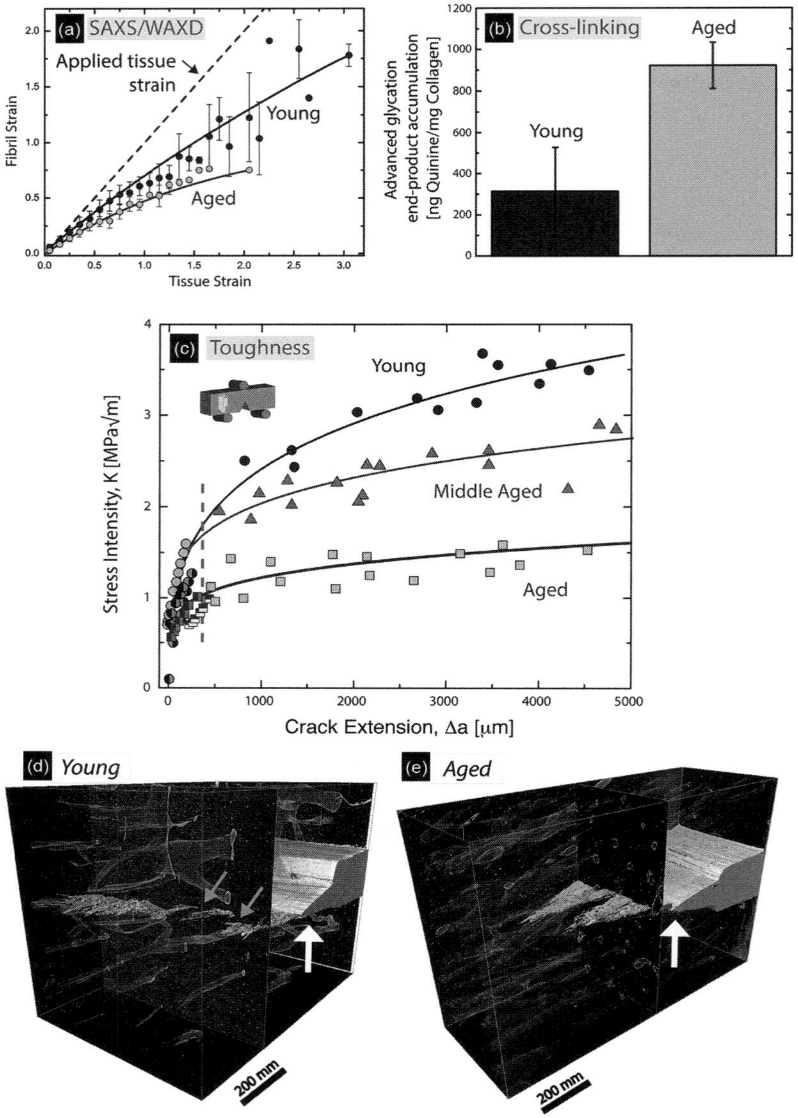

Figure 2.6 *Aging-related changes to bone toughness.* With aging, the structure of bone changes at multiple length-scales, which directly affects the bone's mechanical properties. (a) At small length-scales, mineralized collagen fibrils in aged (85–99-year-old) bone are restricted from deforming as much as in young (34–41-year-old) bone due to (b) the three-fold increase in AGE cross-links restricting the generation of plasticity at the fibrillar scale. (c) Indeed, the initiation and growth toughness of bone decreases dramatically with age. Synchrontron computed microtomography of (d) young and (e) aged fracture toughness samples after crack growth indicates that the large increase in osteon density with age decreases the size and amount of crack bridging allowable in aged bone, and thus contributes to the decrease in toughness with age in human cortical bone. (Adapted from Zimmermann et al. [38].)

(Figure 2.6c). As such, this represents a measure of resistance to crack growth, which can be specifically evaluated in terms of the steepness of the $K_R(a)$ R-curve.

In aged bone, the crack-growth resistance of cortical bone has been found to be severely diminished (Figure 2.6c). For large crack extensions – on the order of millimeters – the crack-growth toughness of young (34–41-year-old) bone is five times higher than aged (85–99-year-old) bone [38,63]. Synchrotron x-ray computed micro-tomography revealed that the aged bone had a three-fold higher osteonal density than young bone, (Figure 2.6d and e). Because the size of the uncracked ligament bridges scales with the osteonal spacing as the microcracks are formed primarily at, or near, the highly mineralized cement lines, the potency of the bridging phenomena becomes degraded with age; indeed, the osteonal density scales inversely with the crack-growth toughness [38,63].

Summary: The aging-related deterioration in mechanical properties is a consequence of combined deleterious alterations to the bone structure at multiple length-scales. Specifically, the aging-related increase in AGEs constrains deformation at small length-scales within the fibril and inhibits the absorption of energy at this scale during mechanical loading. At larger length-scales, the osteon density increases and reduces the toughness as crack bridges, which act to shield the crack tip during crack extension, are now smaller and have a closer spacing [38].

An additional consequence of the loss in intrinsic toughness is that if the fibrils deform less in aged bone, other levels within the structure must be compensating. Indeed, in addition to cross-linking, aging in bone is associated with an increased density/length of microcracks. We theorize that there is a link between the loss in intrinsic fracture toughness, which is associated with an increased AGEs cross-link profile inhibiting fibrillar-level deformation, and the increased microcrack density with age, which must compensate for the loss in plasticity in the fibrils [38].

2.3.2 Role of High Strain Rates

The majority of the literature characterizing the strength and toughness of bone and the mechanisms through which it generates its fracture resistance has been performed at slow strain rates, where it is possible to observe bone fracture. However, most physiological fractures are caused by a traumatic injury and, thus, occur under high strain rate conditions. Therefore, it is necessary to understand whether the strength and toughening mechanisms associated with slow strain rates are also active at the high strain rates associated with physiological activity and fracture incidents.

Fracture toughness experiments at physiological strain rates have shown that the resistance of cortical bone to fracture clearly diminishes at higher strain rates, indicating a change in the bone's extrinsic resistance to fracture (Figure 2.7a) [64–68]. At the microstructural scale (~10–100 μm), the lower toughness is associated with a distinct change in the path of the crack through the bone-matrix structure. At lower strain rates, the crack takes a tortuous path through the bone-matrix structure and deflects at the cement-line boundaries of the osteons (Figure 2.7b). In contrast, at higher strain rates, crack deflection at the cement lines is effectively absent, such that the crack takes a straighter path that penetrates across the face of the osteon (Figure 2.7c). The absence of the crack deflection mechanism at higher strain rates significantly decreases the extrinsic contributions to the bone toughness and suggests that the toughening mechanisms within bone may have time-dependent characteristics.

The change in crack path behavior at different strain rates can be simply explained via the theoretical framework of He et al.[3] [69] for a linear-elastic crack impinging on the interface between two dissimilar materials. Basically, the analysis shows that the crack will change from deflection at the cement lines at slow strain rates to penetration of the cement lines at high strain rates only if the toughness of the bone matrix decreases [68]. The implication of this result is that when loading is applied to bone at a higher rate, the bone-matrix structure ahead of a growing crack has the effect of appearing to be less ductile, i.e., displaying less plasticity or toughness, and as a result, the cement lines are no longer a source of extrinsic toughening through the generation of deflected crack trajectories.[4] Thus the rationale for the lower fracture toughness at high strain rates (i.e., the extrinsic toughness) is potentially tied to the intrinsic behavior.

SAXS has been used to study the intrinsic toughening mechanisms at the fibrillar scale as a function of strain rate [68]. At slower strain rates, the fibril first deforms through elastic stretching of the fibril, which is indicated by the linear relation between the fibrillar and tissue strains, followed by inelastic deformation (i.e., through microcracking, fibrillar sliding, etc.) as indicated by the plateau in the fibril strain (Figure 2.7d). At higher strain rates, the fibril strain vs. tissue strain has a higher slope, especially for the two highest strain rates (Figure 2.7d), which means that more strain or deformation occurs within the fibril. In other words, at higher strain rates the fibril primarily deforms and dissipates energy through elastic stretching and less through inelastic deformation mechanisms (i.e., fibrillar sliding, sacrificial debonding, etc.). Thus the inelastic mechanisms become constrained at higher strain rates due to the viscosity or time-dependent nature of the deformation of the whole mineralized collagen fibril structure [68].

Indeed, while the collagen constituent of bone is commonly regarded as the primary provider of viscoelasticity, recent experimental and computational studies have found that the hierarchical nature of bone is actually responsible for viscoelastic material properties [37,43]. Thus the viscous or inelastic nature of bone may not originate completely from the collagen constituent but from the multiple length-scale architecture, in this case, the fibril. Clearly, the arrangement of collagen and mineral within the fibril, the assembly of fibrils, sacrificial length-scales, and the opening of dilatational

[3] The conditions for a crack penetrating, as opposed to arresting at or delaminating along, an interface between two dissimilar materials, termed 1 and 2, depend on (i) the elastic (Young's) modulus E mismatch of materials across the interface, which is captured with the first Dundurs' parameter = $(E_1 - E_2)/(E_1 + E_2)$, where the subscripts refer to the two materials, and (ii) the ratio of the toughness of the interface, expressed in terms of the strain-energy release rate G_{int}, to the toughness of the material into which the crack will propagate, G_2 [69].

[4] Comparable effects can occur in bone with aging, irradiation damage, and disease where abnormal mineralization, and/or cross-linking profiles within the matrix can reduce the relative inhomogeneity between the bone matrix and the cement lines, again contributing to less deflected crack paths [38,44,73,81].

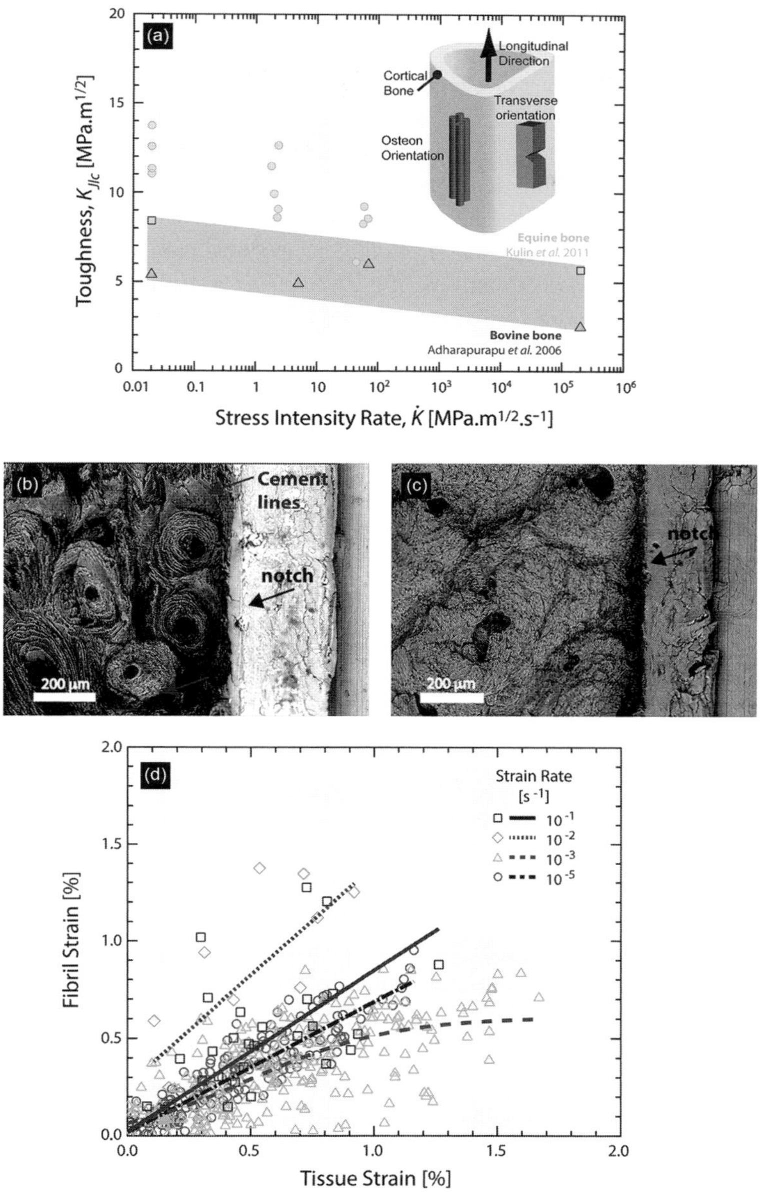

Figure 2.7 *Toughness of bone at high strain rates.* (a) The toughness of human cortical bone decreases as physiologically high strain rates are approached. By analyzing the fracture surfaces of the toughness specimens, (b) the crack clearly takes a tortuous path around the outer boundary of the osteons (i.e., the cement lines) at slow strain rates. (c) However, at the highest strain rate, the crack path takes a path across the face of the osteons. Thus, the extrinsic mechanism of crack deflection is severely diminished at high strain rates, which contributes to the loss in toughness. The loss in crack deflection is most likely a result of the fact that the hierarchical structure of bone inelastically deforms less at high strain rates, which was verified through (d) small angle x-ray scattering measurements at different strain rates, showing more deformation in the fibril and less fibrillar sliding at high strain rates. (Adapted from Zimmermann et al. [68], with permission from Elsevier.)

bands or microcracks all allow bone to deform inelastically, primarily through sliding within and between fibrils [36,38,45,59]. At high strain rates, the fibril has a higher strain because the viscous mechanisms essentially "lock-up", i.e., strain-rate stiffening akin to the behavior of a dashpot at high rates of deformation. Thus, at slower strain rates, the fibrils have a certain amount of plasticity and energy absorption through sliding mechanisms within and between fibrils, while at higher strain rates, the sliding mechanisms lock up, causing the fibrils themselves to stretch further, accounting for less overall plasticity but higher strength in the macroscopic properties.

In conclusion, the effectiveness of the intrinsic and extrinsic mechanisms to generate toughness is compromised at higher strain rates, where the locking up of fibrils restricts their ability to inelastically deform. The reduction in the intrinsic toughness of the bone matrix then influences the extrinsic toughness, such that the lower intrinsic toughness at high strain rates promotes crack penetration across the microstructural interfaces, rather than delamination and crack deflection, as in healthy bone; the more linear crack path reduces the extrinsic toughening and hence lowers the overall resistance to crack propagation.

2.3.3 Bone Disease

There are many bone diseases that biologically cause marked changes in the bone-matrix structure and as such can result in diminished bone fragility. As an example, we discuss here the effect of vitamin D deficiency. Vitamin D is an essential nutrient with many health advantages, but in the case of bone homeostasis, vitamin D plays a key role by aiding the absorption of calcium and phosphate [70,71]. Clinical investigations have linked low levels of vitamin D to lower bone density and a higher risk of fracture [71,72]. The effects of vitamin D deficiency on the composition and structure of human bone that ultimately lead to an increased fracture risk have recently been investigated using a suite of high-resolution experimental techniques to characterize bone quality at multiple length-scales in a set of control and vitamin D deficient bone biopsies [73].

Vitamin D deficiency is characterized by a high degree of osteoid (i.e., unmineralized bone matrix) covering the surfaces within the bone [74]. Therefore, the cause for the increased risk of fracture in vitamin D deficient patients was commonly thought to be due to a lower amount of mineralized bone. However, using high spatial resolution techniques, the bone within the mineralized osteoid frame was found to have the characteristics of aged tissue [73]. Specifically, Fourier-transform infrared spectroscopy revealed a higher cross-linking ratio and a higher carbonate to phosphate ratio in the mineral component (Figure 2.8a and b), which are both associated with aged mineral and collagen. Additionally, synchrotron x-ray computed microtomography showed a higher mineralization in the vitamin D deficient tissue [73]. The tissue ages within the osteoid seams because of a key feature of bone resorption: the osteoclast cells cannot resorb unmineralized bone matrix. Therefore, the tissue underneath the bone osteoid-covered surfaces is essentially trapped.

At higher length-scales, the in situ fracture toughness measurements of the bone were measured in a scanning electron microscope, such that the toughness could be measured while simultaneously observing the crack path. Here, the vitamin D deficient bones had a 21% lower crack initiation toughness and a 31% lower crack growth toughness than the control (healthy) bone (Figure 2.8c) [73]. Crack paths, examined in 3D using synchrotron x-ray computed microtomography (Figure 2.8d and e) indicated that in the vitamin D deficient bone there was significant (30%) change in the crack deflection angle leading to straighter crack trajectories, consistent with the lower toughness of the diseased bone. Additionally, crack paths in vitamin D deficient bone were characterized by far less crack bridging.

Thus the vitamin D deficiency causes a vicious circle, where more bone is resorbed to account for the lack of calcium, but then newly deposited bone tissue cannot be mineralized. This situation leaves an unmineralized layer of osteoid on the surface of bone that in turn cannot be resorbed. In effect, islands of bone are trapped within the osteoid frame and effectively age. The aged material has less intrinsic resistance to crack growth due to the higher cross-linking content and the higher mineralization and a lower extrinsic toughness due to fewer crack bridges and straighter crack paths. Thus, vitamin D deficiency changes the bone at small and larger length-scales to limit, respectively, plasticity (intrinsic toughening) and crack deflection/bridging (extrinsic toughening), all of which serves to diminish bone fragility and increase the risk of fracture [73].

2.4 Emerging Concepts

2.4.1 Microstructural Heterogeneity

In the clinical treatment of osteoporosis, a generation of patients has emerged who have had prolonged osteoporosis treatment with bisphosphonates. While most spontaneous osteoporotic fractures occur in trabecular bone found in the femoral head or the vertebrae, prolonged bisphosphonate treatments have caused a small percentage (~0.2%), but a large number, of patients to develop atypical femoral fractures, which occur within the cortical bone [75]. Bisphosphonates influence the remodeling process by suppressing resorption of bone. This particular treatment has drastically reduced fracture risk in the elderly [76,77], but its prolonged use has been somewhat controversial due to its suppression of bone turnover.

The suppression of bone remodeling allows all the bone packets to age and thus the microstructural heterogeneity of the bone cross-section is reduced [78]. As healthy bone is constantly adapting to new environments by remodeling its structure, the cross-sectional distribution of osteons will likely have different mechanical properties due to different degrees of mineralization and/or the maturity of the collagen and mineral. These osteonal characteristics change as a function of their relative age, termed *tissue age*. A likely consequence of this will be variations in the mechanical properties in cortical bone between individual osteons and between osteons and interstitial tissue; additionally, the mechanical properties will vary between an osteon's lamellae and its cement line.

The heterogeneity of the bone microstructure has been an underappreciated aspect of bone's fracture resistance. Heterogeneity provides a means to induce tortuous crack trajectories, local arrest, and rough irregular crack surfaces, all of which serve to enhance the extrinsic toughness, yet few studies have analyzed the structural

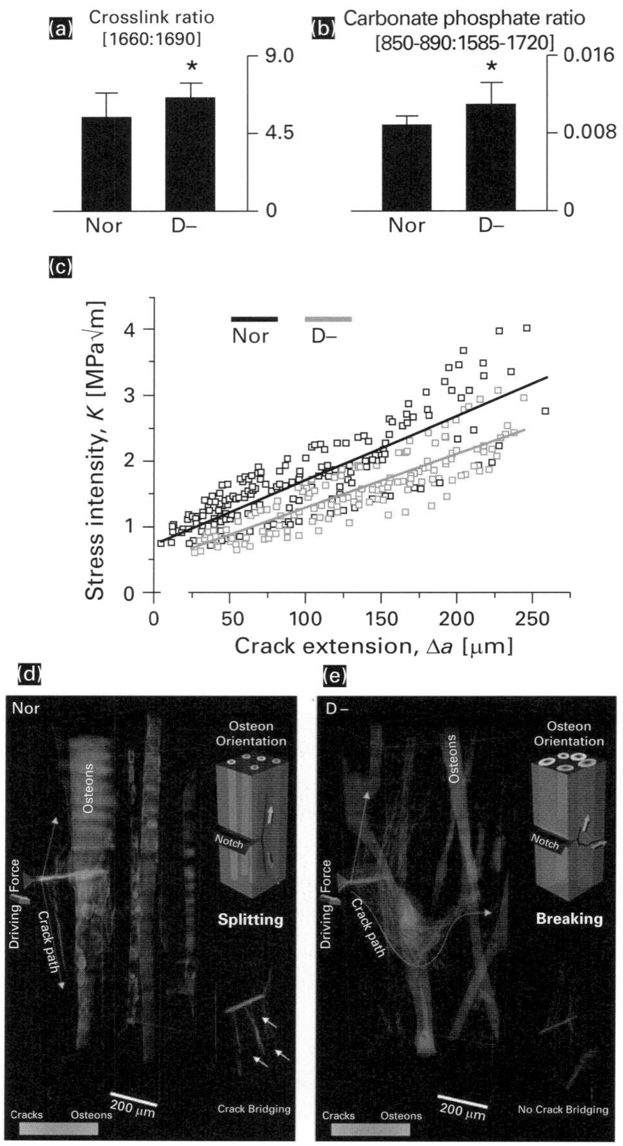

Figure 2.8 *Toughness of bone with vitamin D deficiency.* In vitamin D deficient human bone, the tissue trapped within the characteristic osteoid-covered surfaces ages, as seen through Fourier-transform infrared spectroscopy measurements of (a) the collagen cross-link ratio and (b) the mineral's carbonate to phosphate ratio. (c) The fact that the bone tissue has an aged character results in the loss in the crack initiation and crack growth toughness in vitamin D deficient bone. (d) Indeed, 3D synchrotron computed microtomography of the crack path after toughness testing revealed (d) crack deflection in the control samples and (e) a straighter crack path in the vitamin D deficient samples. (Adapted from Busse et al. [73].)

heterogeneity of bisphosphonate-treated bone. Fracture mechanics studies are underway to determine if such reduced heterogeneity is one of the sources leading to the occurrence of atypical femoral fractures, which have been linked to long-term bisphosphonate use [79,80]. The hypothesis is that the lower heterogeneity would create less deflected crack paths, thereby reducing the toughness and increasing the risk of fracture. Indeed, an increase in the homogeneity of the bone-matrix structure may well be the cause of atypical femoral fractures in individuals after high-dose bisphosphonate use, a notion which is consistent with the smooth nature of these fractures [80].

2.5 Summary and Concluding Remarks

The hierarchical structure of human cortical bone is specifically tailored to resist fracture at multiple length-scales, in particular through intrinsic mechanisms which create plasticity in bone acting at small, submicron length-scales, as well as extrinsic mechanisms at larger, microscopic length-scales which act to shield, and hence inhibit, the growth of cracks. Indeed, changes to the structure at one or more length-scales can severely curtail the effectiveness of these mechanisms leading to a degradation in the overall mechanical properties and increased fracture risk. Specifically, changes to the cross-link profile due to aging play a significant role in limiting plasticity from fibrillar sliding at the molecular level (irradiation can have a similar effect) [38,44,81]. At larger length-scales, increases in the osteon density associated with excessive remodeling with age can have correspondingly negative repercussions to the extrinsic toughness by limiting the creation of crack bridges [38,63]. While a reduction in the heterogeneity of the structure of bone, for example, from the abnormal mineralization associated with certain bone diseases [73,82], in cases of atypical femoral fractures [80] can disrupt the relative degree of mineralization of the cement lines vs. the interstitial bone matrix, causing far less crack deflection and a consequent loss of fracture resistance. The effects can be enhanced when bone is subjected to high strain rates, as the potency of these toughening mechanisms can be diminished [68].

We believe that the understanding developed from studies on the fracture behavior of bone can, not only aid the diagnosis, prevention, and treatment of debilitating bone diseases, which specifically target the key microstructural aspects that impart resistance to bone failure, but also provide valuable insight into the design of new structural engineering materials based on the concept of multiple length-scale hierarchical design.

References

1. Krauss, S., Fratzl, P., Seto, J., et al. (2009). Inhomogeneous fibril stretching in antler starts after macroscopic yielding: Indication for a nanoscale toughening mechanism. *Bone*, 44(6), 1105–1110
2. Launey, M. E., Chen, P. Y., McKittrick, J., & Ritchie, R .O. (2010). Mechanistic aspects of the fracture toughness of elk antler bone. *Acta Biomaterialia*, 6(4), 1505–1514.

3. Yang, W., Chen, I. H., Gludovatz, B., Zimmermann, E. A., Ritchie, R. O., & Meyers, M. A. (2013). Natural flexible dermal armor. *Advanced Materials*, 25(1), 31–48.
4. Zimmermann, E. A., Gludovatz, B., Schaible, E., et al. (2013). Mechanical adaptability of the Bouligand-type structure in natural dermal armor. *Nature Communications*, 4(10), 2634 (doi: http://dx.doi.org/10.1038/ ncomms 3634).
5. Mayer, G. (2005). Rigid biological systems as models for synthetic composites. *Science* 310 (5751), 1144–1147.
6. Lakes, R. (1993). Materials with structural hierarchy. *Nature*, 361(6412), 511–515.
7. Meyers, M. A., McKittrick, J., & Chen, P. -Y. (2013). Structural biological materials: Critical mechanics-materials connections. *Science*, 339(6121), 773–779.
8. Weiner, S., & Wagner, H. D. (1998). The material bone: Structure mechanical function relations. *Annual Review of Materials Science*, 28, 271–298.
9. Currey, J. D. (2006). *Bones: Structure and mechanics*. Princeton: Princeton University Press; p. 456.
10. Robling, A. G., Castillo, A. B., & Turner, C. H. (2006). Biomechanical and molecular regulation of bone remodeling. *Annual Review of Biomedical Engineering*, 8(1), 455–498.
11. Taylor, D., Hazenberg, J. G., & Lee, G. L. (2007). Living with cracks: Damage and repair in human bone. *Nature Materials*, 6, 263–268.
12. Burr, D. B. (2004). Bone quality: Understanding what matters. *Journal of Musculoskeletal & Neuronal Interactions*, 4(2), 184–186.
13. Zimmermann, E. A., Barth, H. D., & Ritchie, R. O. (2012). On the multiscale origins of fracture resistance in human bone and its biological degradation. *JOM*, 64(4), 486–493.
14. Ritchie, R. O. (2011). The conflicts between strength and toughness. *Nature Materials*, 10 (11), 817–822.
15. Launey, M. E., Buehler, M. J., & Ritchie, R. O. (2010). On the mechanistic origins of toughness in bone. *Annual Review of Materials Research*, 40, 25–53.
16. Ritchie, R. O. (1999). Mechanisms of fatigue-crack propagation in ductile and brittle solids. *International Journal of Fracture*, 100(1), 55–83.
17. Evans, A. G. (1990). Perspective on the development of high-toughness ceramics. *Journal of the American Ceramic Society*, 73(2), 187–206.
18. Bailey, A. J. (2001). Molecular mechanisms of ageing in connective tissues. *Mechanisms of Ageing and Development*, 122(7), 735–755.
19. Hodge, A. J., & Petruska, J. A. (1963). Recent studies with the electron microscope on ordered aggregates of the tropocollagen macromolecule. In G. N. Ramachandran (Ed.), *Aspects of protein structure*. New York: Academic Press.
20. Traub, W., Arad, T., & Weiner, S. (1989). 3-Dimensional ordered distribution of crystals in turkey tendon collagen-fibers. *Proceedings of the National Academy of Sciences*, 86(24), 9822–9826.
21. Weiner, S., & Traub, W. (1986). Organization of hydroxyapatite crystals within collagen fibrils. *FEBS Letters* 206(2), 262–266.
22. Landis, W., Hodgens, K., Arena, J., Song, M., & McEwen, B. (1996). Structural relations between collagen and mineral in bone as determined by high voltage electron microscopic tomograph. *Microscopy Research and Technique*, 33(2), 192–202.
23. Landis, W. J., Hodgens, K. J., Song, M. J., et al. (1996). Mineralization of collagen may occur on fibril surfaces: Evidence from conventional and high-voltage electron microscopy and three-dimensional imaging. *Journal of Structural Biology*, 117(1), 24–35.

24. Arsenault, A. L. (1991). Image-analysis of collagen-associated mineral distribution in cryogenically prepared turkey leg tendons. *Calcified Tissue International*, 48(1), 56–62.
25. Maitland, M. E., & Arsenault, A. L. (1991). A correlation between the distribution of biological apatite and amino-acid-sequence of type-i collagen. *Calcified Tissue International*, 48(5), 341–352.
26. Eyre, D. R., Dickson, I. R., & Vanness, K. (1988). Collagen cross-linking in human-bone and articular-cartilage: Age-related-changes in the content of mature hydroxypyridinium residues. *Biochemical Journal*, 252(2), 495–500.
27. Odetti, P., Rossi, S., Monacelli, F., et al. (2005). Advanced glycation end-products and bone loss during aging. *Annals of the New York Academy of Sciences*, 1043(1), 710–717.
28. Saito, M., Marumo, K., Fujii, K., & Ishioka, N. (1997). Single-column high-performance liquid chromatographic fluorescence detection of immature, mature, and senescent cross-links of collagen. *Analytical Biochemistry*, 253(1), 26–32.
29. Sell, D. R., & Monnier, V. M. (1989). Structure elucidation of a senescence cross-link from human extracellular-matrix: Implication of pentoses in the aging process. *Journal of Biological Chemistry*, 264(36), 21597–21602.
30. Martin, R. B., & Burr, D. B. (1989). *Structure, function, and adaptation of compact bone*. New York: Raven Press. p 275.
31. Skedros, J., Holmes, J., Vajda, E., & Bloebaum, R. (2005). Cement lines of secondary osteons in human bone are not mineral-deficient: New data in a historical perspective. *The Anatomical Record A*, 286A, 781–803.
32. Saber-Samandari, S., & Gross, K. A. (2009). Micromechanical properties of single crystal hydroxyapatite by nanoindentation. *Acta Biomaterialia*, 5(6), 2206–2212.
33. Sasaki, N., & Odajima, S. (1996). Stress-strain curve and Young's modulus of a collagen molecule as determined by the X-ray diffraction technique. *Journal of Biomechanics*, 29(5), 655–658.
34. Ritchie, R. O. (1988). Mechanisms of fatigue crack-propagation in metals, ceramics and composites: Role of crack tip shielding. *Materials Science and Engineering A – Structural Materials: Properties, Microstructure and Processing*, 103(1), 15–28.
35. Buehler, M. J. (2007). Molecular nanomechanics of nascent bone: Fibrillar toughening by mineralization. *Nanotechnology*, 18(29), 295102.
36. Gupta, H. S., Wagermaier, W., Zickler, G. A., et al. (2005). Nanoscale deformation mechanisms in bone. *Nano Letters*, 5(10), 2108–2111.
37. Silver, F. H., Christiansen, D. L., Snowhill, P. B., & Chen, Y. (2001). Transition from viscous to elastic-based dependency of mechanical properties of self-assembled type i collagen fibers. *Journal of Applied Polymer Science*, 79(1), 134–142.
38. Zimmermann, E. A., Schaible, E., Bale, H., et al. (2011). Age-related changes in the plasticity and toughness of human cortical bone at multiple length scales. *Proceedings of the National Academy of Sciences of the United States of America*, 108(35), 14416–14421.
39. Nair, A. K., Gautieri, A., Chang, S. W., & Buehler, M. J. (2013). Molecular mechanics of mineralized collagen fibrils in bone. *Nature Communications*, 4.
40. Yuye, T., Ballarini, R., Buehler, M. J., & Eppell, S. J. (2010). Deformation micromechanisms of collagen fibrils under uniaxial tension. *Journal of the Royal Society Interface*, 7(46), 839–850.
41. Siegmund, T., Allen, M. R., & Burr, D. B. (2008). Failure of mineralized collagen fibrils: Modeling the role of collagen cross-linking. *Journal of Biomechanics*, 41(7), 1427–1435.

42. Silver, F. H., Christiansen, D. L., Snowhill, P. B., & Chen, Y (2000). Role of storage on changes in the mechanical properties of tendon and self-assembled collagen fibers. *Connective Tissue Research*, 41(2), 155–164.
43. Gautieri, A., Vesentini, S., Redaelli, A., & Buehler, M. J. (2012). Viscoelastic properties of model segments of collagen molecules. *Matrix Biology*, 31(2), 141–149.
44. Barth, H. D., Zimmermann, E. A., Schaible, E., Tang, S. Y., Alliston, T., & Ritchie, R. O. (2011). Characterization of the effects of X-ray irradiation on the hierarchical structure and mechanical properties of human cortical bone. *Biomaterials*, 32(34), 8892–8904.
45. Fantner, G. E., Hassenkam, T., Kindt, J. H., et al. (2005). Sacrificial bonds and hidden length dissipate energy as mineralized fibrils separate during bone fracture. *Nature Materials*, 4(8), 612–616.
46. Nalla, R. K., Kinney, J. H., & Ritchie, R. O. (2003). Mechanistic fracture criteria for the failure of human cortical bone. *Nature Materials*, 2, 164–168.
47. Thurner, P. J., Chen, C. G., Ionova-Martin, S., et al. (2010). Osteopontin deficiency increases bone fragility but preserves bone mass. *Bone*, 46, 1564–1573.
48. Munch, E., Launey, M. E., Alsem, D. H., Saiz, E., Tomsia, A. P., & Ritchie, R. O. (2008). Tough, bio-inspired hybrid materials. *Science*, 322(5907), 1516–1520.
49. Anderson, T. L. (2005). *Fracture mechanics: Fundamentals and applications*. Boca Raton: CRC Press.
50. Wassermann, N., Brydges, B., Searles, S., & Akkus, O. (2008). In vivo linear microcracks of human femoral cortical bone remain parallel to osteons during aging. *Bone*, 43, 856–861.
51. Norman, T. L., & Wang, Z. (1997). Microdamage of human cortical bone: incidence and morphology in long bones. *Bone*, 20(4), 375–379.
52. Wasserman, N., Yerramshetty, J., & Akkus, O. (2005). Microcracks colocalize within highly mineralized regions of cortical bone tissue. *European Journal of Morphology*, 42(1–2), 43–51.
53. Evans, A. G. (1990). Perspective on the development of high-toughness ceramics. *Journal of the American Ceramic Society*, 73(2), 187–206.
54. Nalla, R. K., Kruzic, J. J., & Ritchie, R. O. (2004). On the origin of the toughness of mineralized tissue: Microcracking or crack bridging? *Bone*, 34, 790–798.
55. Shank, J. K. & Ritchie, R. O. (1989). Crack bridging by uncracked ligaments during fatigue-crack growth in SiC-reinforced aluminum-alloy composites. *Metallurgical Transactions A*, 20A(5), 897–908.
56. Koester, K. J., Ager, J. W., & Ritchie, R. O. (2008). The true toughness of human cortical bone measured with realistically short cracks. *Nature Materials*, 7(8), 672–677.
57. Zimmermann, E. A., Launey, M. E., Barth, H. D., & Ritchie, R. O. (2009). Mixed-mode fracture of human cortical bone. *Biomaterials*, 30(29), 5877–5884.
58. Zimmermann, E. A., Launey, M. E., & Ritchie, R. O. (2010). The significance of crack-resistance curves to the mixed-mode fracture toughness of human cortical bone. *Biomaterials*, 31(20), 5297–5308.
59. Poundarik, A., Diab, T., Sroga, G. E., et al. (2012). Dilational band formation in bone. *Proceedings of the National Academy of Sciences of the United States of America*, 109(47), 19178–19183.
60. Cummings, S. R., Browner, W., Cummings, S. R., et al. (1993). Bone density at various sites for prediction of hip fractures. *The Lancet*, 341(8837), 72–75.
61. Hui, S. L., Slemenda, C. W., & Johnston, C. C. (1988). Age and bone mass as predictors of fracture in a prospective study. *Journal of Clinical Investigation*, 81(6), 1804–1809.

62. Vashishth, D., Gibson, G. J., Khoury, J. I., Schaffler, M. B., Kimura, J., & Fyhrie, D. P. (2001). Influence of nonenzymatic glycation on biomechanical properties of cortical bone. *Bone*, 28(2), 195–201.
63. Nalla, R. K., Kruzic, J. J., Kinney, J. H., Balooch M., Ager J. W., & Ritchie, R. O. (2006). Role of microstructure in the aging-related deterioration of the toughness of human cortical bone. *Materials Science & Engineering C-Biomimetic and Supramolecular Systems*, 26(8), 1251–1260.
64. Adharapurapu, R. R., Jiang, F., & Vecchio, K. S. (2006). Dynamic fracture of bovine bone. *Materials Science & Engineering C-Biomimetic and Supramolecular Systems*, 26(8), 1325–1332.
65. Behiri, J. C., & Bonfield, W. (1984). Fracture-mechanics of bone: The effects of density, specimen thickness and crack velocity on longitudinal fracture. *Journal of Biomechanics*, 17 (1), 25–34.
66. Kulin, R. M., Jiang, F., & Vecchio, K. S. (2011). Effects of age and loading rate on equine cortical bone failure. *Journal of the Mechanical Behavior of Biomedical Materials*, 4(1), 57–75.
67. Kulin, R. M., Jiang, F., & Vecchio, K. S. (2011). Loading rate effects on the R-curve behavior of cortical bone. *Acta Biomaterialia*, 7(2), 724–732.
68. Zimmermann, E. A., Gludovatz, B., Schaible, E., Busse, B., & Ritchie, R. O. (2014). Fracture resistance of human cortical bone across multiple length-scales at physiological strain rates. *Biomaterials*, 35(21), 5472–5481.
69. He, M. Y., & Hutchinson, J. W. (1989). Crack deflection at an interface between dissimilar elastic-materials. *International Journal of Solids and Structures*, 25(9), 1053–1067.
70. DeLuca, H. F. (2004). Overview of general physiologic features and functions of vitamin D. *American Journal of Clinical Nutrition*, 80(6), 1689S–1696S.
71. Lips, P. (2001). Vitamin D deficiency and secondary hyperparathyroidism in the elderly: Consequences for bone loss and fractures and therapeutic implications. *Endocrine Reviews*, 22(4), 477–501.
72. Whyte, M. P., & Thakker, R. V. (2005). Rickets and osteomalacia. *Medicine*, 33(12), 70–74.
73. Busse, B., Bale, H., Zimmermann, E. A., et al. (2013). Vitamin D deficiency induces early signs of aging in human bone, increasing the risk of fracture. *Science Translational Medicine*, 5(193), 193ra188.
74. Priemel, M., von Domarus, C., Klatte, T. O., et al. (2010). Bone mineralization defects and vitamin D deficiency: Histomorphometric analysis of iliac crest bone biopsies and circulating 25-hydroxyvitamin D in 675 patients. *Journal of Bone and Mineral Research*, 25(2), 305–312.
75. Shane, E., Burr, D., Ebeling, P. R., et al. (2010). Atypical subtrochanteric and diaphyseal femoral fractures: Report of a Task Force of the American Society for Bone and Mineral Research. *Journal of Bone and Mineral Research*, 25(11), 2267–2294.
76. Harrington, J. T., Ste-Marie, L. G., Brandi, M. L., et al. (2004). Risedronate rapidly reduces the risk for nonvertebral fractures in women with postmenopausal osteoporosis. *Calcified Tissue International*, 74(2), 129–135.
77. Roux, C., Seeman, E., Eastell, R., et al. (2004). Efficacy of risedronate on clinical vertebral fractures within six months. *Current Medical Research and Opinion*, 20(4), 433–439.
78. Donnelly, E., Meredith, D. S., Nguyen, J. T., et al. (2012). Reduced cortical bone compositional heterogeneity with bisphosphonate treatment in postmenopausal women with intertrochanteric and subtrochanteric fractures. *Journal of Bone and Mineral Research*, 27(3), 672–678.

79. Burr, D. B., Diab, T., Koivunemi, A., Koivunemi, M., & Allen, M. R. (2009). Effects of 1 to 3 years' treatment with alendronate on mechanical properties of the femoral shaft in a canine model: Implications for subtrochanteric femoral fracture risk. *Journal of Orthopaedic Research*, 27(10), 1288–1292.
80. Ettinger, B., Burr, D. B., & Ritchie, R. O. (2013). Proposed pathogenesis for atypical femoral fractures: Lessons from materials research. *Bone*, 55(2), 495–500.
81. Barth, H. D., Launey, M. E., MacDowell, A. A., Ager, J. W., & Ritchie, R. O. (2010). On the effect of X-ray irradiation on the deformation and fracture behavior of human cortical bone. *Bone*, 46(6), 1475–1485.
82. Carriero, A., Zimmermann, E. A., Paluszny, A., et al. (2014). How tough is brittle bone? Investigating osteogenesis imperfecta in mouse bone. *Journal of Bone and Mineral Research*, 29(6), 1392–1401.

3 Bioinspired Design of Multilayered Composites

Lessons from Nacre

Sina Askarinejad and Nima Rahbar

3.1 Background

In the past, significant research has been focused on improving the mechanical properties of lightweight structural materials due to the large demand in bioengineering, aerospace, automotive, armor, and construction applications. This primarily includes advanced structural materials, which are lightweight materials with outstanding mechanical properties such as strength and toughness. Meanwhile, various materials exist in nature that inherently have these exceptional mechanical properties [1]. There is, therefore, a great interest in understanding and analyzing the structure and mechanical behavior of these materials [2]–[7]. Evolution has brought about beautiful, optimized solutions to many problems. Nacre [8], mantis shrimp club [9], bone [10], deep sea sponge [11], bamboo [12,13], and elk antler [14] are just a few of these structural biological materials.

Layered structure and super-strong interface in nacre is the origin of high strength, stiffness, and toughness [15]. Nacre is a natural biocomposite made of aragonite platelets and biological macromolecules. The volume fraction of the organic phase is around 5%. This phase is extremely important in nacre's fracture toughness [16,17]. Previous research on this natural material has shown that its layered "brick-and-mortar" structure is the origin of its remarkable mechanical properties [17,18]. The inorganic phase exists as polygonal tablets with 3–5 μm diameters and the organic phase has a thickness of around 20 nm. The simple brick-and-mortar structure makes the material much tougher than its constituents. The Young's modulus of nacre is reported to be around 70 GPa and its tensile strength has been reported to be around 170 MPa [19]. Nacre is a wonder of nature in its mechanical properties in terms of strength and toughness. This structural material exhibits a toughness (in energy terms) some orders of magnitude higher than that of its primary component ($CaCO_3$), and its strength is among the highest in shell structures [20–22]. Inspired by this structure, there have been significant efforts in the past few decades to synthesize new materials with mechanical performance comparable to nacre [23–32]. The essence of success is to understand the deformation mechanisms inherent in nacre. Hence, mechanical models are needed in order to answer these fundamental questions. The main goal of this study is to explain the basic deformation and toughening mechanisms and the resulting stress distribution in nacre-like multilayered materials through different stages of deformation. Nacre is a natural composite made of 95% brittle aragonite platelets ($CaCO_3$ tablets) [17] and 5% biological macromolecules. This composition has an important effect on nacre's fracture resistance [33]. The elastic modulus of the aragonite is generally assumed to be about $E = 100$ GPa [20,34]. A relatively thick (0.2–0.9 μm) and brittle aragonite is separated by nanoscale interlayers (approximately 20 nm). The mineral layer exists as closely packed polygonal tablets (3–5 μm diameters) separated by a nanoscale organic gap. The tensile strength and Young's modulus of wet nacre are reported to be approximately 140 MPa and 70 GPa, respectively [20,35,36]. The existence and role of water in the structure of nacre has been investigated by many researchers. It has been mentioned that the effect of water was to increase the ductility of nacre and increase the toughness by almost tenfold [37,38].

Wang et al. performed a study investigating the behavior of abalone nacre and pearl oyster nacre under different test setups. The results of four-point bending tests show a Young's modulus of $E \sim 70$ GPa. The tensile curves are highly nonlinear, and the yield stress is in the range of 105–140 MPa [19]. Other researchers have described the origin of toughness by crack deflection, fiber pull-out, and organic matrix bridging [27,39]. Barthelat et al. also performed a numerical and experimental investigation on deformation and fracture of nacre and proposed their results from image correlation, finding the experimental crack resistance curves for nacre with logarithmic function fit [40]. They also claimed that the large breaking strain in the experiments and simulations was due to the waviness of the platelets. However, the numerical model based on the waviness of the platelets overpredicts the strength and underpredicts the ductility [41]. Li et al. studied the nanoscale structure of red abalone shell and suggested that the tablets themselves are made of nanograins [8]. They found that although the tablet strength does not directly affect the deformation mechanism, the integrity of the tablets is due to these nanograins, which prevent the tablets from breaking during the deformation process. Nanoindentation was used by other research groups to study the properties of the individual components of nacre [42]. These studies emphasize that the platelets possess a high fracture stress; i.e., they will remain intact in the deformation process of nacre under low strain rates.

Many researchers believe that the organic matrix in the structure of nacre plays the key role in its high fracture toughness [33]. Therefore, there has been a great interest in understanding the mechanical behavior of the organic matrix. Smith et al. believe that the natural adhesives elongate in a stepwise manner. They claimed that opening of the organic macromolecules' folded domains causes a modular elongation and is the origin of toughness in the natural fibers and adhesives. In that study, a single molecule of protein was pulled, and the behavior of a short molecule and long molecule was investigated. It was mentioned that the short molecules behave more stiffly but required less energy to break [43]. In another study, Mohanty et al. investigated the mechanical response of the organic matrix of nacre. Atomic force microscopy (AFM) was used to pull the organic phase away from the aragonite. The results show a high adhesion force between the proteins and the platelets and are evidence of organic–inorganic interactions [44]. Different experimental approaches have been used to determine the mechanical properties of the organic matrix in different stages of deformation [45–49].

Molecular dynamics simulations have also been used to investigate the role of the organic matrix on the mechanical response of nacre [50–52]. In that study, done by Ghosh et al., the mechanical response of the organic matrix in proximity of aragonite platelets was investigated, and it was indicated that the high elastic modulus of the proteins may be due to the mineral–organic interactions. Barthelat et al. used a two-dimensional finite element model of indentation to fit the experimental nanoindentation load-penetration depth curves of nacre, giving an elastic modulus of around 2.84 GPa [41]. Xu et al. considered the extension of a single biopolymer molecule as a series of helical springs, so that, unfolding of one module increases the stiffness of the biopolymer molecule. In that study, it was shown that a larger spring outer diameter causes a smaller spring constant [53]. Overall, these studies showed that the organic matrix has a high elastic modulus in the range of 4.0 GPa, while unfolding the entangled domains increases its stiffness during the deformation process.

Schaffer et al. claimed that abalone nacre forms by growth through mineral bridges rather than on heteroepitaxial nucleation [54], and many researchers have confirmed this [52,55,56]. Song et al. were the first to consider the role of pillars (mineral bridges) in enhancing the strength of the structure [57]. They believe that the mineral bridges between the tablets increase the fracture strength of the organic matrix interface by a factor of 5. It was also shown that the cracks propagate along the organic matrix. Their transmission electron microscopy (TEM) images showed that the average length of the cracks in each of the organic layers is about 2 μm. This is because of the crack deflection due to the presence of mineral bridges in the biomaterial. In another study, Song et al. studied the properties and performance of the organic matrix and mineral bridges in nacre's structure [58]. In that study, it was shown that the microstructure of nacre should be referred to as "brick-bridge-mortar" structure. Cartwright et al. also studied the dynamics of nacre self-assembly, which indicates the presence of the shear mineral bridges between the platelets in the nacre of gastropods [59]. Their observations show that the mineral tablets start growing through pores in the membrane and initiate the tablet above. They also pointed out how the pore size may influence the rate of growth of mineral bridges (incomplete growth will create asperities on the surface of the platelets). Similar crystalline orientation of platelets in different layers has been shown as strong evidence of the existence of mineral bridges and of a self-assembly mechanism in nacre [60–62]. Checa et al. also used high-resolution imaging techniques to prove the existence of shear mineral bridges and asperities between the layers of platelets in the nacre of gastropods [63]. Furthermore, as Ghosh et al. pointed out, the organic matrix behaves more stiffly in proximity of the minerals [50]. Hence, the existence of pillars and asperities affects the mechanical properties of nacre through two mechanisms: (a) directly, by inclusion of their own properties; and (b) indirectly, by stiffening the organic matrix.

Beyond experimentations and observations, there are different numerical and theoretical approaches to model the mechanical response of nacre [64–70]. Katti and Katti numerically modeled the mechanical behavior of nacre with an elastic 3D finite element model based on brick-and-mortar microstructure, assuming the organic matrix behaves linearly [71,72]. The results show that the organic layer needs to have a high modulus in order to capture the experimental results on nacre. Evans et al. constructed a model consisting of platelets and asperities with a frictionless interface [73]. Their results show that the nanoscale asperities provide a strain-hardening large enough to ensure the formation of multiple dilatation bands but not so large that platelets fracture internally. Shao et al. also studied the size effect on the mechanical response of nacre and proposed a discontinuous crack-bridging model for fracture toughness analysis of nacre [74]. The effect of preexisting structural defects was investigated in the crack-bridging based model in another study by Shao et al. [75]. More recently, Begley et al. studied the brick-and-mortar structure and proposed a model to guide the development of these composites. In that study, the uniaxial response of the brick-and-mortar composites was analyzed and the role of different parameters, such as the bricks' aspect ratio and the volume percentage of the mortar part, were investigated. The proposed analytical model by Begley et al. [76] is used here to validate some of the results. In this section of the chapter, we propose a micromechanics-based model. Using that model, we can investigate the role of platelets, mineral bridges (pillars), nanoscale asperities, and the organic matrix in the overall deformation mechanism of nacre.

3.2 Mechanics of Nacreous Structure

3.2.1 A Micromechanical Model

A detailed investigation on the microstructure of nacre reveals that the structure of abalone nacre is more than a mortar-brick repeating unit, as shown in the increasing scale image Figure 3.1. Nanoasperities, proteins (organic matrix), and mineral pillars (previously referred to as mineral bridges) fill inside the interlayers and gaps. The statistical studies on abalone nacre illustrate that the average length of one platelet on the cross-section is approximately equal to 4.0 μm and the aspect ratio of the platelets is generally around 8.0 [58], measured as L/D, where L is the length of the platelets and D is the thickness. The pillars are placed in the gap between two layers, normal to the longitudinal side of the platelets.

3.2.1.1 Nonlinear Finite Element Models

Aragonite platelets were modeled as beam elements with a bending stiffness set to represent the flexural behavior of hexagonal platelets with E_{tab} = 100 GPa, although they carry minimal bending stresses. The length of a single platelet was considered to be about 4.0 μm, and its thickness was estimated to be about 0.52 μm. The interlayer gap was 24 nm and the gap between two adjacent platelets in a layer was estimated as 20 nm. The density of the platelets was set to be around 3 g/cm^3. The ABAQUS/Explicit software package was used to carry out the finite element simulations.

The fundamental assumption in this model is that the strength of mineral bridges is equal to their theoretical strength = $\sigma \sim E/30$, because the gap between the platelets is less than the critical length scale (30 nm) [67]. Hence pillars were modeled as link elements with stiffness equal to the shear modulus of aragonite $E_{pillars} = G_{tab}$ = 40.0 GPa

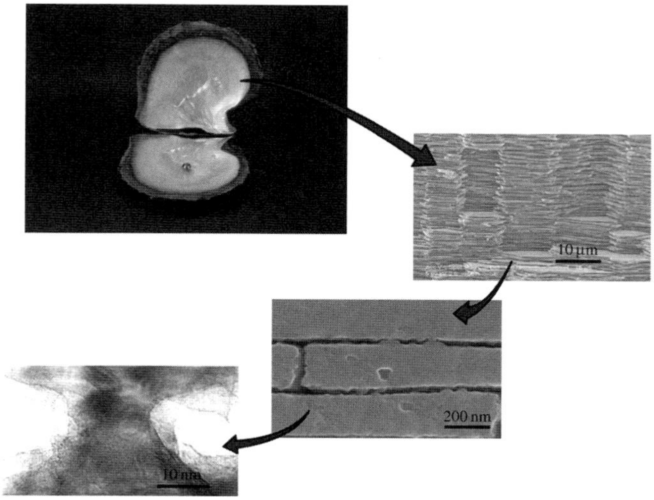

Figure 3.1 Scanning electron microscopy (SEM) images of the microstructure of nacre [19].

and σ_{pillars} = 3.3 GPa. The area fraction of pillars, β_1, was estimated to be about 1.0–4.0%. Moreover, because the density of the pillars in the central region of the platelets is higher than the density of pillars in the outer region of the platelets [58], in our model the pillars modeled as link elements, connecting the middle of the platelets in two adjacent rows in each unit-cell Figure 3.2.

In the proposed model, the behavior of proteins is approximated to be linear elastic before failure. We believe that pillars and asperities confine the organic layer in between the aragonite platelets. This results in a high stiffness value for the organic matrix in the range of ~4.0 GPa. The organic matrix's strength is considered to be near theoretical strength of ~200 MPa. The organic matrix in between the two neighboring platelets was modeled as link elements connecting the edges of the platelets Figure 3.2.

It has been previously shown that the asperities in different layers interpose in many spots. This interlocking affects nacre's behavior by involving more short molecules and making the organic matrix behave [72]. The nanoasperities in between two adjacent platelets provides confinement and entanglement of short molecules in the organic matrix in the proximity of mineral platelets. This is the physical mechanism that provides the high stiffness of the organic matrix. Observations also show nanoscale mineral islands are about 30–100 nm in diameter and 10 nm in amplitude. Nanoasperities show a statistical distribution, with average spacing of 60–120 nm [19]. Accordingly, the area fraction of the asperities, β_2, was estimated to be about 40.0%. Hence the asperities were modeled as link elements to reflect the effects of the interactions of the organic matrix with platelets. This element fails when the displacement between the two platelets in adjacent layers reaches to the point of no interaction between the organic matrix and the platelet, which is when the stress in the organic matrix reaches its strength limit. The asperities are modeled as link elements with stiffness equal to the shear modulus of the organic matrix, $E_{\text{asperities}}$ = 1.6 GPa, and the strength of about 200 MPa.

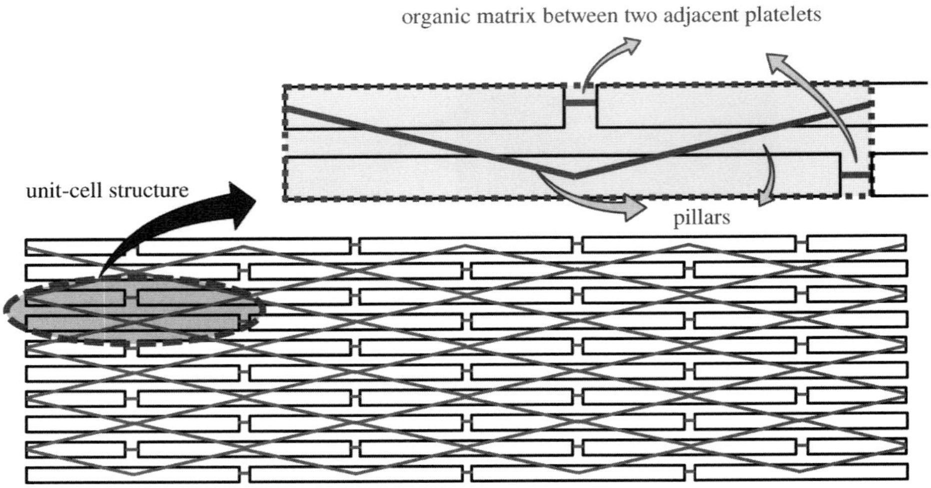

Figure 3.2 Schematic of the arrangement of the link elements representing the organic matrix, pillars, and asperities in the model.

The mechanical behavior of each element in the structure of nacre is shown in Figure 3.3. Using the material properties, geometry, and area fraction of the components, a unit-cell model was created. Displacement was applied in equal time steps to the right-hand side of the system, while the left-hand side of the unit-cell was fixed in the X direction. Figure 3.2 shows a schematic of the unit-cell structure and the way the link elements are arranged to capture the mechanical response of the proposed components. In this system, each element is guaranteed to behave in its natural state; e.g., pillars are modeled as link elements in the direction parallel to the platelets, which translates into shear behavior with no bending.

The unit-cell was used to model the deformation mechanism in a super-cell representing the multilayered structure of nacre. The super-cell was composed of 10 elements in 20 layers for each simulation. Using symmetry, only a quarter of the sample was modeled under a four-point bending experiment. The four-point bending experiment was modeled by applying the linear displacement on the right-hand side of the super-cell as shown in Figure 3.4. For a super-cell, average stress in the layers is plotted versus the average strain of the total cross-section. The simulations were performed for several multilayered models, using the mechanical properties and the prescribed area fractions.

3.2.2 Results: Unit-Cell Behavior and Nacre Micromechanical Model

The simulation result for the mechanical behavior of the unit-cell is presented in Figure 3.5a. The structural components in the unit-cell are schematically presented in Figure 3.5b. All elements carry the load during the early stages of deformation until the stress in the organic matrix under tension in both layers exceeds their strength. Hence, all the elements synergistically contribute to the Young's modulus of the suggested

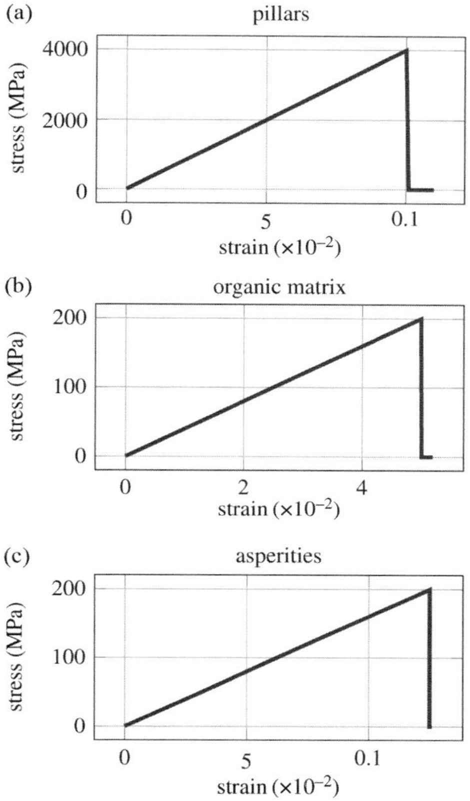

Figure 3.3 Stress–strain behavior of the components present in the unit-cell: (a) pillars, (b) organic matrix, and (c) asperities.

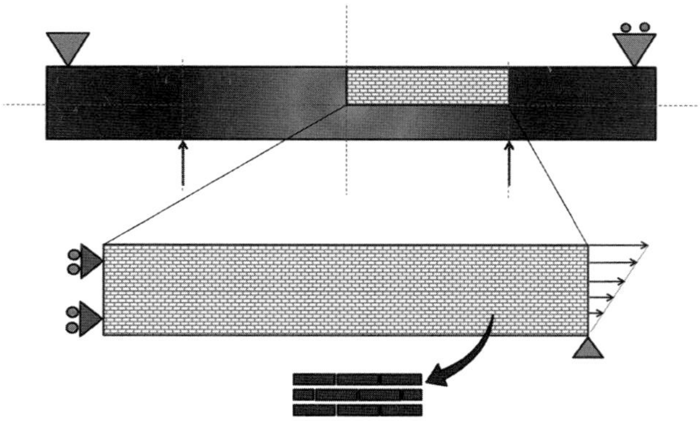

Figure 3.4 Schematic of model used for the simulation of the four-point bend experiment.

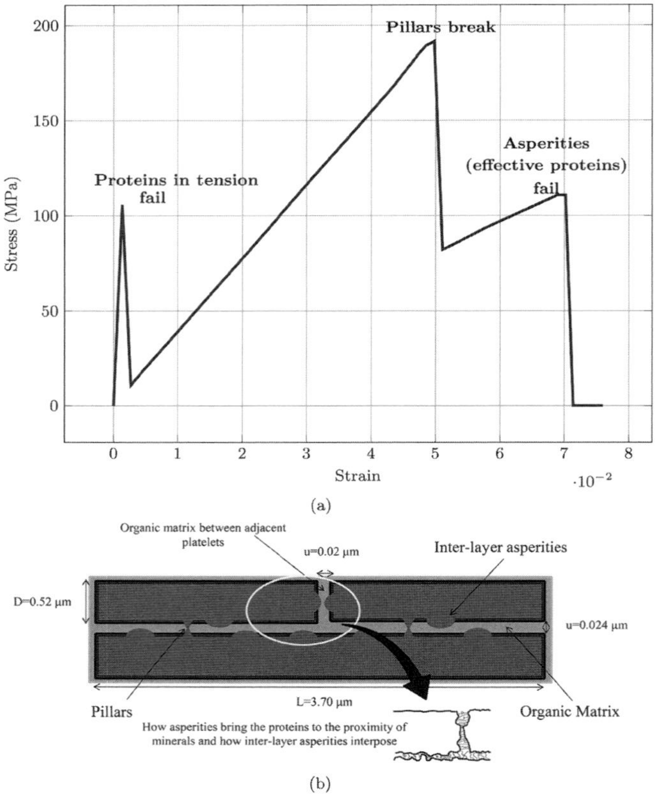

Figure 3.5 (a) Stress–strain plot of the suggested unit-cell and (b) schematic graph of components in the unit-cell structure.

unit-cell. The initial stiffness of the unit-cell with an organic matrix with stiffness value of 4.0 GPa, 40.0% area fraction of asperities, and 2.0% of pillars is about $E = 78.8$ GPa. In the second stage of the loading phase, the pillars and the proteins in shear carry the load. At the end of this stage, pillars break as the stress reaches their strength, and the proteins are the only elements that carry the load. In the third stage, the stiffness of the system comes from the asperities (effective organic matrix). The simulation ends with the failure of the asperity elements, which represents the point of no organic–inorganic interaction between the proteins and the platelets due to the excessive displacement between the two parallel platelets with respect to each other.

The simulation results for a multilayered system similar to the microstructure of nacre is presented in Figure 3.6. Here, average stress in the layers is plotted versus the average strain. Stress averaging is done by dividing the summation of the forces in each layer by the total cross-sectional area in every stage of deformation. The stress–strain curves are in agreement with the previous experimental results [73]. The experimental response of four-point bending test on nacre showed a stiffness of about 70 GPa and a large inelastic deformation area in the stress–strain curve. These values are similar to the simulation

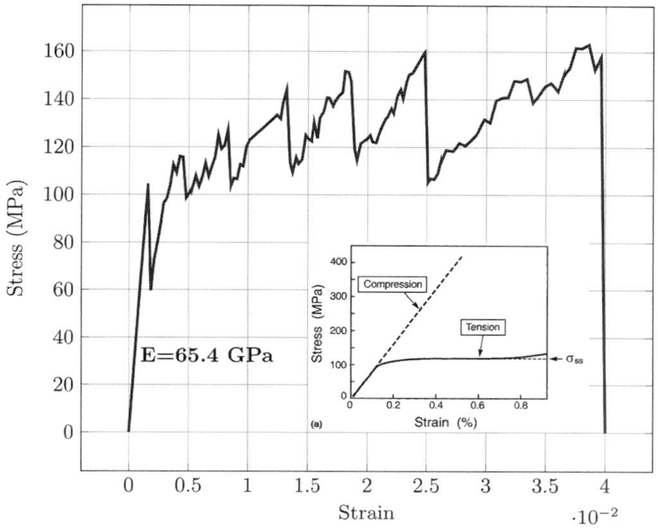

Figure 3.6 Nacre micromechanical model response (compared with the experimental results investigated by Evans et al. [73]).

results for a model with 40.0% area fraction of asperities, 2.0% area fraction of pillars, and an organic matrix with the stiffness of about 4.0 GPa. Additionally, the yield stress of 110 MPa and ultimate strength of 150 MPa is close to the experimentally resulted values of 105 MPa and 140 MPa, respectively. Progressive failures of the organic matrices, pillars, and asperities make the model exhibit significant tensile toughness[1] and a large inelastic deformation. In nacre, tablets sliding spreads throughout the material, and each of the local extensions generated at local sliding zones adds up and causes high strains measured at the macro-scale.

3.3 Mechanics of Nacre-Inspired Ceramic/Polymer Composites

It has been shown that layered structure and super strong interface in nacre is the origin of high strength, stiffness, and toughness [15,77]. Nacre is a natural composite made of 95% aragonite platelets and 5% biological macromolecules. The organic phase is extremely important in nacre's fracture toughness [16,17]. Previous research on nacre has shown that its layered brick-and-mortar structure is the origin of its remarkable mechanical properties [17]. The simple brick-and-mortar structure make the material much tougher than its constituents. Nacre's mechanical properties are strongly dependent on the nanoscale size of the components and hierarchy in its structure [77–80]. One of the important nanoscale features discussed in the literature is the roughness of the tablets; this is called the "asperities." For example, Evans et al. [73] constructed a model

[1] In this paper, tensile toughness is defined as the area under the stress–strain curve.

consisting of platelets and asperities. Their model calculates the stresses needed to displace the plates, resisted by elastic contacts at the asperities. All these studies can be a guide to design and fabricate man-made lamellar composites; however, more theoretical models are needed.

Layered structure can result in high toughness values if properly implemented [81–83]. Salehinia et al. studied the properties of multilayered ceramic/metal composites and investigated the effect of plastic deformation of metals in these materials [84]. There are several analytical studies on the mechanics of multilayered materials [85–89]. An analytical model proposed by Jager and Fratzl [90] estimated the maximum stress and strain of the composite by advancing previously established models of mineralized collagen fibrils [91]. A micromechanical model developed by Kotha et al. [92] derived axial and shear stress distribution in platelets on the assumption that the load carried by the platelets remains constant, and interplatelet load transfer occurs by shear. Shuchun and Yueguang [93] used the classical shear-lag model to study the interdependence of the overall elastic modulus and the number of hierarchical levels in bone-like materials, and compared their results against the tension-shear-chain (TSC) model postulated by Gao et al. [67] and finite element simulations. Investigations by Chen et al. [94] were focused on understanding the characteristic length for efficient stress transfer in staggered biocomposites via derivation of an analytical model followed by numerical simulations. Wei et al. [95] laid down a criterion which reveals the existence of a unique overlap length in biological composites that contributes to an optimization of both strength and toughness. In a study by Begley et al., a micromechanical model is proposed for brick-and-mortar structure with elastic bricks and elastic, perfectly plastic, thin mortar layers [76]. This study was the first to provide a map for designing tough, strong, and stiff materials with a brick-and-mortar structure.

The hierarchical structure is the main inspiration behind numerous attempts to design and fabricate multilayered micro- and nanostructured materials [78,96–98]. Various methods such as electrophoresis [99], dip coating [100], sol–gel [101], and deposition of mineral or clay platelets [102] have been developed to synthesize nacre-like structure. However, in all of these methods the maximum size of the samples that can be fabricated is around several microns, which is not practical for mechanical applications [103]. Freeze casting has proven to be one of the most effective approaches to overcome this limitation where bioinspired multilayered composites are made in larger scales [104–106]. A porous bone-like ceramic was the first structural material made using freeze casting method [107]. The cooling rate and the solution concentration play important roles in improving the mechanical response of these scaffolds [108,109]. Ceramic/polymer and ceramic/metal composites were the first bioinspired composite materials made by freeze casting method [26,109]. However, the effect of the structure on the toughness and the toughening mechanisms, and the role of the components were not fully understood. The main goal of this study is to explain the basic deformation, strengthening, and toughening mechanism and the resulting stress distributions in multilayered materials. The microscale layers and nanoscale features in the samples made by freeze casting and investigating their mechanics distinguish our study from others.

In this section of the chapter, we have also focused on the toughness of multilayered lamellar structures. The effect of lamella thickness on mechanical performance, especially on toughness of the materials, was investigated experimentally and theoretically. The effect of material's architecture, which can be controlled in freeze casting, on the crack-resisting curve of samples, was also studied. Moreover, the effect of mortar properties was assessed by making lamellar composites with different matrix constituents. Furthermore, the importance of imperfect interface in man-made samples, particularly samples made by the freeze casting technique, was investigated. The results of this study can be used as a guide to optimize the mechanics of bioinspired lamellar and brick-and-mortar materials.

3.3.1 Materials and Method

3.3.1.1 Experimental Procedure

The cumbersome freeze casting technique, broadly described as the templating of porous ceramic scaffolds by solidification of a solvent, was used in this study to create the multilayered ceramic/polymer composites. A directional freezing machine with controllable temperature and cooling rate was built and used to fabricate the porous lamellar-structured ceramic samples [110]. All of the steps, such as the slurry preparation, controlled solidification of the slurry, sublimation of the solvent, and densification of the green body or sintering, play important roles in the final structure of the ceramic scaffold and, therefore, mechanical performance of the composites. Hence, in this study, role of solidification rate is explored.

Lamellar alumina scaffolds were prepared by freeze casting of water-based suspensions. The suspensions (solid content 40 wt%) were prepared by dispersing alumina powders with an average diameter of about 150 nm (nano α-Al_2O_3 powder, 99.85%, 40 nm grain size, Inframat Advanced Materials) in deionized water. An ammonium polymethacrylate anionic dispersant (Dar van CN, R. T. Vanderbilt Co., Norwalk, CT, 1.2 wt% of the powder) and an organic binder (polyvinyl alcohol, 1.2 wt% of the powder) were then added to achieve a uniform powder dispersion, and to ensure the integrity of the ceramic structure after removal of the ice. The suspensions were then poured into a cylindrical Teflon mold (25 mm in diameter, 50 mm in height) and placed on a copper cold finger. The temperature of this cold finger was monitored with a temperature sensor. A ring heater and cold bath connected to a liquid nitrogen tank were used to manipulate the temperature of the slurry (Figure 3.7). In order to obtain parallel ceramic lamella aligned over macroscopic dimensions (centimeters), the copper surface, where the ice nucleates, was scratched with a silicon carbide paper and small vibration was applied to the mold during freezing.

The cold finger was then cooled down to –130°C with a cooling rate varying between 1 and 25°C/min. Later, the frozen ice/alumina samples were freeze-dried in order to remove the ice from the structure. Afterwards, the sensitive, porous alumina scaffolds were sintered at 1,500°C for 2 hours to densify the ceramic lamella. This resulted in a bulk ceramic porous scaffold, several centimeters in size, with macroscopically oriented dense lamellar structure. Polydimethylsiloxane (PDMS) and Polyurethane (PU) were

Table 3.1 Mechanical properties of constituent materials

Material	Elastic Modulus (MPa)	Strength (MPa)
Alumina	50,000	150
PDMS	5	20 [111]
PU	300	50 [112]

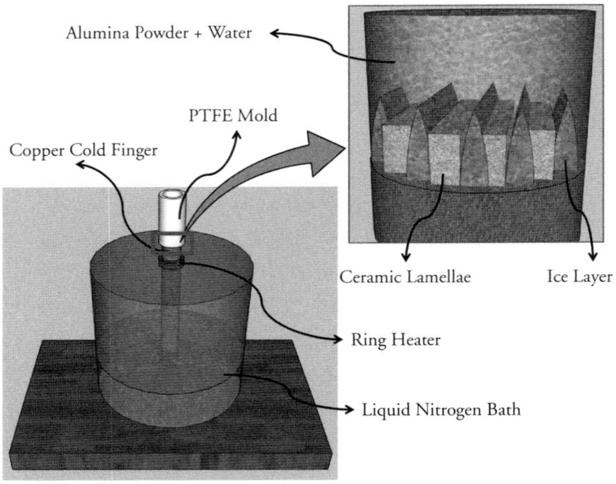

Figure 3.7 Schematic of the freeze casting method used to make samples with microstructures.

infiltrated into the porous sintered scaffold to create the ceramic/polymer composites and study the role of different parameters and the resulting mechanical properties. Finally, vacuum and vibration were applied to soak the polymer into the structure and fill all the gaps and holes. Table 3.1 presents the mechanical properties of constituent materials used to make the samples.

The mechanics of the lamellar-structured ceramic/polymer composites made by the freeze casting method have not been fully explored in the past. We studied the mechanical properties of these samples using tensile tests. Dog-bone samples according to ASTM standard D638 were made and tested using an Instron 8548 micro-tester (Instron, Norwood, MA, USA). We have also investigated the fracture toughness of these samples with dimensions of 22 mm long, 4 mm thick, and 8 mm wide. To perform a single-edge-notch bending experiment, a notch of 0.3 mm depth was created on the samples. The three-point bending test setup was used to perform the experiment. The loading rate of 1 mm/min was used for the experimental procedure. A Canon EOS 600D camera was used to track the crack growth during the experiment as the load was applied and increased on the samples. The recorded raw data were used to calculate the fracture toughness and energy release rate of the composite samples. Energy release rate, J, was calculated from the applied load and instantaneous crack length according to ASTM Standard E1820–06 [113] and was decomposed into elastic and plastic contributions:

$$J = J_{el} + J_{pl}. \tag{3.1}$$

The elastic contribution J_{el} is based on linear elastic fracture mechanics:

$$J_{el} = \frac{K^2}{\bar{E}} \tag{3.2}$$

where K is the stress-intensity factor and $\bar{E}$ is the plane strain Young's modulus,

$$\bar{E} = \frac{E}{1 - \nu^2} \tag{3.3}$$

where ν is Poisson's ratio. The plastic component J_{pl} is given by

$$J_{pl} = \frac{2A_{pl}}{Ba'}, \tag{3.4}$$

where A_{pl} is the plastic area under force vs. displacement curve, B is the specimen thickness, and a is the uncracked ligament length $(W - a)$.

3.3.1.2 The Micromechanical Model: Interlayer, Shear-Lag Analysis

A micromechanical model is also developed to predict the mechanical properties and guide the design of the lamellar-structured composites. In order to discuss the effective properties of a multilayered composite, the balance of forces for the components of the unit-cell, followed by periodic boundary condition should be considered. Hence, the well-known shear-lag theory was employed on a simplified two-dimensional unit-cell of the multilayered composite [114,115]. The previously developed shear-lag models for the layered composites contain several approximations. One of the important simplifications in these models is the assumption of decoupling the x and y directions in the in-plane shear stress, τ_{xy}. This decoupling permits 2D planar elasticity problems to be simplified to a 1D analysis. The in-plane shear stress following infinitesimal theory of deformation is defined as

$$\tau_{xy} = G_{xy}\left(\frac{\partial v}{\partial x} + \frac{\partial u}{\partial y}\right), \tag{3.5}$$

where v is displacement in the y direction, u is the displacement in the x direction, and G_{xy} is in-plane shear modulus. Hence, considering the layers of composite lying down in x direction, the fundamental assumption common to all planar, shear-lag analysis is that

$$\frac{\partial v}{\partial x} = 0 \Rightarrow \tau_{xy} \propto \frac{\partial u}{\partial y}. \tag{3.6}$$

Unidirectional composites can be considered as a layered structure assuming significantly stiff fibers in between soft matrix interlayer. If it is assumed that there are n (an odd number) layers in the geometry, while odd layer numbers are the matrix and the even layer numbers are the fibers, and assuming a linear interpolation between $(u^{(i+1)})$ and $(u^{(i-1)})$, the Hook's law can be written as

$$\tau_{xy}^{(i)} = 0, \quad i = 2m, \tag{3.7}$$

$$\tau_{xy}^{(i)} = \frac{G_{xy}^{(i)}}{t_i}\left(\left\langle u^{(i+1)}\right\rangle - \left\langle u^{(i-1)}\right\rangle\right), \quad i = 2m+1, \tag{3.8}$$

where m is an integer and $\langle\rangle$ indicates averaging over the thickness of the layer and can be defined as

$$\left\langle u^{(i)}\right\rangle = \frac{\frac{1}{t_i}1}{2\pi}\int_{y_{i-1}}^{y_i} u(x,y)dy, \tag{3.9}$$

where $u(y_i)$ and $u(y_{i-1})$ are the axial displacements on the two edges of layer i. Another relation between displacement and shear stress can be derived by using stress equilibrium and Hook's law. The stress equilibrium in layer i can be written as

$$\frac{\partial\sigma_x^{(i)}}{\partial x} + \frac{\partial\tau_{xy}^{(i)}}{\partial y} = 0. \tag{3.10}$$

Integrating over the thickness leads to

$$\frac{d\left(t_i\left\langle\sigma_x^{(i)}\right\rangle\right)}{dx} = \tau(y_{i-1}) - \tau(y_i). \tag{3.11}$$

Later, using the stress–strain relation in the x direction and differentiating with respect to x and averaging over the thickness we have

$$\epsilon_x = \frac{\partial u}{\partial x} = \frac{\sigma_x}{E_x^{(i)}} - \frac{\nu_{xy}^{(i)}\sigma_y}{E_x^{(i)}}, \tag{3.12}$$

$$\frac{d^2\left\langle u^{(i)}\right\rangle}{d^2x} = \frac{1}{t_iE_x^{(i)}}\frac{d\left(t_i\left\langle\sigma_x^{(i)}\right\rangle\right)}{dx} - \frac{\nu_{xy}^{(i)}}{E_y^{(i)}}\frac{d\left(\sigma_y^{(i)}\right)}{dx} \tag{3.13}$$

where, ν_{xy} is the Poisson ratio. Another assumption in the shear-lag theory is that in this equation the second term of the right-hand side is much less than the first term. Hence, the simplifications lead to the following equation:

$$\left\langle u^{(i)}\right\rangle = \frac{1}{t_i}\int_{y_{i-1}}^{y_i} u(x,y)dy. \tag{3.14}$$

Considering these calculations, the shear-lag theory for lamellar-structured stiff fibers and soft matrix with perfect interface can be summarized as

$$\mathbf{A_T}\frac{d^2\tau_T}{d^2x} - \mathbf{B_T}\tau_T = -\tau_{I,0}, \tag{3.15}$$

where A_T is an $\frac{(n-3)}{2} \times \frac{(n-3)}{2}$ diagonal matrix with the ith diagonal element being

$$(A_T)_{i,i} = \frac{t_{2i+1}}{G_{xy}^{(2i+1)}} \tag{3.16}$$

and B_T is an $\frac{(n-3)}{2} \times \frac{(n-3)}{2}$ matrix with the following elements:

$$(B_T)_{i,i-1} = -\frac{1}{E_x^{(2i)} t_{2i}}, \quad (B_T)_{i,i} = -\frac{1}{E_x^{2(i+1)} t_{2(i+1)}} + \frac{1}{E_x^{(2i)} t_{2i}}, \quad (B_T)_{i,i+1} = -\frac{1}{E_x^{2(i+1)} t_{2(i+1)}}. \tag{3.17}$$

A transformation to an equation for average axial stresses in the stiff layers lead to the following equation:

$$\frac{d^2 P_T}{d^2 x} - \mathbf{M}_{\mathbf{I},\sigma} P = \mathbf{M}_{\mathbf{I},\sigma} P_{I,\infty}, \tag{3.18}$$

where P is the vector of forces per unit plate depth on the layers, and can be defined as

$$P = \left(t_1\langle\sigma_x^{(1)}\rangle, t_2\langle\sigma_x^{(2)}\rangle, \ldots, t_{n-1}\langle\sigma_x^{(n-1)}\rangle\right). \tag{3.19}$$

Now, if we assume that the layers with even numbers are the stiff layers, the force vector per unit plate depth, P_T, on these layers is defined as

$$P_T = \left(t_2\langle\sigma_x^{(2)}\rangle, t_4\langle\sigma_x^{(4)}\rangle, \ldots, t_{n-3}\langle\sigma_x^{(n-3)}\rangle\right). \tag{3.20}$$

$P_{I,\infty}$ is the far field or steady state tensile stresses in each layer when there are no shear stress boundary conditions; in other words, the tensile forces in the layers under constant axial stress far away from any ends, discontinuities, or breaks in any layers. Hence, it can be defined as

$$P_T = \left(t_2\langle\sigma_x^{(2)}\rangle, t_4\langle\sigma_x^{(4)}\rangle, \ldots, t_{n-3}\langle\sigma_x^{(n-3)}\rangle\right), \tag{3.21}$$

where $t_T = \sum_{n=2,4}^{n-1} t_i$ and $E_{I,x}^0 = (1/t_T)\sum_{n=2,4}^{n-1} E_x^{(i)} t_i$; that is, the contribution of the stiff layers and $\sigma_0(0)$ is the total applied stress in the x direction when $x = 0$. Moreover, as presented in [116],

$$[M_{I,\sigma}] = [I_L]^{-1}[A_T]^{-1}[B_T][I_L],$$

where $[I_L]$ is an $(n-1) \times (n-1)$ matrix with 1 on the diagonal and in lower half of the matrix zero elsewhere. Hence the elements of $[I_L]^{-1}$ are: $[I_L]_{i,j}^{-1} = 1$ if $i = j$, $[I_L]_{i,j}^{-1} = -1$ if $j = i - 1$, otherwise $[I_L]_{i,j}^{-1} = 0$. Assuming that the fibers are all the same in thickness and properties and the matrix layers have the same thickness and properties, $[M_{I,\sigma}]$ summarizes to

$$\mathbf{M}_{\mathbf{I},\sigma} = \frac{G_m}{E_b t_b t_m} \begin{pmatrix} 1 & -1 & 0 & 0 & \cdots \\ -1 & 2 & -1 & 0 & \cdots \\ 0 & -1 & 2 & -1 & \cdots \\ & & \downarrow & & \\ \cdots & -1 & 2 & -1 & 0 \\ \cdots & 0 & -1 & 2 & -1 \\ \cdots & 0 & 0 & -1 & 1 \end{pmatrix}.$$

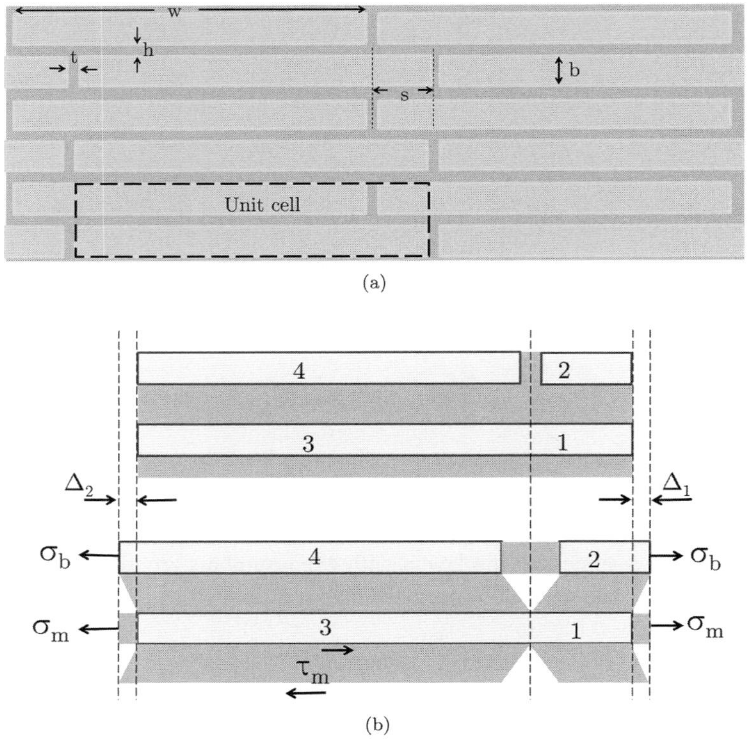

Figure 3.8 (a) Schematic of idealized super-cell structure and the length parameters (t = thickness of the vertical interface, h = thickness of the polymer layer, w = length of the bricks, b = thickness of the ceramic layers, and s = overlap) and (b) unit-cell structure assumed to apply the shear-lag theory.

Here we assumed an idealized composite geometry that is schematically presented in Figure 3.8. The wallpaper symmetry group is *p*mm, with two bricks per unit cell, which are separated by one brick height vertically, and shifted by a distance of s horizontally. The mortar thickness between adjacent plates in each layer (vertical mortar) is assumed to be small in comparison to the brick dimensions. It is assumed that the components deform only in the x-direction (the arrangement of the bricks in the through-thickness direction does not factor into the response).

For the assumed geometry, $M_{I,\sigma}$ would be as follows:

$$\mathbf{M}_{\mathbf{I},\sigma} = \frac{G_m}{E_b t_b t_m} \begin{pmatrix} 1 & -1 \\ -1 & 1 \end{pmatrix}.$$

In other words, the matrix A_T in Eq. (3.16) will be summarized in one term only as $A_T = \frac{t}{G}$. Hence, this notation leads to the following complete set of governing equations:

$$u_1'' = \frac{G_m}{E_b h b}(u_1 - u_2) \rightarrow 0 < x < s, \tag{3.22}$$

$$u_1'' = \frac{G_m}{E_b hb}(u_1 - u_2) \rightarrow 0 < x < s, \tag{3.23}$$

$$u_3'' = \frac{G_m}{E_b hb}(u_3 - u_4) \rightarrow -(w - s) < x < 0, \tag{3.24}$$

$$u_4'' = \frac{G_m}{E_b hb}(u_4 - u_3) \rightarrow -(w - s) < x < 0. \tag{3.25}$$

The corresponding boundary conditions can be written as follows:

$$u_1(0) = u_3(0) = 0, \tag{3.26}$$

$$u_2(s) = \Delta, \tag{3.27}$$

$$u_2(s) - u_3(s - \omega) = \epsilon.\ \omega, \tag{3.28}$$

$$u_3'(0) = u_2'(0) = \alpha[u_2(0) - u_3(0)], \tag{3.29}$$

$$u_1'(s) = u_3'(s - \omega) = 0, \tag{3.30}$$

$$u_1'(0) = u_2'(s) = u_3'(s - \omega), \tag{3.31}$$

where $\alpha = E_m/E_b t$. Using these boundary conditions, the displacement of the different parts of model, u_1, u_2, u_3 and u_4 were calculated.

3.3.1.3 Effect of Imperfect Interface

The quality of the interface plays an important role in the overall mechanical properties of composites [117–123]. In perfectly bonded interfaces, tractions and displacements are continuous across the interface. However, in many cases, the perfect interface assumption is not adequate. The classical approach to an imperfect interface is to relax interfacial continuity and allow a discontinuity in stresses and displacements at the interface [124]. In static loading conditions, stress equilibrium requires the stresses in the interface to be continuous regardless of the quality of the interface. The remaining discontinuities in displacements are functions of interfacial stresses:

$$\begin{aligned} \sigma_{xx}^{(1)} &= \sigma_{xx}^{(2)} = D_n[u_x], & [u_x] &= u_x^{(2)} - u_x^{(4)}, \\ \tau_{xy}^{(1)} &= \tau_{xy}^{(2)} = D_s\left[u_{xy}\right], & \left[u_{xy}\right] &= u_x^{(2)} - u_x^{(1)}, \end{aligned} \tag{3.32}$$

where D_n and D_s are called interface parameters. The parameters represent the ability of the interphase to transfer stress. A perfect interface can be represented by $D_n = D_s = \infty$, which implies vanishing of displacement jumps, and a completely detached interface can be represented by $D_n = D_s = 0$. Intermediate values describe an imperfect interface.

In order to implement the effect of interface into the developed shear-lag models, Eq. (3.33) was modified to

$$(A_T)_{i,i} = \frac{t_{2i+1}}{G_{xy}^{(2i+1)}} + \frac{1}{D_s^{(2i+1)}}, \tag{3.33}$$

where D_s is the interface parameter, as previously described. Using this new assumption, the governing equations were derived and solved.

3.3.2 Results and Discussion

3.3.2.1 Experiments: Microstructure Analysis, Mechanical Response, and Fracture Properties of Lamellar Composites

The results show that changing the cooling rate changes the lamella thickness. During the process of preparing the samples, the temperature of the cold finger was monitored and manipulated by combining a cold bath with ring heaters. The cold finger was cooled to −130°C with the cooling rate varying between 1 and 25°C/min. Controlling the cooling rates under 3°C/min is challenging, and cooling rates above 25°C/min generally cause excessive vibrations in the machine. Figure 3.9 shows an SEM image of a typical ceramic scaffold before polymer infiltration. The ceramic layers can be as thin as 4 μm and as thick as 20 μm, using the presented framework. Figure 3.10 compares the different microstructures of the composite samples made with different cooling rates. It is clear from the images that increasing the cooling rate will decrease the lamella thickness.

An important feature in the structure of nacre that can be mimicked here, in the samples made by the freeze casting method, is the asperities that appear on one side of each ceramic layer. Asperities play an important role in enhancing the polymer/ceramic interfaces [77]. Figure 3.11 shows the asperities in the structure of fabricated composites.

Figure 3.12a compares the stress–strain curves obtained by tensile testing on Al_2O_3/PDMS samples. Four average lamella thicknesses are the result of four average cooling rates. The graphs show that there is a trade-off between strength and ductility of the samples. In these samples, as the lamella thickness decreases, the stiffness and strength increase and the ductility decreases. This is only true for the case of soft polymers, because the weak properties of the polymer causes a low state of stress after the vertical rupture.

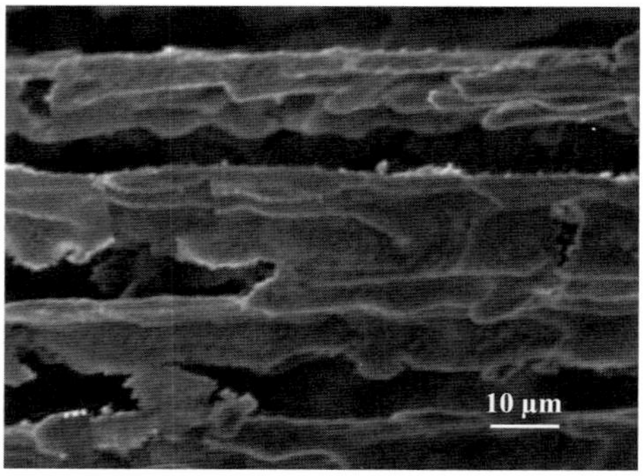

Figure 3.9 SEM image of a bioinspired ceramic scaffold with lamellar structure before polymer infiltration.

(a)

(b)

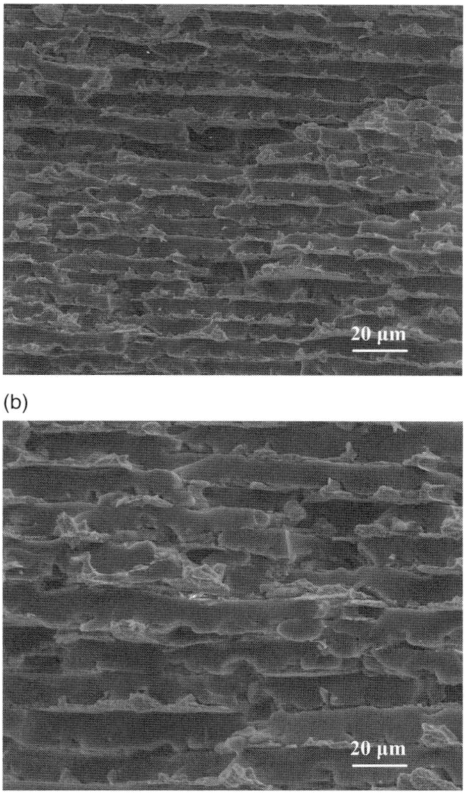

Figure 3.10 SEM image: (a) a lamellar-structured composite with lamella thickness of about 4 μm fabricated with cooling rate of 20°C/min and (b) a lamellar-structured composite with lamella thickness of about 11 μm fabricated with cooling rate of 5°C/min.

Figure 3.12b compares the stress–strain curves of Al_2O_3/PU samples made with different cooling rates. The graphs show that there is an optimum value for the strength. For small lamella thicknesses, the matrix causes the plates to fracture; however, for larger thicknesses, the plates tolerate the stress and the shear stress in the mortar determines the strength. For Al_2O_3/PU samples, the drop in the state of stress is lower, after vertical rupture, than the drop in the state of stress in Al_2O_3/PDMS samples. Comparing these data shows that the stiffness and ultimate strength of Al_2O_3/PU samples are higher than those of the Al_2O_3/PDMS samples. However, this is only due to the higher stiffness and strength of the PU in comparison to PDMS. The larger ultimate strains in samples is due to the fiber pull-out effect, which is more pronounced in the samples with PDMS. The toughness and toughening mechanisms are extensively discussed in Section 3.3.2.3. Single-edge-notch bending tests were performed on the samples, and the stress at each crack length was used to find the R-curve of the fabricated samples [125]. Figure 3.13a and b compares the crack resistance curves for

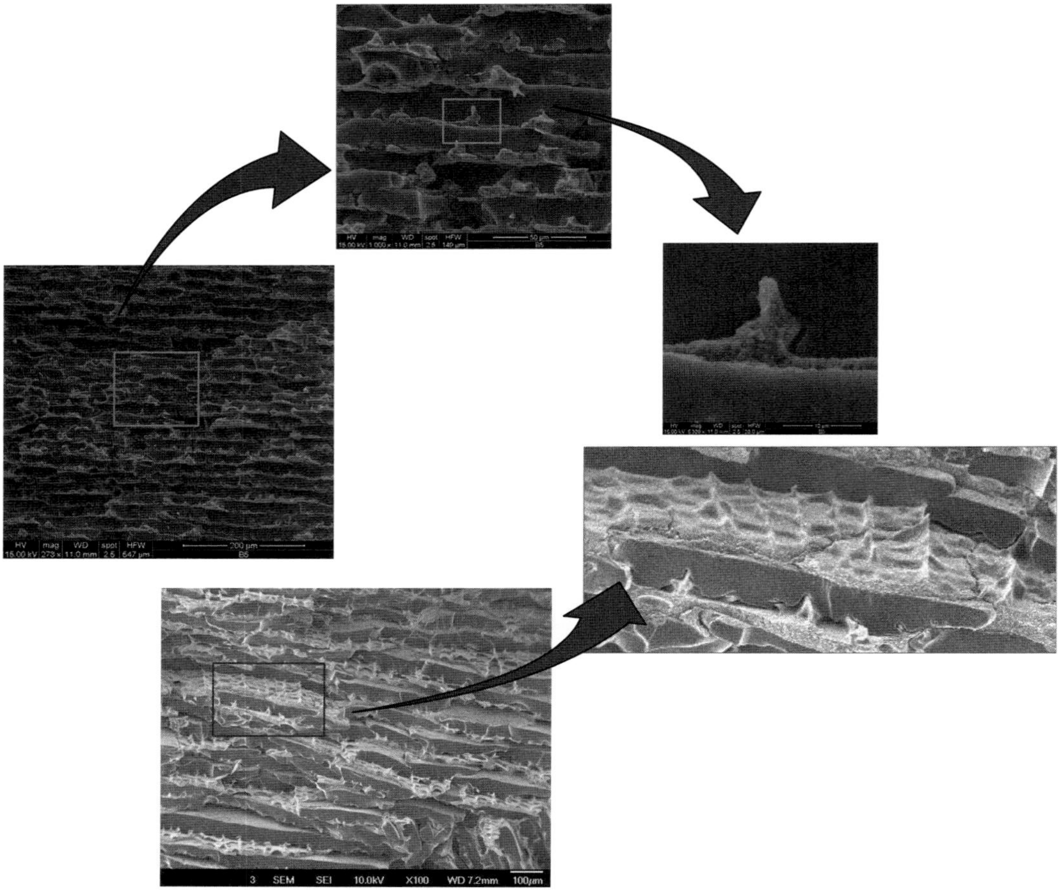

Figure 3.11 Presence of asperities in the structure of samples made by freeze casting plays an important role in their mechanical properties.

Al_2O_3/PDMS and Al_2O_3/PU samples made with different cooling rates, respectively. The influence of ceramic lamella thickness on fracture energy in these samples can be clearly observed in these results.

Figure 3.13a demonstrates that the crack resistance increases as the crack grows in the samples with the larger lamella thicknesses, and this is due to the effective fiber (lamellae) pull-out mechanism that occurs toward the final stage of deformation. This toughening mechanism is less effective in samples with smaller lamella thickness. The same mechanism can be observed in the Al_2O_3/PU samples. However, as the matrix in this case is significantly stiffer than PDMS, the fiber pull-out mechanism is indeed less effective. Hence, the R-curve of samples with small lamella thickness (4 μm) drops significantly. However, it must be noted that all composite samples fabricated here with lamella thicknesses of more than 4 μm outperform nacre with respect to fracture energy. The plane strain critical energy release rate

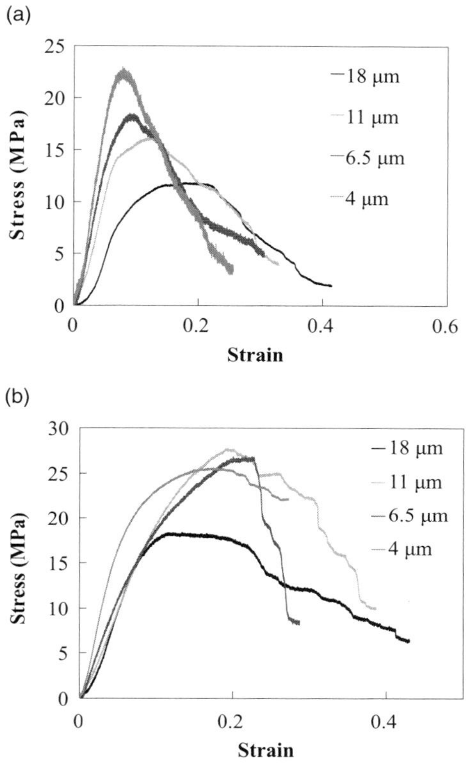

Figure 3.12 Stress–strain relationships for (a) Al_2O_3/PDMS and (b) Al_2O_3/PU samples as a function of lamella thicknesses.

(J_{IC}) for samples with different lamella thickness and different matrix material is also presented in Figure 3.14.

3.3.2.2 Elastic Modulus, Strength, and Effect of Imperfect Interface

The volume fraction of the ceramic phase in the samples made by freeze casting can be assumed to be the same for all the samples, because the alumina to water ratio is the same in all the initial solutions. Analytically, the prescribed shear-lag model was used to derive four second-order differential equations of motion, corresponding to the four parts of the unit cell Figure 3.8. After solving the equations subjected to the boundary conditions, the maximum stress in the bricks σ_b, the maximum shear stress in the horizontal interface τ_i, the stress in the vertical interfaces σ_i, and the composite stress σ_c can be found as

$$\sigma_b = E_b u_2'(0), \tag{3.34}$$

$$\tau_i = \frac{G_m}{h} u_2'(0), \tag{3.35}$$

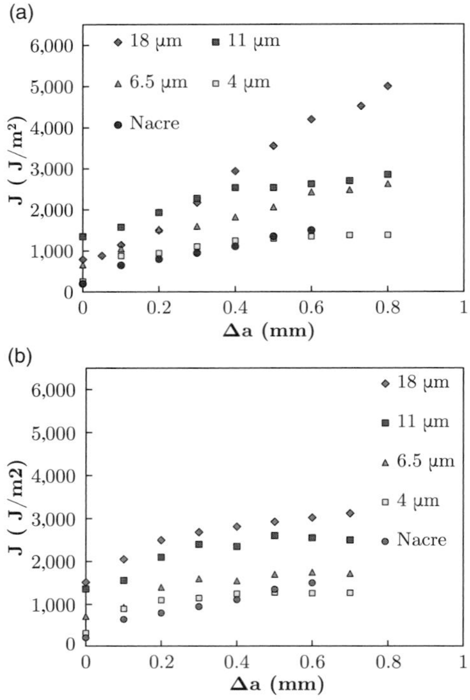

Figure 3.13 Crack resistance curves: Energy release rate as a function of crack growth in (a) Al_2O_3/PDMS samples and (b) Al_2O_3/PU samples. For comparison, the nacre data is extracted from [126].

$$\sigma_i = E_b u_1'(0), \tag{3.36}$$

$$\sigma_c = \frac{1}{2} E_b \left[u_1'(0) + u_2'(0) \right]. \tag{3.37}$$

The effective properties of the composite can then be readily obtained from these equations. Using Eq. (3.37), the stress in the composite can also be found as a function of sample geometry, properties of the components, and the composite strain. Dividing the stress by the strain in the composite leads to the stiffness of the composite as a function of microstructural geometries and properties of the constituents, as

$$E_c = \frac{\omega E_b \left[\left[\frac{1}{2} \sinh(\beta s) + \alpha(\cosh(\beta s) + 1) \right] \sinh(\beta(s-\omega)) - \alpha \sinh(\beta s)[1 + \cosh(\beta(s-\omega))] \right]}{[(1+\alpha\omega)(\cosh(\beta s)+1)] \sinh(\beta(s-\omega)) - \sinh(\beta s) \left[(1+\alpha\omega)(1+\cosh(\beta(s-\omega))) - \frac{\omega}{2}\beta \right]} \tag{3.38}$$

where w is the length of the ceramic layer, s is the overlap of the layers, and α and β parameters are defined as follows:

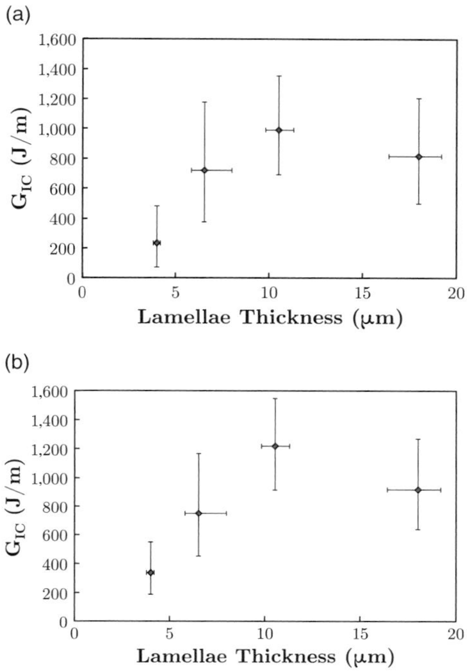

Figure 3.14 Plane strain critical energy release rate, G_c, as a function of lamella thickness in (a) Al_2O_3/PDMS samples and (b) Al_2O_3/PU samples.

$$\beta = 2\sqrt{\frac{G_m}{E_b h b}}, \quad \alpha = \frac{E_m}{E_b t}. \tag{3.39}$$

Using Eqs. (3.34)–(3.36), the stress in the bricks, shear stress in the mortar, and stress in the vertical interfaces can also be obtained:

$$\sigma_i = \frac{\omega E_b \epsilon_c [[2\beta \sinh(\beta s) + \alpha(\cosh(\beta s) + 1)] \sinh(\beta(s-\omega)) - \alpha \sinh(\beta s)[1 + \cosh(\beta(s-\omega))]]}{[(1+\alpha\omega)(\cosh(\beta s)+1)] \sinh(\beta(s-\omega)) - \sinh(\beta s)\left[(1+\alpha\omega)(1+\cosh(\beta(s-\omega))) - \frac{\omega}{2}\beta\right]}, \tag{3.40}$$

$$\sigma_i = \frac{\omega E_b \epsilon_c \alpha [[\cosh(\beta s) + 1] \sinh(2\beta(s-\omega)) - \alpha \sinh(\beta s)[1 + \cosh(\beta(s-\omega))]]}{[(1+\alpha\omega)(\cosh(\beta s)+1)] \sinh(\beta(s-\omega)) - \sinh(\beta s)\left[(1+\alpha\omega)(1+\cosh(\beta(s-\omega))) - \frac{\omega}{2}\beta\right]}, \tag{3.41}$$

$$\tau_i = \frac{\omega E_b \epsilon_c [[\cosh(\beta s) + 1] \sinh(\beta(s-\omega))]}{[(1+\alpha\omega)(\cosh(\beta s)+1)] \sinh(\beta(s-\omega)) - \sinh(\beta s)\left[(1+\alpha\omega)(1+\cosh(\beta(s-\omega))) - \frac{\omega}{2}\beta\right]}. \tag{3.42}$$

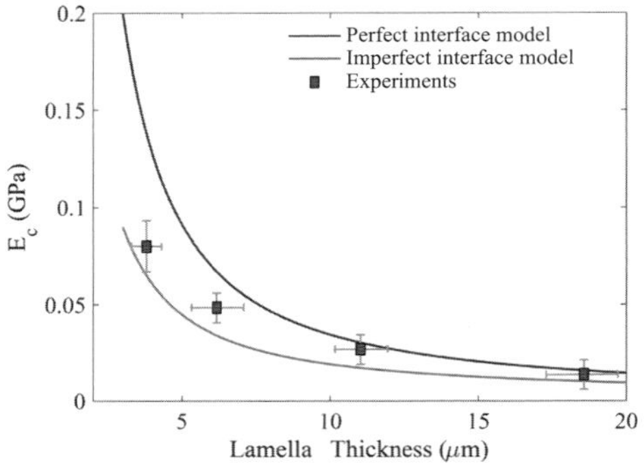

Figure 3.15 Effect of ceramic lamella thickness on the elastic modulus of the bioinspired Al_2O_3/PDMS samples. The mortar thickness is assumed to be 50% larger than the ceramic lamella thickness, according to the SEM images and the volume fraction of the polymer. The error bars represent the standard deviation in the lamella thickness and the elastic moduli.

Finally, α and β parameters can be simply modified to implement the effects of imperfect interfaces into the shear lag model as:

$$\alpha = \frac{D_s b E_m}{E_b t(2E_m + bD_s)}, \quad \beta = 2\sqrt{\frac{G_m D_s}{E_b b(D_s h + E_m)}}.$$

The elastic modulus and the stresses in different elements of the idealized 2D lamellar structure can then be derived as a function of interface property. Figure 3.15 shows the effect of ceramic lamella thickness on the elastic modulus of the samples. The figure also compares the experimental and theoretical results found in this study.

Comparing the theoretical results of models with perfect and imperfect interfaces shows that the effect of imperfection increases as the lamella thickness decreases. Moreover, comparing the experimental results with the theoretical ones shows that the samples with higher lamella thickness behave similarly to the samples with the perfect interface; however, as the lamella thickness decreases, the results lean mostly toward the samples with the imperfect interface. The interface imperfection parameter is assumed to be rather higher for PDMS samples than the ones for PU samples. This is due to two physical phenomena: (1) a molecule agglomerate of PU is larger than PDMS, and this prevents perfect influx of the polymer into the surface voids of the ceramic lamella; and (2) the curing time for PDMS is higher, and this causes more interaction between the ceramic and the polymer layers before the final curing stage. These two different levels of interface imperfection are presented in SEM images obtained from the samples in Figure 3.16.

Using the stresses found in the previous sections, following the shear-lag theory, the composite's strength can now be predicted. The magnitude of stresses in each element at different strain levels defines the failure sequences in the material system. As it can be seen in a unit-cell structure, the vertical interface would fail first if the bricks can tolerate the applied stress. However, the failure in the horizontal interfaces eventually causes the sample's final rupture.

In all these multilayered composites, the ideal scenario is to prevent the failure in the bricks, as it will quickly diminish the strength of the composite. This leads to the use of

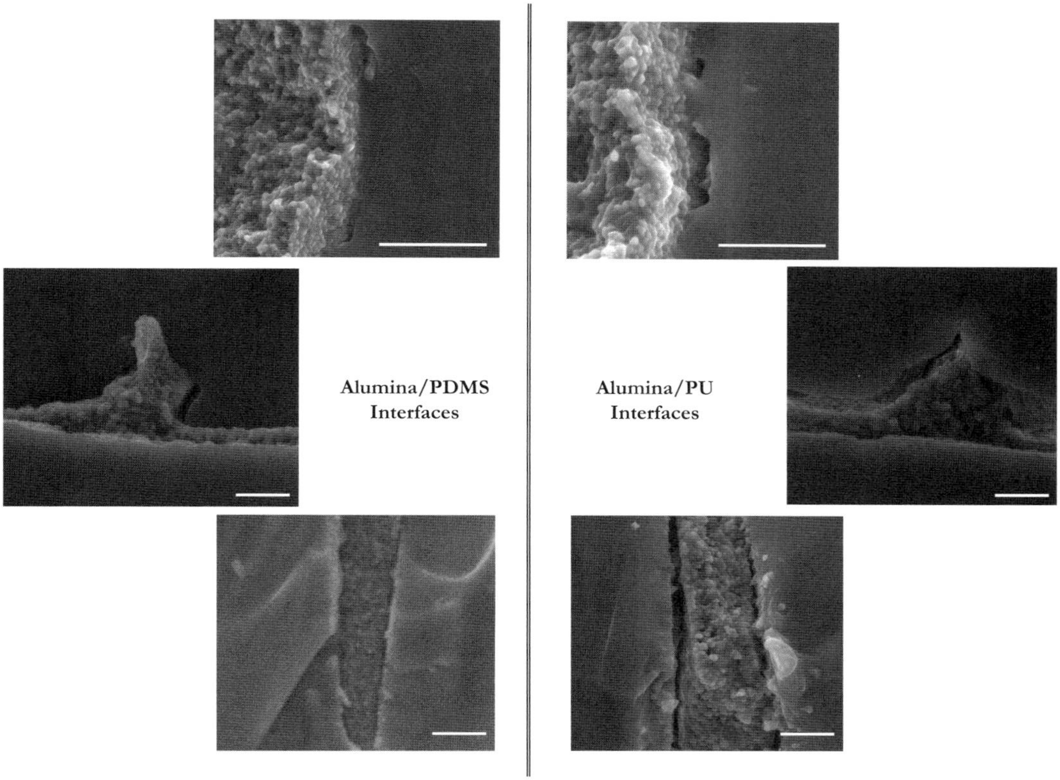

Figure 3.16 The interface property in the samples is a function of different chemical and physical properties of polymers. The imprecations are more dominant at the alumina/PU interfaces.

strong ceramics in these composites. The random flaws in some of the plates does not significantly affect the system as long as there are strong horizontal interfaces. Consequently, the composite can still carry the load after vertical failure, such that the strength is essentially dictated by the stress required to subsequently fail all the horizontal interfaces or fracture the bricks.

The presented analytical framework is used here to find the peak stress (ultimate strength) of the samples. Obviously, the mechanical response of the components as well as the length scales in the model ultimately define the sequence of failure of the elements, and therefore, the peak stress in the system. In the multilayered composites, the vertical rupture happens at the peak stress. If the vertical rupture is delayed due to a mechanism, the peak stress will increase.

The ideal failure scenario that can occur in a multilayered structure is that the vertical interfaces fail first, then some horizontal interfaces start to fail, following by a failure of all vertical interfaces (vertical rupture), and then failure of all horizontal interfaces (horizontal rupture). However, in the lamellar-structured composites with a large volume fraction of polymer, the model shows that vertical rupture happens in a small additional strain after the start of vertical interfaces failure. In Al_2O_3/PU samples, horizontal interfaces start to fail in a higher stress than the Al_2O_3/PDMS samples.

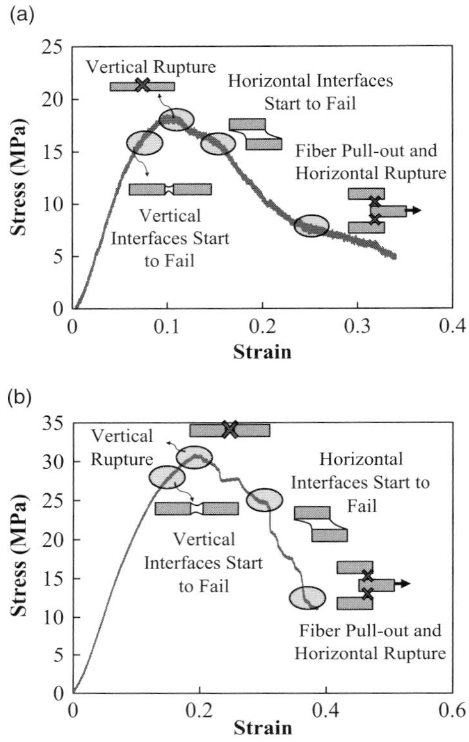

Figure 3.17 Failure sequences in (a) Al_2O_3/PDMS and (b) Al_2O_3/PU samples.

The drop in the composite stress (σ_c) is smaller in Al_2O_3/PU samples after vertical rupture. Hence, Al_2O_3/PU samples exhibit a large strain energy in that step. However, more effective fiber pull-out mechanism can be observed in Al_2O_3/PDMS samples, because the stress in the bricks never actually reaches to their strength during the fiber pull-out process. Figure 3.17 schematically summarizes the failure sequence in the Al_2O_3/PDMS and Al_2O_3/PU samples.

The stress required for a specific failure event (e.g., vertical rupture) is found by equating the associated stress with the material strength and solving for the composite strain E_c required for that event. The composite stress (σ_c) can then be found by multiplying the obtained strain by the composite modulus for that specific event.

In order to find the peak composite stress, failure stress, and failure strain, the following steps need to be taken, since the lamellar structure assumed to be a brick and mortar structure with large mortar fractions [76]: (1) Using the properties of the constituents, calculate the strain in which each of the elements would fail; (2) The minimum strain is defined as the first failure event. If this strain corresponds to either the brick failure or horizontal interface failure, a catastrophic failure will occur, and the corresponding stress is the peak and failure stress; (3) If the minimum strain corresponds to the vertical interface, the next failure event needs to be found. In order to do that, the stress in the bricks and horizontal interface should be found, assuming the failed vertical interface ($\alpha = 0$); (4) If either of stresses in the brick or the horizontal

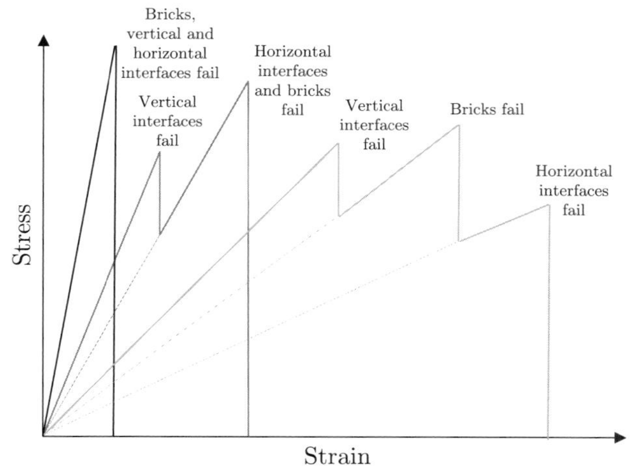

Figure 3.18 Some of the possible scenarios could be predicted using the theoretical analysis. All the parameters are effective on the failure sequences and, subsequently, the maximum strength.

interface (with $\alpha = 0$) is less than their strengths, the sample fails and the stress in vertical interface is the peak and failure stress; (5) If the stresses in the bricks and horizontal interfaces are smaller than their strengths, more strain can be applied to the sample; (6) Using the new modulus and the computed stresses in the elements, the failure strains of the brick and horizontal interface can be found. The stress corresponding to the minimum strain determines the failure stress; and (7) If this stress is more than the vertical interface failure stress, the obtained stress is the peak stress. Figure 3.18 schematically shows some of the possible scenarios that can happen. The highest stress in each of the possible scenarios defines the ultimate strength of the samples.

Figure 3.19a and b shows the calculated analytical strength as a function of ceramic lamella thickness. Similarly, the results of the analytical calculation of strength with the imperfect interface are presented in these figures. It can be observed from these results that the experimental results are similar to the analytical solutions for the imperfect interface problem for the samples with low ceramic lamella thicknesses. Additionally, in Figure 3.19b the theoretical results show that there is an optimum value for the lamella thickness to achieve the highest strength in Al_2O_3/PU samples. This optimum thickness value decreases as the interface gets weaker. The overall drop in the strength value is mainly due to the brick failure, which occurs because of the small aspect ratio of the layers and the high strength of the polymer layers. The effect of interface properties is indeed pronounced in predicting the material properties in these cases. Hence the results clearly show that the effect of interface modification is quite significant on the strength and toughness of samples.

The presented model can also be used to study tensile toughness (area under the stress–strain curve) of the samples by using the values found for the strength of samples. Figure 3.20a and b shows the trend in toughness with respect to the elastic moduli of composites. The toughness calculated from the experimental results is also presented in these figures. Experimental results for all samples are predicted closely with the analytical solutions and the trend in toughness variation as a function of elastic modulus

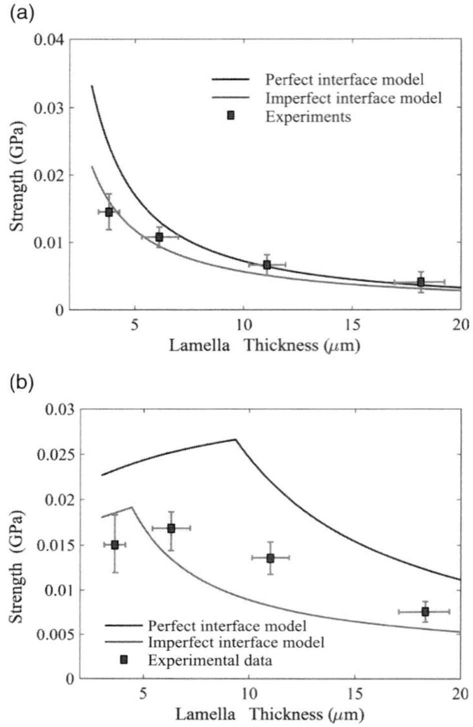

Figure 3.19 Effect of ceramic lamella thickness on the strength of the (a) Al_2O_3/PDMS and (b) Al_2O_3/PU samples. The mortar thickness is assumed to be 50% larger than the ceramic lamella thickness according to the SEM images. The error bars represent the standard deviation in the lamella thickness and the strength.

matches well with the presented imperfect interface model. The differences between the experimental and modeling results can be listed as: (1) The thickness and length of the plates are not uniform throughout the samples and the limitations in manufacturing method (such as excessive vibration in higher cooling rates) cause some minor anomalies in the microstructure; (2) The behavior of the polymer layer may be different at different thicknesses while we assume the same number for the specific material in the theoretical model; (3) The interactions of polymer with ceramic layer at the final stage of deformation that is important in toughening mechanisms is not included in the shear-lag model. When the polymers are soft, the toughness increases as the elastic modulus increases; however, when the polymer gets stiff and strong, the toughening mechanisms at the later stages of the deformation get less effective, and the toughness does not increase as the elastic modulus increases.

3.3.2.3 Fracture Toughening Mechanisms

The tensile toughness added to materials due to the presence of fibers in the system can be calculated by the following equation [127]:

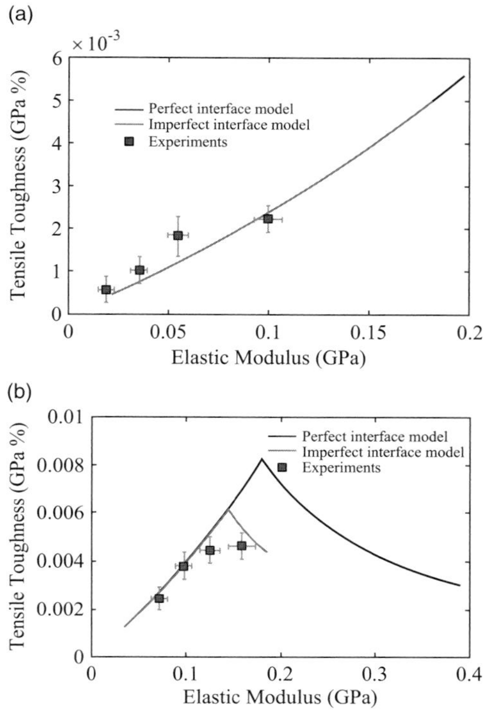

Figure 3.20 Variation of tensile toughness in (a) Al_2O_3/PDMS and (b) Al_2O_3/PU samples. Tensile toughness increases as the elastic modulus increases in samples with the soft matrix (such as PDMS). For PU samples, there is an optimum value for the toughness as the elastic modulus increases. The effect of imperfect interface is significant in these samples.

$$\Delta K = 2f \int_0^{u_*} \bar{\sigma}(u) du, \tag{3.44}$$

where K is the toughness, f is the volume fraction of fibers, u is the crack opening, whichever is smaller, $u*$ is the minimum of two crack opening values, opening at the crack mouth and opening at the edge of the bridging zone (Figure 3.21). $\bar{\sigma}$ is the net traction applied on the matrix crack by fibers in the crack wake. In fiber/matrix systems, both intact fibers and fibers that fail away from the crack plane increase the stress needed for crack growth [128]. Unbroken fibers toughen the system by bridging effect and the broken ones resist crack opening by frictional sliding as the fibers pull out of the matrix [129].

In the studies about the composites with ceramic fibers, they usually assign a specific number for the strength of the fibers (S); however, this is not what happens in reality. If we assume a certain number for the strength of the fibers, they must fail at the location of the highest stress, which eliminates the effect of fiber pull-out in the analysis. Here, the contribution of both fiber bridging and pull-out on the crack growth resistance is taken to account by considering the statistical nature of failure in brittle ceramic fibers.

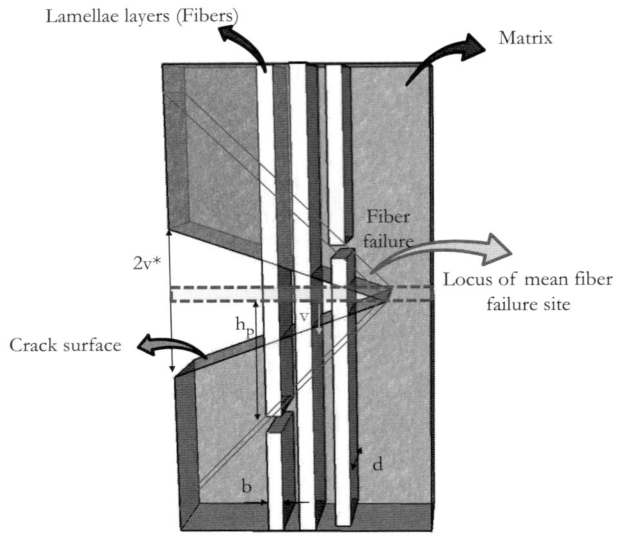

Figure 3.21 A schematic indicating crack bridging by intact and fractured lamellae layers. Different dimensions and the locus of mean fiber failure site are shown. Fiber pull-out length can also be found.

The fibers are assumed to have a highly imperfect interface so that they slip with respect to the matrix in the crack wake at a constant shear resistance, τ, which can be found by analyzing the matrix shear strength and the interface properties [130]. The resulting slip length, l, is presumed to be large enough to allow all of the interfacial shear to be accommodated along the zone of sliding. The axial stress in each intact fiber varies within the slip length

$$\sigma = T\left(1 - \frac{z}{l}\right), \qquad l = \frac{TA}{M\tau}, \tag{3.45}$$

where T is the stress in the fiber at the crack plane, $A = b \times d$ is the fiber cross section area (b is the fiber thickness and d is the lamellae layer depth), $M = 2(b + d)$ is the fiber circumference exposed to the matrix, and z is the distance from the crack plane. A further assumption of the analysis is the use of a fiber strength distribution that satisfies weakest-link statistics. The probability density function, P, for fiber failure as a function of the peak stress, T, and the distance from the crack plane, z, can be derived as

$$P = \frac{2md}{A_0\sigma_0^m}\left(T - \frac{2z\tau}{b}\right)^{m-1} \exp\left(-\frac{2bdT^{m+1}}{(m+1)\tau A_0\sigma_0}\right), \tag{3.46}$$

where d is the depth of the lamellae layers, m is the shape parameter, and σ_0 is the scale parameter. Eq. (3.46) is the basic formula that governs several statistical parameters such as pull-out length, h, the mean strength, S, and the net stress, $\bar{\sigma}$. These parameters play important roles in toughness of the lamellar-structured composites. The average

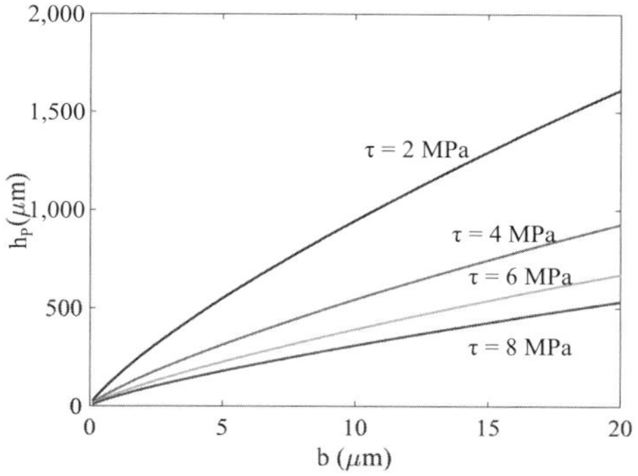

Figure 3.22 Fiber pull-out length as a function of lamellae thickness for different shear resistance of the mortar. Decreasing the shear resistance and increasing the lamellae thickness increase the fiber pull-out length.

failure position for all fibers that have failed at a stress less than T can be found and simplified as

$$\bar{h} = \frac{A_0}{4(b+d)}\left(\frac{\sigma_0}{\eta}\right)^m \Gamma\left(\frac{m+2}{m+1}\right), \tag{3.47}$$

where Γ is the complete gamma function and

$$\eta = \left[\frac{A_0 \sigma_0^m \tau(m+1)}{2A}\right]^{\frac{1}{m+1}}. \tag{3.48}$$

As can be seen in this equation, the fiber pull-out length can be calculated as a function of the lamellae thickness and the shear stress in the mortar. Figure 3.22 shows the fiber pull-out length versus the lamellae thickness for different shear resistance of mortar. As the shear resistance increases, the pull-out length decreases.

Therefore, the corresponding solution for the mean strength of all the fibers is

$$\bar{S} = \eta\Gamma\left(\frac{m+2}{m+1}\right). \tag{3.49}$$

In order to find the effect of fibers on the toughness of composites, the fraction of fibers that fail at each location behind the crack tip and the associated failure site should be determined. The fraction of failed fibers near the matrix crack front can be assumed to be zero. As the matrix crack grows, the weaker fibers begin to fail in the immediate crack wake, near the crack plane. Thereafter, upon additional crack extension, further fibers fail in the wake, at locations further from the crack plane. Hence, for failed fibers, the stress at the crack plane is:

$$\sigma_p = \frac{2\pi\tau}{(b+d)}(h-u) \qquad (h > u) \tag{3.50}$$

$$\sigma_p = 0 \quad (h < u), \tag{3.51}$$

where u is the crack opening displacement. Moreover, for intact fibers, the peak stress, T, can be determined by the following equation:

$$T=\left(\frac{4\pi E_f\tau(1+\zeta)u}{b+d}\right)^{\frac{1}{2}}, \tag{3.52}$$

where ζ is

$$\zeta=\frac{E_f f}{E_m(1-f)}. \tag{3.53}$$

Using this analysis, the average stress, σ, on the fibers at an opening, u, can be written as

$$\bar{\sigma}=(1-q)T+q\langle\sigma_p\rangle, \tag{3.54}$$

where q is the fraction of fibers that have failed at opening, u, and $\langle\sigma_p\rangle$ is the average value of the pull-out stress associated with all prior failures. An important point in using this equation is that the number of fibers assumed to be sufficient to allow use of average values, which is quite practical in microstructured lamellar composites. The fraction of failed fibers at an opening, u, can be found using the proposed probability function, P, by finding the cumulative fiber failure probability at $S = T$

$$q\equiv\int_0^S\int_0^l P(z,T)dzdT. \tag{3.55}$$

Hence, the average stress in the fibers in the lamellar-structured composite with respect to the crack opening, geometric parameters, and mechanical properties of constituents can be derived as

$$\begin{aligned}\bar{\sigma}=\eta\sqrt{\frac{u}{v}}\exp\left[-\left(\frac{u}{v}\right)^{\frac{m+1}{2}}\right]+\frac{\eta}{(1+\zeta)(1+m)}\left\{1-\exp\left[-\left(\frac{u}{v}\right)^{\frac{m+1}{2}}\right]\right\}\\ \times\left\{\gamma\left[\frac{m+2}{m+1},\left(\frac{u}{v}\right)^{\frac{m+1}{2}}\right]-\frac{\eta(m+1)}{2E_f}\left(\frac{u}{v}\right)\right\}.\end{aligned} \tag{3.56}$$

The average stress for a crack opening length is a function of geometry, shear resistance, and Weibull parameters. Figure 3.23 shows the dependence of average stress on the prescribed parameters.

Using this analysis, the details on toughness can be highlighted by separately considering the contributions of fiber bridging and fiber pull-out. Hence, the bridging and pull-out components of the toughness can be found and simplified as

$$\begin{aligned}\Delta K&=2f\int_0^{u_*}\bar{\sigma}(u)du=2f\int_0^{u_*}\left[(1-q)T+\langle q\sigma_p\rangle\right]du\\&=2f\Big(\underbrace{\int_0^{u_*}[(1-q)T]du}_{\Delta K_b}+\underbrace{\int_0^{u_*}q\langle\sigma_p\rangle du}_{\Delta K_p}\Big),\end{aligned} \tag{3.57}$$

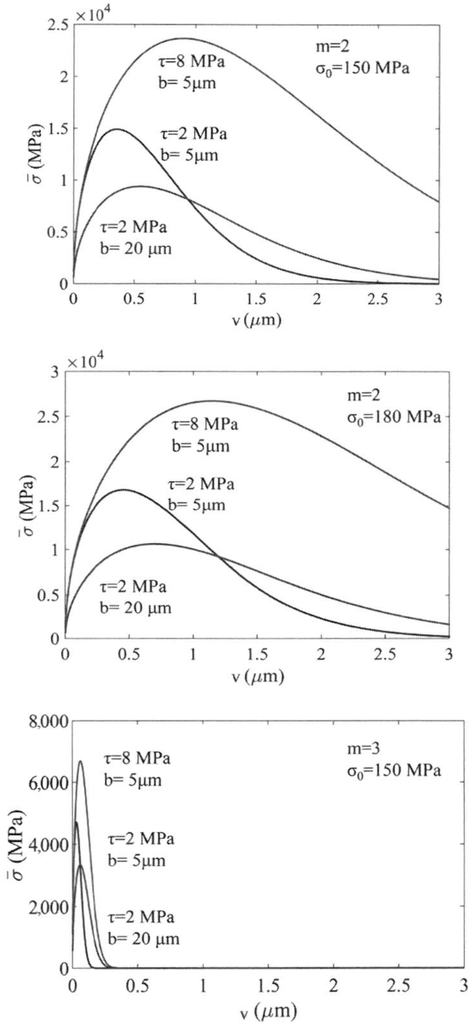

Figure 3.23 Stress/displacement characteristics for a crack surface bridged by fibers. The purpose of these graphs is to illustrate the dependence of toughness on the shear resistance and thickness of the lamellae at particular amount of m and σ_0.

$$\Delta K_b = \frac{4f\nu\eta}{m+1}\gamma\left[\frac{3}{m+1}, \left(\frac{u_*}{\nu}\right)^{\frac{m+1}{2}}\right] \tag{3.58}$$

$$\Delta K_p \sim \langle h\rangle^2\left(\frac{(b+d)\tau}{bd}\right), \tag{3.59}$$

where $\gamma(r,\ s)$ is the incomplete gamma function and

$$\nu = \frac{\eta^2 bd}{4E_f\tau(1+\zeta)(b+d)}. \tag{3.60}$$

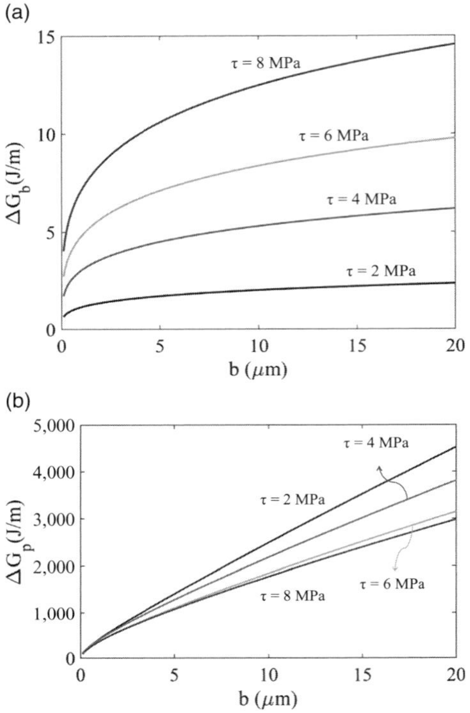

Figure 3.24 The energy release rate gradient due to the bridging and pull-out toughening mechanisms. As the results show, in the lamellar-structured composites, the effect of fiber pull-out is more significant than the bridging effect.

As it is shown, the crack mouth opening, u_*, has an important effect on toughening mechanisms [131]. If we assume that the crack has an elliptical shape, u_* can be found with the following equation:

$$u_* = \frac{2K_c}{\bar{E}}\sqrt{\frac{a}{\pi}}, \tag{3.61}$$

where $\bar{E}$ is the plane strain Young's modulus and a is the length of crack.

Moreover, selection of lamella's mean strength, m, and lamella's strength distribution function, σ_0, within the typical range obtained experimentally [132], enables prediction of specific stress/displacement characteristics for a crack surface bridged by lamella. Here, we used the values of m and σ_0, previously found experimentally for alumina samples [133,134]. For our samples made by freeze casting and sintered in 1,500°C for 2 hours, both these values are assumed to be less than the regular alumina samples. Moreover, in the freeze casting process, larger numbers of flaws appear in the ceramic lamella of samples fabricated with a high cooling rate. Hence for samples with thinner lamella, a smaller value of m should be assumed. Considering all these assumptions, the extreme cases for our bioinspired composite corresponds to Al_2O_3/PDMS samples with lamella thickness of 18 μm with the highest fiber pull-out toughening, and Al_2O_3/PU

Table 3.2 Mechanical properties of constituent materials

Lamellae thickness (μm)	$\Delta J_p + \Delta J_b$ (J/m^2)	ΔJ_{exp} (J/m^2)
4	1,015	550 ± 428
6.5	1,623	1,920 ± 121
11	1,786	2,120 ± 254
18	4,361	4,820 ± 322

samples with lamella thickness of 4 μm with the highest fiber bridging effect. These findings can also be observed in the R-curves found by fracture experiments presented in Figure 3.13.

The contributions to overall fracture energy from the toughening mechanisms, $\Delta J = \Delta J_b + \Delta J_p$, in PDMS samples are compared in Table 3.2. These data clearly validate the presented theoretical framework in predicting fracture energies of bioinspired lamellar composites for the range of lamella thicknesses investigated.

3.3.3 Implications

The implications of these results are quite significant in the design of biologically inspired super-tough ceramic composites. Man-made brick-and-mortar and lamellar ceramic/polymer composites with optimized structure and components' mechanical properties can outperform nacre in toughness. However, microstructural requirements to improve strength, stiffness, and toughness are different for lamellar structures. Decreasing the lamella thickness increases the elastic modulus in the layered structure and the detailed properties of the matrix play the main role in the maximum amount of stress that the composite can tolerate. The results presented here would, therefore, suggest that there are different material design parameters that need to be taken into account to achieve the optimal material properties for a specific application using the freeze casting technique. There are also some design and fabrication limitations for lamellar-structured samples made by this method: (1) the thickness of the layers are limited to 4–18 μm, (2) there are limitations in material selection with specific desired properties, and (3) there are limitations in matrix material selection due to the complexity of matrix infiltration. There is also a significant need to investigate the effect of weak interface properties in these bioinspired composites in order to compute the optimum microstructure and materials selection. Interface properties, which depend on the chemical and mechanical bonding between ceramic lamella and matrix, impose another important design parameter into the multilayered materials mechanics models.

3.4 Conclusions

We have investigated the mechanical response of bioinspired structural composites made by the freeze casting method. Al_2O_3/PDMS and Al_2O_3/PU lamellar composites were fabricated, characterized, and experimented. The main goal of this study is to understand the deformation and toughening mechanisms of the lamellar-structured composites. For this purpose, an experimental setup was built and used to fabricate multilayered ceramics/polymer composite using the cumbersome freeze casting technique. A shear-lag model is also developed to study the effective properties of these composites.

The overall performance of multilayered bioinspired materials depends on four important factors: (1) geometry of the microstructure which can be controlled by the cooling rate in freeze casting method, (2) properties of the ceramic lamella, (3) properties of the polymer (matrix), and (4) the ceramic/polymer interface properties. The results show that the toughness of composite decreases as the ceramic layer thickness decreases. Hence the optimum thickness depends on the overall mechanical properties of the components. Fiber bridging and pull-out processes were theoretically and experimentally investigated as the main toughening mechanisms and correlated with the increase in the crack resistance curve of the samples.

In this chapter, the imperfect interface model was also implemented in the shear-lag model describing the deformation mechanisms in multilayered composites. The analytical results using the imperfect interface model match well with the experimental results obtained from the samples fabricated by the freeze casting technique. Hence the findings of this study allow us to optimize and improve the mechanical performance of lamellar-structured composites made by freeze casting. Moreover, the results of this study emphasize the effect of interface properties (which is the key feature in biological materials) and further enable us to focus on nature's strategy to create superb biological composites such as nacre.

Studying the strength and stiffness of natural composites has a profound impact on the development of man-made materials. Designs in nature are incredible for their combination of building blocks and amplification of properties [3,5]. One of these composites is nacre, which possesses impressive mechanical properties in terms of its strength and stiffness. Nacre is made up of 95% aragonite ($CaCO_3$) and 5% organic materials [77]. These components are weak individually, with protein as soft as skin and minerals as brittle as chalk; however, they come together to form a remarkably strong composite [17]. The high mineral content in nacre makes it one of the stiffest natural composites [3]. The arrangement of nacre structure allows it to be twice as strong and a thousand times tougher than its constituent materials. For example, while ceramics are superior to metals because they are stiffer, harder, and can be used at higher temperature, their brittleness is a major drawback. However, replicating the mechanisms in nacre could lead to ceramic-based composites that are significantly tougher than typical ceramics, broadening their potential uses [5]. There is much effort focused on mimicking the structure of nacre using various methods. Even so, the deformation and toughening mechanisms of these materials need more in-depth investigation in order to design and fabricate nature-inspired composites.

References

1. Ortiz, C., & Boyce M. C. (2008). Bioinspired structural materials. *Science*, 319(5866), 1053–1054.
2. Weaver, J. C., Milliron, G. W., Miserez, A., et al. (2012). The stomatopod dactyl club: A formidable damage-tolerant biological hammer. *Science*, 336(6086), 1275–1280.
3. Barthelat, F. (2010). Nacre from mollusk shells: A model for high-performance structural materials. *Bioinspiration & Biomimetics*, 5(3), 035001.
4. Woesz, A., Weaver, J. C., Kazanci, M., et al. (2006). Micromechanical properties of biological silica in skeletons of deep-sea sponges. *Journal of Materials Research*, 21(08), 2068–2078.
5. Espinosa, H. D., Juster, A. L., Latourte, F. J., Loh, O. Y., Gregoire, D., & Zavattieri, P. D. (2011). Tablet-level origin of toughening in abalone shells and translation to synthetic composite materials. *Nature Communications*, 2, 173.
6. Mayer, G., & Sarikaya, M. (2002). Rigid biological composite materials: structural examples for biomimetic design. *Experimental Mechanics*, 42(4), 395–403.
7. Currey, J. D., Landete-Castillejos, T., Estevez, J., et al. (2009). The mechanical properties of red deer antler bone when used in fighting. *Journal of Experimental Biology*, 212(24), 3985–3993.
8. Li, X., Chang, W. C., Chao, Y. J., Wang, R., & Chang, M. (2004). Nanoscale structural and mechanical characterization of a natural nanocomposite material: The shell of red abalone. *Nano Letters*,4(4), 613–617.
9. Grunenfelder, L. K., Suksangpanya, N., Salinas, C., et al. (2014). Bio-inspired impact-resistant composites. *Acta Biomaterialia*, 10(9), 3997–4008.
10. Prabhakaran, M. P., Venugopal, J., & Ramakrishna, S. (2009). Electrospun nanostructured scaffolds for bone tissue engineering. *Acta Biomaterialia*, 5(8), 2884–2893.
11. Johnson, M., Walter, S. L., Flinn, B. D., & Mayer, G. (2010). Influence of moisture on the mechanical behavior of a natural composite. *Acta Biomaterialia*, 6(6), 2181–2188.
12. Askarinejad, S., Kotowski, P., Shalchy, F., & Rahbar, N. (2015). Effects of humidity on shear behavior of bamboo. *Theoretical and Applied Mechanics Letters*, 5(6), 236–243.
13. Askarinejad, S., Kotowski, P., Youssefian, S., & Rahbar, N. (2016). Fracture and mixed-mode resistance curve behavior of bamboo. *Mechanics Research Communications*, 78, 79–85.
14. Launey, M. E., Chen, P. Y., McKittrick, J., & Ritchie, R. O. (2010). Mechanistic aspects of the fracture toughness of elk antler bone. *Acta Biomaterialia*, 6(4), 1505–1514.
15. Song, Z. Q., Ni, Y., Peng, L. M., Liang, H. Y., & He, L. H. (2016). Interface failure modes explain non-monotonic size-dependent mechanical properties in bioinspired nanolaminates. *Scientific Reports*, 6, 23724.
16. Xu, Z. H., & Li, X. (2011). Deformation strengthening of biopolymer in nacre. *Advanced Functional Materials*, 21(20), 3883–3888.
17. Ji, B., & Gao, H. (2004). Mechanical properties of nanostructured biological materials. *Journal of Mechanics and Physics of Solids*, 1963–1990.
18. Askarinejad, S., Choshali, H. A., Flavin, C., & Rahbar, N. (2018). Effects of tablet waviness on the mechanical response of architected multilayered materials: Modeling and experiment. *Composite Structures*, 195, 118–125.

19. Wang, R. Z., Suo, Z., Evans, A. G., Yao, N., & Aksay, I. A. (2001). Deformation mechanisms in nacre. *Journal of Materials Research*, 16(09), 2485–2493.
20. Jackson, A. P., Vincent, J. F. V., & Turner, R. M. (1988). The mechanical design of nacre. *Proceedings of the Royal Society of London Series B*, 234, 415–440.
21. Meyers, M. A., Chen, P. Y., Lin, A. Y. M., & Seki, Y. (2008). Biological materials: Structure and mechanical properties. *Progress in Materials Science*, 53(1), 1–206.
22. Espinosa, H. D., Rim, J. E., Barthelat, F., & Buehler, M. J. (2009). Merger of structure and material in nacre and bone: Perspectives on de novo biomimetic materials. *Progress in Materials Science*, 54(8), 1059–1100.
23. Meyers, M. A., McKittrick, J., & Chen, P. Y. (2013). Structural biological materials: Critical mechanics-materials connections. *Science*, 339(6121), 773–779.
24. Porter, M. M., Yeh, M., Strawson, J., et al. (2012). Magnetic freeze casting inspired by nature. *Materials Science and Engineering: A*, 556, 741–750.
25. Munch, E., Launey, M. E., Alsem, D. H., Saiz, E., Tomsia, A. P., & Ritchie, R. O. (2008). Tough, bio-inspired hybrid materials. *Science*, 322(5907), 1516–1520.
26. Launey, M. E., Munch, E., Alsem, D. H., Saiz, E., Tomsia, A. P., & Ritchie, R. O. (2010). A novel biomimetic approach to the design of high-performance ceramic/metal composites. *Journal of the Royal Society Interface*, 7(46), 741–753.
27. Wang, R. Z., Wen, H. B., Cui, F. Z., Zhang, H. B., & Li, H. D. (1995). Observations of damage morphologies in nacre during deformation and fracture. *Journal of Materials Research*, 30(9), 2299–2304.
28. Ritchie, R. O. (2011). The conflicts between strength and toughness. *Nature Materials*, 10(11), 817–822.
29. Hunger, P. M., Donius, A. E., & Wegst, U. G. (2013). Platelets self-assemble into porous nacre during freeze casting. *Journal of the Mechanical Behaviour of Biomedical Materials*, 19, 87–93.
30. Bonderer, L. J., Feldman, K., & Gauckler, L. J. (2010). Platelet-reinforced polymer matrix composites by combined gel-casting and hot-pressing. Part I: Polypropylene matrix composites. *Composite Science and Technology*, 70(13), 1958–1965.
31. Tang, Z., Kotov, N. A., Magono, S., & Ozturk, B. (2003). Nanostructured artificial nacre. *Nature Materials*, 2(6), 413–418.
32. Bouville, F., Maire, E., Meille, S., Van de Moortle, B., Stevenson, A. J., & Deville, S. (2014). Strong, tough and stiff bioinspired ceramics from brittle constituents. *Nature Materials*, 13(5), 508–514.
33. Jackson, A. P., Vincent, J. F. V., & Turner, R. M. (1989). A physical model of nacre. *Composite Science and Technology*, 36(3), 255–266.
34. Bass, J. D. (1995). Elasticity of minerals, glasses, and melts. In *Mineral physics and crystallography: A handbook of physical constants*, vol. 2, 45–63.
35. Currey, J. D., & Taylor, J. D. (1974). The mechanical behavior of some molluscan hard tissues. *Journal of Zoology, London*, 173, 395–406.
36. Currey J. D. (1976). Further studies on the mechanical properties of mollusc shell material. *Journal of Zoology, London* 180, 445–453.
37. Verma, D., Katti, K., & Katti, D. (2007). Nature of water in nacre: A 2D Fourier transform infrared spectroscopic study. *Spectrochimica Acta A*, 67(3), 784–788.
38. Mohanty, B., Katti, K. S., Katti, D. R., & Verma, D. (2006). Dynamic nanomechanical response of nacre. *Journal of Materials Research*, 21(08), 2045–2051.

39. Feng, Q. L., Cui, F. Z., Pu, G., Wang, R. Z., & Li, H. D. (2000). Crystal orientation, toughening mechanisms and a mimic of nacre. *Materials Science and Engineering C*, 11(1), 19–25.
40. Barthelat, F., & Espinosa, H. D. (2007). An experimental investigation of deformation and fracture of nacre/mother of pearl. *Experimental Mechanics*, 47(3), 311–324.
41. Barthelat, F., Li, C. M., Comi, C., & Espinosa, H. D. (2006). Mechanical properties of nacre constituents and their impact on mechanical performance. *Journal of Materials Research*, 21(8), 1977–1986.
42. Bruet, B. J. F, Qi, H. J., Boyce, M.C., et al. (2005). Nanoscale morphology and indentation of individual nacre tablets from the gastropod mollusc *Trochus niloticus*. *Journal of Materials Research*, 20(9), 2400–2419.
43. Smith, B. L., Schaffer, T. E., Viani, M., et al. (1999). Molecular mechanistic origin of the toughness of natural adhesives, fibres and composites. *Nature*, 399(6738), 761–763.
44. Mohanty, B., Katti, K. S., & Katti, D. R. (2008). Experimental investigation of nanomechanics of the mineral-protein interface in nacre. *Mechanics Research Communications*, 35(1), 17–23.
45. Katti, K. S., Mohanty, B., & Katti, D. R. (2006). Nanomechanical properties of nacre. *Journal of Materials Research*, 21(5), 1237–1242.
46. Moshe-Drezner, H., Shilo, D., Dorogoy, A., & Zolotoyabko, E. (2010). Nanometer-scale mapping of elastic modules in biogenic composites: The nacre of mollusk shells. *Advanced Functional Materials*, 20(16), 2723–2728.
47. Xu, Z. H., Yang, Y., Huang, Z., & Li, X. (2011). Elastic modulus of biopolymer matrix in nacre measured using coupled atomic force microscopy bending and inverse finite element techniques. *Materials Science and Engineering C*, 31(8), 1852–1856.
48. Stempfle, P., Pantale, O., Njiwa, R. K., Rousseau, M., Lopez, E., & Bourrat, X. (2007). Friction-induced sheet nacre fracture: Effects of nano-shocks on cracks location. *International Journal of Nanotechnology*, 4(6), 712–729.
49. Stempfle, P. H., Pantale, O., Rousseau, M., Lopez, E., & Bourrat, X. (2010). Mechanical properties of the elemental nanocomponents of nacre structure. *Materials Science and Engineering C*, 30(5), 715–721.
50. Ghosh, P., Katti, D. R., & Katti, K. S. (2007). Mineral proximity influences mechanical response of proteins in biological mineral-protein hybrid systems. *Biomacromolecules*, 8(3), 851–856.
51. Barthelat, F., Tang, H., Zavattieri, P. D., Li, C. M., & Espinosa, H. D. (2007). On the mechanics of mother-of-pearl: A key feature in the material hierarchical structure. *Journal of Mechanics and Physics of Solids*, 55(2), 306–337.
52. Meyers, M. A., Lin, A. Y. M., Chen, P. Y., & Muyco, J. (2008). Mechanical strength of abalone nacre: Role of the soft organic layer. *Journal of the Mechanical Behaviour of Biomedical Materials*, 1(1), 76–85.
53. Xu, Z. H., & Li, X. (2011). Deformation strengthening of biopolymer in nacre. *Advanced Functional Materials*, 21(20), 3883–3888.
54. Schaffer, T. E., Ionescu-Zanetti, C., Proksch, R., et al. (1997). Does abalone nacre form by heteroepitaxial nucleation or by growth through mineral bridges? *Chemistry of Materials*, 9(8), 1731–1740.
55. Weiner, S., & Lowenstam, H. (1986). Organization of extracellularly mineralized tissues: A comparative study of biological crystal growth. *Critical Reviews in Biochemistry*, 20(4), 365–408.

56. Addadi, L., & Weiner, S. (1997). Biomineralization: A pavement of pearl. *Nature*, 389 (6654), 912–915.
57. Song, F., Soh, A. K., & Bai, Y. L. (2003). Effect of nanostructures on the fracture strength of the interfaces in nacre. *Journal of Materials Research*, 18(8), 1741–1744.
58. Song, F., Soh, A. K., & Bai, Y. L. (2003). Structural and mechanical properties of the organic matrix layers of nacre. *Biomaterials*, 24, 3623–3631.
59. Cartwright, J. H., & Checa, A. G. (2007). The dynamics of nacre self-assembly. *Journal of the Royal Society Interface*, 4(14), 491–504.
60. Hou, W. T., & Feng, Q. L. (2003). Crystal orientation preference and formation mechanism of nacreous layer in mussel. *Journal of Crystal Growth*, 258(3), 402–408.
61. Checa, A. G., & Rodriguez-Navarro, A. B. (2005). Self-organisation of nacre in the shells of Pterioida (Bivalvia: Mollusca). *Biomaterials*, 26(9), 1071–1079.
62. Checa, A. G., Okamoto, T., & Ramrez, J. (2006). Organization pattern of nacre in Pteriidae (Bivalvia: Mollusca) explained by crystal competition. *Proceedings of the Royal Society B: Biological Sciences*, 273(1592), 1329–1337.
63. Checa, A. G., Cartwright, J. H., & Willinger, M. G. (2011). Mineral bridges in nacre. *Journal of Structural Biology*, 176(3), 330–339.
64. Dimas, L. S., Bratzel, G. H., Eylon, I., & Buehler, M. J. (2013). Tough composites inspired by mineralized natural materials: Computation, 3D printing, and testing. *Advanced Functional Materials*, 23(36), 4629–4638.
65. Dimas, L. S., Buehler, M. J. (2014). Modeling and additive manufacturing of bio-inspired composites with tunable fracture mechanical properties. *Soft Matter*, 10(25), 4436–4442.
66. Currey, J. D. (1977). Mechanical properties of mother of pearl in tension. *Proceedings of the Royal Society B*, . 196, 443–463.
67. Gao, H., Ji, B. H., Jager, I. L., Arzt, E., & Fratzl, P. (2003). Materials become insensitive to flaws at nanoscale: Lessons from nature. *Proceedings of the National Academy of Sciences*, 100, 5597–5600.
68. Katti, D. R., Katti, K. S., Sopp, J. M., & Sarikaya, M. (2001). 3D finite element modeling of mechanical response in nacre-based hybrid nanocomposites. *Computational and Theoretical Polymer Science*, 11, 397–404.
69. Katti, K. S., Katti, D. R., Pradhan, S. M., & Bhosle, A. (2005). Platelet interlocks are the key to toughness and strength in nacre. *Journal of Materials Research*, 20(05), 1097–1100.
70. Sen, D., & Buehler, M. J. (2011). Structural hierarchies define toughness and defect-tolerance despite simple and mechanically inferior brittle building blocks. *Scientific Reports*, 1, 35.
71. Katti, D. R, & Katti, K. S. (2001). Modeling microarchitecture and mechanical behavior of nacre using 3d finite element techniques. *Journal of Materials Science*, 36, 1411–1417.
72. Katti, D. R., Pradhan, S. M., & Katti, K. S. (2004). Modeling the organic-inorganic interfacial nanoasperities in a model bio-nanocomposite, nacre. *Reviews on Advanced Materials Science*, 6(2), 162–168.
73. Evans, A. G., Suo, Z., Wang, R. Z., Aksay, I. A., He, M. Y., Hutchinson, J. W. (2001). Model for the robust mechanical behavior of nacre. *Journal of Materials Research*, 16(9), 2475–2484.
74. Shao, Y., Zhao, H. P., Feng, X. Q., & Gao, H. (2012). Discontinuous crack-bridging model for fracture toughness analysis of nacre. *Journal of the Mechanics and Physics of Solids*, 60(8), 1400–1419.

75. Shao, Y., Zhao, H. P., & Feng, X. Q. (2014). On flaw tolerance of nacre: A theoretical study. *Journal of the Royal Society Interface*, 11(92), 20131016.
76. Begley, M. R., Philips, N. R., Compton, B. G., et al. (2012). Micromechanical models to guide the development of synthetic brick and mortarcomposites. *Journal of the Mechanics and Physics of Solids*, 60(8), 1545–1560.
77. Askarinejad, S., & Rahbar, N. (2015). Toughening mechanisms in bioinspired multilayered materials. *Journal of the Royal Society Interface*, 12(102), 20140855.
78. Dimas, L. S., & Buehler, M. J. (2012). Influence of geometry on mechanical properties of bio-inspired silica-based hierarchical materials. *Bioinspiration & Biomimetics*, 7(3), 036024.
79. Zhu, T. T., Bushby, A. J., & Dunstan, D. J. (2008). Size effect in the initiation of plasticity for ceramics in nanoindentation. *Journal of the Mechanics and Physics of Solids*, 56(4), 1170–1185.
80. Dunstan, D. J., & Bushby, A. J. (2013). The scaling exponent in the size effect of small scale plastic deformation. *International Journal of Plasticity*, 40, 152–162.
81. Li, N., Nastasi, M., & Misra, A. (2012). Defect structures and hardening mechanisms in high dose helium ion implanted Cu and Cu/Nb multilayer thin films. *International Journal of Plasticity*, 32, 1–16.
82. Salehinia, I., Wang, J., Bahr, D. F., & Zbib, H. M. (2014). Molecular dynamics simulations of plastic deformation in Nb/NbC multilayers. *International Journal of Plasticity*, 59, 119–132.
83. Salehinia, I., Shao, S., Wang, J., & Zbib, H. M. (2015). Interface structure and the inception of plasticity in Nb/NbC nanolayered composites. *Acta Materialia*, 86, 331–340.
84. Salehinia, I., Shao, S., Wang, J., & Zbib, H. M. (2014). Plastic deformation of metal/ceramic nanolayered composites. *JOM*, 66(10), 2078–2085.
85. Dutta, A., Tekalur, S. A., & Miklavcic, M. (2013). Optimal overlap length in staggered architecture composites under dynamic loading conditions. *Journal of the Mechanics and Physics of Solids*, 61(1), 145–160.
86. Yao, H., Song, Z., Xu, Z., & Gao, H. (2013). Cracks fail to intensify stress in nacreous composites. *Composites Science and Technology*, 81, 24–29.
87. Zhang, Z. Q., Liu, B., Huang, Y., Hwang, K. C., & Gao, H. (2010). Mechanical properties of unidirectional nanocomposites with non-uniformly or randomly staggered platelet distribution. *Journal of the Mechanics and Physics of Solids*, 58(10), 1646–1660.
88. Liu, G., Ji, B., Hwang, K. C., & Khoo, B. C. (2011). Analytical solutions of the displacement and stress fields of the nanocomposite structure of biological materials. *Composites Science and Technology*, 71(9), 1190–1195.
89. Dimas, L. S., Giesa, T., & Buehler, M. J. (2014). Coupled continuum and discrete analysis of random heterogeneous materials: elasticity and fracture. *Journal of the Mechanics and Physics of Solids*, 63, 481–490.
90. Jager, I., & Fratzl, P. (2000). Mineralized collagen fibrils: A mechanical model with a staggered arrangement of mineral particles. *Biophysical Journal*, 79(4), 1737–1746.
91. Wagner, H. D., & Weiner, S. (1992). On the relationship between the microstructure of bone and its mechanical stiffness. *Journal of Biomechanics*, 25(11), 1311–1320.
92. Kotha, S. P., Kotha, S., & Guzelsu, N. (2000). A shear-lag model to account for interaction effects between inclusions in composites reinforced with rectangular platelets. *Composites Science and Technology*, 60(11), 2147–2158.

93. Shuchun, Z., & Yueguang, W. (2007). Effective elastic modulus of bone-like hierarchical materials. *Acta Mechanica Solida Sinica*, 20(3), 198–205.
94. Chen, B., Wu, P. D., & Gao, H. (2009). A characteristic length for stress transfer in the nanostructure of biological composites. *Composites Science and Technology*, 69(7), 1160–1164.
95. Wei, X., Naraghi, M., & Espinosa, H. D. (2012). Optimal length scales emerging from shear load transfer in natural materials: Application to carbon-based nanocomposite design. *ACS Nano*, 6(3), 2333–2344.
96. Li, X., Gao, H., Scrivens, W. A., et al. (2005). Structural and mechanical characterization of nanoclay-reinforced agarose nanocomposites. *Nanotechnology*, 16(10), 2020.
97. Morits, M., Verho, T., Sorvari, J., et al. (2017). Toughness and fracture properties in nacremimetic clay/polymer nanocomposites. *Advanced Functional Materials*, 27 (10), 1605378.
98. Niebel, T. P., Bouville, F., Kokkinis, D., & Studart, A. R. (2016). Role of the polymer phase in the mechanics of nacre-like composites. *Journal of the Mechanics and Physics of Solids*, 96, 133–146.
99. Long, B., Wang, C. A., Lin, W., Huang, Y., & Sun, J. (2007). Polyacrylamide-clay nacre-like nanocomposites prepared by electrophoretic deposition. *Composites Science and Technology*, 67(13), 2770–2774.
100. Bonderer, L. J., Studart, A. R., & Gauckler, L. J. (2008). Bioinspired design and assembly of platelet reinforced polymer films. *Science*, 319(5866), 1069–1073.
101. Mammeri, F., Le Bourhis, E., Rozes, L., & Sanchez, C. (2005). Mechanical properties of hybrid organicinorganic materials. *Journal of Materials Chemistry*, 15(35–36), 3787–3811.
102. Chen, R., Wang, C. A., Huang, Y., & Le, H. (2008). An efficient biomimetic process for fabrication of artificial nacre with ordered-nanostructure. *Materials Science and Engineering: C*, 28(2), 218–222.
103. Corni, I., Harvey, T. J., Wharton, J. A., Stokes, K. R., Walsh, F. C., & Wood, R. J. K. (2012). A review of experimental techniques to produce a nacre-like structure. *Bioinspiration & Biomimetics*, 7(3), 031001.
104. Valashani, S. M. M., & Barthelat, F. (2015). A laser-engraved glass duplicating the structure, mechanics and performance of natural nacre. *Bioinspiration & Biomimetics*, 10 (2), 026005.
105. Zlotnikov, I., Gotman, I., Burghard, Z., Bill, J., & Gutmanas, E. Y. (2010). Synthesis and mechanical behavior of bioinspired ZrO2-organic nacre-like laminar nanocomposites. *Colloids and Surfaces A: Physicochemical Engineering Aspects*, 361(1), 138–142.
106. Bai, H., Walsh, F., Gludovatz, B., et al. (2016). Bioinspired hydroxyapatite/poly (methyl methacrylate) composite with a nacre mimetic architecture by a bidirectional freezing method. *Advanced Materials*, 28(1), 50–56.
107. Deville, S., Saiz, E., & Tomsia, A. P. (2006). Freeze casting of hydroxyapatite scaffolds for bone tissue engineering. *Biomaterials*, 27(32), 5480–5489.
108. Launey, M. E., Munch, E., Alsem, D. H., et al. (2009). Designing highly toughened hybrid composites through nature-inspired hierarchical complexity. *Acta Materialia*, 57(10), 2919–2932.
109. Wegst, U. G., Schecter, M., Donius, A. E., & Hunger, P. M. (2010). Biomaterials by freeze casting. *Philosophical Transactions of the Royal Society of London A: Mathematical, Physical and Engineering Sciences*, 368(1917), 2099–2121.

110. Askarinejad, S., & Rahbar, N. (2018). Mechanics of bioinspired lamellar structured ceramic/polymer composites: Experiments and models. *International Journal of Plasticity*, 107, 122–149.
111. Liu, M., Sun, J., Sun, Y., Bock, C., & Chen, Q. (2009). Thickness-dependent mechanical properties of polydimethylsiloxane membranes. *Journal of Micromechanics and Microengineering*, 19(3), 035028.
112. Qi, H. J., & Boyce, M. C. (2005). Stressstrain behavior of thermoplastic polyurethanes. *Mechanics of Materials*, 37(8), 817–839.
113. ASTM International. (2006). *ASTM E1820–06: Standard test method for measurement of fracture toughness annual book of ASTM standards, Vol. 03.01: metals – mechanical testing; elevated and low-temperature tests; metallography.* West Conshohocken, PA: ASTM International.
114. Hedgepeth, J. M. (1961). Stress concentrations in filamentary structures.
115. Nairn, J. A. (1988). Fracture mechanics of unidirectional composites using the shear-lag model I: Theory. *Journal of Composite Materials*, 22(6), 561–588.
116. Nairn, J. A., & Mendels, D. A. (2001). On the use of planar shear-lag methods for stress-transfer analysis of multilayered composites. *Mechanics of Materials*, 33(6), 335–362.
117. Benveniste, Y., & Miloh, T. (1986). The effective conductivity of composites with imperfect thermal contact at constituent interfaces. *International Journal of Engineering Science*, 24(9), 1537–1552.
118. Cheng, Z. Q., He, L. H., & Kitipornchai, S. (2000). Influence of imperfect interfaces on bending and vibration of laminated composite shells. *International Journal of Solids and Structures*, 37(15), 2127–2150.
119. Wang, Z., Zhu, J., Jin, X. Y., Chen, W. Q., & Zhang, C. (2014). Effective moduli of ellipsoidal particle reinforced piezoelectric composites with imperfect interfaces. *Journal of the Mechanics and Physics of Solids*, 65, 138–156.
120. Hashin, Z. (1991). Composite materials with interphase: Thermoelastic and inelastic effects. In *Inelastic deformation of composite materials*. New York: Springer; pp. 3–34.
121. Gu, S. T., He, Q. C., & Pense, V. (2015). Homogenization of fibrous piezoelectric composites with general imperfect interfaces under anti-plane mechanical and in-plane electrical loadings. *Mechanics of Materials*, 88, 12–29.
122. Nairn, J. A., & Liu, Y. C. (1997). Stress transfer into a fragmented, anisotropic fiber through an imperfect interface. *International Journal of Solids and Structures*, 34(10), 1255–1281.
123. Nairn, J. A. (2007). Numerical implementation of imperfect interfaces. *Computational Materials Science*, 40(4), 525–536.
124. Martin, P. A. (1992). Boundary integral equations for the scattering of elastic waves by elastic inclusions with thin interface layers. *Journal of Nondestructive Evaluation*, 11(3–4), 167–174.
125. Tada, H., Paris, P. C., & Irwin, G. R. (2000). *The analysis of cracks handbook* (pp. 58–58). New York: ASME Press. Chicago.
126. Munch, E., Launey, M. E., Alsem, D. H., Saiz, E., Tomsia, A. P., & Ritchie, R. O. (2008). Tough, bio-inspired hybrid materials. *Science*, 322(5907), 1516–1520.
127. Thouless, M. D., & Evans, A. G. (1988). Effects of pull-out on the mechanical properties of ceramic-matrix composites. *Acta Metallurgica*, 36(3), 517–522.
128. Marshall, D. B., Evans, A. G., Drory, M., et al. (1986). Fracture mechanics of ceramics.
129. Phillips, D. C. (1972). The fracture energy of carbon-fibre reinforced glass. *Journal of Materials Science*, 7(10), 1175–1191.

130. Marshall, D. B., & Evans, A. G. (1985). Failure mechanisms in ceramic-fiber/ceramic-matrix composites. *Journal of the American Ceramic Society*, 68(5), 225–231.
131. Cox, B. N., & Marshall, D. B. (1994). Concepts for bridged cracks in fracture and fatigue. *Acta Metallurgica et Materialia*, 42(2), 341–363.
132. Stewart, R. L., Chyung, K., Taylor, M. P., et al. (1986). *Fracture mechanics of ceramics*. New York: Plenum Press, 33.
133. Curkovic, L., Bakic, A., Kodvanj, J., & Haramina, T. (2010). Flexural strength of alumina ceramics: Weibull analysis. *Transactions of FAMENA*, 34(1), 13–19.
134. Evans, A. G., & McMeeking, R. M. (1986). On the toughening of ceramics by strong reinforcements. *Acta Metallurgica*, 34(12), 2435–2441.

4 Bamboo-Inspired Materials and Structures*

Ting Tan and Wole Soboyejo

4.1 Introduction

Bamboo is a group of perennial grasses in the family Poaceae, subfamily Bambusoideae, tribe Bambuseae [1]. One estimation classified bamboo into 75 genera and approximately 1,500 species [1]. The ordinary species of giant bamboo includes *Phyllostachys heterocycla pubescens* (Moso), *Bambusa stenostachya* (Tre Gai), *Guadua angustifolia* (Guadua), and *Dendrocalamus giganteus* (Dendrocalamus). Moso is the most widely distributed bamboo for utilization [2]. This species is native to China, and was introduced to Japan in about 1736 and to Europe before 1880 [2]. The word, *moso* (in Japanese) or *mao zhu* (in Chinese), means hairy culm sheaths and this bamboo is named for the pubescent down at the bottom of its new culms.

Bamboo consists of a culm above the ground and a rhizome system under the ground (Figure 4.1). The culm is a hollow cylinder with structural nodes in between. In addition to supporting the bamboo structure, there are channels to exchange water and nutrients with the earth. The rhizomes grow laterally with nodes and internodes [3]. There are two basic forms of rhizomes, i.e., leptomorphs (runners) and pachymorphs (clumpers). The runners grow rapidly and send internodes far away from their parent plants, whereas the clumpers cannot expand more than few inches per year. The ordinary method of bamboo propagation is transplanting rhizomes, not planting seeds [4]. Moso is a running bamboo, consisting of three primary life stages, i.e., the timber development, the culm stabilization, and the culm aging [5]. The first stage ranges from the new culm to the fifth year. Juvenile Moso leaves are larger than those found on mature plants. Lignification of the culm is completed at the end of this stage. The second stage spans from the fifth to the tenth years. Metabolism of the culm stabilizes with physiological activities. The final, declining stage is the culm aging, during which the culm weight, mechanical strength, and nutrient degrade significantly.

Bamboo grows in many tropical and temperate regions, with the exception of Europe and Antarctica (Figure 4.1). The world bamboo forest is approximately 65,520 square kilometres [6]. Asia possesses 65% of the global bamboo forest. Bamboo is among the

* The authors wish to acknowledge the University of Vermont and Princeton University for their support. The authors acknowledge financial support from UVM clean energy fund (028613) and Vermont Venture Innovation Fund (031365).

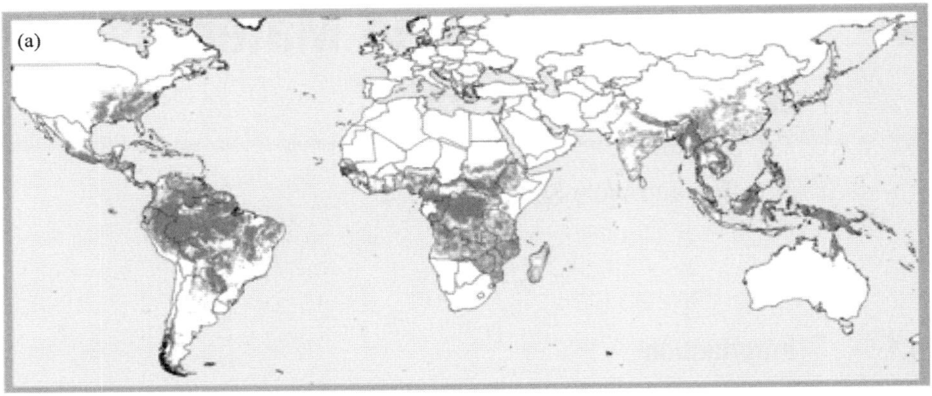

Figure 4.1 Global distribution of bamboos in existing forests. (Reproduced from Zhao et al. [11] with permission from Oxford University Press.)

fastest-growing plants in the world. The primary method of bamboo growth is vegetative reproduction [7–9]. New rhizomes penetrate the ground to produce a secondary culm until it reaches a stable size [10]. Large bamboo can reach the height of 15–30 m in about 2–4 months, whereas the smallest bamboos reach only several inches in height [7]. Data from the United States shows that the typical height of Moso bamboo was 7.6 m in Oregon, and the maximum height was 17 m in South Carolina [2]. The minimum temperature for Moso bamboo growth is –17°C for most cultivars.

Bamboo has been used to build human houses for many centuries [6,12–14]. This application includes both ordinary houses and the scaffolding used in the construction industry [15]. Other large-scale structures have also been constructed using bamboo. For example, a bamboo bridge has been built in the Guangdong province of China [16]. In Columbia, Simon Velez demonstrated his architectural talent by building a bamboo church, which synergizes grace and sustainability. For Expo Hanover 2000, Velez built a bamboo pavilion in Germany after making a full-scale model structure that survived various tests in Columbia [17]. Bamboo is a promising material to build shelters after disasters [18,19]. A test bamboo house survived loads equivalent to the 7.8 Kobe earthquake on a shake table [20] with little obvious damage. In April 1991, approximately 20 bamboo houses survived near the epicenter of a 7.5 Richter scale earthquake in Costa Rica [8]. Bamboo can be used to build disaster shelters in hazard regions within a short time due to its abundance.

There have been more than 1,000 documented uses of bamboo [21]. In ancient China, bamboo books were crafted from culm strips to record history [22]. Bamboo has been a popular theme in Chinese paintings and writings [4]. A wide variety of applications are made using bamboo around the world, including furniture, handicrafts, indoor flooring, textiles, energy sources, and musical instruments [19]. Meanwhile, bamboo has been successfully used to fabricate racing bicycles [23], due to its high strength to weight ratio and shape factors [24–26], In addition, bamboo can be a food source. The winter shoots of Moso bamboo are tasty food, primarily due to their size and the abundance of water and nutrients absorbed during their growth [27].

4.2 Bamboo Properties

4.2.1 Bamboo Cross Sections

Numerous researchers have studied the microstructural features in bamboo cross sections [15,28–34]. The culm includes approximately 40% cellulose fibers, 52% parenchyma tissue, and 8% conduction tissue (vessels, sieve tubes, companion cells), which may vary with species and bamboo growth stages [1]. Sclerenchyma fibers are the primary supporting tissues within the vascular bundles of bamboo. The phloem vessels transport sugars and nutrients, whereas the metaxylem vessels transport water for the main culm [35,36]. The tissue surrounding the vascular bundles is parenchyma. It has been shown that the ratio between the vascular bundles and the parenchyma decreases radially from the outside to the inside surface (Figure 4.2a and b). In the longitudinal direction, the narrowing of the bamboo results in a reduction of the parenchyma toward the culm head [1].

In addition to the functionally graded microstructure in the cross section, the cell structure of fibers and parenchyma in natural bamboo also exhibits remarkable mechanical features. The primary chemical contents of bamboo fibers (Figure 4.2c and d) are cellulose, hemicellulose, and lignin [37]. One feature of bamboo fibers is the thick polylamellate secondary walls, which consist of alternating broad and narrow layers with varying microfibril orientation. The cell walls of bamboo fibers comprise stiff cellulose fibrils and pliant amorphous matrices with various biopolymers [38–44]. Their performance is affected by cell wall thickness, cellulose orientation, degree of crystallization, cell wall fraction, matrix composition, chemical bonding, and water content [38]. The microfibril angles are in the range of 3–10° in the broad layers, whereas the angles are in the range of 30–90° in the narrow layers [45]. The alternating layers contribute substantially to the remarkable strength and toughness of bamboo fibers [45]. A statistical model was created to simulate the drying fracture process of the bamboo *Guadua angustifolia* [46]. A network of flexible hexagonal cells was connected by brittle joints that break at a certain tensile strength. A power law distribution was found between the numbers and sizes of the avalanche breaking events, whereas different geometries were not critical to the breaking statistics. A more direct observation of cracked bamboo fiber cells was obtained by Schott [47] when heating the bamboo culm. Intergranular cracks formed along the fiber cell boundaries, and radial cracks occurred within single bamboo fiber cells.

There are two types of bamboo parenchyma cells, i.e., longitudinally elongated cells and short tub-like cells, which are immersed in between [1] and are separated by thick walls that are lignified in the early stages of intermodal development (Figure 4.3a). The biological roles of these two different types of parenchyma cells have been an interesting research topic. He et al. [50] performed experiments to characterize the lignification of these two cell types at various life stages. Based on their results, the short and long parenchyma cells were lignified in 2-month-old bamboo culms. However, for the short cells, the lignified regions were limited to the contact regions with the long ones. The walls of the short cells at the corner

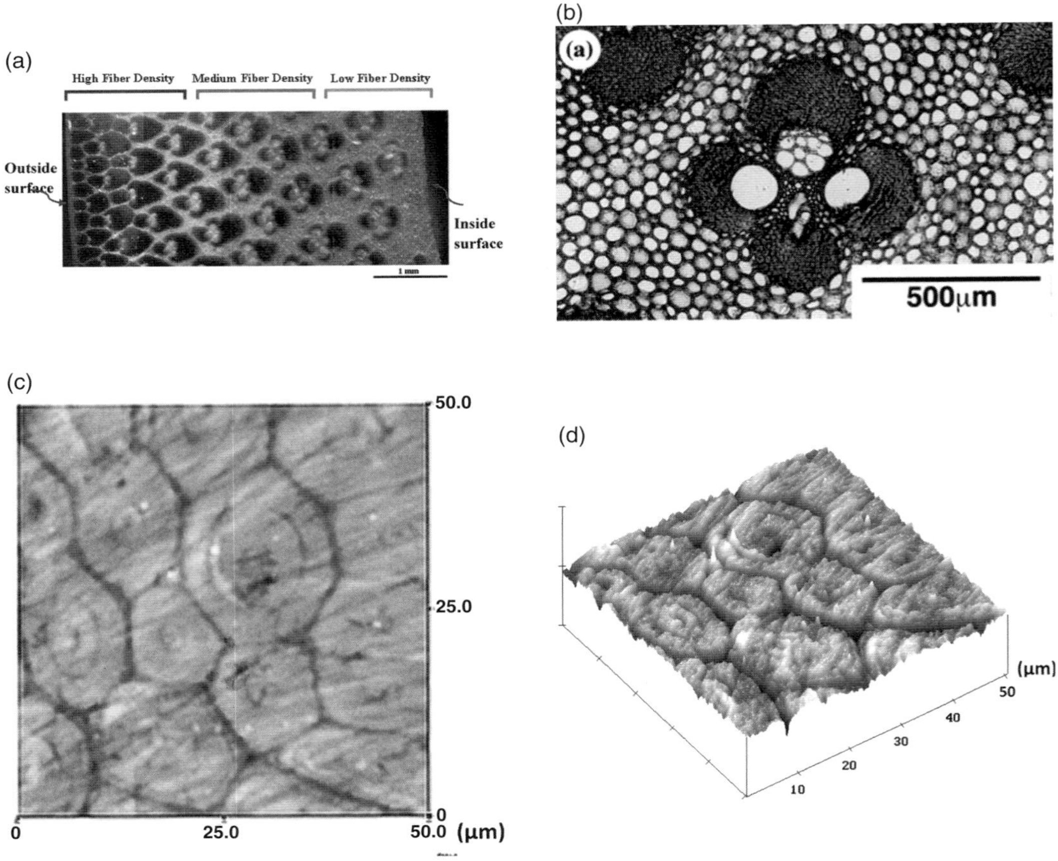

Figure 4.2 (a) An optical image of the functionally graded mesostructure of Moso bamboo. (b) Vascular bundles with protoxylem and metaxylem vessels in Moso bamboo, (reproduced from Abe and Yano [48] with permission from Springer.) Atomic Force Microscopy images of bamboo fibers (c) 2D, and (d) 3D images. (Parts a, c, and d reproduced from Tan et al. [49] with permission from Elsevier.)

regions never lignified – even in 7-year-old culms [50]. In another observation [47], residual fiber bridging was found between heated long parenchyma cells, whereas the corner regions of the short parenchyma cells kept their geometries without excessive deformation. A conceptual model is proposed by the authors to explain the toughening mechanism of the parenchyma. Once cracked, microfibril bridging within the long cell walls resists crack growth. When the crack reaches the corner regions of the short cell, the unlignified portions deform to dissipate more energy, which diverts the crack into other directions (Figure 4.3b). Otherwise, cracks could grow along the lignified contact regions, causing the bamboo structure to fail catastrophically. More work is needed to further elucidate the toughening mechanisms of bamboo parenchyma cells.

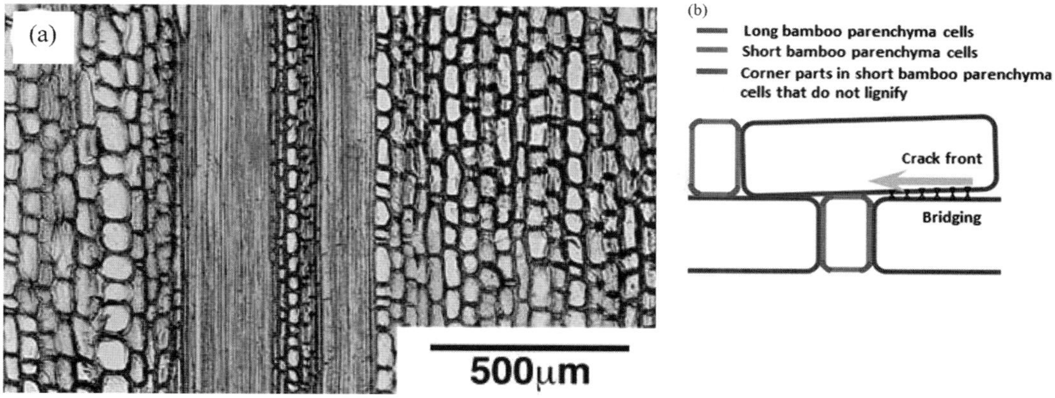

Figure 4.3 (a) Bamboo parenchyma consists of longer and shorter cells. (Reproduced from Abe and Yano [48] with permission from Springer.) (b) A conceptual model of toughening mechanisms in bamboo parenchyma.

4.2.2 Young's Modulus

The Young's modulus of bamboo depends on many factors, such as age, species, and material orientation. Ideally, bamboo is treated as a unidirectional, fiber-reinforced composite. The rule of mixture is used to characterize the Young's modulus [24],

$$E = fE_f + (1 - f)E_m, \tag{4.1}$$

where f is the fiber volume ratio, and E_f and E_m are the Young's moduli of the fiber and matrix, respectively. Nogata and Takahashi [51] performed unidirectional tests to measure the Young's modulus of Moso bamboo using bulk specimens along the longitudinal axis, from which the modulus is ~15 GPa. Amada et al. [29] reported that the Young's modulus of bamboo matrix is ~2 GPa, whereas the fiber modulus is ~46 GPa. Prior studies also measured the Young's modulus of bamboo strips varying from the outside to the inside surface, along which the fiber volume ratio decreases [31,51,52]. Li [31] used the bending test to measure the Young's modulus of 1-, 3-, and 5-year-old Moso bamboo. The reported results ranged from 5 to 16 GPa, and similar results (4–16 GPa) were obtained by Ma et al. [52]. Tan et al. [49] used nanoindentations to measure the Young's moduli radially, in which the values ranged from 7.5 to 13 GPa. All studies showed that the Young's moduli decrease from the outside to the inside, consistent with the fiber gradient in Figure 4.2a and 4.4a.

4.2.3 Tensile Strength

Bamboo has high tensile strength, which is partially due to the remarkable mechanical properties of the bamboo fiber cells [53]. Nogata and Takahashi [51] reported that the

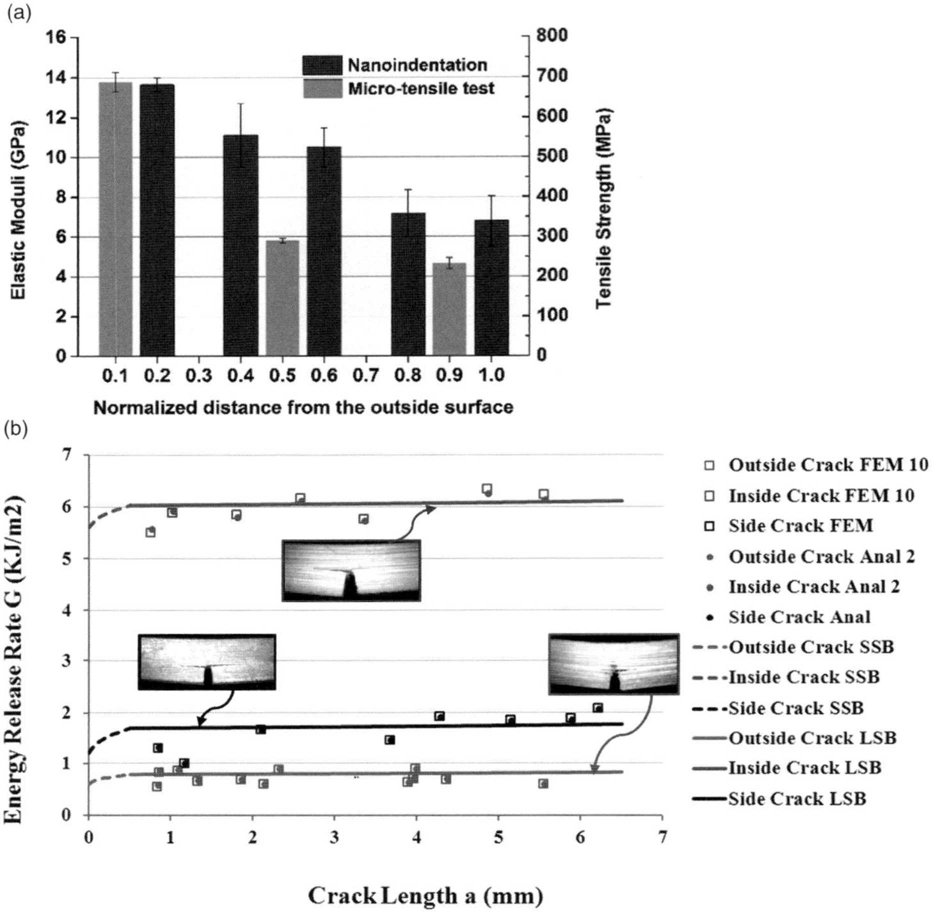

Figure 4.4 (a) The radial distributions of elastic moduli and tensile strength obtained from nanoindentations. (b) Fracture resistance-curve results for Moso bamboo structure. (Reproduced from Tan et al. [49] with permission from Elsevier.)

tensile strength of the matrix (parenchyma) is ~50 MPa, whereas tensile strength of the fiber is ~810 MPa. Amada et al. [29] measured the tensile strength of Moso culm bamboo along the axial direction, which ranged between 100 and 400 MPa from the inside to outside. Meanwhile, the tensile strength of Moso culm bamboo measured by Nogata and Takahashi [51] varied from 100 to 750 MPa along the same direction. Tan et al. [49] performed microtensile tests using Moso strips, and the tensile strength ranged from 200 to 600 MPa. All studies showed that the tensile strength degrades from the outside to inside surface, which is consistent with the fiber gradient shown in Figure 4.2a and 4.4a.

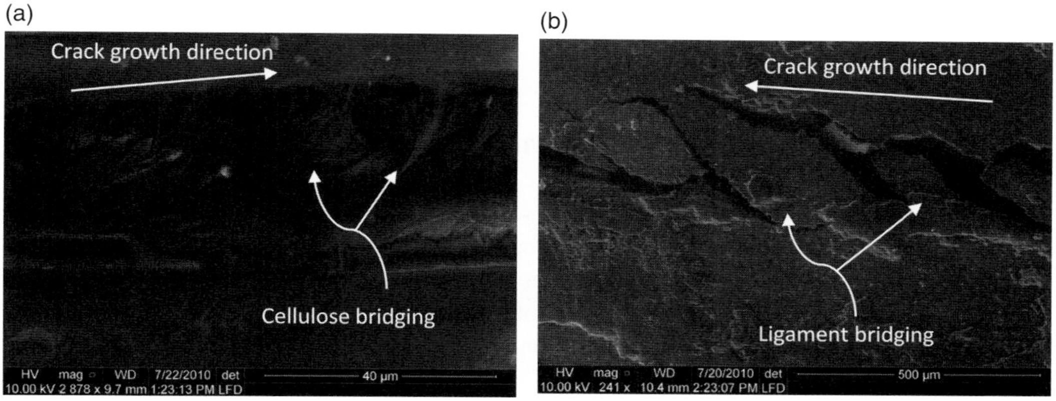

Figure 4.5 SEM images of crack bridging in Moso bamboo: (a) cellulose bridging in the outside notched SENB sample and (b) ligament bridging in the side-notched SENB sample. (Reproduced from Tan et al. [49] with permission from Elsevier.)

4.2.4 Fracture Resistance

Due to the fiber gradient in the microstructure, bamboo exhibits extraordinary fracture behavior. Shao et al. [54] used the double cantilever test to measure the energy release rate, i.e., G_{IC}, of Moso bamboo by cracking the bamboo along the longitudinal–radial direction, and the fracture resistance was ~0.36 kJ/m^2. Amada and Untao [55] performed tensile tests using the single-edge notched dog bone specimen to measure the fracture toughness, i.e., K_{IC}, of Moso bamboo, and the values ranged from ~20 to 100 MPa$\sqrt{\text{m}}$. In order to characterize fracture along with the fiber gradient, Tan et al. [49] performed four-point bending tests to measure the fracture resistance of Moso bamboo. Specimens were fabricated with notches from the inside, outside, and side surfaces. In all cases, cracks deflect horizontally to both ends instead of propagating along the notch direction. A good agreement was achieved between fracture resistance predictions from the bimaterial analytical fracture model [56] and the finite element model [49]. Crack bridging was the primary toughening mechanism in the crack deflection. For samples with notches from the outside and inside surfaces, cellulose fiber bridging is the major toughening mechanism (Figure 4.5a). Lower cellulose bridging fiber density was observed for samples with notches from the outside surface than for specimens with notches from the inside. For samples with notches from the side surface, ligament bridging is the primary toughening mechanism (Figure 4.5b), resulting in the modest tensile strength. Both small- and large-scale bridging models [57–59] were used to estimate the slightly elevated fracture resistance of Moso bamboo. It is shown that the fracture resistance of Moso bamboo increases from the outside to the inside, which is the opposite of the trends for strength and toughness (Figure 4.4b). This is because less energy is invested to split fiber arrays closer to the outside than to break lignin webs closer to the inside regions. This synergy of strength and fracture resistance imparts superior mechanical properties to bamboo.

4.3 Bamboo Fiber-Reinforced Polymeric Composites

Bamboo has been used as fiber reinforcement for various engineering composites due to its excellent mechanical properties and abundance. Conventional bamboo composites include chipboard, plybamboo, strand-woven bamboo, and hybrid laminates [60]. For bamboo chipboards, the reinforcements are made from small bamboo culms or bamboo wastes [12], exhibiting the irregular size and geometries. To fabricate plybamboo, hybrid bamboo laminates, raw bamboo culms, or strips are compressed with other plant fibers or adhesives to produce laminate boards. Substantial effort has been devoted to producing bamboo fiber-reinforced polymeric (BFRP) composites [61,62]. Two major components in BFRP composites are bamboo fibers and polymer matrices [37,61] the fabrication of which are detailed in the following sections.

4.3.1 Fabrication Process

Bamboo fibers are obtained using chemical or mechanical methods from natural bamboo culms [37]. Initially, natural bamboo culms are spilt into small strips. In chemical methods, crushed bamboo strips are pre-alkali treated to isolate fibers, which then go through the multistep bleaching process. This method has been widely used by manufacturing industries due to its rapidness. In mechanical methods, the crushed bamboo strips are treated by enzymes to produce spongy structures for the fiber extraction. Rough fibers are prepared using a process of cutting, separation, boiling, and enzyme fermentation, whereas finer fibers are prepared using a process of boiling, enzyme fermentation, washing, drying, acid treatment, oil soaking, and air drying [63].

The polymeric matrices for BFRP composites include polypropylene [64–69], polyester [70–73], epoxy [71,74–76], phenolic resin [70,77], polyvinyl chloride [78], polystyrene, and phenol formaldehyde [21]. After drying, bamboo fibers were alkali-treated in NaOH solution before mixing with polymer matrices [37]. Extra chemical modifications were performed on bamboo fibers to increase the adhesion between fibers and matrices [68]. Then, bamboo fibers and matrices were melt-blended above the ambient temperature before injection molding [68,74,79,80]. The composites cure at ambient temperature for fabrication, which can be tailored into different geometries based on needs. A summary of the fabrication methods is included in Table 4.1.

4.3.2 BFRP Composites

Polypropylene-based BFRP composites are widely used due to their good mechanical properties [64–69,81], in which the tensile strength exceeds that of commercial wood-pulp composites. Chattopadhyay et al. [69] produced short bamboo fiber reinforced polypropylene composites containing various percentages of chemically modified

Table 4.1 A summary of representative bamboo fiber-reinforced polymer composites

	Treatment	Matrix	Bamboo
Chen et al. [64]	Bamboo fibers up to 2 mm were vacuum dried. Fibers and matrix were oven dried at 80°C and 180 mmHg for 2 h. The mixture was injection molded at 210°C and 5 MPa pressure. The boards were heated 20 min and then under ambient cooling.	Polypropylene and maleated polypropylene	*Bambusa paravariabilis*
Thwe and Liao [79]	Short bamboo fibers and 3 mm E-Glass fibers 3 mm were oven dried. Constituents underwent melt mixing at 190°C and 40 rpm for 8–10 min before injection molding.	Polypropylene and maleated polypropylene	
Lee et al. [68]	Bamboo fibers were air dried at 80°C for 24 h. After NaOH pretreatment, fibers were chemically modified using two silane (AS and TMOS). The mixture was then input into an extruder at 180°C and 100 rpm for composite production.	Polypropylene	
Kushwaha and Kumar [71,72]	Dried bamboo mats were treated in NaOH solution at room temperature and then neutralized. Polyester resin with 2% hardener and 2% accelerator. Matrix was spread onto the seven-layer composite before being pressed in the hydraulic press to 170 kN for 6 h. The composites were cured at 80°C for 1.5 h.	Polyester	
Rajulu et al. [74]	Bamboo fibers of 10 mm were treated in 1% NaOH solution for 30 min and then sun-dried for 2 weeks. The composites were cured at 65°C for 24 h in an oven.	Epoxy	*Dendrocalamus strictus*
Das and Chakraborty [77]	Bamboo strips dipped into NaOH solution with different concentrations for 1 h at ambient temperature. The strips were washed using distilled water and subsequently neutralized with 2% H_2SO_4. By placing the strips within a resin bed, molding was performed at 180°C and 15 atm for 5 min.	Novolac resin	*Bambusa balcooa*
Bansal and Zoolagud [21]	Air-dried bamboo strips of 0.6–1.0 mm length were used to weave mats, whose slivers were at an angle of 45° to the edge. Dried resin-coated mats were assembled up to 5 plies and hot pressed to produce bamboo mat boards.	Phenol formaldehyde	

bamboo fibers. It was found that both fiber volume ratios and modification methods were critical to mechanical properties of the composites. SEM characterization exhibited that a better bonding was achieved in maleic anhydride modified polypropylene matrix, based on which good fiber-matrix adhesion is achieved [65,66]. Other work [79,80] studied the mechanical behavior of hybrid bamboo–glass fiber reinforced polypropylene composites (BGRP). Specimens with short bamboo fibers exhibited better tensile strength and modulus than specimen made from plain polypropylene. The composites prepared at 30% fiber with 2% maleic anhydride-grafted polypropylene (MAPP) concentrations showed optimum performance [80].

For polyester-based BFRP, the effect of the fiber modification from different concentrations of alkali treatments on tensile and flexural strengths were studied [71,72,82]. It was found that alkali treatments reduced the water uptake in the manufacturing process, resulting in more durable composites. Other treatment methods, such as permanganate and benzoylation, also enhanced the mechanical properties of polyester-based composites [71]. Different silane-treated bamboo fibers were studied for their water absorption in these composites [71]. Excellent wear resistance of epoxy-based BFRP was reported compared to that of plain epoxy, which was attributed to the high shear resistance provided by the bamboo fibers [75]. It was also found that the amino-functional silane reduced the water absorption of epoxy-based BFRP composite, resulting from the better adhesion between fibers and matrices [71]. Another work, by Rajulu et al. [74], reported that the void content is critical to the composite performance, which was affected by the content of short fibers.

There is a growing need for use of bamboo composites in various applications (Figure 4.6), including housing, transportation, and industrial products [60,83,84]. For example, more than 200,000 m^2 of plybamboo were used to build the roof at the Madrid–Barajas airport. This award-winning design project combined the use of

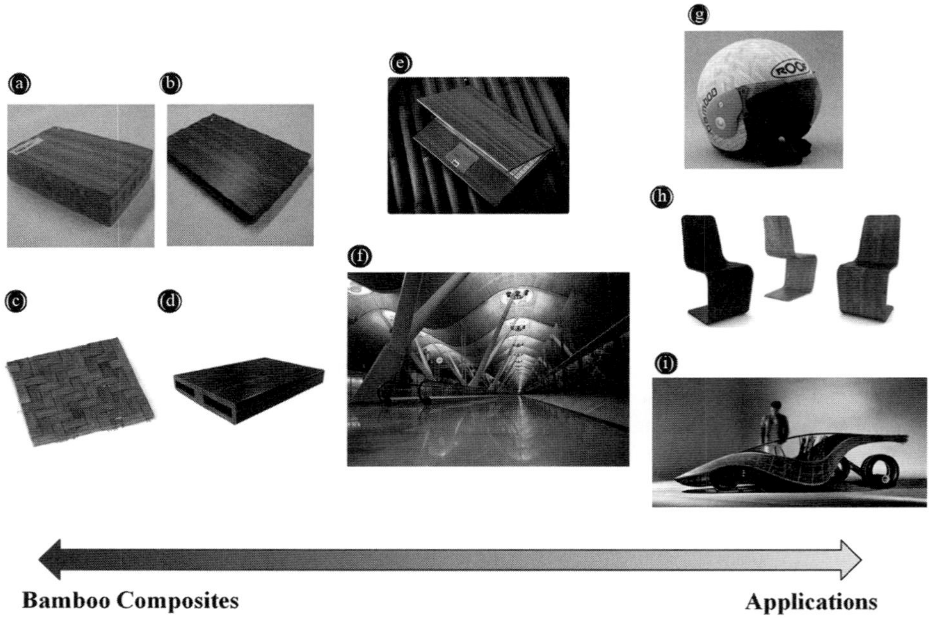

Figure 4.6 Bamboo reinforced composites and their applications: (a) Plybamboo. (b) Strand-woven bamboo. (c) Woven bamboo mats. (d) Bamboo fiber reinforced composites. (e) Bamboo laptop case. (f) Plybamboo roof. (g) Bamboo helmet. (h) Bamboo spring chair. (i) Bamboo concept car.

sustainable materials with the aesthetic use of daylight. BFRP composites have been used to make sport products, including bamboo helmets, surfboards, boat paddles, and vehicle components [83]. For example, Mitsubishi has fabricated interior doors using bamboo fibers and plant-based urethane resins. Also, a concept car has been created using bamboo and other natural fiber-reinforced materials [83]. Furthermore, bamboo has been used to make electronic packaging and renewable energy infrastructure. More effort is needed to develop technical standards that will control the quality of the BFRP products, particularly on the long-term behavior of these composites. Meanwhile, a balance needs to be established between the increasing demand for bamboo products and the bamboo plantations [18].

4.4 Bamboo-Inspired Materials and Systems

4.4.1 Bamboo-Inspired Functionally Graded Materials

Hierarchical materials in nature have always been the prototypes for innovative material research and development [85–93]. One feature of natural bamboo is the gradation between vascular bundles and parenchyma from the outside and inside surface. This functionally graded structure gives bamboo excellent mechanical properties, such as strength and fracture resistance [49,94]. Functionally graded materials (FGM) are effective in reducing the stress concentrations between different layers, which is critical to the performance of many devices and connecters [95,96]. Inspired by the microstructure of natural bamboo, innovative FGMs have been synthesized to meet different mechanical and thermal needs, including the ceramic–metal joints [97], piezoelectric actuators [88], and biomimetic cylindrical structures [52,98].

4.4.1.1 Ceramic–Metal Joints

Ceramic–metal joints are critical to structures at elevated temperatures. Their performance is affected by a combination of chemical (thermodynamics and kinetics at interface, wetting, adhesion, etc.) and mechanical factors (residual stresses, defects, fracture energy, etc.). Thus, the design methodologies are critical to the resulting products [97,99,100]. Aiming at reducing the stresses during process, Bruck et al. [95] developed a functionally graded nickel–alumina joint with the gradual transition in compositions. Powder mixtures of different fractions of nickel and alumina were blended with the binder in a ball mill to form slurry [100]. Initially, uniform layers were fabricated by cold-pressing powders uniaxially and they were then sintered. The multiinterlayer specimen was fabricated using the hot isostatic press, during which the temperature and pressure were ramped, held, and cooled. In the FGM, the volume fraction of nickel was graded by

$$V_{Ni} = \left(\frac{x}{t}\right)^p, \tag{4.2}$$

where p is the gradient exponent, x is the distance from the layer of pure alumina, and t is the layer thickness (Figure 4.7). A series of mechanical tests were performed to

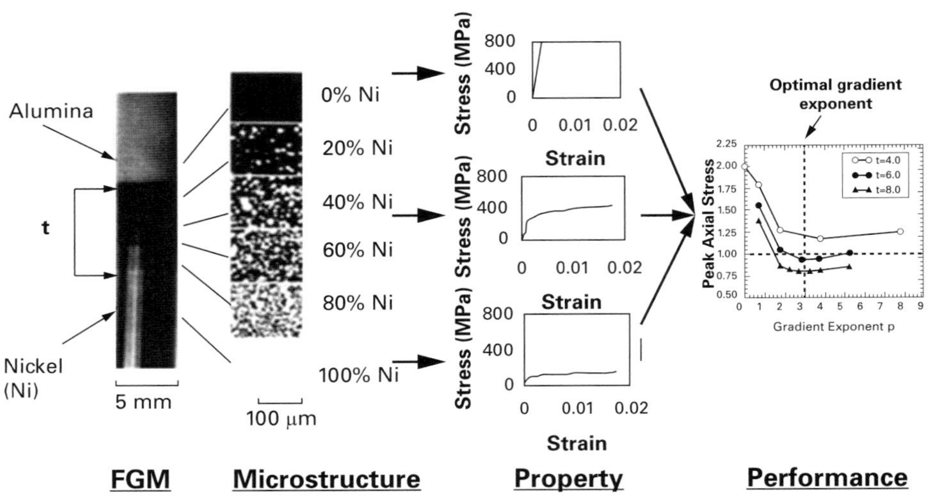

Figure 4.7 Relationship between the microstructure, mechanical properties, and performance of a functionally graded ceramic–metal joint. (Reproduced from Bruck et al. [95] with permission from Springer.)

characterize the mechanical behavior of each single layer. The results showed that higher strength occurred in layers with higher factions of alumina, which is consistent with the gradient microstructure. To find the optimum fraction of nickel composition, different patches of FGM specimens were measured for their peak axial stresses in the service condition. The results showed that an optimum gradient exponent, p, of ~3.2 could be obtained for specimens with different layer thicknesses.

4.4.1.2 Piezoelectric Actuators

Piezoelectric actuators are important to the performance of microelectromechanical (MEMS) systems. Usually, the degradation of the bonding strength between the piezoelectric plates and metal shims is detrimental to conventional unimorph and monomorph actuators [96,101]. Inspired by the microstructure of Moso bamboo, a novel piezoelectric actuator was developed using graded porous zirconate titanate (PZT). A mixture of commercial PZT and stearic acid powders was compacted by die and cold isotatic pressing [88]. Mixtures of layers with different compositions were sintered in air using a series of heating rates. The FGM samples were prepared by compacting different uniform layers and then sintering them together. The graded porosity distribution is given by

$$\left(1 - \frac{p_i}{p_n}\right) = \left(\frac{h_i}{h_n}\right)^m, \tag{4.3}$$

where p_i and p_n are the porosities of the ith and the top (nth) layers, h_i and h_n are the corresponding layer thickness, and m is the thickness parameter. By performing experiments on a series of FGM specimens, the optimum m that provides the maximum bending curvature is ~1.92. Further effort showed that a good agreement was obtained between these measurements and predictions from a laminate model [101]. The functionally graded piezoelectric actuator successfully alleviated the stress concentrations between different layers. Numerical simulations are needed to reveal the optimal gradient that to meet the functional needs [102].

4.4.1.3 Cylindrical Structures

The functionally graded bamboo structure is good at resisting both bending and buckling loads. de Vries [98] used 3D printing techniques to produce functionally graded cylindrical structures. Two polymers were used as the fiber and the matrix, respectively. A series of fiber–matrix composite models were created to simulate the bamboo cross section, including a model with two concentric layers representing the fiber and matrix materials, and two models with equal-size or varying-size fibers based on the bamboo fiber gradient. Four-point bending tests were performed to characterize the flexural capabilities of these cylindrical specimens. Even though the relationship between the number of fiber layers and the enhanced mechanical performance has not yet been elucidated [98], the 3D printing techniques exhibit opportunities to create advanced hierarchical materials in the future.

Along with its excellent bending resistance, natural bamboo is stable under compression, even with a ratio of diameter to length of 1/150~1/250. Ma et al. [52] created a cylindrical shell model numerically using pipes and ribs. The thin-walled cylindrical shells simulated the vascular bundles, whereas the varied pipes and ribs simulated the parenchyma cells. Nonlinear buckling analysis was performed to characterize the load bearing efficiency, i.e., the ratio between the critical buckling stress and the mass of the cylinder. The results showed that the load-bearing efficiency in the functionally graded cylinder was about 125% higher than that of the homogeneous cylinder [52]. Instead of exhibiting a local buckling for conventional cylinders, the functionally graded cylinder exhibited a more favorable global buckling.

4.4.1.4 Fiber-Reinforced Composites

Li et al. [103] designed cylinders using glass fiber reinforced epoxy composites by simulating the bamboo microfibrils. Four types of cylinders were fabricated using the same fiber–matrix ratio with different reinforcement patterns. The two control types were solid and hollow cylinders reinforced by axially aligned fibers. The third type was a hollow single-layer cylinder reinforced by right helical fibers at 15°. The forth type was a double-layered cylinder reinforced by 80% right helical fibers at 15° and 20% left helical fibers at 30°. Various mechanical tests, including bending, compression, and tension tests, were performed to evaluate the mechanical performance of these cylinders. The results showed that the double-helical structures exhibit the optimum performance in terms of strength, ductility, and stability [103]. Multilayered glass fiber reinforced composite panels were created with and without the transitional zones to simulate the bamboo fiber layers. In order to characterize their mechanical performance, double-side grooved specimens and three-point bending tests were applied to measure the interlaminar shear strength [104]. In both tests, specimens possessing the transition zones exhibited higher interlaminar strength than those without transition zones. This demonstrated that the functionally graded feature reduced stress concentrations between different cell wall layers.

4.4.2 Bamboo-Inspired Renewable Energy Systems

Because bamboo has evolved to resist wind loads over thousands of years, it is a sustainable candidate for wind turbine blades. Tan et al. [105–107] worked with UVM undergraduate students to create a bioinspired, renewable energy harvesting system consisting of a bamboo wind turbine and solar panels (Figure 4.8). First, wind turbine blades were made by wrapping bamboo laminates around 3D-printed skeletons, and LED lights were installed on the blade edges (Figure 4.8c and d). Then, a microprocessor was used to store the harnessed wind and solar energy in a battery to power the lights (Figure 4.8e). The system is off-grid, like an individual plant. When installed, a sustainable relationship is established between the bamboo wind turbine and the natural environment. Given the abundance of bamboo, bamboo wind turbines could be used in many places around the world, particularly in developing regions. Meanwhile, lights on the blades can be tailored to meet different needs, such as warning, entertainment, and advertisement (Figure 4.8g). If LED lights were replaced by ultraviolet or ultrasound sensors, the signals could warn bats and birds to help prevent collisions. In short, the goal is to not simply build wind turbines, but to plant them.

4.5 Summary

This chapter presents the recent advances in bamboo-inspired materials and structures. As a fast-growing grass, bamboo consists of a culm with multiple nodes above the

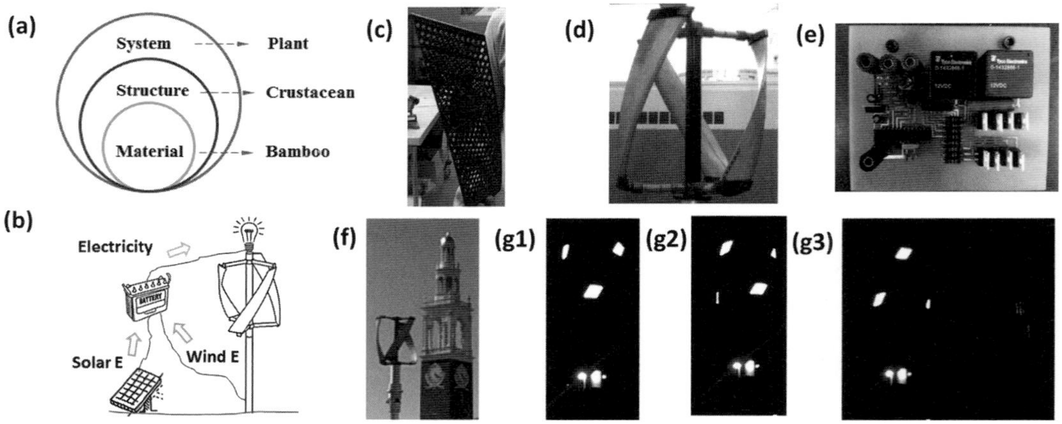

Figure 4.8 (a) A multiscale, bioinspired concept. (b) Energy paths in the hybrid wind and solar energy harvesting system (E stands for energy). (c) A 3D-printed helical blade skeleton. (d) A close view of the vertical bamboo wind turbine with LED lights attached to trailing edges of blades. (e) The integrated microprocessor control system. (f) A field image in daytime. (g1-g3) Snapshots at night. The background is the Ira Allen Chapel at UVM.

ground, and a rhizome system growing laterally under the ground. Moso is the most widely distributed bamboo, consisting of three primary life stages, i.e., the timber development, the culm stabilization, and the culm aging. Bamboo has been used by human beings for various applications over thousands of years The culm includes cellulose fibers and parenchyma tissues. The graded ratio between the vascular bundles and the parenchyma enables bamboo to perform as a functionally graded material, with remarkable strength, modulus, and fracture resistance.

Bamboo fiber reinforced polymeric composites have been developed using natural bamboo fibers and various polymeric matrices, such as polypropylene, polyester, and epoxy. Different methods have been used to modify increase the fiber–matrix adhesion. The good qualities of BFRPs ensure a variety of home and industrial applications. Inspired by the bamboo microstructure, novel FGMs have been developed to meet different mechanical and thermal needs, including the ceramic–metal joints, piezoelectric actuators, and biomimetic cylinders. Future effort could be devoted to the creation of additional bamboo-inspired materials and structures.

1. Technical standards are needed to guide the fabrication of BFRP products, particularly on the long-term behavior. A balance needs to be achieved between the increasing need for bamboo products and the sustainable utilization of bamboo plantations.
2. 3D printing techniques could contribute significantly to the creation of innovative hierarchical materials in the future.
3. More work is needed to understand the intercellular toughening mechanisms of bamboo fiber layers and alternating parenchyma cells. Multiscale experiments are critical to understand the fundamental mechanical behavior of hierarchical natural materials.
4. Bamboo is a good candidate to make wind turbines because of its intrinsic wind-resistant properties. Due to the abundance of bamboo plantations, bamboo wind turbines could be used over the world for various applications.

References

1. Liese, W. (1998). *The anatomy of bamboo culms* [Technical report]. Beijing, China: International Network for Bamboo and Rattan.
2. Lewis, D., & Miles, C. (2007). *Farming bamboo*. Raleigh, NC: Lulu Press.
3. Clark, L. G., & Pohl, R. W. (1996). *Agnes Chase's first book of grasses: The structure of grasses explained for beginners*. Washington, DC: Smithsonian Institution Press.
4. McClure, F. (1966). *The bamboos: A fresh perspective*. Cambridge, MA: Harvard University Press.
5. Fu, J. (2001). Chinese Moso bamboo: Its importance. *Bamboo*, 22, 5–7.
6. Lobovikov, M., Ball, L., Guardia, M., & Russo, L. (2007). *World bamboo resources: A thematic study prepared in the framework of the global forest resources assessment*. Rome, Italy: Food and Agriculture Organization of the United Nations Press.
7. Farrelly, D. (1996). *The book of bamboo: A comprehensive guide to this remarkable plant, its uses, and its history*. San Francisco: Sierra Club Books Press.

8. Janssen, J. J. (2000). *Designing and building with bamboo* [Technical report]. Beijing, China: International Network for Bamboo and Rattan.
9. Banik, R. L. (1995). *A manual for vegetative propagation of bamboos* [Technical report]. Beijing, China: International Network for Bamboo and Rattan.
10. Chapman, G. P. (1996). *The biology of grasses*. Wallingford, UK: CABI Press.
11. Zhao, H., Gao, Z., Wang, L., et al. (2018). Chromosome-level reference genome and alternative splicing atlas of Moso bamboo (*Phyllostachys edulis*). *GigaScience*, 7, 115.
12. van der Lugt, P. (2008). *Design interventions for stimulating bamboo commercialization-Dutch design meets bamboo as a replicable model* [PhD thesis]. Delft, Netherlands: Delft University of Technology.
13. Flander, K. D., & Rovers, R. (2009). One laminated bamboo-frame house per hectare per year. *Construction and Building Materials*, 23, 210–218.
14. Paudel, S. K., & Lobovikov, M. (2003). Bamboo housing: Market potential for low-income groups. *Journal of Bamboo and Rattan*, 2, 381–396.
15. Chung, K., & Yu, W. (2002). Mechanical properties of structural bamboo for bamboo scaffoldings. *Engineering Structures*, 24, 429–442.
16. Shen, S. (2007). The art of survival: Case study of Crosswaters Ecolodge, Nankunshan, Guangdong. *Urban Space Design*, 3, 64–73 (in Chinese).
17. Dethier, J., Liese, W., Otto, F., Schaur, E., & Steffans, K. (2000). *Grow your own house-Simon Velez and bamboo architecture*. Weil am Rhein, Germany: Vitra Design Museum Press.
18. Richard, M. J. (2013). *Assessing the performance of bamboo structural components* [PhD thesis]. Pittsburgh, PA: University of Pittsburgh.
19. van der Lugt, P., Vogtländer, J., & Brezet, H. (2009). *Bamboo, a sustainable solution for western Europe-design cases, LCAs and land-use* [Technical report]. Beijing, China: International Network for Bamboo and Rattan.
20. Vengala, J., Jagadeesh, H. N., & Pandey, C. N. (2008). *Development of bamboo structures in India: Modern bamboo structures*. London, UK: Taylor & Francis Press; pp. 51–63.
21. Bansal, A. K., & Zoolagud, S. (2002). Bamboo composites: Material of the future. *Journal of Bamboo and Rattan*, 1, 119–130.
22. Loewe, M., & Shaughnessy, E. L. (1999). *The Cambridge history of Ancient China: From the origins of civilization to 221 BC*. Cambridge, UK: Cambridge University Press.
23. Johnson, S. (2008). Reinventing the wheel. *The Daily Princeton*. Retrieved from www.dailyprincetonian.com/2008/04/24/20982/
24. Soboyejo, W. (2002). *Mechanical properties of engineered materials*. New York: CRC Press.
25. Rahbar, N., Jorjani, M., Riccardelli, C., et al. (2010). Mixed mode fracture of marble/adhesive interfaces. *Materials Science and Engineering A*, 527, 4939–4946.
26. Wang, J. J. A., Ren, F., Tan, T., & Liu, K. (2015). The development of in situ fracture toughness evaluation techniques in hydrogen environment. *International Journal of Hydrogen Energy*, 40, 2013–2024.
27. Burros, M. (1987). Winter bamboo shoots. *New York Times*. Retrieved from www.nytimes.com/recipes/2689/winter-bamboo-shoots.html
28. Gibson, L., Ashby, M., Karam, G., Wegst, U., & Shercliff, H. (1995). The mechanical properties of natural materials. II. Microstructures for mechanical efficiency. *Proceedings of the Royal Society A: Mathematical, Physical and Engineering Sciences.*, 450, 141–162.
29. Amada, S., Ichikawa, Y., Munekata, T., Nagase, Y., & Shimizu, H. (1997). Fiber texture and mechanical graded structure of bamboo. *Composites Part B: Engineering*, 28, 13–20.

30. Ghavami, K., Rodrigues, C., & Paciornik, S. (2003). Bamboo: Functionally graded composite material. *Asian Journal of Civil Engineering (Building and Housing)*, 4, 1–10.
31. Li, X. B. (2004). *Physical, chemical, and mechanical properties of bamboo and its utilization potential for fiberboard manufacturing* [Master thesis]. Baton Rouge: Louisiana State University.
32. Ray, A. K., Das, S. K., Mondal, S., & Ramachandrarao, P. (2004). Microstructural characterization of bamboo. *Journal of Materials Science*, 39, 1055–1060.
33. Ray, A. K., Mondal, S., Das, S. K., & Ramachandrarao, P. (2005). Bamboo: A functionally graded composite-correlation between microstructure and mechanical strength. *Journal of Materials Science*, 40, 5249–5253.
34. Silva, E. C. N., Walters, M. C., & Paulino, G. H. (2006). Modeling bamboo as a functionally graded material: Lessons for the analysis of affordable materials. *Journal of Materials Science*, 41, 6991–7004.
35. Lo, T. Y., Cui, H., & Leung, H. (2004). The effect of fiber density on strength capacity of bamboo. *Materials Letters*, 58, 2595–2598.
36. Lo, T. Y., Cui, H., Tang, P., & Leung, H. (2008). Strength analysis of bamboo by microscopic investigation of bamboo fibre. *Construction and Building Materials*, 22, 1532–1535.
37. Abdul Khalil, H., Bhat, I., Jawaid, M., Zaidon, A., Hermawan, D., & Hadi, Y. (2012). Bamboo fibre reinforced biocomposites: A review. *Materials & Design*, 42, 353–368.
38. Burgert, I., & Keplinger, T. (2013). Plant micro-and nanomechanics: Experimental techniques for plant cell-wall analysis. *Journal of Experimental Botany*, 64, 4635–4649.
39. Fratzl, P., Burgert, I., & Gupta, H. S. (2004). On the role of interface polymers for the mechanics of natural polymeric composites. *Physical Chemistry Chemical Physics*, 6, 5575–5579.
40. Nishida, M., Tanaka, T., Miki, T., Shigematsu, I., Kanayama, K., & Kanematsu, W. (2014). Study of nanoscale structural changes in isolated bamboo constituents using multiscale instrumental analyses. *Journal of Applied Polymer Science*, 131, 9.
41. Zou, L. H., Jin, H., Lu, W. Y., & Li, X. D. (2009). Nanoscale structural and mechanical characterization of the cell wall of bamboo fibers. *Materials Science and Engineering C*, 29, 1375–1379.
42. Youssefian, S., & Rahbar, N. (2015). Molecular origin of strength and stiffness in bamboo fibrils. *Scientific Reports*, 5, 11116.
43. Yu, Y., Tian, G., Wang, H., Fei, B., & Wang, G. (2011). Mechanical characterization of single bamboo fibers with nanoindentation and microtensile technique. *Holzforschung*, 65, 113–119.
44. Habibi, M. K., & Lu, Y. (2014). Crack propagation in bamboo's hierarchical cellular structure. *Scientific Reports*, 4, 1–7.
45. Wai, N., Nanko, H., & Murakami, K. (1985). A morphological study on the behavior of bamboo pulp fibers in the beating process. *Wood Science and Technology*, 19, 211–222.
46. Villalobos, G., Linero, D. L., & Muñoz, J. D. (2011). A statistical model of fracture for a 2d hexagonal mesh: The cell network model of fracture for the bamboo *Guadua angustifolia*. *Computer Physics Communications*, 182, 188–191.
47. Schott, W. (2005). Bamboo under the microscope. Retrieved from www.powerfibers.com/Bamboo_under_the_Microscope.pdf
48. Abe, K., & Yano, H. (2010). Comparison of the characteristics of cellulose microfibril aggregates isolated from fiber and parenchyma cells of Moso bamboo (*Phyllostachys pubescens*). Cellulose, 17, 271–277.

49. Tan, T., Rahbar, N., Allameh, S., et al. (2011). Mechanical properties of functionally graded hierarchical bamboo structures. *Acta Biomaterialia*, 7, 3796–3803.
50. He, X. Q., Suzuki, K., Kitamura, S., Lin, J. X., Cui, K. M., & Itoh, T. (2002). Toward understanding the different function of two types of parenchyma cells in bamboo culms. *Plant and Cell Physiology*, 43, 186–195.
51. Nogata, F., & Takahashi, H. (1995). Intelligent functionally graded material: Bamboo. *Composites Engineering*, 5, 743–751.
52. Ma, J. F., Chen, W. Y., Zhao, L., & Zhao, D. H. (2008). Elastic buckling of bionic cylindrical shells based on bamboo. *Journal of Bionic Engineering*, 5, 231–238.
53. Laroque, P. (2007). *Design of a low cost bamboo footbridge* [Master thesis]. Cambridge, MA: Massachusetts Institute of Technology.
54. Shao, Z. P., Fang, C. H., & Tian, G. L. (2009). Mode I interlaminar fracture property of Moso bamboo (*Phyllostachys pubescens*). *Wood Science and Technology*, 43, 527–536.
55. Amada, S., & Untao, S. (2001). Fracture properties of bamboo. *Composites Part B: Engineering*, 32, 451–459.
56. Charalambides, P., Lund, J., Evans, A., & Mcmeeking, R. (1989). A test specimen for determining the fracture resistance of bimaterial interfaces. *Journal of Applied Mechanics*, 56, 77–82.
57. Budiansky, B., Amazigo, J. C., & Evans, A. G. (1988). Small-scale crack bridging and the fracture toughness of particulate-reinforced ceramics. *Journal of the Mechanics and Physics of Solids*, 36, 167–187.
58. Bloyer, D., Ritchie, R., & Rao, K. V. (1998). Fracture toughness and R-curve behavior of laminated brittle-matrix composites. *Metallurgical and Materials Transactions A*, 29, 2483–2496.
59. Bloyer, D., Ritchie, R., & Rao, K. V. (1999). Fatigue-crack propagation behavior of ductile/brittle laminated composites. *Metallurgical and Materials Transactions A*, 30, 633–642.
60. Suhaily, S. S., Khalil, H. A., Nadirah, W. W., & Jawaid, M. (2013). Bamboo based biocomposites material, design and applications. *Materials science-advanced topics*. Rijeka, Croatia: InTech Press; pp. 489–517.
61. Rassiah, K., & Megat Ahmad, M. (2013). A review on mechanical properties of bamboo fiber reinforced polymer composite. *Australian Journal of Basic and Applied Sciences*, 7, 247–253.
62. Deshpande, A. P., Bhaskar Rao, M., & Lakshmana Rao, C. (2000). Extraction of bamboo fibers and their use as reinforcement in polymeric composites. *Journal of Applied Polymer Science*, 76, 83–92.
63. Yao, W., & Zhang, W. (2011). Research on manufacturing technology and application of natural bamboo fibre. In *Intelligent Computation Technology and Automation (Vol. 2)*. Shenzhen, Guangdong, China: International Conference on IEEE; pp. 143–148.
64. Chen, X., Guo, Q., & Mi, Y. (1998). Bamboo fiber-reinforced polypropylene composites: A study of the mechanical properties. *Journal of Applied Polymer Science*, 69, 1891–1899.
65. Shito, T., Okubo, K., & Fujii, T. (2002). *Development of eco-composites using natural bamboo fibers and their mechanical properties*. Transactions on The Built Environment: WIT Press, 4, pp. 175–182.
66. Sano, O., Matsuoka, T., Sakaguchi, K., & Karukaya, K. (2002). *Study on the interfacial shear strength of bamboo fibre reinforced plastics*. Transactions on The Built Environment: WIT Press, 4, 147–156.

67. Okubo, K., Fujii, T., & Yamamoto, Y. (2004). Development of bamboo-based polymer composites and their mechanical properties. *Composites Part A: Applied Science and Manufacturing*, 35, 377–383.
68. Lee, S. Y., Chun, S. J., Doh, G. H., Kang, I. A., Lee, S., & Paik, K. H. (2009). Influence of chemical modification and filler loading on fundamental properties of bamboo fibers reinforced polypropylene composites. *Journal of Composite Materials*, 43, 1639–1657.
69. Chattopadhyay, S. K., Khandal, R., Uppaluri, R., & Ghoshal, A. K. (2011). Bamboo fiber reinforced polypropylene composites and their mechanical, thermal, and morphological properties. *Journal of Applied Polymer Science*, 119, 1619–1626.
70. Das, M., & Chakraborty, D. (2009b). The effect of alkalization and fiber loading on the mechanical properties of bamboo fiber composites, part 1: Polyester resin matrix. *Journal of Applied Polymer Science*, 112, 489–495.
71. Kushwaha, P. K., & Kumar, R. (2010). Studies on the water absorption of bamboo-epoxy composites: The effect of silane treatment. *Polymer-Plastics Technology and Engineering*, 49, 867–873.
72. Kushwaha, P. K., & Kumar, R. (2011). Influence of chemical treatments on the mechanical and water absorption properties of bamboo fiber composites. *Journal of Reinforced Plastics and Composites*, 30, 73–85.
73. Wong, K., Zahi, S., Low, K., & Lim, C. (2010). Fracture characterisation of short bamboo fibre reinforced polyester composites. *Materials & Design*, 31, 4147–4154.
74. Rajulu, A. V., Chary, K. N., Reddy, G. R., & Meng, Y. (2004). Void content, density and weight reduction studies on short bamboo fiber–epoxy composites. *Journal of Reinforced Plastics and Composites*, 23 127–130.
75. Nirmal, U., Hashim, J., & Low, K. (2012). Adhesive wear and frictional performance of bamboo fibres reinforced epoxy composite. *Tribology International*, 47, 122–133.
76. Osorio, L., Trujillo, E., Van Vuure, A., & Verpoest, I. (2011). Morphological aspects and mechanical properties of single bamboo fibers and flexural characterization of bamboo/epoxy composites. *Journal of Reinforced Plastics and Composites*, 30, 396–408.
77. Das, M., & Chakraborty, D. (2009a). Processing of the uni-directional powdered phenolic resin-bamboo fiber composites and resulting dynamic mechanical properties. *Journal of Reinforced Plastics and Composites*, 28, 1339–1348.
78. Kim, J. Y., Peck, J. H., Hwang, S. H., et al. (2008). Preparation and mechanical properties of poly (vinyl chloride)/bamboo flour composites with a novel block copolymer as a coupling agent. *Journal of Applied Polymer Science*, 108, 2654–2659.
79. Thwe, M. M., & Liao, K. (2003). Durability of bamboo-glass fiber reinforced polymer matrix hybrid composites. *Composites Science and Technology*, 63, 375–387.
80. Samal, S. K., Mohanty, S., & Nayak, S. K. (2009). Polypropylene-bamboo/glass fiber hybrid composites: Fabrication and analysis of mechanical, morphological, thermal, and dynamic mechanical behavior. *Journal of Reinforced Plastics and Composites*, 28, 2729–2747.
81. Tan, T., Santos, S. F. D., Savastano, H., & Soboyejo, W. (2012). Fracture and resistance-curve behavior in hybrid natural fiber and polypropylene fiber reinforced composites. *Journal of Materials Science*, 47, 2864–2874.
82. Kushwaha, P., & Kumar, R. (2009). Enhanced mechanical strength of BFRP composite using modified bamboos. *Journal of Reinforced Plastics and Composites*, 28, 2851–2859.
83. Koren, G. (2010). *New bamboo product for the global market* [Master thesis]. Delft, Netherlands: Delft University of Technology.

84. Boobicycles (2013). Boo fixie. Retrieved from http://boobicycles.com/gallery/?id=7215762702420521 8&title=Boo%20Fixie
85. Meyers, M. A., Chen, P. Y., Lin, A. Y. M., & Seki, Y. (2008). Biological materials: Structure and mechanical properties. *Progress in Materials Science*, 53, 1–206.
86. Gibson, L. J. (2012). The hierarchical structure and mechanics of plant materials. *Journal of the Royal Society Interface*, 9, 2749–2766.
87. Wegst, U. G., Bai, H., Saiz, E., Tomsia, A. P.,, & Ritchie, R. O. (2015). Bioinspired structural materials. *Nature Materials*, 14, 23–36.
88. Li, J. F., Takagi, K., Ono, M., et al. (2003). Fabrication and evaluation of porous piezoelectric ceramics and porosity–graded piezoelectric actuators. *Journal of the American Ceramic Society*, 86, 1094–1098.
89. Zhou, B. (2000). Bio-inspired study of structural materials. *Materials Science and Engineering C*, 11, 13–18.
90. Ribbans, B., Li, Y., & Tan, T., 2016. A bioinspired study on the interlaminar shear resistance of helicoidal fiber structures. *Journal of the Mechanical Behavior of Biomedical Materials*, 56, 57–67.
91. Tan, T., & Ribbans, B. (2017). A bioinspired study on the compressive resistance of helicoidal fibre structures. *Proceedings of the Royal Society A: Mathematical, Physical and Engineering Sciences*, 473, 20170538.
92. Alghamdi, S., Tan, T., Hale-Sills, C., et al. (2017). Catastrophic failure of nacre under pure shear stresses of torsion. *Scientific Reports*, 7, 13123.
93. Alghamdi, S., Du, F., Yang, J., & Tan, T. (2018). The role of water in the initial sliding of nacreous tablets: Findings from the torsional fracture of dry and hydrated nacre. *Journal of the Mechanical Behavior of Biomedical Materials*, 88, 322–329.
94. Schulgasser, K., & Witztum, A. (1992). On the strength, stiffness and stability of tubular plant stems and leaves. *Journal of Theoretical Biology*, 155, 497–515.
95. Bruck, H., Evans, J., & Peterson, M. (2002). The role of mechanics in biological and biologically inspired materials. *Experimental Mechanics*, 42, 361–371.
96. Taya, M. (2003). Bio-inspired design of intelligent materials. In *Smart structures and materials, electroactive polymer actuators and devices (Vol. 5051)*. Bellingham, WA: International Society for Optics and Photonics; pp. 54–66.
97. Rabin, B. H., Williamson, R. L., Bruck, H. A., et al. (1998). Residual strains in an Al_2O_3-Ni joint bonded with a composite interlayer: Experimental measurements and FEM analyses. *Journal of the American Ceramic Society*, 81, 1541–1549.
98. de Vries, D. V. W. M. (2010). *Biomimetic design based on bamboo* [Master thesis]. Eindhoven, Netherlands: Eindhoven University of Technology.
99. Winter, A. N., Corff, B. A., Reimanis, I. E., & Rabin, B. H. (2000). Fabrication of graded nickel-alumina composites with a thermal-behavior-matching process. *Journal of the American Ceramic Society*, 83, 2147–2154.
100. Williamson, R., Rabin, B., & Drake, J. (1993). Finite element analysis of thermal residual stresses at graded ceramic-metal interfaces. Part I. Model description and geometrical effects. *Journal of Applied Physics*, 74, 1310–1320.
101. Almajid, A. A., & Taya, M. (2001). 2D-elasticity analysis of FGM piezo-laminates under cylindrical bending. *Journal of Intelligent Material Systems and Structures*, 12, 341–351.
102. Rubio, W. M., Vatanabe, S. L., Paulino, G. H., & Silva, E. C. N. (2011). Functionally graded piezoelectric material systems – A multiphysics perspective. *Advanced computational*

materials modeling: From classical to multi-scale techniques. Weinheim, Germany: Wiley Press; pp. 301–339.

103. Li, S., Zeng, Q., Xiao, Y., Fu, S., & Zhou, B. (1995). Biomimicry of bamboo bast fiber with engineering composite materials. *Materials Science and Engineering: C*, 3, 125–130.
104. Markham, M., & Dawson, D. (1975). Interlaminar shear strength of fibre-reinforced composites. *Composites*, 6, 173–176.
105. Tan, T., Ren, F., Wang, J. J. A., et al. (2013). Investigating fracture behavior of polymer and polymeric composite materials using spiral notch torsion test. *Engineering Fracture Mechanics*, 101, 109–128.
106. Tan, T., Xia, T., O'Folan, H., et al. (2014). Sustainability in beauty: A review and extension of bamboo inspired materials. *Blucher Material Science Proceedings*, 1, 18–21.
107. Tan, T., Xia, T., O'Folan, H., et al. (2015). Sustainability in beauty: An innovative proposing-learning model to inspire renewable energy education. *The Journal of Sustainability Education*, 8, 1–7.

Part II

Structures

5 Bioinspired Underwater Propulsors

Tyler Van Buren, Daniel Floryan, and Alexander J. Smits

5.1 Introduction

In recent years, there has been considerable interest in developing novel underwater vehicles that use propulsion systems inspired by biology [1,2]. Such vehicles have the potential to uncover new mission capabilities and improve maneuverability, efficiency, and speed [3,4]. Here we will explore the physical mechanisms that govern the performance – especially swimming speed and efficiency – of propulsive techniques inspired by biology. We will also show that we can translate the understanding we have gained from biology to the design of a new generation of underwater vehicles.

Many aquatic animals are capable of great speed, high efficiency, and rapid maneuvering. Engineers have been able to mimic their techniques by constructing robotic imitations, with considerable success. One of the best-known examples is Robotuna, an eight-link tendon- and pulley-driven robot, whose external shape has the form of a bluefin tuna and which is capable of emulating the swimming motion of a live tuna [5]. This project evolved into the Ghostswimmer, a prototype Navy vehicle that swims by manipulating its dorsal (back), pectoral (chest), and caudal (tail) fins [6]. However, many biological features do not exist for the purpose of swimming alone – they could exist for survivability or reproductive purposes. As a design paradigm, therefore, it may be better to abandon biomimetic designs for ones that are inspired by biology but not constrained by it.

In general, we can identify four major types of swimmers, with examples illustrated in Figure 5.1:

Oscillatory: these animals propel themselves primarily using a semi-rigid caudal fin or fluke that is oscillated periodically. Examples include salmon, tuna, and dolphin.

Undulatory: these animals utilize a traveling wave along their body or propulsive fins to push fluid backward. Examples include eels, lampreys, and rays.

Pulsatile: these animals periodically "inhale" a volume of water and then discharge it impulsively as a jet, producing thrust in the direction opposite the jet. Examples include jellyfish, squid, and some mollusks.

Drag-based: these animals force a bluff body such as a rigid flipper through the water to generate thrust by reaction. Examples include humans, turtles, and ducks.

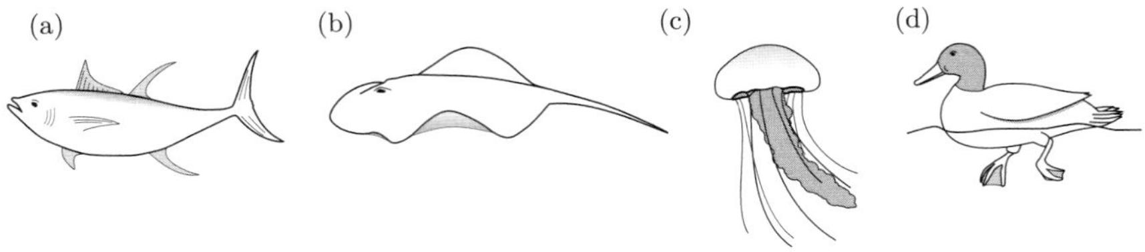

Figure 5.1 Examples of four swimming types: (a) oscillatory – tuna, (b) undulatory – ray, (c) pulsatile jet – jellyfish, and (d) drag-based – duck.

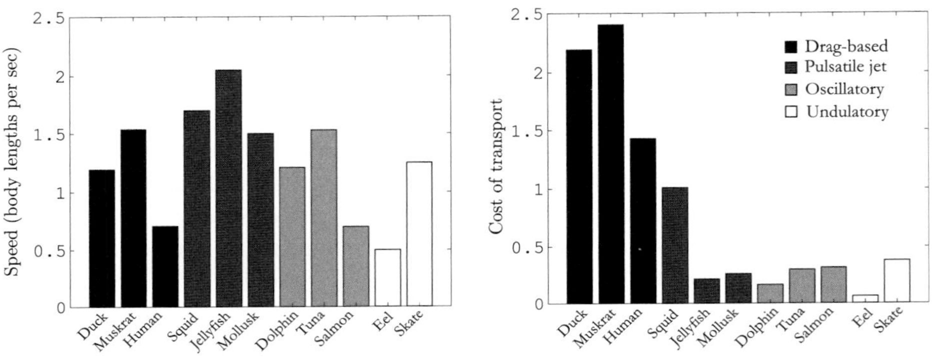

Figure 5.2 Relative swimming speed and cost of transport for oscillatory, undulatory, pulsatile jet, and drag-based swimmers. (Data summarized from Fish [8].)

Some swimmers use more than one swimming technique. For example, the sea turtle has been reported to use the *power stroke* [7], where the first half of the stroke is drag-based and the second half is oscillatory. In this chapter, we will treat the different swimming techniques in isolation, although we encourage readers to think about possible designs that could incorporate positive aspects of each.

It is not surprising that different swimmers show differences in their performance characteristics. Typical swimming speed (normalized by body length) and cost of transport of the four types of swimmers are shown in Figure 5.2. The cost of transport quantifies the energy efficiency of transporting an animal or vehicle from one place to another. In biology, it is often expressed as the distance traveled per unit energy cost (similar to miles per gallon). Despite large differences in size and swimming mechanisms, we see that most organisms swim between 0.5 and 1.5 body lengths per second. This is a typical cruising speed; the maximum swimming speed can be very different among different swimmer types. However, drag-based swimmers have a notably higher cost of transport than the others. This makes sense, as most drag-based swimmers do not necessarily solely live in water, and many have evolved to also walk or fly.

We now examine the basic mechanisms that these different types of swimmers employ, as they might be implemented on a vehicle.

5.2 Mechanics of Underwater Propulsion

In steady swimming, where there is no acceleration or deceleration, the thrust produced by the propulsive system is balanced exactly by the drag on the vehicle, in the time-average. For underwater vehicles, the drag force has two major components: the friction drag due to the viscous shear stresses acting on the surface of the vehicle, and the pressure, or form drag due to the pressure losses in the wake. For streamlined vehicles, such as those shaped like fish, the viscous drag component tends to dominate, whereas for bluff bodies, exemplified by more boxy shapes, the form drag dominates. An important parameter is the Reynolds number *Re*, which is a measure of the importance of inertial forces to viscous forces, and is defined by $Re = \rho U_\infty L/\mu$, where U_∞ is the speed, L is a characteristic dimension of the vehicle such as its length, and ρ and μ are the density and viscosity of the fluid, in our case fresh or salt water. At large Reynolds numbers, typical of most fish and all underwater vehicles, the form drag is almost independent of Reynolds number, but the viscous drag always remains a function of Reynolds number.

For thrust, most modern human-designed propulsors utilize some sort of continuous rotation (think propellers), which is not a motion natural to biology. Fish and mammals such as dolphins and whales use fins and flukes to propel themselves in combined pitching and heaving motions, turtles use a paddling motion, while squid eject jets of fluid. We find that there are four major sources of thrust: (1) drag-based thrust, (2) lift-based thrust, (3) added mass forces, and (4) momentum injection, as illustrated in Figure 5.3. Drag is the force acting opposite to the direction of motion of the body, lift is the force produced normal to the direction of motion, added mass forces are due to the inertia of the water that is put in motion by the body, and momentum injection is the force produced by jetting fluid from the body, as used by squid and jellyfish. These propulsion types can be, and often are, combined in practical systems, but we will first consider them separately.

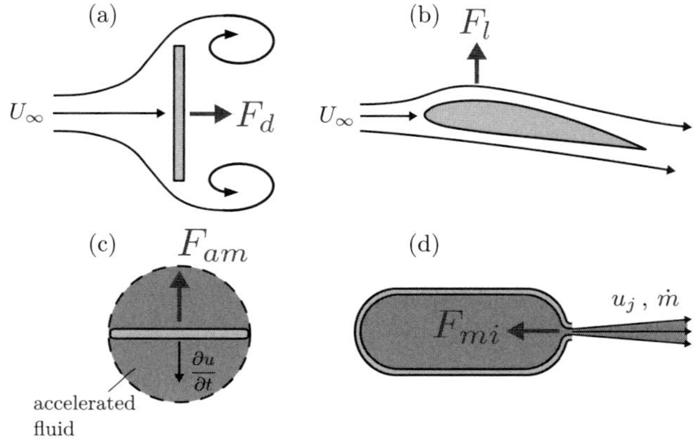

Figure 5.3 Typical thrust generation mechanisms: (a) drag-based, (b) lift-based, (c) added mass, and (d) momentum injection.

Drag-based thrust actually uses the form drag experienced by a bluff body to generate thrust. Human swimmers make extensive use of drag-based thrust. For example, a swimmer doing the breaststroke will spread her hands, push water rearwards, and so by action–reaction, propels herself forward. The (steady) drag-based thrust F_d is given by

$$F_d = \frac{1}{2}\rho U_\infty^2 A C_D, \tag{5.1}$$

where A is the frontal area of the body, U_∞ is the speed, and the drag coefficient C_D is defined by the body shape (see Figure 5.3a). The drag coefficient of a hand is about 1, so it takes about 10 N of force to move a hand at 1 m/s through water.

As to lift-based thrust, we noted that lift was defined as the force acting normal to the direction of motion. For a simple airfoil or hydrofoil in steady motion, the lift force is given by

$$F_l = \frac{1}{2}\rho U_\infty^2 A C_L, \tag{5.2}$$

where C_L is the lift coefficient, which depends on the shape of the object and the angle of attack α (see Figure 5.3b). For a thin hydrofoil, C_L depends only on the angle of attack ($=2\pi\alpha$). Because the lift is always perpendicular to the body motion, getting thrust from lift requires the propulsor to move laterally as well as translationally, so that the effective (local) velocity seen by the propulsor is such that a portion of the lift acts in a direction that produces thrust (that is, a forward velocity). This is sometimes referred to as the Knoller–Betz effect.

Thrust due to added mass may be illustrated by considering the differences between suddenly moving your hand in air, as compared to doing it in water. In both cases, there would be a drag force, as described previously, due to form drag, but there is an additional force required to move the surrounding fluid. This "added mass" force F_{am} depends on the mass of fluid put into motion, as well as the acceleration required (Newton's Second Law), as in

$$F_{am} = m_f \frac{\partial u}{\partial t}, \tag{5.3}$$

where m_f is the fluid mass and $\partial u/\partial t$ is the fluid acceleration. It is relatively small in air but much larger in water, because the density of water is about 800 times that of air. The reaction force generated by this impulsive motion in water can generate considerable thrust. The actual force can be difficult to assess accurately because it is difficult to precisely estimate the volume of fluid that is accelerated by a particular motion, but for simple motions and shapes, some reasonable estimates can be given. There is often a misconception that added mass is more important in water than in air because water is heavier, but all of these forces scale with the fluid density, thus the weight of the fluid does not dictate the relative importance of thrust generation mechanisms.

Lastly, we consider the direct injection of momentum into the surrounding fluid, most commonly via a jet. If you were to inflate a balloon and then release it before tying the bottom opening, it would erratically dart around the room while deflating. This is because the pressure inside the balloon is released through the small opening, creating a jet of air. This released air has momentum, and so it creates a reaction force that causes the balloon to move. Similarly, if you were to let go of a running hose it would whip around dramatically. In the steady case, the momentum of the jet governs the force generated according to

$$F_{mi} = \dot{m} v_j, \tag{5.4}$$

where $\dot{m}$ is the mass flow rate through the jet opening and v_j is the jet velocity. When properly harnessed, this force can be utilized as a controlled thrust generation mechanism.

Thus far, we have identified four possible sources of thrust that are important for underwater propulsion. In addition, we need to consider the efficiency, which is typically defined by

$$\eta = \frac{F_x U_\infty}{\mathcal{P}}, \tag{5.5}$$

where $\mathcal{P}$ is the power input into the propulsor, F_x is the thrust generated by the propulsor, and U_∞ is the velocity of the vehicle. This definition is termed the "Froude efficiency," and it is the fraction of power input into the propulsor that is used to propel the vehicle forward, ignoring the question of how efficiently the input power was generated to begin with. We emphasize that the numerator is the work that the system needs to do against the drag experienced by the vehicle; that is, for steady swimming, F_x balances the vehicle drag such that the vehicle moves at a constant speed. The denominator is the power input into the propulsor, or the power that the propulsor has at its disposal in order to do work against the drag experienced by the vehicle. The form of the numerator is as written here for all steadily swimming vehicles, whereas the form of the input power in the denominator differs for different modes of propulsion.

With these basic concepts in place, we can now examine the traits of different swimmers (oscillatory, undulatory, pulsatile, and drag-based) to identify elements that could inspire propulsor design, with a special focus on thrust and efficiency. Due to the similarities in thrust production mechanisms and wake characteristics, we will combine oscillatory and undulatory swimmers into a single section. For each swimming style, we will present our current understanding of the fundamental mechanisms that govern propulsive performance. Then, we will propose a simplified robotic swimmer concept for each swimming type where we use aquatic swimmers as inspiration, without constraining ourselves to mimic biology.

5.3 Oscillatory and Undulatory

Oscillatory and undulatory swimmers make up a large fraction of the aquatic life synonymous with high swimming speed and efficiency, and so they have become the

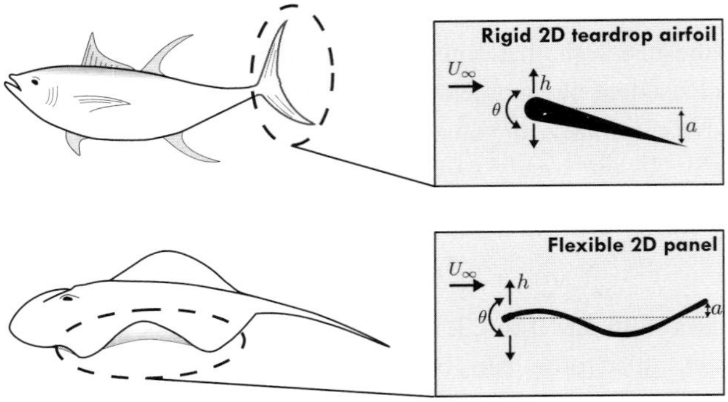

Figure 5.4 Example of how one could simplify an oscillatory body/caudal fin (top) and undulatory median/paired fins (bottom) swimmers. Motion inputs: heave $h(t)$ and pitch $\theta(t)$, and output: trailing edge amplitude $a(t)$.

focus of propulsion inspiration. These swimmers use lateral motion of their propulsive surfaces to generate thrust. *Thunniform* swimmers, such as tuna, tend to be more oscillatory swimmers – the primary propulsion is achieved using a flapping motion of their caudal fin or fluke and the body is fairly rigid. *Anguilliform* swimmers such as eels and snakes are more undulatory – the body itself is the primary propulsor. Rays also use undulatory motion, but they use their elongated pectoral fins instead of a caudal fin.

Figure 5.4 shows examples of how to model an oscillatory or undulatory swimmer. For example, an oscillatory swimmer's propulsor can be modeled using a pitching and heaving foil, or an undulatory swimmer's propulsor could be modeled as a flexible panel. Although these simple models may not look much like fish, they can reveal much of the underlying physics and provide simple designs for possible application to underwater vehicles.

5.3.1 Wake Characteristics

The wake of a propulsor is like its footprint, in that it can reveal the physical mechanisms by which thrust is produced. For example, the flow over a stationary cylinder periodically separates, resulting in an oscillating drag force. In the wake, this unsteadiness is evinced by a train of alternating sign vortices, known as a von Kármán vortex street, shown in Figure 5.5a. If you were to measure the time-averaged velocity field downstream of the cylinder, you would see a mean momentum deficit in the wake, indicating that a net drag is acting on the body. A similar wake is exhibited by a pitching or heaving foil, but here the vortices are of the opposite sign, forming a reverse von Kármán vortex street, shown in Figure 5.5b. This orientation results in a time-average velocity field that has a mean momentum excess in the wake, indicating thrust production.

When the propulsor has a finite width, the wake becomes highly three-dimensional, as shown in Figure 5.5c. The spanwise vortices seen in a two-dimensional propulsor

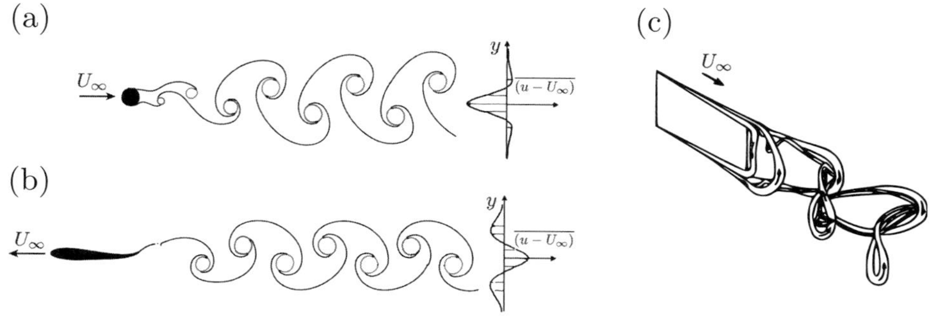

Figure 5.5 Vortex wake and time-averaged velocity profiles of a two-dimensional (a) drag-producing cylinder and (b) thrust-producing foil. (c) Three-dimensional vortex skeleton of a pitching panel. ((a–b) reprinted from Eloy [9] with permission from Elsevier, (c) from Buchholz and Smits [10] reproduced with permission.)

become interconnected loops, and the interaction among these vortex loops compresses the wake in the spanwise direction while spreading it in the panel-normal direction. Van Buren et al. [11] altered the shape of the trailing edge to manipulate the vortex dynamics of the wake. They found that they delayed or enhanced the wake compression and breakdown, impacting the thrust and efficiency of the propulsor. Hence, it may be possible to control the performance of a propulsor by altering the vortex structure in the wake.

Care must be taken when solely using the wake for assessing performance, which is something that has gained popularity since the advent of flow measurement techniques, like particle image velocimetry, that make the wake easily accessible. There is no strong correlation between the wake structure, specifically vortex spacing and pattern, and the performance characteristics of an unsteady propulsor [12]. The wake does have thrust and power information, but to properly access it one would also need to know the full unsteady pressure field, which is difficult to obtain experimentally.

5.3.2 Motion Type

For our simplest model of a propulsor, we will neglect the influence of the main body of the vehicle and consider a sinusoidally oscillating rigid foil in isolation. Its motion can be deconstructed into a time-varying pitching component $\theta(t)$ (twisting about the leading edge), and heaving component $h(t)$ (pure lateral translation, or plunging), that together produce a trailing edge motion $a_t(t)$ (see Figure 5.4). The foil is assumed to be rectangular, with a chord length c and a span s.

The thrust generated by pitching motion is due purely to added mass forces, since the time-average lift force is zero [13]. We therefore expect the thrust to scale as the product of the streamwise component of the added mass ($\sim \rho c^2 s$) and the acceleration ($\sim c\ddot{\theta}$), where the symbol $\sim$ denotes "it varies as." That is,

$$F_x \sim \rho s c^3 \ddot{\theta}\theta,$$

and so the time-averaged thrust scales according to

$$\bar{F}_x \sim \rho s c^3 f^2 \theta_0^2 \approx \rho s c f^2 a^2, \tag{5.6}$$

where f is the frequency of oscillation and $a \approx c\theta_0$ is the trailing edge amplitude for small pitching motions.

The thrust produced by heaving motions is primarily due to lift-based forces, and added mass forces in the thrust direction are typically small [13]. Thus, we expect the thrust to scale as the streamwise component of the instantaneous lift force. That is,

$$F_x \sim L(\dot{h}/U^*),$$

where L is the lift force, $\dot{h}$ is the heave velocity, and U^* is the effective velocity seen by the foil. If we assume that the contribution to the lift is quasi-steady, and that the angle of attack is small, so that $\alpha \approx \dot{h}/U^*$, then

$$F_x \sim \frac{1}{2}\rho U^{*2} sc\,(2\pi\alpha)\left(\dot{h}/U^*\right) \sim \pi\rho sc\,\dot{h}^2,$$

so that the mean thrust scales as

$$\bar{F}_x \sim \rho s c f^2 h_0^2 = \rho s c f^2 a^2, \tag{5.7}$$

where $a = h_0$ for heaving motions. Interestingly, at this level of approximation, the thrust generated through pitch and heave scale similarly, and have no dependence on velocity (which has been confirmed experimentally by [14]). A more detailed analysis is offered in Floryan et al. [13], which includes unsteady and nonlinear effects.

Typically, the performance and motions are presented nondimensionally. The thrust and power are then normalized using aerodynamic convention, and so we obtain the thrust and power coefficients as well as the resulting efficiency, defined by

$$C_T = \frac{F_x}{\frac{1}{2}\rho U_\infty^2 sc}, \qquad C_P = \frac{F_y \dot{h} + M_z \dot{\theta}}{\frac{1}{2}\rho U_\infty^3 sc}, \qquad \eta = C_T/C_P, \tag{5.8}$$

where F_y is the side force and M_z is the spanwise moment. We also introduce the Strouhal number,

$$St = \frac{2fa}{U_\infty}, \tag{5.9}$$

which can be thought of as a ratio of the wake width to the spacing between shed vortices, or as the ratio of the lateral velocity of the trailing edge to the freestream velocity. Another parameter is the reduced frequency, defined by

$$f^* = \frac{fc}{U_\infty}, \tag{5.10}$$

which is the time it takes a particle to pass from the leading to the trailing edge of the propulsor compared to the period of its motion. Note that f^* does not depend on the motion amplitude.

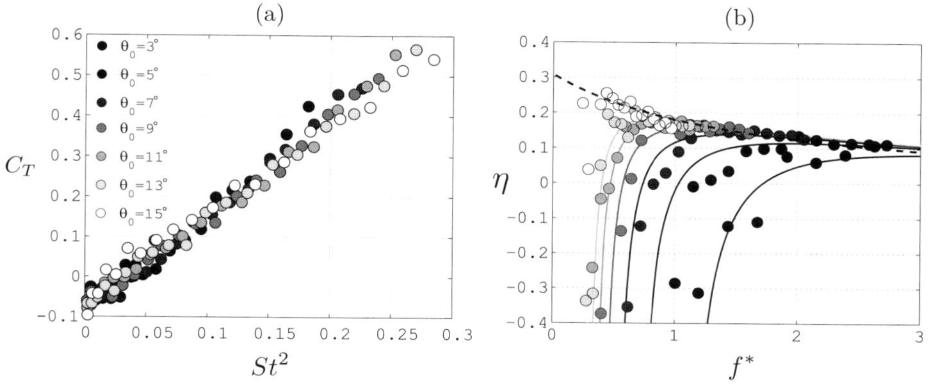

Figure 5.6 Time-averaged (a) thrust and (b) efficiency for a pitching two-dimensional teardrop foil. Lines represent the the analytical solutions, where the solid lines include viscous drag, and the dashed line is the inviscid approximation. (Taken from Floryan et al. [13].)

In much of the current literature, the thrust, power, and efficiency are shown as a function of Strouhal number to define the performance envelope (Reynolds number effects are often neglected, but more about that later). Figure 5.6 shows the thrust and efficiency of a two-dimensional pitching propulsor for a range of frequencies and amplitudes. The thrust goes as $C_T \sim St^2$ for all of the pitch amplitudes, which is in accord with Eq. (5.6), originally proposed by Floryan et al. [13]. They also showed that for inviscid flow the efficiency should follow a decaying curve with f^*, which is in accord with experiments for large values of f^* (see Figure 5.6b). At lower frequencies, the efficiency decreases dramatically due to the viscous drag, which drives the thrust – and therefore the efficiency – negative.

On their own, pitching and heaving motions are not very fish-like in appearance, and they only reach 20–30% peak propulsive efficiency. When these motions are combined, however, they can resemble fish-like motions much more closely, and achieve significant improvements in performance. Figure 5.7 exemplifies how simply adding heave to a pitching motion can dramatically increase the thrust and efficiency of the propulsor. The phase difference ϕ between the pitch and heave motions governs the motion path of the foil, and to be most fish-like, the trailing edge of the foil must lag the leading edge such that the foil slices through the water. This reduces the maximum angle of attack of the foil and the likelihood of flow separation. Many researchers suggest that $\phi = 270°$ maximizes the propulsive efficiency [15–17]. The scaling arguments for pitching or heaving motions presented by Floryan et al. [13] have since been successfully extended to simultaneously pitching and heaving motions in Van Buren et al. [17].

There are other ways to manipulate the propulsor performance. For instance, Van Buren et al. [18] showed that the thrust production is strongly correlated to the trailing edge velocity of the propulsor in both pitch and heave. For a sinusoidal motion, the trailing edge velocity can be changed by adjusting the frequency or amplitude of motion. However, a motion profile that is more like a square wave can have a much

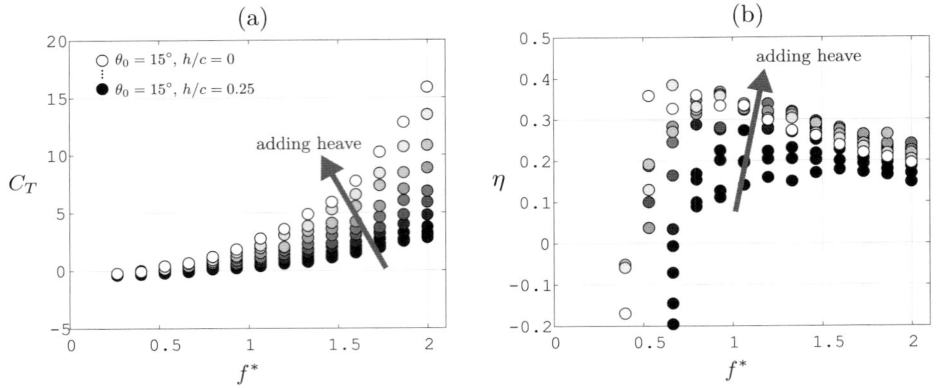

Figure 5.7 Time-averaged (a) thrust and (b) efficiency for a two-dimensional teardrop foil with incremental increases in heave amplitude to a pitching foil. Phase between pitch and heave is $\phi = 270°$. (Taken from Van Buren et al. [17], reprinted by permission of the American Institute of Aeronautics and Astronautics, Inc.)

higher peak trailing edge velocity than a sinusoidal motion of equal amplitude and frequency. By making the motion more square-like, the thrust production could be increased by over 300%.

If energy expenditure is more important than thrust, you might consider intermittent motions. Floryan et al. [19] showed that by swimming intermittently, the energy expenditure decreased linearly with duty cycle when compared to continuous swimming, although the average swimming speed is reduced accordingly. In general, other factors such as the metabolic rate will need to be taken into account, and these will generally reduce the benefits of intermittent swimming.

Lastly, one can achieve a significant performance boost by swimming near a boundary, or in the wake of another swimmer. Quinn et al. [20,21] showed experimentally that when a propulsor swims near a solid boundary it experiences a thrust increase due to ground effect. Similarly, two propulsors can significantly improve performance by swimming side-by-side [22]. The maximum thrust is found for a phase difference of 180°, and the maximum efficiency is found for 0°. In both instances, two paired foils could produce higher thrust and efficiency than a single foil. Experiments and simulations have also demonstrated the advantages of fish schooling, indicating that one swimmer can benefit from being in the wake of the other [23,24].

5.3.3 Flexibility

So far, we have considered only rigid propulsors, but most fish exhibit some form of passive/active flexibility in their propulsive fins while swimming. This flexibility turns out to be a valuable asset in increasing the propulsive performance. For example, Dewey et al., Quinn et al., and Floryan and Rowley [25–28] separately explored the influence of adding flexibility to simple pitching and heaving propulsors. They found that flexibility adds a hierarchy of resonances to the system, which can be modeled using the linear beam equation. That is,

$$\rho sc \frac{\partial^2 e}{\partial t^2} + EI \frac{\partial^4 e}{\partial x^4} = F_{ext}, \tag{5.11}$$

where e is a small panel deflection, EI is the panel stiffness, and F_{ext} is the force from the fluid. In the usual beam equation, the first term uses properties of the beam, but here it is expected that the virtual mass forces due to the water surrounding the panel will dominate (panel mass = virtual mass), and so fluid properties are used instead (see also [29]). This yields a new nondimensional frequency

$$\hat{f}^* = f\left(\rho sc^5/(EI)\right)^{1/2}, \tag{5.12}$$

which is the ratio of the frequency of the driving motion to the first resonant frequency of the panel when added mass is much larger than panel mass.

Let us consider a simple panel in fish-like motion (pitching and heaving separated by $\phi = 270°$) of varying flexibility (Π_1) and compare that to a stiff panel ($\Pi_1 \sim \infty$), where

$$\Pi_1 = \frac{Et^3}{12\left(1 - \nu_p^2\right)\rho U_\infty^2 c^3}.$$

Here ν_p is the Young's modulus of the panel, and t is the panel thickness. The thrust and efficiency of these panels, covering the range of flexural stiffnesses seen in biological propulsors, are shown in Figure 5.8 for a large parameter space of leading edge motions. For certain flexibilities, there is a clear envelope where the thrust production is up to 1.5 times higher than the thrust of a rigid panel. This is because the frequency of the panel

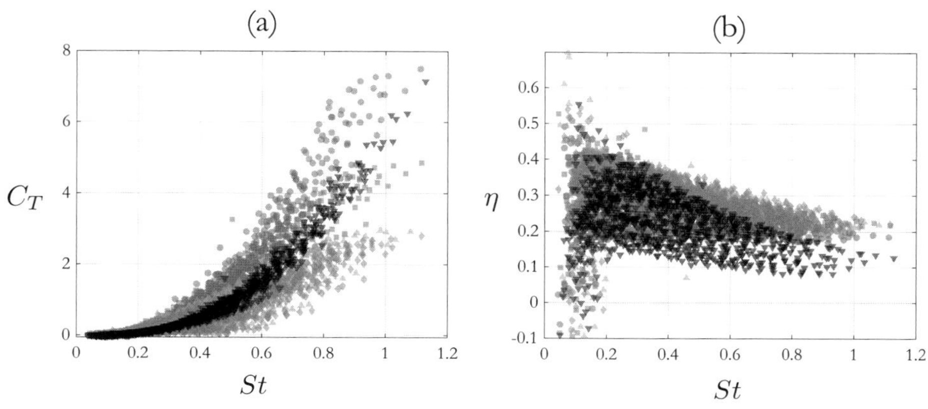

Figure 5.8 Effects of flexibility on time-averaged (a) thrust and (b) efficiency for pitching and heaving panels separated by phase $\phi = 270°$. Pitch amplitudes $\theta = \{6°, 9°, \ldots, 15°\}$; heave amplitudes $h_0/c = \{0.083, 0.167, \ldots, 0.33\}$; and frequencies $f = \{0.2, 0.25, \ldots, 1\text{ Hz}\}$. The rigid panel corresponds to $\Pi_1 \sim \infty$ (upsidedown triangles), and flexible panels range from $\Pi_1 = 0.5$ (diamonds), 0.7 (triangles), 1.4 (squares), and 2.8 (circles).

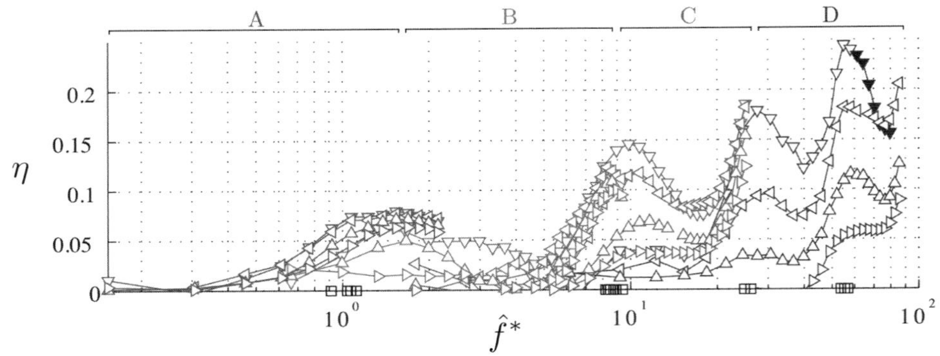

Figure 5.9 Efficiency peaks of heaving flexible panels at multiple resonance modes. Panels A, B, C, and D have stiffnesses $EI = 3.2 \times 10^{-1}$, 1.1×10^{-2}, 8.1×10^{-4}, 6.9×10^{-5}, respectively. (Taken from Quinn et al. [30], reproduced with permission.)

motion is close to the resonance of the system, and the trailing edge amplitude becomes amplified. The peak efficiency is less influenced by flexibility than thrust, but the efficiency tends to remain higher over a larger Strouhal number range.

The work of Dewey et al. [25] found similar trends to these for pitching panels of varying aspect ratio, over the same range of flexibilities. Quinn et al. [30] extended this work to very flexible panels in heave, where more than one resonance mode could be excited (similar to undulatory swimmers). They found that efficiency peaks occurred close to each resonant frequency, as shown in Figure 5.9, indicating that a simple analysis based on linear beam theory can sometimes be a valuable tool in predicting the performance of flexible propulsors.

5.3.4 Concept Design

Consider the concept design using oscillatory and undulatory motions shown in Figure 5.10. The long tube-shaped body is similar to many underwater vehicles in use today, with control surfaces for stability and maneuvering located near the stern.

The propulsors are a series of rigid and flexible pitching and heaving panels whose motions can be individually controlled. The first pair of propulsors near the bow are rigid, which can be used to provide thrust through lift- and added mass-based forces, and maneuvering through drag-based forces. The other pairs of propulsors are flexible to maximize thrust and efficiency of swimming. Multiple propulsors are intended to exploit fin–fin interactions. The motion profiles of all pairs of propulsors can be individually tuned and may take advantage of nonsinusoidal motions to achieve the thrust benefits associated with higher trailing edge velocities.

This type of vehicle is likely to be used where high efficiency and quiet swimming are valued. For long missions, the surface of the rigid panels could house flexible solar panels, so that the vehicle can occasionally recharge its batteries at the water surface, making it fully independent.

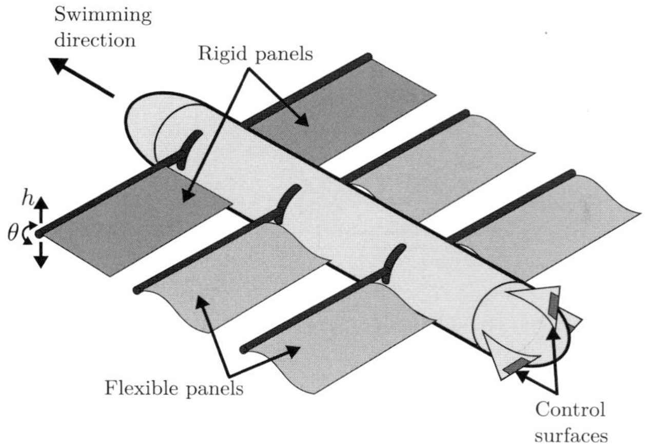

Figure 5.10 Concept design that uses undulatory and oscillatory propulsion, taking advantage of fin–fin interaction and flexibility performance benefits.

5.4 Pulsatile Jet

Pulsatile jet swimmers include squid, jellyfish, and mollusks. They propel themselves forward by impulsively injecting momentum into the surrounding fluid, producing forward thrust. Pulsatile jet swimmers have been widely studied in relation to jellyfish (see, for example [31]), but our general understanding is enriched by research on starting jets, vortex ring evolution, and synthetic jets (a flow control device that creates a train of vortex rings [32]).

We can model a pulsatile jet propulsor as a driver (e.g., a piston or a piezoelectric membrane) that displaces a fluid within a cavity that has a small opening to produces a jet. If the fluid driver oscillates periodically, the jet becomes pulsatile. This type of simplification is shown schematically in Figure 5.11, where a piston with diameter d_p oscillates at peak velocity $u_p(t)$ and peak amplitude Δ, displacing a fluid volume V through an orifice with diameter d_o, creating a time-varying jet with exit velocity $u_j(t)$.

5.4.1 Wake Characteristics

As we did in the previous section, we can examine the wake to give us insight into the thrust mechanisms of pulsatile swimmers. Dabiri et al. [31] directly measured the velocity field downstream of jellyfish and showed that their wakes are primarily made up of a train of vortex rings produced by the periodic expulsion of water. Figure 5.12 shows a sample flow measurement of a jellyfish wake (three-dimensional) and compares it to a snapshot of the flow produced by a synthetic jet (two-dimensional). The wakes are qualitatively similar in that they create successive vortex rings or vortex pairs that convect away from the orifice at their own induced velocity. The vortices generate a

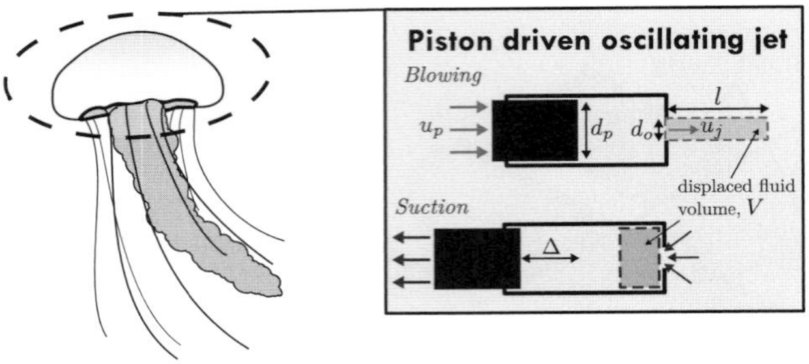

Figure 5.11 Example of how one could simplify a pulsatile swimmer into a piston-driven oscillating jet.

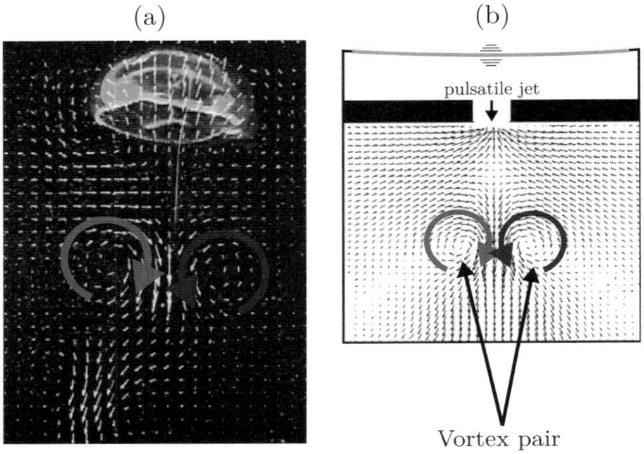

Figure 5.12 Velocity field measurement of (a) jellyfish and (b) synthetic jet. Both show similar vortex pairs generated at the orifice and propagating away at their own induced velocity. (Adapted from Dabiri et al. [31] and Van Buren et al. [33].)

region of high velocity on the centerline that represents the bulk of the momentum that is injected into the fluid to produce thrust.

The vortex ring generated by a circular orifice (like a jellyfish) is stable and simply diffuses as it evolves downstream. A vortex loop generated by a nonaxisymmetric orifice (like a scallop or clam) is unstable, and it can exhibit periodic axis-switching downstream until it reaches an axisymmetric equilibrium [34,35]. This instability can induce vortex pinch-off and an earlier breakdown, which can affect the thrust production.

5.4.2 Vortex Formation and Evolution

A vortex ring is formed when a volume of fluid is driven impulsively through a circular orifice in the cavity, and the boundary layers on the walls of the cavity separate at the

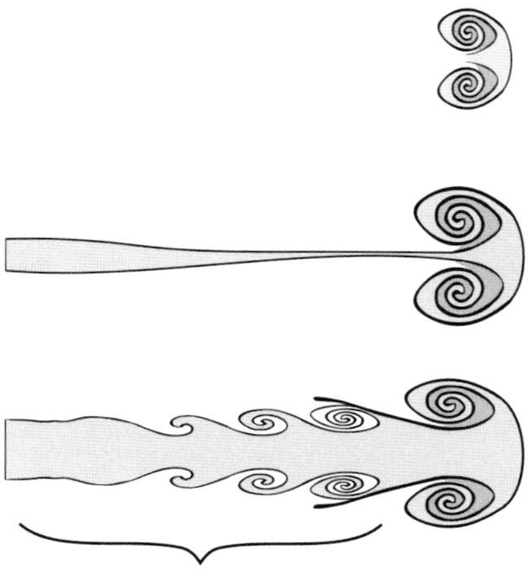

Figure 5.13 Illustration of the vortex formation from a piston-driven cavity with a circular orifice at stroke ratios (a) $l/d_o = 2$, (b) 3.8, and (c) 14.5. After the critical stroke ratio of $l/d_o = 4$, the appearance of a trailing jet is observed. (Vortex structure inspired from Gharib et al. [36].)

orifice edges, resulting in a growing roll-up of vorticity. The volume of fluid that exits the orifice can be estimated as a column with length l and diameter d_o. The aspect ratio of this fluid column is called the stroke ratio l/d_o, and Gharib et al. [36] showed that there is a critical stroke ratio $l/d_o \sim 4$, after which the circulation in the main vortex stops growing and a trailing jet forms behind the vortex. This occurs when the fluid driver cannot provide energy fast enough to continue the vortex growth. Figure 5.13 shows a sketch of a vortex formed using $l/d_o = 2$, 3.8, and 14.5 exhibiting the trailing jet that forms after the critical stroke ratio is reached.

The trailing jet formation can dramatically influence swimming performance in jellyfish. Dabiri et al. [31] studied seven jellyfish species of various sizes, and tracked their vortex formation while swimming. Three of the jellyfish species produced vortices below the critical stroke ratio, while the other four exceeded the critical stroke ratio so that they produced vortex rings with a trailing jet. Figure 5.14 shows that the jellyfish that commonly exceeded the critical stroke ratio swam faster than their counterparts, but much less efficiently. From a design perspective, this means that by changing the stroke ratio one could either favor thrust production or efficiency.

Now, for a zero-net-mass-flux condition, any vortex created through exhalation needs an equal and opposite inhalation (or suction) to refill the cavity. For example, Figure 5.12b shows the moment of the suction phase of a synthetic jet where fluid is inhaled into the cavity. The suction comes from all directions, and with relatively low peak velocities. Hence the negative thrust produced during the suction phase is relatively small compared to that seen in the blowing phase which directs all of the momentum in one direction.

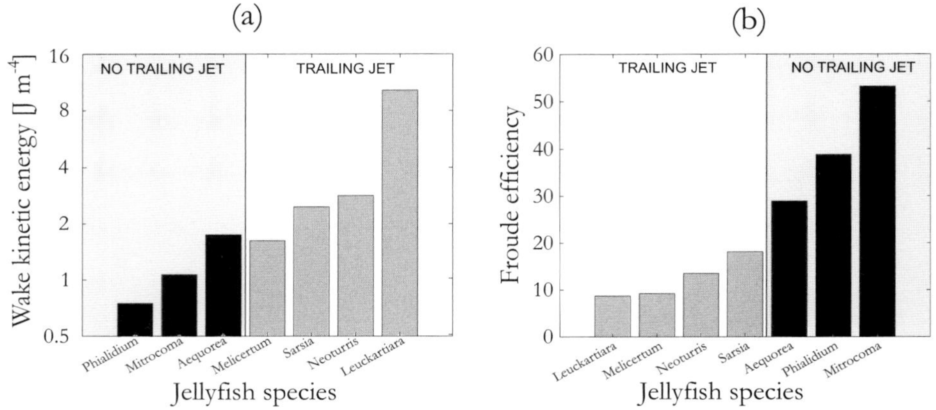

Figure 5.14 Various species of jellyfish (a) wake kinetic energy and (b) propulsive efficiency while swimming. Four jellyfish commonly exhibit a vortex with a trailing jet, resulting in higher thrust but much lower efficiency. (Data taken from Dabiri et al. [31].)

In addition, the vortex spacing in a series of pulses can strongly impact the resulting velocity field. For synthetic jets, the Strouhal number is defined as

$$St = \frac{f_j d_o}{U_j}, \tag{5.13}$$

where f_j is the jet frequency and U_j is some characteristic velocity of the jet, either the time-average of the absolute streamwise velocity or the peak velocity. This is effectively the inverse of the stroke ratio if we recognize that $U_j/f_j \sim l$. Holman et al. [37] showed that there is a minimum stroke ratio that must be achieved to create a jet. If the stroke ratio is too small, the vortex pairs become too close together, and the vortices created by one blowing cycle could be pulled back into the orifice by the next suction cycle. For two-dimensional flows, the minimum stroke ratio is $l/d_o \approx 1$ (the actual number depends on orifice geometry). Thus, we can define a range of stroke ratios, $1 \geq l/d_o \leq 4$, for which a train of trailing jet-free vortices can be formed.

5.4.3 Thrust Production

Consider the force required to generate a vortex with circulation $\Gamma = \oint v \cdot dS$ where v is the fluid velocity on closed surface S. The circulation can be thought of as a measure of the rotation rate of a vortex, and we can relate it to the equivalent impulse (integral of force over time) according to [38]:

$$I = \int F_x dt = \frac{1}{2}\pi d_o^2 \rho \Gamma. \tag{5.14}$$

Consequently, the mean thrust force per pulse is $\overline{F_x} = I/t_p$ where t_p is the time of a single pulse.

The circulation is an output of the system, and it would be helpful to be able to predict the force based upon known inputs. We can start by using a simple slug model [39,40] to relate the time-change in circulation with the jet velocity. That is,

$$\frac{d\Gamma}{dt} \approx \frac{1}{2} u_j^2. \tag{5.15}$$

A correction to u_j needs to be applied to account for the growth of the boundary layer within the orifice, which is especially important for high stroke ratios. Hence,

$$u_j^* = u_j\left(1 + 8\sqrt{\frac{l/d_o}{\pi Re}}\right), \tag{5.16}$$

where Re is the Reynolds number based on U_j and d_o. A simple conservation of mass relation can relate the input piston velocity to the jet velocity, where

$$u_j = u_p\left(\frac{d_p}{d_o}\right)^2 \tag{5.17}$$

for a circular piston and orifice. Using these simplified relations, the thrust generated by a series of vortex rings can be estimated. These relations stem from connecting the circulation estimates presented by various researchers [39,40] to the force that would be required to create that circulation [38]. They can be used to understand how the thrust force might vary with input parameters, although, to our knowledge, they have not yet been verified against experiment.

Krueger and Gharib [41] directly measured the impulse and thrust production of a piston-driven vortex generator, varying the peak acceleration of the piston. Their results are shown in Figure 5.15. The thrust coefficient was defined by

$$C_T = \frac{F_x}{\frac{1}{2}\rho U_j^2 \pi (d_o/2)^2}. \tag{5.18}$$

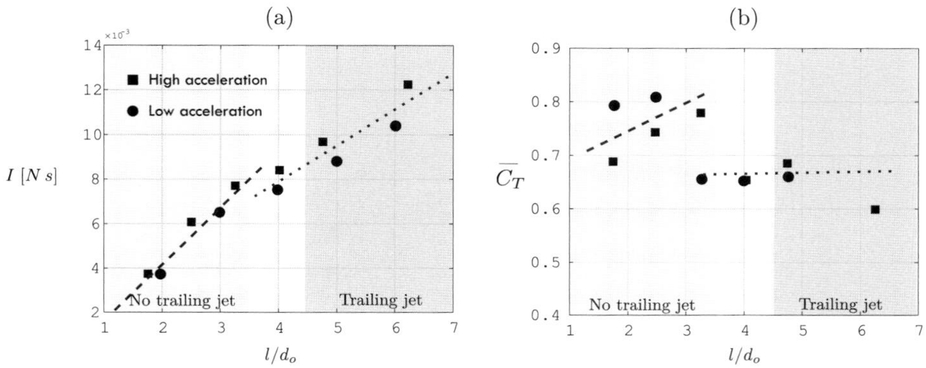

Figure 5.15 Impact of stroke ratio on the (a) impulse and (b) thrust coefficient. High and low accelerations considered for vortex pairs with and without trailing jets. (Data taken from Krueger and Gharib [41] with rough trend lines added.)

We have added some lines to highlight trends in the data. For high and low accelerations, the impulse grows linearly with stroke ratio, because for higher stroke ratios the jet vortex is formed over a longer period of time. Note the change in offset and slope of the trend lines at the point where a trailing jet is produced ($l/d_o \geq 4$). Similar trends are seen for the average thrust per pulse, where a penalty occurs at the production of a trailing jet. These simple experiments help explain the performance measurements of Dabiri et al. [31] on swimming jellyfish.

Much still needs to be studied on pulsatile jets as a viable propulsion system. This includes further parametric studies on the force generated by vortex ring generators and defining and measuring the efficiency. We saw that the stroke ratio is a critical parameter and that by changing the stroke ratio one could either favor thrust production or efficiency. In future work, compliant materials that make up the cavity and orifice could be considered. Also, elegant and compact designs would be necessary to propel an operating aquatic vehicle.

5.4.4 Concept Design

Figure 5.16 presents a concept design of an aquatic vehicle that uses pulsatile jet propulsion. There are two main oscillating piston-driven cavities, located near the bow and stern. Taking advantage of one-way flow valves, we can ensure that thrust is produced during both halves of the actuation cycle. During the first half-cycle, flow is inhaled at the bow and exhaled at the stern. During the second half-cycle, flow is directed to the surface of the vehicle to either apply blowing or suction within the boundary layer. Through oscillatory blowing and suction on the surface, we can add momentum to the boundary layer to make it less susceptible to separation and mitigate turbulence [42,43].

To provide steering, we can incorporate controlled, actuated flexiblity at mid-body to maneuver. As is demonstrated by agile swimmers like the sea lion [44], excellent maneuvering ability can be achieved from body distortion alone.

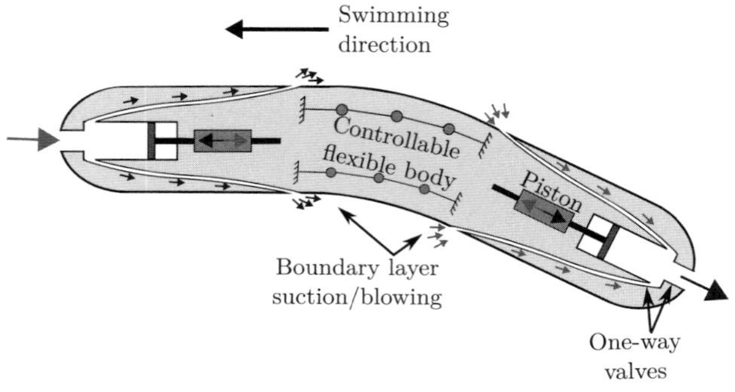

Figure 5.16 Concept design that uses pulsatile jet propulsion, taking advantage of one-way valves so that both strokes produce thrust, soft robotics to use the body to maneuver, and flow control concepts to limit flow separation over the body.

This type of design may be capable of high swimming speeds. The torpedo-like body could be made exceptionally streamlined, minimizing drag, thereby helping to increase efficiency. The thrust and efficiency would be primarily dictated by the design of the pistons. Also, this design concept could be made relatively small, which may be ideal for remote sensing devices.

5.5 Drag-Based Propulsion

Drag-based propulsion is primarily used by amphibious creatures such as turtles and ducks who have evolved to be proficient in land and water travel. As we saw in Figure 5.2, drag-based swimmers are not very efficient, but they are competitive in relative swimming speed. We will show that they still have much to offer as inspiration for propulsion systems.

A simple model for a drag-based propulsor is shown in Figure 5.17. Various bluff body shapes provide different drag characteristics for similar size and flow velocity, so the planform and aspect ratio are important. For application to underwater propulsion, the bluff body needs to be moved periodically, but for modest reduced frequencies, the forces and flow fields may be estimated from steady flow considerations.

5.5.1 Wake Characteristics

The wake of a stationary bluff body has been a topic of interest in aerodynamics and hydrodynamics for centuries.

As an incoming flow approaches a bluff body, it is redirected around it. The fluid near the surface of the body is decelerated due to viscous friction (viscous drag), and a boundary layer develops. The boundary layer experiences a streamwise pressure gradient which is favorable where the external flow is accelerating and adverse where it is decelerating. In an adverse pressure gradient, there is a resultant streamwise force acting on the flow inside the boundary layer. If this force is large enough, it can force the fluid to actually reverse direction, and the boundary layer is said to separate. Separation is accompanied by a recirculating flow in the wake, and large pressure losses which appear as form drag. Form drag due to separation is typically much larger than the viscous drag due to the boundary layer, and so the shape of the body determines the drag since the shape dictates the point of separation and the size of the wake.

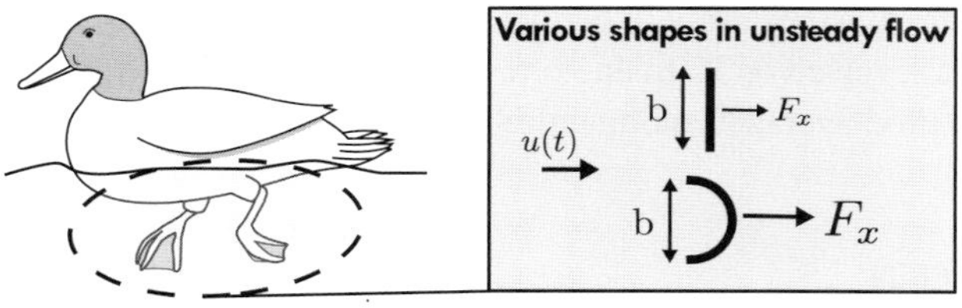

Figure 5.17 Modeling a drag-based swimmer for fundamental study. Various shapes have different drag characteristics for similar projected lengths b.

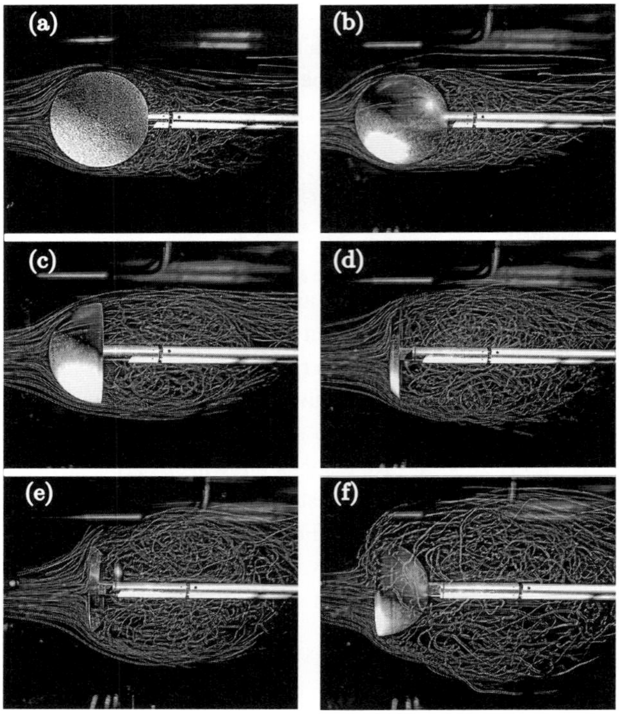

Figure 5.18 Helium bubble flow visualization of air flow around three-dimensional bluff bodies: (a) rough sphere, (b) smooth sphere, (c) forward-facing cup, (d) forward-facing chamfered plate, (e) backward-facing chamfered plate, and (f) reverse-facing cup. (Images taken in the Rensselaer Polytechnic Institute undergraduate subsonic wind tunnel.)

This process is illustrated in Figure 5.18, where the flow around various three-dimensional shapes is visualized using helium bubbles in a wind tunnel. Long exposure photographs reveal the path-lines of the helium bubbles. We see that in each case the incoming flow first bends around the body, and then at some point separation occurs, and a wake is formed. The size of the wake dramatically changes for the various shapes, where the spheres have the smallest wake and the reverse-facing cup has the largest wake. The drag force generally correlates with wake size, where the larger wake corresponds to the larger drag. The wake can be unstable, like the cylinder exhibited previously in Figure 5.5a, resulting in periodic separation on the body producing a time-varying drag and a reverse von Kármán vortex street. The drag can be directly estimated by measuring the time-averaged momentum deficit in the wake.

5.5.2 Thrust and Efficiency

In discussing the thrust and efficiency of drag-based propulsors, we will often use the term "drag" interchangeably with "thrust." Although this may seem confusing, for this class of swimmers, the drag produced by the propulsor is, by action–reaction, the thrust of the system.

As we have indicated earlier, the drag generated by bluff bodies is generally split into viscous and pressure, or form drag. For a long, thin flat plate oriented parallel to the incoming flow, the major drag component would be viscous drag because the frontal area of the plate is small and there is considerable area for fluid friction to act on the body. When the plate is perpendicular to the flow, however, there is large form drag, because of the pressure losses due to separation and the formation of a large wake. For drag-based propulsion, we will focus mainly on pressure drag because it is easier to produce in large quantity and the flows we are considering have relatively low viscous forces (that is, high Reynolds numbers).

The drag coefficient of a body moving steadily through a fluid is usually defined as

$$C_D = \frac{F_x}{\frac{1}{2}\rho U_p^2 sb}, \tag{5.19}$$

which is just another form of Eq. (5.1), where U_p is the mean velocity of the propulsor with respect to the surrounding fluid (not to be confused with the vehicle velocity, U_∞), b is the projected width of the body, and s is the span-length. At high Reynolds numbers, the drag coefficient for most bluff body shapes is a constant independent of Reynolds number, and it is generally obtained empirically. Table 5.1 shows the drag coefficients for various two- and three-dimensional shapes. The shape clearly has a large impact, with C_D varying from 1.17 for a cylinder and 2.30 for a forward facing cup (empty hemicylinder), which coincides well with the wake sizes shown in Figure 5.18. Generally the drag decreases by around half for three-dimensional shapes of similar cross-section. As designers, we see that for quasi-steady flow (low reduced frequencies) we can maximize thrust by increasing the speed and size of the propulsor, and optimizing its shape.

Table 5.1 Drag coefficients for various two- and three-dimensional shapes in a flow going left to right (Data taken from [45])

Cross-section	Shape (2D)	C_D	Shape (3D)	C_D
	Cylinder	1.17	Sphere	0.47
	Square rod	2.05	Cube	1.05
	Flat plate	1.98	Circular plate	1.17
	Hemicylinder (full)	1.16	Hemisphere (full)	0.42
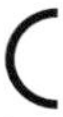	Hemicylinder (empty)	1.20	Hemisphere (empty)	0.38
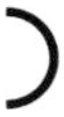	Hemicylinder (empty)	2.30	Hemisphere (empty)	1.42

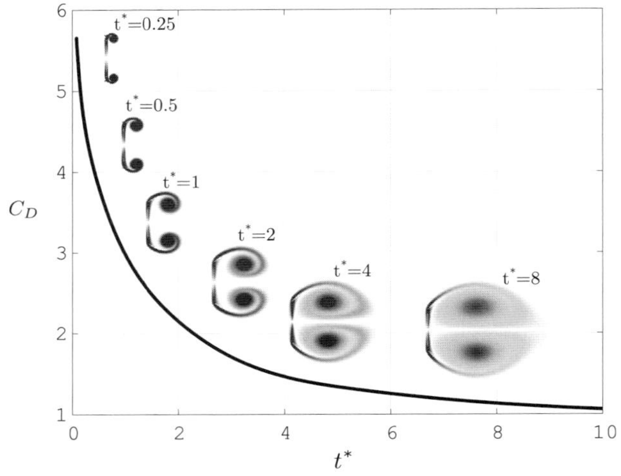

Figure 5.19 Drag coefficient for an impulsively started flat plate where the flow velocity is normal to the plate at Reynolds number Re = 500. Inlaid graphics represent the qualitative vorticity field at six points in time. (Simulations using immersed boundary code [46,47].)

If we now consider unsteady flows, we can identify additional ways to increase thrust. Figure 5.19 shows a simulation of the drag coefficient and wake of an impulsively started two-dimensional flat plate at Reynolds number $Re = 500$. Here, we nondimensionalize the time by the time it would take a fluid particle to travel a distance X, where X is the height of the plate, thus $t^* = t\ (U_p/X)$. There is a dramatic decrease in drag between $0 \leq t^* \leq 10$ as the vortices that shed from the plate edges grow to their maximum size. This means that if one could periodically and impulsively use a drag-based propulsor to generate thrust and limit the development time to $t^* \leq 2$, on average the thrust production would be four times higher than that produced by a quasi-steady drag-based propulsor. Although the Reynolds number of these simulations is relatively small, the broad conclusions will carry over to higher Reynolds numbers since, for a thin plate normal to a flow, there can be no net viscous drag in the flow direction, thus the form drag dominates.

The efficiency of the drag-based propulsor cannot be considered in isolation as we did for lift-based propulsors, because efficiency is the ratio of an output to an input, and there is no clear output for a drag-based propulsor. However, we can define an efficiency if we include the entire vehicle. Consider a human paddling a canoe at constant velocity (for details see [48]). We define a system including the boat and the paddle but not the human. An energy balance of this system is

$$W_{ext} = \Delta E_K + \Delta E_P + W_{int}, \tag{5.20}$$

where ΔE_K and ΔE_P are the kinetic and potential energy changes of the system which are zero for constant velocity and unchanging height, and W_{int} is the internal work of the system, which we will neglect. The external work W_{ext}, consists of the work done by the human on the paddle and the boat (W_h), the work done by the water on the paddle (W_p), and the work done by the water on the boat (W_b). The work of the human balances the resistance of the water, thus

$$W_h - W_p - W_b = 0, \tag{5.21}$$

and we can define a system efficiency that compares the useful output energy (boat moving forward) to the input energy (human)

$$\eta_s = \frac{W_b}{W_h} = \frac{W_b}{W_p + W_b}. \tag{5.22}$$

If the work is constant, we can substitute power for work, and power is given by the product of force and velocity, and so the efficiency is given by

$$\eta_s = \frac{F_b U_b}{F_p (U_p - U_b) + F_b U_b}, \tag{5.23}$$

where F_b and F_p are the drag on the boat and paddle, U_b is the velocity of the boat, and U_p is the velocity of the paddle relative to the boat. If the velocity is constant we have a balance of forces $F_b = F_p$, and the efficiency just becomes a velocity ratio

$$\eta_s = \frac{U_b}{U_p}. \tag{5.24}$$

From Eq. (5.1) combined with our balance of forces we obtain

$$U_b^2 A_b C_{D_b} = (U_p - U_b)^2 A_p C_{D_p}. \tag{5.25}$$

Solving for U_p in terms of U_b, applying the constraint that $U_p > U_b$ because to produce thrust the paddle must move faster than the boat, and using Eq. (5.24) gives

$$\eta_s = \frac{1}{1 + \sqrt{\xi}}, \tag{5.26}$$

where ξ is the ratio $C_{D_b} A_b / (C_{D_p} A_p)$. This result, obtained for paddling a canoe, can be used for any vehicle propelled using drag-based propulsor. The theoretical system efficiency η_s is shown in Figure 5.20 as a function of ξ, which indicates that for a high-efficiency drag-based propulsion system it is best to have a large and high-drag propulsor and a small and low-drag vehicle. Interestingly, these are all things that also result in higher thrust and speed, thus coupling high thrust and efficiency. It is clear that drag-based propulsors are not inherently inefficient, and with proper design, they could be competitive with other propulsors.

5.5.3 Concept Design

Our concept vehicle, like most animals that utilize drag-based propulsion, will be amphibious. It can swim fully submerged, at the water surface, or traverse on land. It will be suitable for tasks that require versatility and large payloads.

Two tank-like tracks will be used to facilitate motion on both land an water. Inside the tracks are spring-loaded half-cylinders individually driven via solenoid, and can

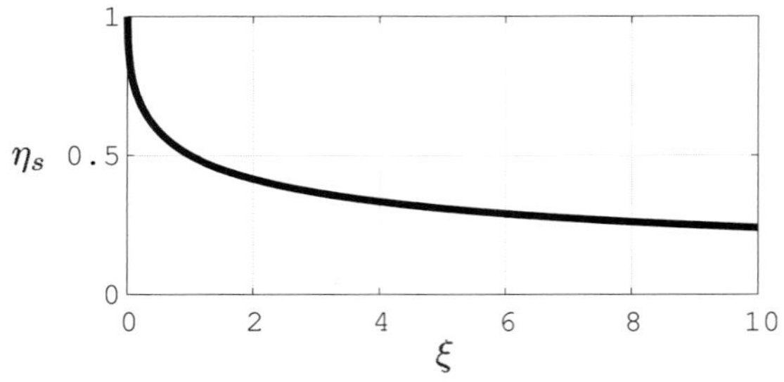

Figure 5.20 Drag-based propulsion system efficiency. Here, $\xi = C_{D_b}A_b/(C_{D_p}A_p)$.

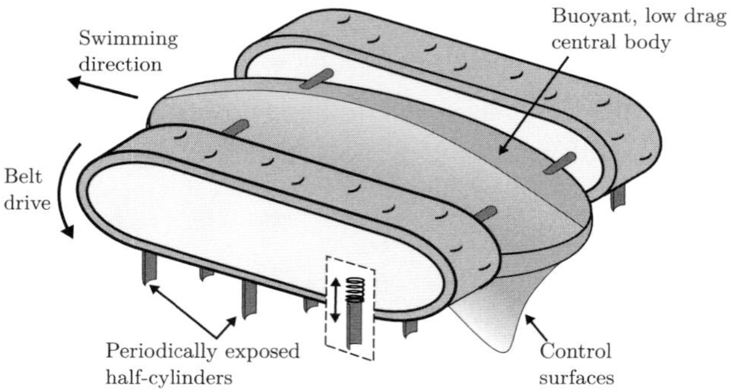

Figure 5.21 Concept design of an amphibious vehicle that uses drag-based propulsion, with open half-cylinders periodically exposed to the flow for limited time. The vehicle can be flipped to travel on the ground like a tank.

be impulsively exposed to the surrounding fluid. This maximizes our available thrust for drag-based propulsion, with the highest drag shape possible being periodically impulsively started. The spacing of the propulsors will promote drag via propulsor interaction, much like the interaction of schooling fish or flocking birds. The half-cylinders will only be exposed on a single side of each track, which will allow water travel on one side and land travel on the other (requiring the vehicle to flip in transition). Note, if the vehicle was designed for submerged water travel only, the tracks could be turned on their side and shrouded to expose only the propulsors to be more hydrodynamic.

The central body will carry a dry payload and trap enough air for neutral buoyancy. Control surfaces will steer the vehicle during water travel.

5.6 Concluding Remarks

We have attempted to provide an informative and intriguing summary of how aquatic swimmers can inspire biological propulsion systems. We identified three major groups of aquatic swimmers for inspiration. First were oscillatory and undulatory swimmers (for example, tuna or manta ray), that periodically oscillate their flukes, or median and caudal fins to create lift-based and added mass propulsion. We showed that for high thrust and efficiency, pitching and heaving motions must be combined at a phase offset of about 270°, a limitation set by the need to reduce the peak angle of attack of the motion. Adding flexibility was paramount to further increasing the thrust and efficiency of the propulsor.

Second were pulsatile swimmers (for example, jellyfish), which create unsteady jets to inject momentum into the flow to produce thrust. The stroke ratio l/d_o, was identified as one of the key governing parameters of performance. A minimum stroke ratio of ≈ 1 is required to ensure the formation of a jet, otherwise the pulse of the previous stroke would be sucked back into the orifice. A maximum stroke ratio of ≈ 4 was required to prohibit the formation of a trailing jet, which negatively impacts the mean thrust of each pulse.

Lastly, drag-based swimmers (for example, turtle) that create thrust from the pressure-drag produced by moving a bluff body through the water. The propulsor shape is critical in maximizing thrust for a given size and velocity. Quasi-two-dimensional shapes produced nearly twice as much drag as low aspect ratio three-dimensional shapes. Also, by limiting the flow development time from an impulsive start, the average thrust could be significantly increased. Drag-based propulsors appear to offer some previously unexplored potential for fast and efficient swimming.

We hope that our proposed concept vehicles, based on what we currently know about biological propulsors, provide inspiration to future engineers and scientists to explore the design of future underwater vehicles.

This work was supported by ONR Grant N00014–14-1-0533 (Program Manager Robert Brizzolara).

References

1. Colgate, J. E., & Lynch, K. M. (2004). Mechanics and control of swimming: A review. *IEEE Journal of Oceanic Engineering*, 29(3), 660–673.
2. Bandyopadhyay, P. R. (2005). Trends in biorobotic autonomous undersea vehicles. *IEEE Journal of Oceanic Engineering*, 30(1), 109–139.
3. Fish, F. E., Lauder, G. V., Mittal, R., et al. (2003). Conceptual design for the construction of a biorobotic AUV based on biological hydrodynamics. In *13th International Symposium on Unmanned Untethered Submersible Technology, Durham, NH*. 24–27.
4. Fish, F. E., Smits, A. J., & Bart-Smith, H. (2010). Biomimetic swimmer inspired by the manta ray. In Y. Bar-Cohen (Ed.), *Biomimetics: Nature-based innovation*, Boca Raton, FL: CRC Press.
5. Triantafyllou, M. S., & Triantafyllou, G. S. (1995). An efficient swimming machine. *Scientific American*, 272(3), 64–71.
6. Rufo, M., & Smithers, M. (2011). Ghostswimmer? AUV: Applying biomimetics to underwater robotics for achievement of tactical relevance. *Marine Technology Society Journal*, 45(4).

7. Lutz, P. L., Musick, J. A., & Wyneken, J. (2002). *The biology of sea turtles* (Vol. 2). CRC Press, Boca Raton, FL.
8. Fish, F. E. (2000). Biomechanics and energetics in aquatic and semiaquatic mammals: platypus to whale. *Physiological and Biochemical Zoology*, 73(6), 683–698.
9. Eloy, C. (2012). Optimal Strouhal number for swimming animals. *Journal of Fluids and Structures*, 30, 205–218.
10. Buchholz, J. H. J., & Smits, A. J. (2006). On the evolution of the wake structure produced by a low-aspect-ratio pitching panel. *Journal of Fluid Mechanics*, 546, 433–443.
11. Van Buren, T., Floryan, D., Brunner, D., Senturk, U., & Smits, A. J. (2017). Impact of trailing edge shape on the wake and propulsive performance of pitching panels. *Physical Review Fluids*, 2(1), 014702.
12. Floryan, D., Van Buren, T., & Smits, A. J. (2019). Swimmers' wakes are not reliable indicators of swimming performance. arXiv:1906.10826.
13. Floryan, D., Van Buren, T.,Rowley, C. W., & Smits, A. J. (2017). Scaling the propulsive performance of heaving and pitching foils. *Journal of Fluid Mechanics*, 822, 386–397.
14. Van Buren, T., Floryan, D., Wei, N., & Smits, A. J. (2018). Flow speed has little impact on propulsive characteristics of oscillating foils. *Physical Review Fluids*, 3(1), 013103.
15. Quinn, D. B., Lauder, G. V., & Smits, A. J. (2015). Maximizing the efficiency of a flexible propulsor using experimental optimization. *Journal of Fluid Mechanics*, 767, 430–448.
16. Xu, M., & Wei, M. (2016). Using adjoint-based optimization to study kinematics, deformation of flapping wings. *Journal of Fluid Mechanics*, 799, 56–99.
17. Van Buren, T., Floryan, D., & Smits, A. J. (2018). Scaling and performance of simultaneously heaving and pitching foils. *AIAA Journal*, 1–12.
18. Van Buren, T., Floryan, D, Quinn, D., & Smits, A. J. (2017). Nonsinusoidal gaits for unsteady propulsion. *Physical Review Fluids*, 2, 053101.
19. Floryan, D., Van Buren, T., & Smits, A. J. (2017). Forces and energetics of intermittent swimming. *Acta Mechanica Sinica*, 33(4), 725–732.
20. Quinn, D. B., Moored, K. W., Dewey, P. A., & Smits, A. J. (2014). Unsteady propulsion near a solid boundary. *Journal of Fluid Mechanics*, 742, 152–170.
21. Quinn, D. B., Lauder, G. V., & Smits, A. J. (2014). Flexible propulsors in ground effect. *Bioinspiration & Biomimetics*, 9(3), 036008.
22. Dewey, P. A., Quinn, D. B., Boschitsch, B. M., & Smits, A. J. (2014). Propulsive performance of unsteady tandem hydrofoils in a side-by-side configuration. *Physics of Fluids*, 26(4), 041903.
23. Boschitsch, B. M., Dewey, P. A., & Smits, A. J. (2014). Propulsive performance of unsteady tandem hydrofoils in an in-line configuration. *Physics of Fluids*, 26(5), 051901.
24. Maertens, A. P., Gao, A., & Triantafyllou, M. S. (2017). Optimal undulatory swimming for a single fish-like body and for a pair of interacting swimmers. *Journal of Fluid Mechanics*, 813, 301–345.
25. Dewey, P. A., Boschitsch, B. M., Moored, K. W., Stone, H. A., & Smits, A. J. (2013). Scaling laws for the thrust production of flexible pitching panels. *Journal of Fluid Mechanics*, 732, 29–46.
26. Dewey, P. A. (2013). *Underwater flight: Hydrodynamics of the manta ray* [PhD thesis]. Princeton, NJ: Princeton University.
27. Quinn. D. B. (2015). *Optimizing the efficiency of Batoid-inspired swimming* [PhD thesis]. Princeton, NJ: Princeton University.
28. Floryan, D., & Rowley, C. W. (2018). Clarifying the relationship between efficiency and resonance for flexible inertial swimmers. *Journal of Fluid Mechanics*, 853, 271–300.

29. Allen, J. J., & Smits, A. J. (2001). Energy harvesting eel. *Journal of Fluids and Structures*, 15 (3–4), 629–640.
30. Quinn, D. B., Lauder, G. V., & Smits, A. J. (2014). Scaling the propulsive performance of heaving flexible panels. *Journal of Fluid Mechanics*, 738, 250–267.
31. Dabiri, J. O., Colin, S. P., Katija, K., & Costello, J. H. (2010). A wake-based correlate of swimming performance and foraging behavior in seven co-occurring jellyfish species. *Journal of Experimental Biology*, 213(8), 1217–1225.
32. Glezer, A., & Amitay, M. (2002). Synthetic jets. *Annual Review of Fluid Mechanics*, 34, 503–529.
33. Van Buren, T., Whalen, E., & Amitay, M. (2014). Vortex formation of a finite-span synthetic jet: high Reynolds numbers. *Physics of Fluids*, 26, 014101–21.
34. Dhanak, M. R., & Bernardinis, B. (1981). The evolution of an elliptic vortex ring. *Journal of Fluid Mechanics*, 109, 189–216.
35. Van Buren, T., Whalen, E., & Amitay, M. (2014). Vortex formation of a finite-span synthetic jet: effect of rectangular orifice geometry. *Journal of Fluid Mechanics*, 745, 180–207.
36. Gharib, M., Rambod, E., & Shariff, K. (1998). A universal time scale for vortex ring formation. *Journal of Fluid Mechanics*, 360, 121–140.
37. Holman, R., Utturkar, Y., Mittal, R., Smith, B., & Cattafesta, L. (2005). Formation criterion for synthetic jets. *AIAA Journal*, 43(10), 2110–2115.
38. Huggins, E. R. (1966). Impulse and vortices. *Physical Review Letters*, 17(26), 1284.
39. Shusser, M., Gharib, M., Rosenfeld, M., & Mohseni, K. (2002). On the effect of pipe boundary layer growth on the formation of a laminar vortex ring generated by a piston/cylinder arrangement. *Theoretical and Computational Fluid Dynamics*, 15(5), 303–316.
40. Dabiri, J. O., & Gharib, M. (2004). A revised slug model boundary layer correction for starting jet vorticity flux. *Theoretical and Computational Fluid Dynamics*, 17(4), 293–295.
41. Krueger, P. S., & Gharib, M. (2003). The significance of vortex ring formation to the impulse and thrust of a starting jet. *Physics of Fluids*, 15(5), 1271–1281.
42. Choi, H., Jeon, W. P., & Kim, J. (2008). Control of flow over a bluff body. *Annual Review of Fluid Mechanics*, 40, 113–139.
43. Cattafesta, L., & Sheplak, M. (2011). Actuators for active flow control. *Annual Review of Fluid Mechanics*, 43, 247–272.
44. Fish, F. E., Hurley, J., & Costa, D. P. (2003). Maneuverability by the sea lion zalophus californianus: turning performance of an unstable body design. *Journal of Experimental Biology*, 206(4), 667–674.
45. Hoerner, S. F. (1965). *Fluid-dynamic drag: Practical information on aerodynamic drag and hydrodynamic resistance*. Brick Town, NJ: Hoerner Fluid Dynamics.
46. Taira, K., & Colonius, T. (2007). The immersed boundary method: a projection approach. *Journal of Computational Physics*, 225(2), 2118–2137.
47. Colonius, T., & Taira, K. (2008). A fast immersed boundary method using a nullspace approach and multi-domain far-field boundary conditions. *Computer Methods in Applied Mechanics and Engineering*, 197(25), 2131–2146.
48. Cabrera, D., & Ruina, A. (2006). *Propulsive efficiency of rowing oars* [Technical report]. Ithaca, NY: Cornell University.

6 Bioinspired Design of Dental Functionally Graded Multilayer Structures

Jing Du, Xinrui Niu, and Wole Soboyejo

6.1 Introduction

Oral health is of great importance to people's general health. Dental disease is more prevalent than most people imagine. For example, caries, which can lead to partial or total loss of teeth, affect almost 100% adults and 60–90% of schoolchildren [1]. To correct the dental malfunction caused by tooth loss due to various reasons, such as caries, aging, injury, etc, dental crowns have been adopted as a common treatment.

However, the strength and fatigue life of crowns are unsatisfactory for many reasons. First, the choice of materials is limited. Metals are durable, but heavy and not biocompatible. Ceramics and their composites are light, comfortable, and biocompatible, but fragile. Thus, none of the available materials can satisfy all of the requirements of dental crowns, which include many engineering aspects such as strength, longevity, biocompatibility, and aesthetics.

Second, because no single class of materials can meet all of the requirements for dental crowns, multilayered architectures are often used to achieve the balance of properties that is needed for applications. Hence, these dental crowns use layered structures, which are subjected to contact loads due to occlusal activity in the oral cavity. Under such conditions, the contact loads can give rise to subsurface stresses that can lead ultimately to subcritical cracking and final failure of the multilayered crown structures. Such cracking, which is often enhanced by the abrupt changes in the elastic properties from one layer to the other, can be managed using functionally graded multilayer structures that are inspired by the structure of natural teeth.

In contrast to synthetic dental structures (including dental crowns, bridges, implants, and so on), natural teeth exhibit superior strength and durability. Prior research has shown that natural teeth exhibit excellent mechanical resilience, even though the mechanical properties such as Young's moduli and hardness of the constituents of natural teeth, are not greater than those of the synthetic materials that are used in dental crowns. The magic is that nature has constructed teeth with astonishing sophistication in the "processing method" and the underlying microstructures. Hence, the superior overall performance of natural teeth is the result of the fabrication method and structural design [2].

In the case of natural teeth, nature provides the bioinspiration for the processing and structural design of layered structures with good resistance to cracking. For example, the functionally graded interfacial layers in the dentoenamel junction (DEJ) provide the inspiration for the design of multilayered dental structures that reduce the overall stresses in the top ceramic layers of dental ceramic crown structures. This will be explored in this chapter, using a combination of mechanics and materials models.

This chapter is divided into five sections. Following the introduction in Section 6.1, some basic dental and mechanics concepts are presented in Section 6.2 for the design of layered structures. This is followed by Section 6.3, in which a review of prior work on conventional crowns and bioinspired functional graded crowns is provided. Future directions are elucidated in Section 6.4, before summarizing the insights from this chapter in Section 6.5.

6.2 Basic Concepts

6.2.1 Dental Multilayered Model Structures

The strong interest in improving the life-time and performance of posterior all-ceramic crowns has stimulated considerable research efforts on the contact damage of dental multilayers [3–11]. Dental multilayers are widely recognized as engineering idealizations of dental crowns. They simplify the curved complex structures of dental crowns into flat multilayered structures with circular cross-sections and similar sizes to real human teeth [3,10–13]. Hertzian contact is commonly used to simulate occlusal contact [3,6,10–15].

In most of the basic research, ceramic crown structures are idealized as flat multilayered structures that are deformed under Hertzian contact loading, as shown in Figure 6.1 [14]. The flat multilayered structures investigated in the literature usually consist of a top layer of ceramic or glass of about 0.5–1.5 mm thick, a substrate layer of polymer-based composite or polymer of about 4–12 mm thick, with an adhesive layer of about 100 μm thick in between. The diameter of the Hertzian indenter varies from 0.8 mm to 20 mm [14]. In most dental crowns, the Young's modulus of top ceramic layer is 70–400 GPa. The modulus of the substrate polymeric layer is ~18 GPa, while that of dental adhesive is generally between 3 and 5 GPa.

6.2.2 Contact Damage of Dental Multilayers

Figure 6.2 shows some damage and cracking modes in the top ceramic layer of dental multilayered structures subjected to Hertzian contact load.

R represents the subsurface radial cracking that often initiates at the center of the bottom surface (a.k.a. subsurface) of ceramic layers [3,4,11–13,16]. These are complex three-dimensional cracks that propagate not only along the radial directions but also in thickness direction, upward toward the top surface of dental ceramic and downward through adhesive layer toward substrate layer. Fracture of substrate layer causes bulk failure of the multilayer structure [9, 17–19].

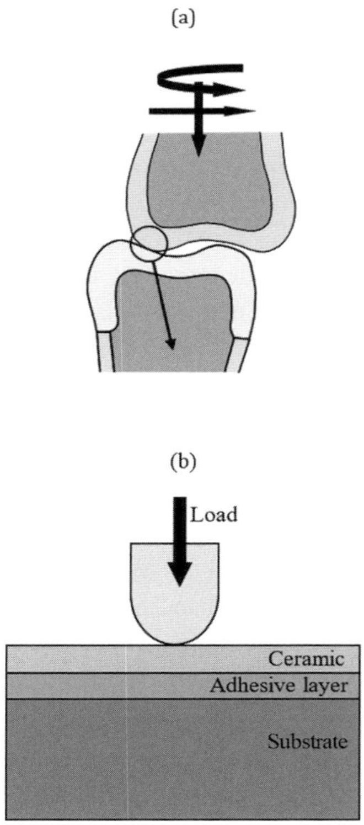

Figure 6.1 Real occlusal activities and model. (a) Schematic of real occlusal activities of teeth/crowns. (b) Simplified multilayered model structure under Hertzian contact.

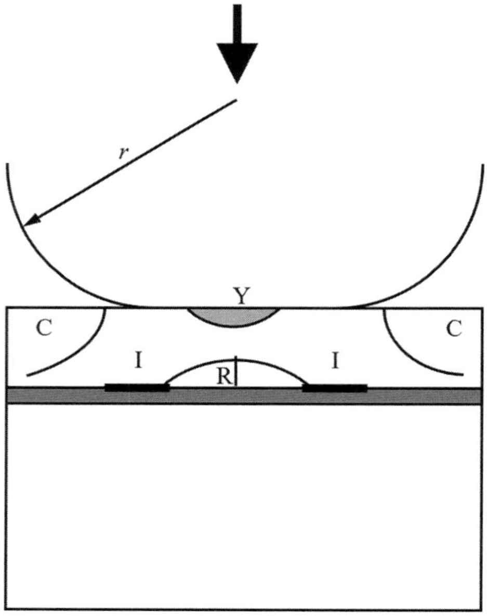

Figure 6.2 Multiple failure methods in dental multilayer structures under Hertzian contact [57].

(a)

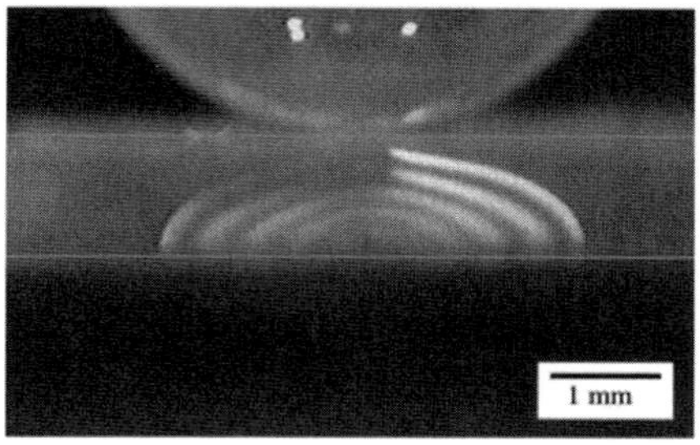

(b)

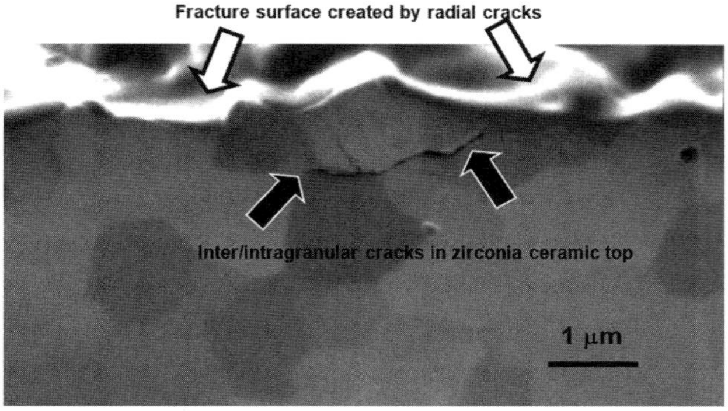

Figure 6.3 (a) Image of radial cracks initiated at the subsurface of the top glass layer in a glass/adhesive/substrate dental multilayer structure [20]. (b) Focused ion beam (FIB) image shows inter/intragranular cracks in zirconia ceramic top layer. The microcracks are associated with radial cracks [13].

Figure 6.3a shows a typical picture of radial cracking (at initiation stage) in a glass/adhesive/substrate structure [20]. This cracking is due to the stress concentrations in the subsurface regime in the top ceramic layer that is caused by Young's moduli mismatch, as shown in the finite element simulations by Huang et al. [21]. The failure mode is consistent with the major clinical failure mode of ceramic dental crowns, as reported by Kelly [4].

The morphology of the radial cracks is presented in Figure 6.3b, in which the solid arrows point to inter- and intragranular cracks, while the hollow arrows point to fracture surfaces created by the radial cracks. The image presented in Figure 6.3b was taken with a dual-beam focused ion beam (FIB)/scanning electron microscope (SEM) system with a 30 kV ion beam. The image shows that the radial crack mainly propagates along grain boundaries, with some incidence of transgranular cracking in some of the grains.

In Figure 6.2, C represents the cone cracks that initiate from the top surface of the ceramic layer and propagate toward the interfaces between ceramic and adhesive layers. Observing from the top view, their location is just outside of the contact area between the indenter ball and the ceramic layer. Figure 6.4 shows some typical images of these cracks in a glass/adhesive/substrate structure [14].

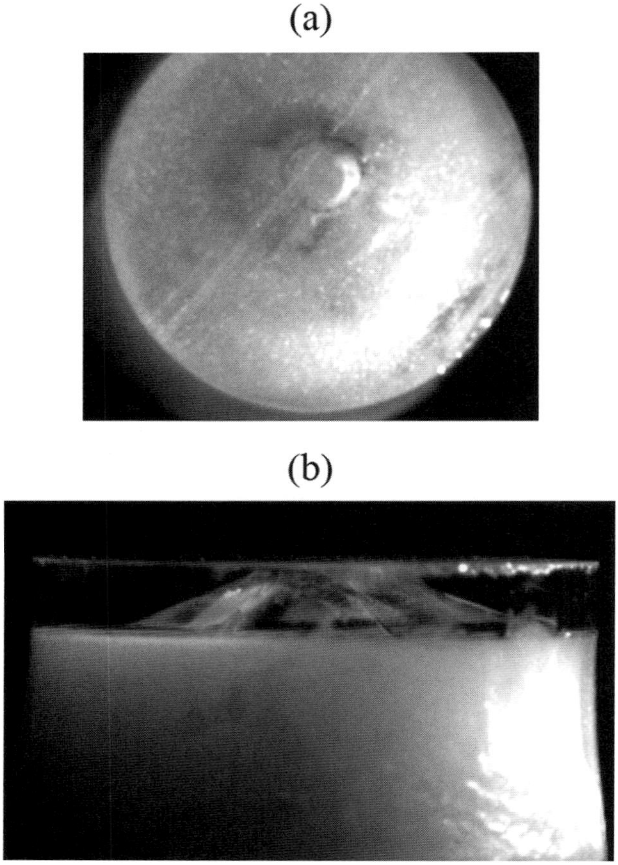

Figure 6.4 Images of cone cracks in the top glass layer in a glass/adhesive/substrate dental multilayer structure: (a) top view and (b) side view [14].

In Figure 6.2, Y represents the quasi-plastic damage that occurs right underneath the loading indenter. It is a form of yield deformation in response to the shear stresses [22,23]. Figure 6.5 shows the typical images of the quasi-plastic damage in indented glass-ceramics [24].

In Figure 6.2, I represents the interfacial debonding between the ceramic and adhesive layers. This is often associated with the propagation of radial cracks. Figure 6.6 shows the typical images of interfacial cracks between zirconia ceramic and dental cement adhesive layers [13].

Other minor failure modes may also develop, including median, outer, and inner cone cracks and partial cone cracks. These depend on loading and environmental conditions. However, they are not the primary failure modes in the experiments and theories presented in this chapter.

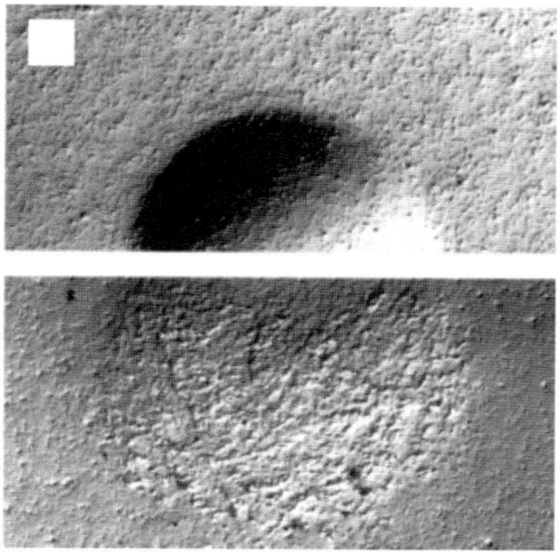

Figure 6.5 Top and section views of the quasiplastic damage in bonded-interface specimens of coarse-grain micaceous glass-ceramic, from indentation with WC sphere of radius 3.18 mm at load 1,000 N [24].

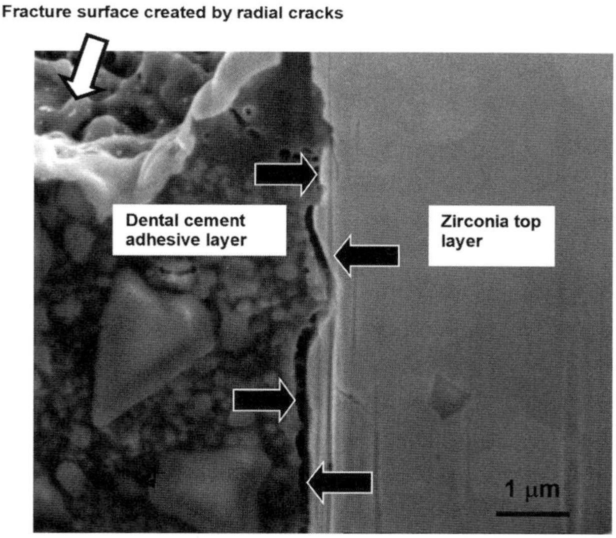

Figure 6.6 SEM image showing interfacial cracking between zirconia ceramic and dental cement adhesive layers, where solid arrows indicate interfacial cracks and hollow arrows indicate fracture surface created by radial cracks. Picture was taken by the dual-beam FIB/SEM system with 5 kV electron beam [13].

6.2.3 Viscoelasticity

Many biomaterials exhibit time-dependent viscoelastic or "creep" behavior. For metallic biomaterials, viscous behavior is often associated with diffusion or dislocation motion. However, for polymeric biomaterials, viscosity is mainly associated with chain stretching, unfolding, or sliding. Creep-induced fracture may also occur by the formation and growth of voids at grain or crystalline boundaries [25].

The mechanical response due to viscous flow could be represented by combinations of spring and dashpot models. Figure 6.7a shows the Maxwell model that has spring and dashpot in series. The solution of the Maxwell model can be expressed as [25]

$$\frac{d\varepsilon}{dt} = \frac{\sigma}{\eta} + \frac{1}{E}\frac{d\sigma}{dt}, \tag{6.1}$$

where ε is the strain, σ is the stress level, E is the Young's modulus, η is the viscosity, and t is the time.

Figure 6.7b shows the Voigt model, which combines the spring and dashpot in parallel. The response of the Voigt model in creep is expressed as [25]

$$\varepsilon = \frac{\sigma_0}{E}\left[1 - \exp\left(-\frac{t}{\tau}\right)\right] \tag{6.2}$$

where σ_0 is the initial stress level and $\tau = \eta/E$ is the relaxation time.

For the problems with complex material behaviors, the three-element model (Figure 6.7c), the four-element model, or the even more complex Kelvin model are widely used [25]. The three-element model (also known as the Zener creep model), shown in Eq. (6.3), will be interpreted and implanted with an elastic model, the slow crack growth model, to explain the deformation and cracking mechanism of contact damage in bioinspired dental multilayered structures. Thus,

$$\frac{1}{E} = \frac{1}{k_1}\left\{1 - \frac{k_2}{k_1 + k_2}\exp\left[-\frac{k_1 k_2 t}{\eta(k_1 + k_2)}\right]\right\}, \tag{6.3}$$

where t is the time, k_1 and k_2 are the spring constants, and η is the dashpot constant, as shown in Figure 6.7c.

6.2.4 Viscosity Assisted Slow Crack Growth Model

Developed in the 1970s, the power law slow crack growth (SCG) model is an elastic model that describes stable subcritical crack growth behavior [26]. It was adopted by researchers to explain the occupancy of subsurface radial cracks in the top ceramic layers [10,12,27]. For the dental multilayer structure shown in Figure 6.1b, when the middle adhesive layer is very thin, it can be simplified as a bilayered structure. For the monotonic loading condition, when the bilayer structure is loaded at a fixed loading rate $\dot{P}$, the failure condition could be expressed as [7]

$$P_c = \left[A(N+1)\,\dot{P}\right]^{1/(N+1)}, \tag{6.4}$$

where P_c is the failure load; A is a load, time, and thickness independent quantity, and N is the crack velocity exponent.

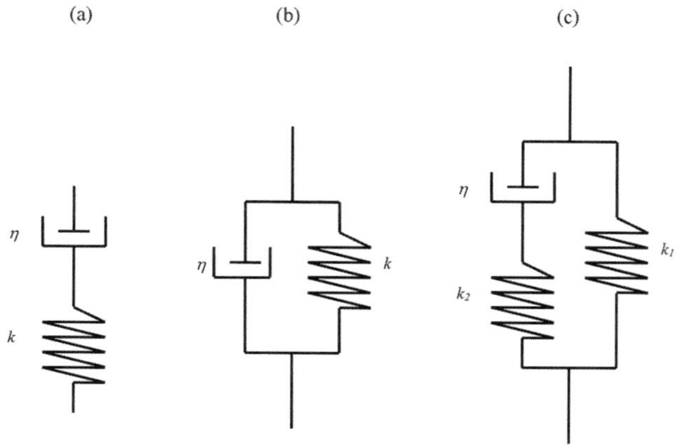

Figure 6.7 Schematics of viscoelastic spring-dashpot models: (a) Maxwell model, (b) Voigt model, and (c) three-element (Zerner) model [25].

For the fatigue loading with constant frequency, the failure condition is given as [7]

$$P_{\max}^{N} t = 2AN^{0.47}, \tag{6.5}$$

where $P_{\max}$ is the maximum load of the sinusoidal cyclic loading, and t is the time to failure.

The SCG model captures the contact damage of dental multilayer structures solely contributed by the slow crack growth in the top ceramic layer. However, it neglects the middle adhesive layer and treats the top ceramic and substrate polymeric layers as linearly elastic materials. The viscous flow in the substrate and middle layers is neglected. In an effort to consider the combined effects of slow crack growth in the top ceramic layer and the viscosity-induced loading rate effects in adhesive and substrate layers, the rate dependent Young's modulus assisted SCG (RDEASCG) model, was developed for prediction of failure loads in three-layered dental structures under monotonic loading [12]. The model is explained by the flowchart shown in Figure 6.8.

To model the viscosity-assisted fatigue behavior, a substrate creep effect (SCE) fatigue model was developed [11] to provide relationship between the maximum load and the failure time that corresponds to the pop-in of the subsurface radial crack in the top glass layer. The pop-in loads can be predicted from Eq. 6.6 shown here:

$$P_m^N\left[\log\left(\frac{CE_c}{k_1}\right)\left\{1-\frac{k_2}{k_1+k_2}\exp\left[-\frac{k_1k_2}{\eta(k_1+k_2)}\right]\right\}\right]^N t_f = H, \tag{6.6}$$

where t_f is the time to failure, E_c is the Young's moduli of the top ceramic layer, C is a dimensionless coefficient, and H is a material- and geometry-dependent constant.

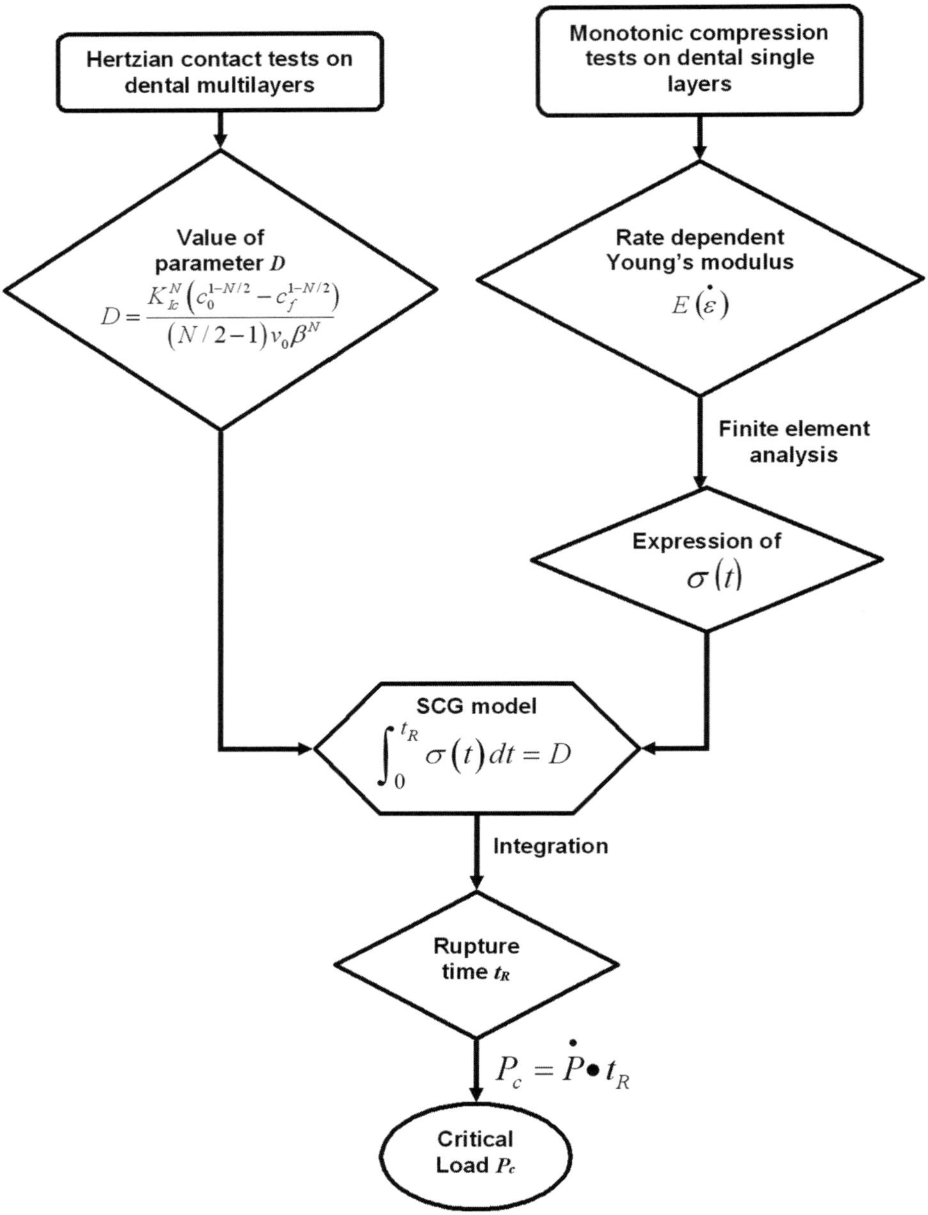

Figure 6.8 Flowchart of rate dependent Young's modulus effected SCG (RDEASCG) model [12,13].

6.3 Review

6.3.1 History of Dental Ceramics

Full or partial tooth replacement has been used in human history for a very long time [28]. A few thousand years ago, materials such as ivory, bone, wood, metals/alloys, or even animal teeth, were used because they "looked" or "felt" like teeth. In 1774, ceramics were first used successfully in dentistry. They effectively solved problems such as stained, decayed, and terminally malodorous dentures [4]. More importantly, they are "strong" and biocompatible.

However, despite these advantages, the widespread application of ceramics in prostheses has not lived up to the original expectations [9]. While initially appreciated for their hygienic qualities, early ceramics did not faithfully replicate the optical characteristics of natural teeth. They were also awkward to fit as prosthetic replacements, and they had high failure rates. Hence, over the next 190 years after 1774, efforts were made to improve the aesthetics, particularly with respect to translucency, and fabrication methods in support of dental practice [4].

In the 1960s, dental ceramics were formulated for routine fusion onto metal substructures. This greatly broadened the use of ceramics, but failed to match the aesthetic qualities of natural teeth. In subsequent decades, attention has been paid to improving structural properties of ceramic systems, net-shape processing, machining as a forming method, and analyzing clinical failure [4]. Innovations in ceramic science have also resulted in the creation of structural materials with improved mechanical performance [9].

Ceramics are brittle, and they generally have poor crack growth resistance, especially when initial defects are introduced by the shaping process. Dental crowns continue to fail at a rate of approximately 3% each year [29], with the highest fracture rates occurring on posterior crowns and bridges, where stresses are greatest [9]. Prior research has also shown that almost 20% of posterior all-ceramic dental crowns with resin-retained ceramics fail within the first 5 years of service in the oral cavity [8].

Motivated by the need for improved dental multilayers, a number of new materials and technologies have been developed for dental restorative crowns and bridges [2,30,31]. These materials must meet a number of mechanical, chemical, thermal, and optical requirements in order to withstand the rigors of dental applications. The key properties include strength, fracture toughness, wear and hardness, compatibility with an oral environment, and aesthetics [2].

At the present stage, the strengths of synthetic crowns appear to be comparable with or greater than those of natural materials (Table 6.1) [31,32]. However, their strength values do not reflect their crack growth resistance. The reported values [33,34] of fracture toughness for all-ceramic dental materials are noticeably below those of natural teeth [2]. Also, some of the ceramic top layers have been shown to be aggressive on opposing natural teeth [35].

On the other hand, the financial drivers for developing damage-resistant crown are high. Dental implants (which require a crown) account for about $1 billion of revenues

Table 6.1 Physical and mechanical properties of the oral hard tissues of human teeth [31,32]

Property	Dentin	Enamel
Density ($g{\cdot}cm^{-3}$)	2.1	2.96
Compressive elastic modulus (GPa)	18–24	60–120
Compressive strength (MPa)	230–370	94–450
Tensile elastic modulus (GPa)	11–19	
Tensile strength (MPa)	30–65	8–35
Shear strength (MPa)	138	90
Flexural strength (MPa)	245–280	60–90
Hardness (GPa)	0.13–0.51	3–6
Fracture toughness ($MPa{\cdot}m^{1/2}$)	3.08	0.8

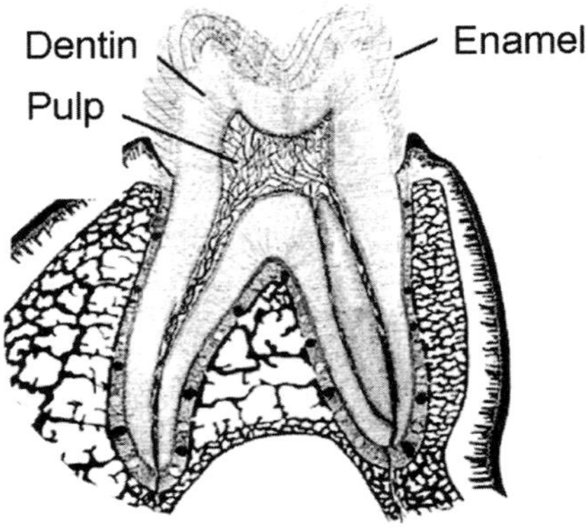

Figure 6.9 Schematic of the natural tooth structure [14].

each year. Also, the market is estimated to be growing at a rate of 18% per year. Furthermore, dental crowns generate over $2 billion each year in revenues, with 20% of the units being all-ceramic [9]. The aging population is likely to drive the demand for dental crowns [9].

6.3.2 Resilience of Natural Teeth

The natural teeth show superior overall properties to artificial crowns [2]. Hence, the knowledge of the structure of the human tooth is very important for the design of artificial dental crowns. The natural tooth is a remarkable example of nature's ability to design a complex and functional composite [2]. A human tooth consists of pulp, enamel, and dentin, as shown in Figure 6.9 [14]. Pulp is an organic and soft connective tissue that is beyond the scope of this chapter. Dentin and enamel are hard tissues that are composed of inorganic minerals and organic materials.

Dentin is a complex connective tissue that is composed of ~50 vol% inorganic hydroxyapatite crystals, 23 vol% organic material (mostly collagen fibers) and

27 vol% water [36–38]. The majority of the dentin is made up of a highly cross-linked collagen phase mineralized with hydroxyapatite (HAP) crystallites [36]. The main structural function of dentin is to provide a mechanically tough support for the enamel. The cross-linked and mineralized collagen phase resists the deformation in the structure. The highly mineralized walls, dentinal tubules, and their short lateral extensions also contribute to the stiffening of the structure. The fracture toughness is enhanced by the composite microstructure, which contains dispersed inorganic particles in a collagen matrix and hence multiple interfaces to interfere with crack propagation. An investigation of the fracture properties of dentin revealed that dentin is tougher through its thickness than it is laterally [32].

Enamel is the most highly mineralized tissue of the body, roughly 90 vol% HAP. Mature enamel is a complex inorganic-organic composite [36,37]. The organic matrix is composed mainly of noncollagenous glycoproteins [36]. The HAP phase has a highly organized structure composed of long rods of packed crystallites extending from the outer enamel surface downward to the DEJ. The function of enamel is to provide a hard, wear-resistant surface. The hardness of enamel is roughly five times that of dentin. However, enamel is a brittle material. It fractures along the weak paths parallel to the rods [39]. Nevertheless, the well-designed microstructure and strong attachment to the underlying dentin resists crack propagating.

Though very different in their structures and properties, as listed in Table 6.1 [31,32], dentin and enamel work together harmoniously through an interface known as the DEJ [2]. The microstructure of the DEJ has been studied by Huang et al. [21] using scanning electronic microscopy, as shown in Figure 6.10. It is a unique area in the crown of a tooth that is distinct from either the contiguous enamel or dentin [40]. The natural DEJ supplies the bond between dentin and enamel. It also resists cracks that originate in enamel from penetrating into the dentin. To provide these functions, nature has evolved a complex interface in which components from enamel and dentin interpenetrate and intermingle [17,39,40]. Figure 6.11 shows a schematic of the "microscallops" formed by the collagen bundles within the major scallops at the DEJ [40]. These coarse collagen bundles merge with other fibrils before or after entering the enamel matrix [40]. Likewise, the common mineral phase (HAP) is continuous across the junction. The interface is not smooth, but instead is a series of linked semicircles (or scallops) that increase contact area, and thus reduces the likelihood of dentin and enamel separating during function. It also imparts strength through stress transfer [2].

In contrast to artificial bonds between crowns and dentin, the DEJ rarely fails, except when it is affected by inherited disorders. Knowledge of DEJ toughening mechanisms is important in designing artificial crowns and teeth [41].

In 1994, Lin and Douglas set up an experiment to determine the fracture behavior of the DEJ. A fracture test was designed to initiate a crack at the vertex of the chevron in the enamel, across the DEJ zone, and propagated into the bulk dentin [17]. The results showed that there was an extensive plastic deformation (83 $\pm$ 12%) collateral to the fracture process at the DEJ zone. The fractography revealed that the deviation of the crack path involved an area that was approximately 50–100 μm deep. The parallel-oriented coarse collagen bundles with diameters of 1–5 μm at the DEJ zone may play a

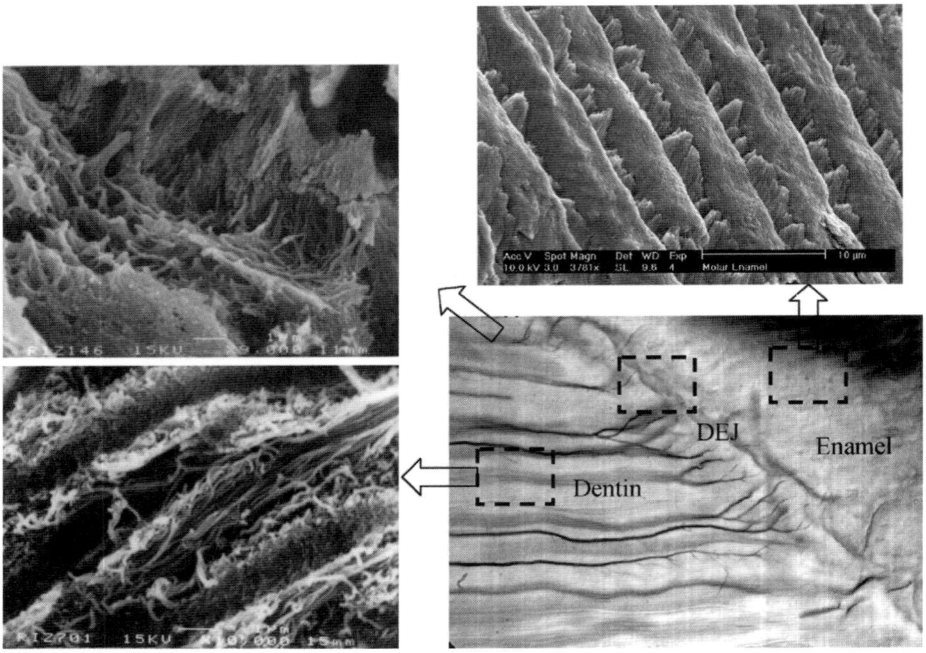

Figure 6.10 Microstructure of dentoenamal junction (DEJ) under SEM [21].

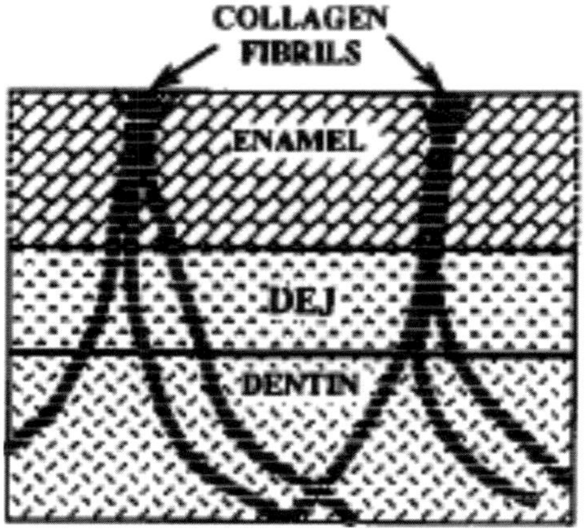

Figure 6.11 Schematic of microscallops formed by the collagen bundles within the major scallops at the DEJ [40].

significant role in resisting the enamel crack [40]. White et al. [41] also investigated the DEJ failure mechanisms by performing microindentation tests at three sites across the DEJ zone of 10 human incisor teeth. The results showed that the DEJ did not undergo catastrophic interfacial delamination, and the damage was distributed over a broad zone instead. DEJ-zone damage occurred primarily within the adjacent layer of specialized

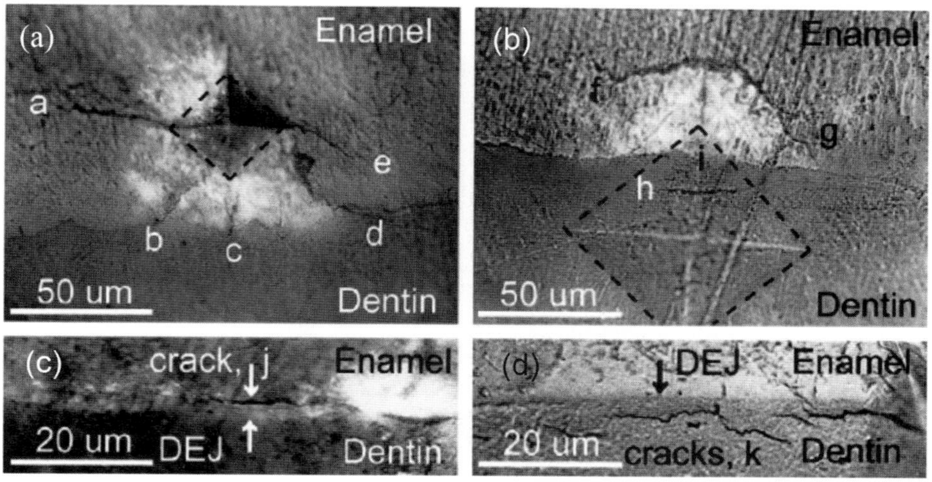

Figure 6.12 Light micrographs of the DEJ zone after indentation. Dashed rhombi indicate footprints of indentations, where letters a–k denotes cracks. (a) Indentation centered on enamel side of DEJ; (b) Indentation centered on dentin side of optical DEJ; (c) Crack within enamel is close to and parallel to DEJ but does not involve DEJ itself; (d) Series of interrupted cracks or partial tearing within dentin that are close to and slightly oblique to DEJ but do not involve DEJ itself k [41].

first-formed enamel, and the optical DEJ interface resisted delamination as shown in Figure 6.12 [41]. All of these factors may explain the fact that in the intact tooth, the multiple full-thickness cracks commonly found in enamel do not typically cause total failure of the tooth by crack extension into dentin [17].

Prior researchers [42,43] performed atomic force microscope (AFM) based nanoindentation tests to measure the Young's modulus of the natural DEJ area. Their results in Figure 6.13 showed that, within the DEJ region, mechanical properties such as Young's modulus and hardness change gradually from that of hard, brittle enamel to that of the softer durable dentin layer. The fracture results [43] once again demonstrated that it is extremely difficult to initiate cracks in dentin at the DEJ or propagate cracks from enamel to dentin across the DEJ [44,45].

6.3.3 Design of DEJ-Like Dental Interfaces

Posterior all-ceramic crowns show short lifetime [9]. The reason could be poor interfaces between the layers. They normally consist of a ceramic layer (mimicking enamel), supported by a ceramic-filled polymeric layer (mimicking dentin) held in place with a glass-filled, resin-based cement layer. In most dental crowns, the Young's modulus of the top ceramic layer is between 70 and 400 GPa. The modulus of the substrate polymeric layer is around 18 GPa and that of dental cements are generally between 3 and 5 GPa [4]. The mismatch in Young's moduli results in high stresses in the top

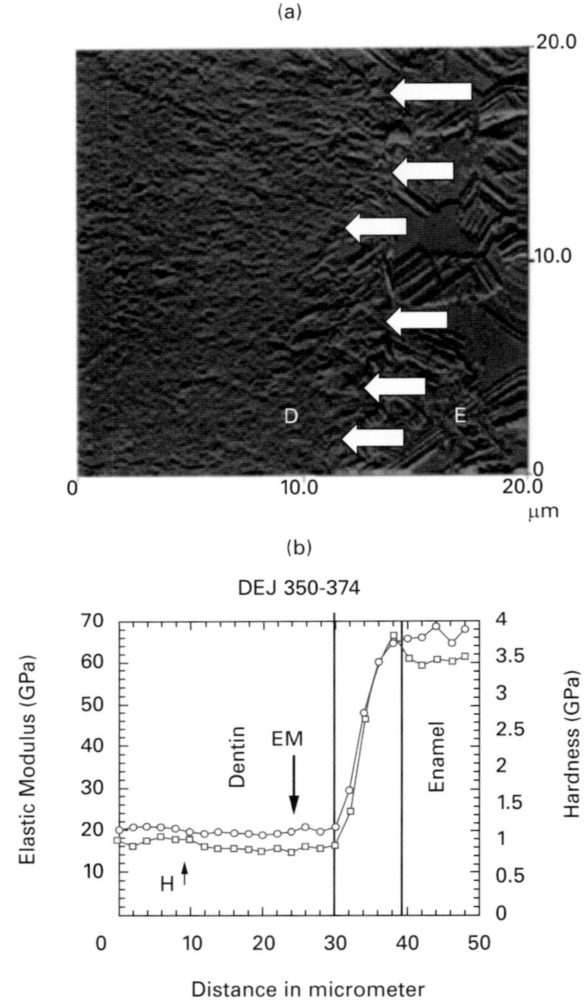

Figure 6.13 (a) AFM images of a DEJ region. Arrows indicate the approximate position of the interface; E: enamel, and D: dentin [42]; (b) Gradually changed hardness and elastic modulus (EM) over DEJ region [43].

ceramic layer. These stresses ultimately result in subsurface radial cracks in the top ceramic layer [21].

Making better adhesive layers remains a challenging problem in the design of dental crowns. Efflandt and coworkers [46] explored the interaction between bioactive glasses and dentin that were extracted from human teeth under simulated oral conditions. Their research shows the exciting possibility of developing better bioactive glass/polymer bonding agents. Although using bioactive glasses in their bulk form was not a feasible option, the presence of bioactive glass may be of benefit to interfaces in new dental adhesives or tissue-engineering restorations [46].

Zhang et al. [47] produced a porous composite consisting of polysulfone (or cellulose acetate). They produced bioactive glass particles by using phase separation techniques. The composites asymmetric structures with dense top layers and porous structures underneath. These porous composites may have potential applications as interfacial materials between soft and hard tissues, such as the artificial cartilage/bone interface [47].

Francis et al. [2] described a procedure to produce a DEJ-like interface and enamel coating involved depositing slurries of oxide or glass powder by a draw-down blade method, drying at 100°C for 15 minutes followed by higher temperature heating. Microstructural analysis showed that the interpenetration in this DEJ-like interface originated from a solidified melt phase penetrating into the man-made dentin-like layer below. The $CaO\text{-}Al_2O_3\text{-}SiO_2$ (CAS) composition provided strong bonding [2]. Compared with the natural DEJ, the DEJ-like interface developed in this study needed to go a step further, with collagen projections across the interface. Also, fracture properties of the DEJ-like interface and the ability to interfere with or arrest crack propagation have not been investigated [2].

6.3.4 Concepts of Bioinspired Dental FGM

The concept of using functionally graded materials (FGMs) arose in the mid 1980s as a way of improving the thermomechanical performance of aerospace structures. Material gradation offers a way of maintaining the high strength of an inner core and high thermal resistance of an outer layer, while eliminating the deleterious effects of a sharp interface [48].

Inspired by both the prior work on FGM and natural teeth structure, Huang et al. [21] proposed to use a bioinspired FGM layer as a DEJ-like dental adhesive in the dental crown structures. In their computational study, Huang et al. added a FGM layer with its Young's modulus varying gradually from the dental cement layer to the enamel-like dental ceramic layer (Figure 6.14; [21]). Finite element simulations of the structure showed that adding an FGM adhesive layer can significantly reduce the stress concentrations in the subsurface of the top ceramic layer. This increased the resistance of the structure to radial cracking. Moreover, the estimated critical crack lengths were much greater than those estimated in both existing dental crowns and the natural tooth

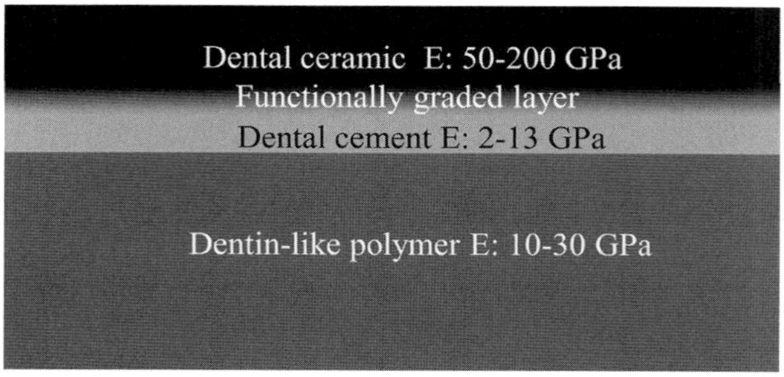

Figure 6.14 Schematic of a model dental multilayer structure with functionally graded layer between the cement and the ceramic layers [21].

Table 6.2 Mechanical properties of the human enamel obtained from the indentation analyses [49]

Material	HV_c (GPa)	*E* (GPa)
Human enamel (young)		
Inner	3.1	75
Middle	3.5	82
Outer	4.1	87
Human enamel (old)		
Inner	3.0	79
Middle	3.4	90
Outer	4.0	100

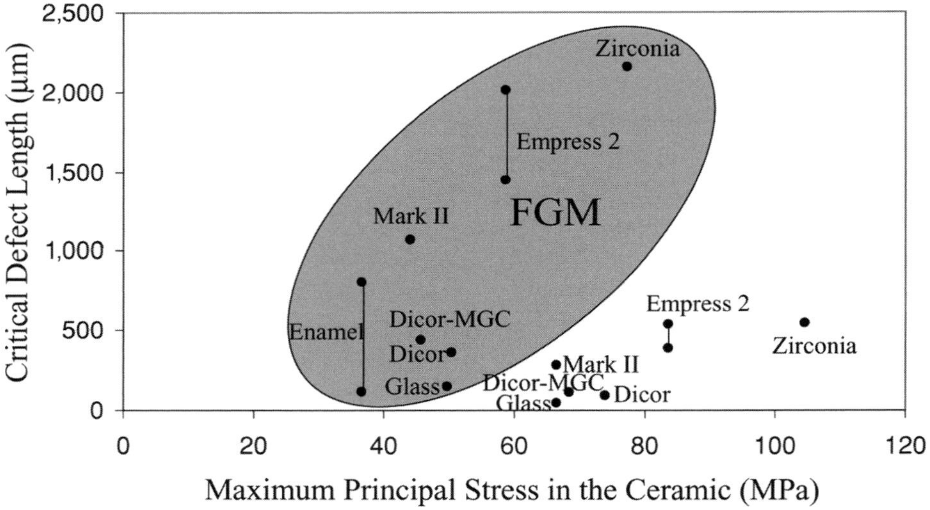

Figure 6.15 Critical crack lengths versus maximum principal stresses in graded and nongraded multilayer structures [21].

enamel–dentin complex (Figure 6.15; [21]). The result suggested the possibility of building synthetic, bioinspired, functionally graded dental multilayers that have comparable or better durability than those of natural teeth.

The promise of these results stimulated the current effort to develop bioinspired functionally graded dental multilayers. It is also important to know that FGMs may improve the current single-layer dental materials, especially dental ceramics. Park et al. [49] characterized the hardness, elastic modulus, and apparent fracture toughness as a function of distance from the DEJ using indentation approaches. The results showed that the hardness and the elastic modulus increased with distance from the DEJ (Table 6.2; [49]). Huang et al. also suggested that the idea of FGM needs to be extended to two- and three-dimensional geometries, instead of only along the thickness direction [21].

Prior work has been done to study the effects of FGM architectures on the performance of the dental multilayer structures. The maximum principal stress at the subsurface center of the top ceramic layer is associated with the major clinical failure mode, the subsurface radial crack. Simulations have shown that as the thickness of the FGM increased, the stress concentration in the structure was reduced, and the maximum principal stress in the subsurface center of the top ceramic layer decreased. Thicker FGM provided better support to the top ceramic layer, because it had higher stiffness. However, when the thickness of the single dental adhesive layer increased, the opposite was true [50]. Simulations [21,50] have also shown that a linear Young's modulus gradient will provide a balanced maximum principal stress in the top ceramic layer and the FGM layer. The nonlinear Young's modulus gradient in the FGM increased the maximum principal stress in the top ceramic layer if the higher gradient was near the top ceramic layer. It increased the maximum principal stress in the FGM if the higher gradient was near the substrate. This suggested that the optimal design of dental FGM architecture was around linear gradation, which was consistent with the Young's modulus distribution in the DEJ.

FGMs are also of interest for increasing the adhesion between dental implants and oral cavity bone. The biomechanical behavior of a functional graded biomaterial (FGBM) dental implant was investigated [51] by using a three-dimensional finite element method, and taking into account the interaction between the implant and the surrounding bone. The computational results showed that the use of an FGBM implant effectively reduced the stress difference at the implant–bone interfaces where maximum stresses occured. At the same time, it had relatively small effects on the natural frequencies of the whole system. Furthermore, Traini et al. [52] used a laser metal sintering technique to construct a titanium alloy dental implant incorporating a gradient of porosity, from the inner core to the outer surface. The FGMs were proven to give better adaptation to the elastic properties of the bone.

As for the FGMs, there are a number of in-depth studies on crack growth resistance and fracture behavior. Rousseau et al. [48] studied crack tip deformations in FGMs parallel to the gradient direction and subjected to low-velocity, symmetrical impact loading. Paulino and coworkers [53–55] have developed accurate graded finite elements for the modeling of FGM. Such models could clearly form the basis for future work to compute crack driving forces and develop better understanding of fracture processes in FGM. They could also be applied to the design of future bioinspired structures with nonlinear material laws that have more advanced scope. There is, therefore, a need for further research on FGMs for future applications in dentistry.

6.3.5 Fabrication and Characterization of Bioinspired FGM Structure

Based on the previous concepts of the bioinspired FGM structure design, researchers developed a method to fabricate the bioinspired FGM structures using a ceramic particle reinforced, polymer-based composite material [56]. A steel plate mold containing holes was filled with a dentin-like composite material, which was a clinically used dental material, and then cured with UV light to become the substrate of the dental multilayer

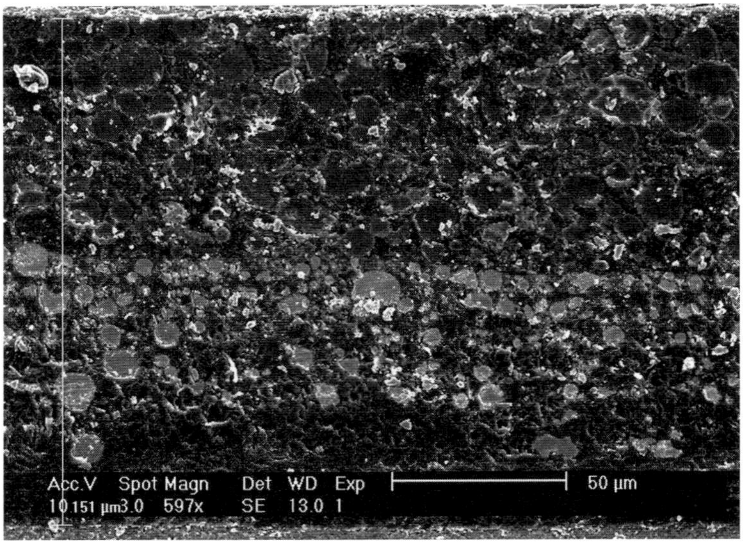

Figure 6.16 SEM image of the microstructure of functionally graded material [56].

model structure. The FGM were produced using nanocomposite materials – a mixture of zirconia or alumina nanoparticles and an epoxy matrix. After mixing, a wire-wound wet-film applicator rod was then used to spread the nanocomposite material across the steel plate. When pulled across the steel plate, with a fluid film in front of the rod, the applicators could control the thickness of the fluid film. After each layer had been deposited and spread, the plate was cured in a vacuum oven. The deposition and curing process was then repeated to build up the multilayered structures. From bottom to top, the functionally graded nanocomposite layers contained different weight percentages of ceramic particles to change the stiffness of the layers. The crown-like dental ceramic layer on top was fabricated with a medical grade 3 mol% yttria-stabilized zirconia material. It was pressed onto the last layer of the FGM before curing [50,56].

Figure 6.16 [56] shows the microstructure of the fabricated dental FGM structure. A total of 10 layers with microscale thicknesses were used to produce the graded structure. Also, with each graded layer, either alumina or zirconia particles were used to achieve the desired modulus. The higher moduli alumina particles were used to achieve the desired moduli at lower particle volume fractions. This minimized problems of particle clustering during the mixing of composite layers with high zirconia particle fractions. The microsized ceramic "particles" were clusters of nanosized powders.

Nanoindentation experiments were carried out on the polished cross sections of the fabricated dental FGM structure to provide inputs regarding the architecture and the actual Young's modulus distributions of the structure as shown in Figure 6.17 [50]. The measurements suggest an almost sinusoidal variation in Young's modulus across

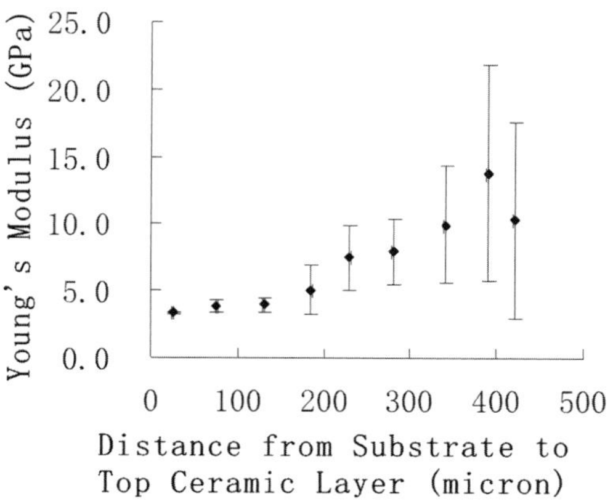

Figure 6.17 Distribution of Young's moduli across the fabricated dental FGM layer [50].

the FGM. The variabilities in the Young's moduli were attributed partly to the clustering and nonuniform particle distributions. The dental composite substrate was found to have a modulus of ~20 GPa, which mimicked the Young's modulus of the remaining dentin in natural teeth, whereas the top zirconia ceramic layer had a modulus >200 GPa. The measured Young's modulus distribution was incorporated into the mechanics models to study the effects on the performance of the dental structures for their applications in dentistry.

6.3.6 Performance of the Dental FGM Structures

Hertzian contact experiments were performed on dental multilayer structures with and without FGM. The tests were conducted at clinically relevant loading rates between 1 N/s and 1,000 N/s. The critical load was determined from the displacement discontinuity in the load-displacement curves obtained from the measurement. Figure 6.18 [56] shows that the FGM structures had ~20–30% higher critical pop-in loads than flat conventional dental multilayers without FGM for any given loading rates.

Following the fracture under Hertzian tests, the cross sections of the tested samples were examined under optical microscope to discover the failure modes of the dental FGM structure [50]. Figure 6.19 shows a typical subsurface radial crack observed in the ceramic top layer. Evidence of the interfacial cracking between the top zirconia ceramic layer and the FGM layer was also apparent in this image. The results, therefore, suggest that critical load corresponded to the pop-in of radial cracks in the center of the subsurface of the top zirconia layer. The pop-in was associated with radial crack propagation into the top zirconia ceramic layer, along with partial interfacial cracking between the top zirconia ceramic layer and the FGM.

The slow crack growth (SCG) model was used to study the crack growth and predict the critical loads. The differences of critical loads of the structure with and without FGM were greater for slower loading rates than they were for faster loading rates. These

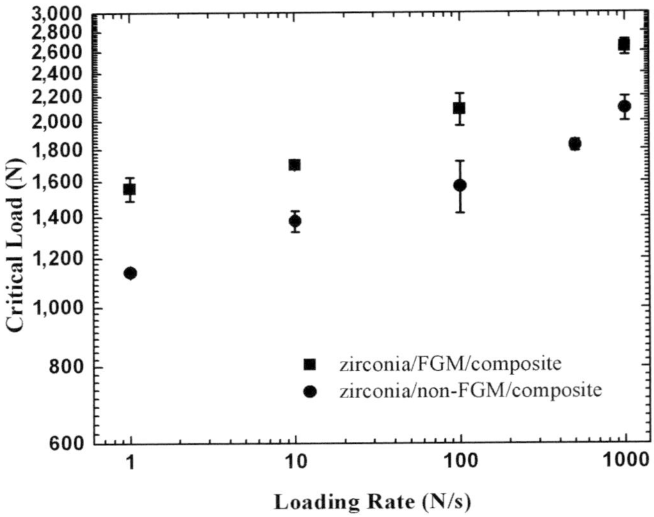

Figure 6.18 Comparison of the critical loads under Hertzian contact loading at various clinically relevant loading rates of the dental multilayer structures with and without FGM [56].

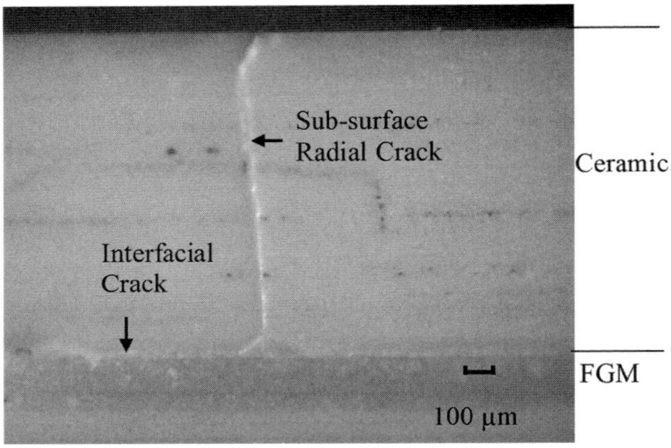

Figure 6.19 Failure modes of the dental FGM structure revealed by optical images of the cross-section after Hertzian contact tests [50].

suggested a loading rate dependence of the layer properties should be considered. Rate dependent material properties were incorporated to consider the effects of the viscosity of the polymeric materials using a load rate dependent Young's modulus assisted slow crack growth (RDEASCG) model [12,56]. The predictions of the critical load were made by inputting the stresses obtained from FEM simulation into the RDEASCG model. The results show that the predictions from the RDEASCG model were comparable with those obtained from the experiments.

6.4 Future Directions

The study of bioinspired design of dental multilayers showed that bioinspired, functionally graded dental multilayers significantly increased the strength of dental multilayers under monotonic loading. However, there are still some challenges that must be resolved before potential applications of bioinspired functionally graded dental multilayers can be fully explored in actual dental crowns.

Further work is clearly needed to explore the extent to which the FGM adhesives can improve fatigue lives. The clinical performance of such structures under cyclic loading conditions needs to be investigated. The fatigue lives of the dental multilayer structures with and without the FGM should be measured and compared.

Research is also needed to study the possible effects of other FGM architectures beyond those examined in this chapter. For example, the gradient direction of the FGM could be not only along the Herzian contact load axis, but also perpendicular to the load axis. The idea of FGM needs to be extended to two- and three-dimensional geometries, instead of only along the thickness direction [21]. The different FGM architectures could be fabricated and compared.

Also, in reality, tooth to tooth contacts (shown in Figure 6.1a) are far more complex than the contact between the Hertzian ball and idealized flat dental multilayers that were discussed in this chapter. Therefore, we have the following suggestions for the future work: First, curvature effects on contact damage under both monotonic and cyclic loading need further investigation. Then, effects of complex loading (shear contact loading and normal + shear loading), both in air and in a wet environment, should be studied. Finally, a mechanistically based framework for predicting contact lives of real dental structures with complex geometries under both monotonic and cyclic loadings in water is yet to be established.

Figure 6.6 suggests that the pop-in of radial cracks is accompanied by interfacial cracking between the ceramic and cement layers. Hence, for the development of durable dental crown systems, future work is clearly needed. Interfacial cracking between top ceramic and middle adhesive layers (both traditional dental cement and bioinspired functionally graded structures) should be considered as possible cracking modes in the future analyses and design of dental multilayers. Studies of interfacial fracture could be analogous to those in orthopedic interfaces; e.g., bone/tissue interface.

In addition, environmental exposures that are relevant to occlusal conditions need to be explored. The effects of humidity and acidity on the fabricated structure should be examined. The FGM structure could be tested under different humidity and acidity conditions. In the prior work [16], a model was developed to understand the mechanism of water absorption induced cracking in dental multilayered structures. This also stimulated our interests in several issues in the future. The interaction between stress and diffusion are coupled. Hence, the overall stress state and SCG phenomena cannot be estimated simply by linear superposition. Further work is clearly needed to develop physics-based mechanics models that incorporate the effects of water diffusion and contact loading for the prediction of the fatigue lives of dental crowns. Future work is also needed to study SCG phenomena in the aqueous environments that are closer to those in the oral cavity; i.e., beyond distilled water.

6.5 Summary

This chapter presents the results of a combined experimental and theoretical/computational study of the bioinspired design of the dental FGM structures. The need for dental crowns generates a huge market every year, which is increasing due to the steady growth of the aging population. The lifetime of dental crowns, however, does not live up to expectations. Hence, there is a need to improve the current design of dental crowns.

On the other hand, natural teeth have different features that distinguish them from artificial restoration crowns. The DEJ has a complex microstructure, which resists the crack propagation, and a graded Young's modulus distribution, which provide a buffer to the stress distribution. Artificial ceramic dental crowns, instead of having a DEJ, are bonded to the remaining dentin using a single layer of dental adhesive material. The Young's modulus mismatch of the dental adhesive material and crown causes high stress concentrations in the subsurface regime of the top ceramic layer in dental multi-layer structures. This results in subsurface cracking in the top ceramic layer. This failure mode is consistent with the major clinical failure mode of ceramic dental crowns.

Inspired by natural teeth, the design of DEJ-like graded dental interfaces was proposed using various methods. Bioinspired dental FGM structure is a design using a DEJ-like graded dental adhesive to replace the single layer adhesive in the dental crown structures. Simulation efforts have been put into the design of the bioinspired dental FGM structures. It has been shown that the FGM can reduce the stress concentration in the structure and reduce the maximum principal stress in the top ceramic layer, which was relevant to the subsurface cracks. The effects of various architectures, including the thickness and the Young's modulus gradient, were also studied using simulation.

Using a ceramic particle reinforced nanocomposite material, the bioinspired dental FGM structures were fabricated. The Young's modulus of the FGM was controlled by modulating the component and composition of the ceramic reinforcement. The critical loads of these structures under Hertzian contact were ~20–30% higher than those of conventional structures using single-layer adhesive, at any clinically relevant loading rate.

The SCG model provided fracture mechanics measurements of the crack growth behavior in the structured. The viscosity of the materials was considered by incorporating the rate-dependent Young's moduli into the SCG model. The predictions provided by the RDEASCG had good agreement with the experimental measurements.

For the future application of bioinspired dental FGM structures in dentistry, further work is clearly needed to assess the clinical performance of such structures under cyclic loading conditions and environmental exposures that are relevant to occlusal conditions. The other FGM architectures also need to be explored. The complex geometries and loading conditions of the real dental crowns should also be considered.

References

1. World Health Organization (WHO). (2012). Oral health fact sheet. Retrieved from www.who.int/news-room/fact-sheets/detail/oral-health.
2. Francis, L. F., Vaidya, K. J., Huang, H. Y., & Wolf, W. D. (1995). Design and processing of ceramic-based analogs to the dental crown. *Materials Science and Engineering: C*, 3(2), 63–74. doi:10.1016/0928-4931(95)00088-7
3. Huang, M., Niu, X., Shrotriya, P., Thompson, V., Rekow, D., & Soboyejo, W. O. (2005). Contact damage of dental multilayers: Viscous deformation and fatigue mechanisms. *Journal of Engineering Materials and Technology*, 127(1), 33. doi:10.1115/1.1836769
4. Kelly, J. R. (1997). Ceramics in restorative and prosthetic dentistry. *Annual Review of Materials Science*, 27(1), 443–468. doi:10.1146/annurev.matsci.27.1.443
5. Lawn, B. R., Lee, K. S., Chai, H., et al. (2000). Damage-resistant brittle coatings. *Advanced Engineering Materials*, 2(11), 745–748. doi:10.1002/1527-2648(200011)2:11<745::AID-ADEM745>3.0.CO;2-E
6. Lawn, B. R., Pajares, A., Zhang, Y., et al. (2004). Materials design in the performance of all-ceramic crowns. *Biomaterials*, 25(14), 2885–2892. doi:10.1016/j.biomaterials.2003.09.050
7. Lee, C.-S., Kim, D. K., Sanchez, J., Pedro, M., Antonia, P., & Lawn, B. R. (2002). Rate effects in critical loads for radial cracking in ceramic coatings. *Journal of the American Ceramic Society*, 85(8), 2019–2024.
8. Malament, K. A., & Socransky, S. S. (1999). Survival of Dicor glass-ceramic dental restorations over 14 years: Part I. Survival of Dicor complete coverage restorations and effect of internal surface acid etching, tooth position, gender, and age. *The Journal of Prosthetic Dentistry*, 81(1), 23–32.
9. Rekow, D., & Thompson, V. P. (2007). Engineering long term clinical success of advanced ceramic prostheses. *Journal of Materials Science: Materials in Medicine*, 18(1), 47–56. doi:10.1007/s10856-006-0661-1
10. Zhang, Y., Lawn, B. R., Rekow, E. D., & Thompson, V. P. (2004). Effect of sandblasting on the long-term performance of dental ceramics. *Journal of Biomedical Materials Research. Part B, Applied Biomaterials*, 71(2), 381–386. doi:10.1002/jbm.b.30097
11. Zhou, J., Huang, M., Niu, X., & Soboyejo, W. O. (2007). Substrate creep on the fatigue life of a model dental multilayer structure. *Journal of Biomedical Materials Research. Part B, Applied Biomaterials*, 82(2), 374–382. doi:10.1002/jbm.b.30742
12. Niu, X., & Soboyejo, W. (2006). Effects of loading rate on the deformation and cracking of dental multilayers: Experiments and models. *Journal of Materials Research*, 21(04), 970–975. doi:10.1557/jmr.2006.0114
13. Niu, X., Yang, Y., & Soboyejo, W. (2008). Contact deformation and cracking of zirconia/cement/foundation dental multilayers. *Materials Science and Engineering: A*, 485(1–2), 517–523. doi:10.1016/j.msea.2007.09.014
14. Shrotriya, P., Wang, R., Katsube, N., Seghi, R., & Soboyejo, W. O. (2003). Contact damage in model dental multilayers: An investigation of the influence of indenter size. *Journal of Materials Science: Materials in Medicine*, 14(1), 17–26.
15. Zhang, Y., Pajares, A., & Lawn, B. R. (2004). Fatigue and damage tolerance of y-tzp ceramics in layered biomechanical systems. *Journal of Biomedical Materials Research. Part B, Applied Biomaterials*, 71(1), 166–171. doi:10.1002/jbm.b.30083
16. Huang, M., Thompson, V. P., Rekow, E. D., & Soboyejo, W. O. (2007). Modeling of water absorption induced cracks in resin-based composite supported ceramic layer structures. *Journal of Biomedical Materials Research*, 1, 124–130. doi:10.1002/jbmb

17. Lin, C. P., & Douglas, W. H. (1994). Structure-property relations and crack resistance at the bovine dentin-enamel junction. *Journal of Dental Research*, 73(5), 1072–1078. doi:10.1177/00220345940730050901
18. Tsai, Y. L., Petsche, P. E., Anusavice, K. J., & Yang, M. C. (1998). Influence of glass-ceramic thickness on Hertzian and bulk fracture mechanisms. *The International Journal of Prosthodontics*, 11(1), 27–32.
19. Thompson, J. Y., Anusavice, K. J., Naman, A., & Morris, H. F. (1994). Fracture surface characterization of clinically failed all-ceramic crowns. *Journal of Dental Research*, 73(12), 1824–1832. doi:10.1177/00220345940730120601
20. Lawn, B., Deng, Y., Miranda, P., Pajares, A., Chai, H., & Kim, D. K. (2002). Overview: Damage in brittle layer structures from concentrated loads. *Journal of Materials Research*, 17 (12), 3019–3036. doi:10.1557/JMR.2002.0440
21. Huang, M., Rahbar, N., Wang, R., Thompson, V., Rekow, D., & Soboyejo, W. O. (2007). Bioinspired design of dental multilayers. *Journal of Materials Science: Materials in Medicine*, 18(1), 57–64. doi:10.1007/s10856-006-0662-0
22. Lawn, B. R., Padture, N. P., Cait, H., & Guiberteau, F. (1994). Making ceramics "ductile." *Science*, 263(5150), 1114–1116. doi:10.1126/science.263.5150.1114
23. Lawn, B. R. (2005). Indentation of ceramics with spheres: A century after Hertz. *Journal of the American Ceramic Society*, 81(8), 1977–1994. doi:10.1111/j.1151-2916.1998.tb02580.x
24. Peterson, I. M., Wuttiphan, S., Lawn, B. R., & Chyung, K. (1998). Role of microstructure on contact damage and strength degradation of micaceous glass-ceramics. *Dental Materials*, 14(1), 80–89.
25. Soboyejo, W. (2003). *Mechanical properties of engineered materials*. CRC Press.
26. Wiederhorn, S. M. (1974). Subcritical crack growth in ceramics. In R. C. Bradt, D. P. H. Hasselman, & F.F. Lange (Eds.), *Fracture mechanics of ceramics*. Springer. doi:10.1007/978-1-4615-7014-1_12
27. Zhang, Y., Kim, J.-W., Bhowmick, S., Thompson, V. P., & Rekow, E. D. (2009). Competition of fracture mechanisms in monolithic dental ceramics: Flat model systems. *Journal of Biomedical Materials Research. Part B, Applied Biomaterials*, 88(2), 402–411. doi:10.1002/jbm.b.31100
28. Ratner, B. D., Hoffman, A. S., Schoen, F. J., & Lemons, J. E. (2004). *Biomaterials science: An introduction to materials in medicine*. Elsevier.
29. Burke, F. J. T., Fleming, G. J. P., Nathanson, D., & Marquis, P. M. (2002). Are Adhesive technologies needed to support ceramics? An assessment of the current evidence. *The Journal of Adhesive Dentistry*, 4(1), 7–22.
30. McLean, J. W. (1979). *The science and art of dental ceramics. Volume I: The nature of dental ceramics and their clinical uses.*
31. Craig, R. G. (1989). *Restorative dental materials*. Mosby
32. Mowafy, O. M. El., & Watts, D. C. (1986). Fracture toughness of human dentin. *Journal of Dental Research*, 65(5), 677–681. doi:10.1177/00220345860650050901
33. Rosenstiel, S. F., & Porter, S. S. (1989). Apparent fracture toughness of all-ceramic crown systems *The Journal of Prosthetic Dentistry*, 62(5), 529–532. doi:10.1016/0022-3913(89)90073-5
34. Taira, M., Nomura, Y., Wakasa, K., Yamaki, M., & Matsui, A. (1990). Studies on fracture toughness of dental ceramics. *Journal of Oral Rehabilitation*, 17(6), 551–563.
35. DeLong, R., Sasik, C., Pintado, M. R., & Douglas, W. H. (1989). The wear of enamel when opposed by ceramic systems. *Dental Materials*, 5(4), 266–271.

36. Cate, A. R. (1980). Ten. Oral histology: Development, structure and function.
37. Miles, A. E. W. (1967). Structural and chemical organization of teeth.
38. Linde, A. (1984). *Dentin and dentinogenesis*. CRC Press.
39. Rasmussen, S. T., Patchin, R. E., Scott, D. B., & Heuer, A. H. (1976). Fracture properties of human enamel and dentin. *Journal of Dental Research*, 55(1), 154–164. doi:10.1177/00220345760550010901
40. Lin, C. P., Douglas, W. H., & Erlandsen, S. L. (1993). Scanning electron microscopy of type i collagen at the dentin-enamel junction of human teeth. *Journal of Histochemistry & Cytochemistry*, 41(3), 381–388. doi:10.1177/41.3.8429200
41. White, S. N., Miklus, V. G., Chang, P. P., et al. (2005). Controlled failure mechanisms toughen the dentino-enamel junction zone. *The Journal of Prosthetic Dentistry*, 94(4), 330–335. doi:10.1016/j.prosdent.2005.08.013
42. Fong, H., Sarikaya, M., White, S. N., & Snead, M. L. (2000). Nano-mechanical properties profiles across dentin–enamel junction of human incisor teeth. *Materials Science and Engineering: C*, 7(2), 119–128. doi:10.1016/S0928-4931(99)00133-2
43. Marshall, G. W., Balooch, M., Gallagher, R. R., Gansky, S. A., & Marshall, S. J. (2001). Mechanical properties of the dentinoenamel junction: AFM studies of nanohardness, elastic modulus, and fracture. *Journal of Biomedical Materials Research*, 54(1), 87–95.
44. Rasmussen, S. T. (1984). Fracture properties of human teeth in proximity to the dentinoenamel junction. *Journal of Dental Research*, 63(11), 1279–1283. doi:10.1177/00220345840630110501
45. Xu, H. H. K., Smith, D. T., Jahanmir, S., et al. (1998). Indentation damage and mechanical properties of human enamel and dentin. *Journal of Dental Research*, 77(3), 472–480. doi:10.1177/00220345980770030601
46. Efflandt, S. E., Magne, P., Douglas, W. H., & Francis, L. F. (2002). Interaction between bioactive glasses and human dentin. *Journal of Materials Science. Materials in Medicine*, 13 (6), 557–565.
47. Zhang, K., Ma, Y., & Francis, L. F. (2002). Porous polymer/bioactive glass composites for soft-to-hard tissue interfaces. *Journal of Biomedical Materials Research*, 61(4), 551–563. doi:10.1002/jbm.10227
48. Rousseau, C.-E., & Tippur, H. V. (2001). Dynamic fracture of compositionally graded materials with cracks along the elastic gradient: Experiments and analysis. *Mechanics of Materials*, 33(7), 403–421. doi:10.1016/S0167-6636(01)00065-5
49. Park, S., Quinn, J. B., Romberg, E., & Arola, D. (2008). On the brittleness of enamel and selected dental materials. *Dental Materials*, 24(11), 1477–1485. doi:10.1016/j.dental.2008.03.007
50. Du, J., Niu, X., Rahbar, N., & Soboyejo, W. (2013). Bio-inspired dental multilayers: Effects of layer architecture on the contact-induced deformation. *Acta Biomaterialia*, 9(2), 5273–5279. doi:10.1016/j.actbio.2012.08.034
51. Yang, J., & Xiang, H.-J. (2007). A three-dimensional finite element study on the biomechanical behavior of an FGBM dental implant in surrounding bone. *Journal of Biomechanics*, 40 (11), 2377–2385. doi:10.1016/j.jbiomech.2006.11.019
52. Traini, T., Mangano, C., Sammons, R. L., Mangano, F., Macchi, A., & Piattelli, A. (2008). Direct laser metal sintering as a new approach to fabrication of an isoelastic functionally graded material for manufacture of porous titanium dental implants. *Dental Materials*, 24 (11), 1525–1533. doi:10.1016/j.dental.2008.03.029

53. Paulino, G. H., Jin, Z.-H., & Dodds, R. H. (2007). 2.13-failure of functionally graded materials. In *Comprehensive structural integrity* (pp. 607–644). Elsevier Ltd.
54. Kim, J.-H., & Paulino, G. H. (2003). An accurate scheme for mixed-mode fracture analysis of functionally graded materials using the interaction integral and micromechanics models. *International Journal for Numerical Methods in Engineering*, 58(10), 1457–1497. doi:10.1002/nme.819
55. Walters, M. C., Paulino, G. H., & Dodds, R. H. (2004). Stress-intensity factors for surface cracks in functionally graded materials under Mode-I thermomechanical loading. *International Journal of Solids and Structures*, 41(3–4), 1081–1118. doi:10.1016/j.ijsolstr.2003.09.050
56. Niu, X., Rahbar, N., Farias, S., & Soboyejo, W. (2009). Bio-inspired design of dental multilayers: Experiments and model. *Journal of the Mechanical Behavior of Biomedical Materials*, 2(6), 596–602. doi:10.1016/j.jmbbm.2008.10.009
57. Niu, X. (2008). *Contact damage of dental multilayers*. Princeton University.

7 Bionic Organs

Kiavash Kiaee, Yasamin A. Jodat, and Manu Sebastian Mannoor

7.1 Introduction

Defined as the interface of biology and electronics, "bionics" is the science of integrating electronic devices with biological systems to construct hybrid systems that can restore the full functionality of an impaired biological organ or provide additional features and augmented capabilities (Figure 7.1). Indeed, the main goal of designing bionic organs is to restore the original functionality or replace the anatomical defects with enhanced abilities, such that the resulted hybrid would be able to assist humans in highly complex or hazardous tasks. Despite common artificial organs with merely mechanical and electronic elements, bionic organs consist of both mechanical and cellular components coupled in order to regenerate organ architecture and function, and tissue regrowth [1].

The idea of bionic systems goes back to ancient history, when primitive versions of implants included plates, screws, and metallic materials that were used to replace an amputated limb or provide fixation of a bone after a fracture. This idea has been evolved drastically since the introduction of novel fabrication methods, electronics, and biocompatible materials. The new approach to bionic organs is to create customized human organs that potentially outperform previous organ capabilities by taking advantage of engineered nanomaterials and tissue engineering techniques, as well as recent advanced technologies such as 3D bioprinting. These new capabilities enable a new generation of functional electronics that are tailored specifically for integration with biological systems and present tremendous opportunities in many fields, including regenerative medicine, medical monitoring, human–machine interfaces, prosthetics, and augmented bionics.

However, unsuitability of traditional electronic elements for biological integration, fundamental differences between structure and formation of man-made versus biological entities, and our limited understanding of advantageous or adverse interactions between these two systems pose challenges that we will further discuss in this chapter.

7.1.1 Integration Challenges

To restore or augment a human capability, a bionic organ should be designed in such a way that the assimilation with the human body does not result in any side effects, discomforts, or inflammations, nor should it provoke a foreign-object rejection response

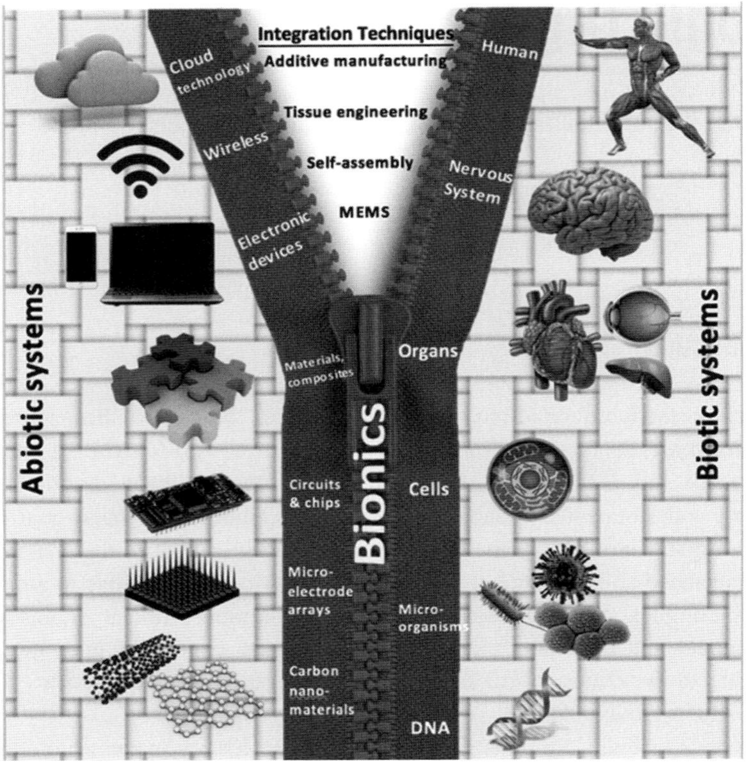

Figure 7.1 The field of bionics develops a multidimensional interaction between engineered materials and biological entities to address human needs. The degrees of integration and levels of achievement in bionic organs and devices are illustrated here.

from the body [2–4]. Due to various disparities and structural gaps between the biological and nonbiological systems, reaching a seamless integration is still a major challenge in designing bionic systems and devices. A major disparity between biological systems and nonbiological elements stems from the elemental structure and properties. In fact, biological systems are dominated by light elements such as carbon, nitrogen, oxygen, hydrogen, and calcium. Unlike in inorganic systems, where elements such as iron, chromium, and nickel are abundant, biological systems are mainly limited to metallic ions. For instance, iron found in red blood cells (RBCs) exists in ionic forms that are bound to hemoglobin, where it has a chemical functionality rather than a mechanical one. Moreover, the structure of biological systems consists of mainly polymers or composites of polymers and ceramic particles.

Another key difference between abiotic engineered and biological systems is the formation process [5–7]. Biological systems and materials are *grown* (not made) into a whole organism (plant or animal) via self-assembly, guided by the principles of developmental biology. As a consequence, biological systems are, in general, able to reconfigure and adapt to environmental changes and self-heal when damaged. In addition, the biologically controlled self-assembly driven growth takes place at or near room temperature and at atmospheric pressure under common physiological conditions [8]. In contrast, abiotic electronic or mechanical systems are *fabricated* from selected materials based on secure engineering design and considering extreme conditions. This design-driven fabrication process often involves high temperatures, pressure, and what are considered to be "biologically harsh" chemical conditions.

Moreover, the design of high-quality electronic materials is highly impacted by the mechanical deficiencies and lifetime issues such as fatigue and failure. The biological systems, however, are capable of regeneration and reconstruction owing to the ability of constant renewal and growth encoded in the system. This limitation arises from the considerable disparity in mechanical properties between the two categories. For instance, the Young's modulus in electronics range from 1 to 100 GPa [9], whereas the Young's modulus of organics such as skin ranges between 0.1 and 1 MPa [10,11].

Furthermore, despite the two-dimensional and planar structures of electronics and inorganics, the natural growth in biological systems allows for extremely complicated three-dimensional integration of various functional materials into a single system. Organs of animals and parts of plants such as leaves represent very good examples of such hierarchical integration of materials and structures achieved via the growth process. As a consequence, biological systems can also be functionally graded with hierarchical structures, a property that is relatively difficult to achieve via conventional fabrication methods.

Figure 7.2 summarizes some of the key differences in properties and functionalities between the two classes of materials and systems originating from the fundamental disparity in the *growth* and *fabrication* processes [12].

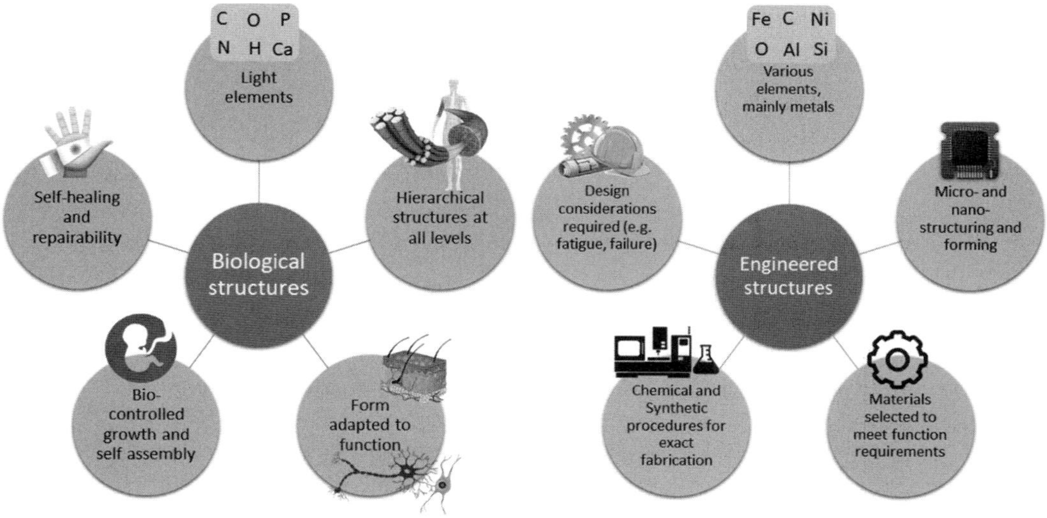

Figure 7.2 Dichotomy between biological and engineered systems stemming from material properties, composition, and formation.

7.1.2 Addressing the Challenges with Innovative Approaches

To address the aforementioned issues and integration challenges, a variety of strategies have been established. Self-assembly is one of the biomimetic methods that bridges the gap between the top-down and bottom-up approaches by programming autonomous assembly of building blocks when the parts are either too small or too complex to be compiled conventionally. Other approaches include the constantly improving attempts to achieve biocompatible materials and nanoscale electronics by various techniques of chemical functionalization or modification. Carbon nanotubes (CNTs), graphene, and quantum dots (QD) are examples of the nanomaterials that have exhibited excellent size- and shape-dependent electronic and mechanical properties [13]. Herein, we will focus on 3D bioprinting, a new technique that has revolutionized the world of electronic implants by offering higher levels of organ mimicry.

7.2 3D Bioprinting: One Step Closer to Reality

These strategies utilize the combined strength of modern imaging and 3D modeling techniques, advanced materials and cell culture systems, and a plethora of 3D bioprinting processes in a workflow that can be broadly categorized into four stages: (i) Computer-assisted imaging and design, (ii) preparation of biomaterial and biological components, (iii) bioprinting process, and (iv) tissue maturation and conditioning (Figure 7.3).

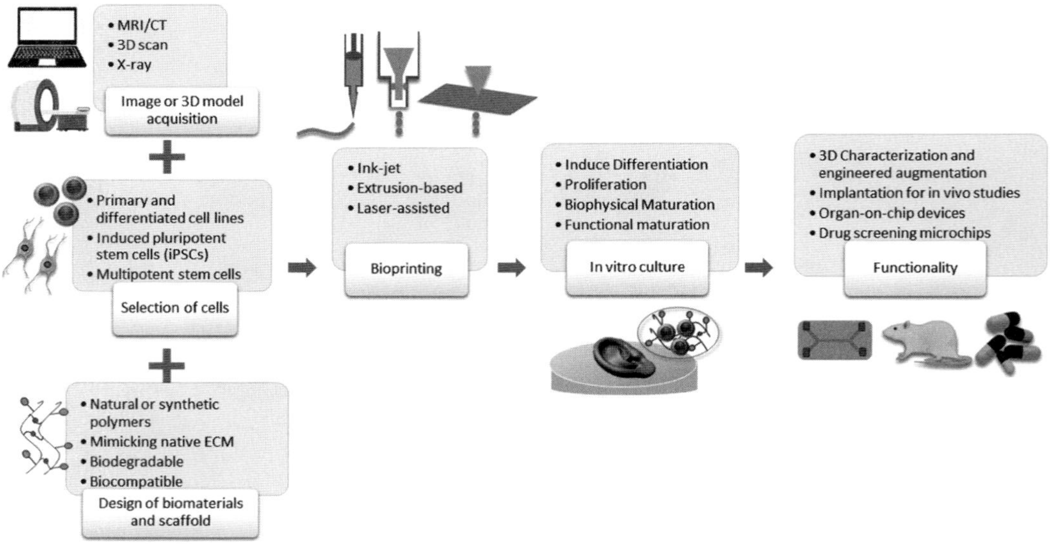

Figure 7.3 A typical bioprinting procedure, includes different steps: (i) Computer-assisted imaging and model design, (ii) engineering and design of biomaterials and scaffolding, (iii) bioprinting process, (iv) tissue culture and maturation, (v) functional testing.

7.2.1.1 Computer-Assisted Imaging and Design

A key component to successful bioprinting is obtaining an accurate and functional design. The irregular and complex 3D structure of tissues and organs on different scales represent challenges that can only be surmounted using advanced medical imaging and computer aided design and manufacturing (CAD-CAM) techniques. Widely used imaging modalities for this purpose include magnetic resonance imaging (MRI) and computed tomography (CT), although alternatives such as ultrasound and laser scanning and software-based image processing techniques are abundant and are becoming increasingly more attractive considering their affordability, flexibility, and improved accuracy [14,15]. After acquisition of imaging data, 2D cross-sections and various 3D representations including wire-frame, solid objects, contours, and meshes can be generated utilizing CAD-CAM techniques [16]. These flexible and powerful models are then used in clinical applications to design the desired structures for bioprinting.

7.2.2 Preparation of Biomaterials and Biological Components

7.2.2.1 Required Material Properties of an Ideal Bioink

Most 3D bioprinting processes require a meticulously selected and engineered mix of biomaterials and biological components called bioink. Before diving into the realm of many materials and chemicals from various origins that are used in this process, however, it is helpful to understand what constitutes a suitable biomaterial and what requirements are recognized. Most importantly, biomaterials are biocompatible. Unfortunately, biocompatibility is often used as a broad and at times ambiguous term, so we will attempt to provide a concise but clear definition. Biocompatibility is granted when adverse local and systematic reactions from the living target organism or tissue is limited under a certain threshold even after long-term exposure. In practice, this is assessed through vigorous testing of local, systematic, tumorigenic, reproductive, and developmental effects of the candidate material employing different methods and assays, including analytical chemistry, genetic toxicology, and cytotoxicity assays. These methods test for irritation reactivity, immunogenicity, pyrogenicity, subacute or subchronic toxicity, genetic toxicology, and implantation and hemocompatibility. Numerous regulatory guidelines govern this process with an important example being ISO 10993.

However, since in many bioprinting applications biomaterials are coupled with living cells, conditions other than biocompatibility should also be satisfied, including: (i) adequate porosity to allow cell migration as well as transfer of oxygen, signal molecules, nutrients and waste material to and away from cells; (ii) degradability at a suitable rate, optimally at the same rate as regeneration, to allow new tissue formation; and (iii) mechanical properties comparable to native tissue, allowing prints to hold the established shape while still being flexible and stretchable enough to avoid implantation and long-term problems.

Lastly is the issue of printability. As we will see when bioprinting processes are discussed in more detail in the next section, biomaterials are often passed through a nozzle during printing and are laid down in precise and predetermined patterns, layer after layer to form the final structure. The first step, passing through a nozzle, requires the material to be in liquid form, whereas after being laid down, a more permanent form is required, otherwise we are left with a shapeless blob of liquid.

In regular 3D printing processes, this is achieved by raising the printing temperature slightly above the melting temperature of the material, which yields a fluid that can be drawn from the nozzle but is still viscous enough to retain its shape before it is rapidly cooled and transitions into solid form. This is not possible in bioprinting, however, since in many processes cell-laden biomaterials need to be printed, and cells are extremely sensitive to temperature and will not endure high temperatures. A routine strategy to overcome this problem is adding hydrogel materials to the bioink. These polymers can be prepared as liquids with tunable viscosity but can also be induced to transition into hydrogel form, through a process called cross-linking. This can be achieved through different mechanisms based on the material, including physical, ionic, chemical, thermal, or stereo complex cross-linking. Another advantage is that mechanical properties of hydrogel materials are often highly tunable and can be made to closely approximate that of the target tissue.

7.2.2.2 Materials

So far, we have established certain requirements for materials to be used in bioprinting, namely, biocompatibility, porosity, biodegradability, suitable mechanical properties, and printability. Through the years, researchers have identified and isolated various materials for this purpose and most, if not all, are polymers with two major categories, natural polymers and synthetic polymers.

Natural polymers are naturally occurring and are often isolated from plant and animal products. They include proteins such as collagen, gelatin, fibrin, fibronectin, and thrombin, as well as polysaccharides such as alginate, hyaluronic acid, chitin, chitosan, and cellulose. Synthetic polymers include polyethylene glycol (PEG), polycaprolactone (PCL), and polyglycolic acid (PGA).

Nevertheless, no single material has been identified as the gold standard for 3D bioprinting, and this is usually because of the very specific and application-based requirements. For instance, alginate, a natural biopolymer often isolated from cell walls of brown algae, can be mixed with water to form a viscous printable solution that will undergo gelation upon exposure to calcium ions. The gel properties can be easily tuned to mimic many native tissues and the gelation process is fast and easy. Moreover, the material is porous, biocompatible, and biodegradable, making it an excellent choice for bioprinting. However, cell adhesion and growth is usually very poor in alginate [17], thus chemical modification or additive substances are required. One successful strategy has been to functionalize alginate with immobilized arginylglycylaspartic acid (RGD) peptides to promote cell adhesion and growth [18,19]. This is a recurring theme, as material selection is an area of ongoing research in bioprinting and researchers often mix and modify different natural and synthetic polymers to achieve excellent mechanical properties and favorable biological interactions at the same time.

7.2.3 3D Bioprinting Processes

After preparation of bioink, the material has to be formed and cross-linked into the designed geometry with sufficient special resolution.

7.2.3.1 Ink-Jet Printing

Inkjet printing is a rapid, widely used, and cost-effective bioprinting method where droplets of bioink are ejected from a nozzle. This is achieved by one of two mechanisms, thermal or piezoelectric. In thermal systems, a filament is rapidly heated by an electric pulse which creates pulses of vapor pressure that drive the bioink droplets. Since the heating is localized and momentary, lasting only a few microseconds, even mammalian cells that are extremely sensitive to heat and pressure have been printed without significant adverse effects or loss of viability [20–23]. In piezoelectric systems, droplet formation and ejection is achieved via rapid volume changes in a piezoelectric crystal induced by electrical pulses [24]. Advantages of using piezoelectric systems include higher control over several important factors such as stresses experienced by cells during droplet generation and impact, droplet size and velocity, and generation rate. Potential challenges include cell agglomeration and sedimentation during the printing process, so bioink stability has to be ensured for the duration of the print [24], and limited range of bioink viscosities that can be processed.

7.2.3.2 Laser-Assisted Bioprinting

Another approach for 3D biofabrication is termed *laser-assisted biofabrication*. This method is based on laser-induced forward transfer, a process proposed in 1986 by Bohandy et al. [25] for direct writing of metals from thin-film substrates using laser pulses. In the format that is used today for bioprinting, a slide is covered with a thin film of conductive material, usually metal thin films. This film is then covered with cell-laden bioink. Spatially and temporally controlled laser pulses on this donor slide will cause a sharp local temperature rise that leads to pressure pulses of evaporated metal, trapped air, or water vapor from ink. This pressure drives a very small droplet of bioink from the donor slide to another slide, where the printed structure is being formed. Although the need for precisely controlled laser pulses means that these systems are often more expensive than alternative bioprinting processes, its many advantages have led to increased use of this method in laboratory settings. These benefits include very high resolution, the ability to process bioinks of a wide range of viscosities, and most importantly, nozzle-free application, which means there is no potential for clogging, which is a major challenge with other bioprinting methods, including extrusion-based and inkjet printing.

7.2.3.3 Extrusion-Based Bioprinting

After the success of extrusion-based 3D printing systems, also known as fused deposition modeling or FDM, an ever-increasing pallette of materials are being 3D printed using similar extrusion-based methods, and to date, biological materials such as hydrogels, cell aggregates, synthetic polymers, and more have been successfully printed using this method. The heart of the system consists of a robotically controlled nozzle with precisely controlled movements in three dimensions and controlled rates of material deposition. In many systems, especially those designed to process synthetic polymers, the nozzle temperature is also controlled and can be raised to about 300°C, but this is normally avoided when treating heat-sensitive biological materials. The extrusion mechanism can be either pneumatic or mechanical, using pistons and screws [16]. The simplicity and low cost of these systems have made them widely popular among researchers. Other important considerations leading to success of extrusion-based bioprinting include high spatial resolution and the ability to process bioink with viscosities ranging from 30 mPa/s to 6 × 107 mPa/s and very high cell densities [16]. However, currently all process parameters, including nozzle diameter, printing speed, and dispensing rate have to be optimized for every new bioink formulation, which is a burdensome endeavor. Also, frequent clogging of the nozzle, high shear stresses on cells which can lead to lower cell viability compared to other methods, and limited cartridge capacity disrupting continuous prints are other challenges that need to be addressed before these bioprinting methods find widespread applications beyond research laboratories.

7.3 Bionic Organs

The following section will discuss the structure, properties, and challenges of some of the bionic organs developed to date with a focus on the sensory organs.

7.3.1 Bionic Eye

A major target of bionics is treating vision impairments such as glaucoma, age-related macular degeneration (AMD), diabetic retinopathy, cataracts, and retinitis pigmentosa. Since the 1950s, great efforts have been dedicated to treating visual impairments in several different ways. A common pathway is using bionic devices and prostheses to induce the neuronal activity by compensating for the function of the disabled area to propagate the signal across the remaining visual pathway, resulting in visual perception. The structure of the human eye is designed such that the light can be received and processed through visual contact. Specifically, light enters the eye through the cornea, a transparent layer over the eye which refracts the light, and travels through the pupil to the lens and is eventually projected on the retina, an area containing photoreceptor cells. These comprise rod-shaped and cone-shaped cells, and they are responsible for conscious light perception and color differentiation. Next, the visual signals are transduced to neuronal signals and processed inside the retinal neuronal networks. Finally, these signals are transferred to the visual cortex through the ganglion cells axons in the optic nerve. To date, several attempts have been made to address eye problems based on the extent and cause of the impairment. Thus, the site of stimulation in these devices varies by illness and the specific section that is affected. Some possible sites of the stimulation include the retina, optic nerve, lateral geniculate nucleus, and visual cortex.

7.3.1.1 Retinal Implants

In patients with AMD or hereditary retinal degeneration, such as retinitis pigmentosa, the retina degenerates gradually, often leading to full blindness in adults; however, the visual pathway beyond the retina is often intact and remains chiefly functional. In these cases, a retinal implant would be sufficient to restore functional levels of vision. Some types of retinal implants include: Argus II, Boston Retinal Implant Project, Epi-Ret 3, Intelligent Medical Implants (IMI) and Alpha-IMS (Retina Implant AG). In the first four devices, the image is captured from the environment using external cameras. Next, the cameras transmit the visual information to the implanted microelectrode array consisting of 25–100 electrodes, the number and density of which determines the pixel resolution [26]. The last type of retinal implant (Alpha-IMS) does not rely on external cameras to acquire images. Instead, the device consists of a multiphotodiode array. Each photodiode is sensitive to the amount of light entering the retina and relays the visual signal to the amplification circuit and electrodes, which then transmit the charge to the adjacent retinal layers. Propagating the signal towards the bipolar cells, the surviving visual pathway will take on the signal processing afterwards [27]. Based on the type of defect, the stimulation site can target different types of cells, and thus, the electrodes can be placed in different retinal sites, such as epiretinal, intrascleral, suprachoroidal, and subretinal.

Each of the aforementioned implants have different clinical availability, ranging from merely animal availability (Boston Retinal Implant), to clinical trials (Epi-Ret 3, IMI), to FDA approval (Angus II) [26]. A major advantage of retinal implants is the employment of the surviving cells of the visual pathway to accommodate visual perception. However, because this procedure is highly dependent on the viability of the inner retinal neuronal networks, defective or lost inner retinal system (e.g., glaucoma and optic neuropathy) would impede the effectiveness of the retinal implants and demand other practices [28].

7.3.1.2 Cortical Implants

In deeper levels of visual impairment, the damage to the inner retina or optic nerve disrupts the visual pathway in such a way that a retinal prosthesis will not be effective, and cortical stimulation is required to reach levels of vision functionality. Since the first cortical implant in 1968 by Brindley and Lewin [29], studies have attempted to use cortical penetrating electrodes to stimulate the occipital pole and the surrounding area in the visual cortex [30,31]. Despite effectiveness, the clinical practice of cortical implantation still remains limited [32–34] since, not only is the procedure highly invasive, but it also contains major infection and inflammation risks as well as requiring careful postimplantation visual rehabilitation [35].

7.3.1.3 Challenges and Novel Approaches

Aside from invasiveness and recovery challenges, a major challenge in visual implants is maximizing the visual acuity and the resulting perception of the image. People with healthy visual systems have the ability to differentiate between small changes in light density and spatial patterns as a result of the small receptive field of a single photoreceptor [36]. On the other hand, in implant-based stimulations, the image acuity is highly dependent on the number of electrodes. Moreover, the cross-talk between adjacent electrodes as well as the possibility of tissue damage are other challenges in visual implants [35,36]. Recent research has been focused on employing other methods to restore levels of visual impairment by neural stimulation, some of which include optogenetics, alternating magnetic fields, neurotransmitter injection, and altering ionic gradients across neural cell membranes [28]. Although still very young, these novel methods could take the visual bionics to the next level by better imitating human vision in the future. Recently, Ruiters et al. [37] proposed the first customized ocular prosthesis designed with the aid of a 3D-printed impression-free mold of the anophthalmic cavity. During this process, after acquiring a cone beam CT scan (CBCT), a 3D model of the anophthalmic cavity was designed and 3D printed and fitted in a 68-year-old male suffering from a long-standing sore blind eye. Although demonstrating preliminary results in a single case report, the 3D printing technology is known to have ample potential to improve and customize ocular prosthetics.

7.3.2 Bionic Ear

Targeting deafness and auditory impairments for a long time, bionic devices have made a great contribution toward treatment of these disorders. This section will describe these improvements by first discussing the hearing mechanism and the reasons behind hearing impairments. Next, a brief review of cochlear implants will be provided, followed by a novel approach, which has moved further toward achieving a fully functional bionic ear.

When sound impulses strike the ear, the resulting mechanical vibration travels through the ear canal to the tympanic membrane whose vibration conveys the sound waveform to the inner ear through the oscillation of three bones in the middle ear called auditory ossicles. The oscillation is then transferred to the inner ear which consists of a coil-shaped fluid-filled membrane known as cochlea. Propagation of vibrations through the cochlear fluid is then followed along the basilar membrane (BM), a tonotopic structure with varied width and stiffness along the cochlea which acts as a base for auditory receptor hair cells. Equipped with "stereocilia" mechanosensors, the hair cells produce an action potential upon deflection as a result of BM movement. The signal is then transmitted to the brain via the auditory nerve for further processing.

In most cases of the hearing loss, the hair cells are damaged or destroyed, whereas the remaining auditory pathway is often intact and functional. Auditory impairment can stem from a variety of causes, including genetic mutations and hereditary deafness, aging, consumption of drugs (e.g., gentamicin, cisplatin), physical trauma, long-term exposure to noise, and infectious diseases and disorders (e.g. meningitis, herpes viruses, or syphilis). As shown in Figure 7.4a, partial or complete absence of hair cells in a defective ear leads to the loss of connection to the rest of the auditory nerve and the central nervous system. The neuronal section between the spiral ganglion cells and the inner ear usually undergoes retrograde degeneration and gradually loses its functionality [38].

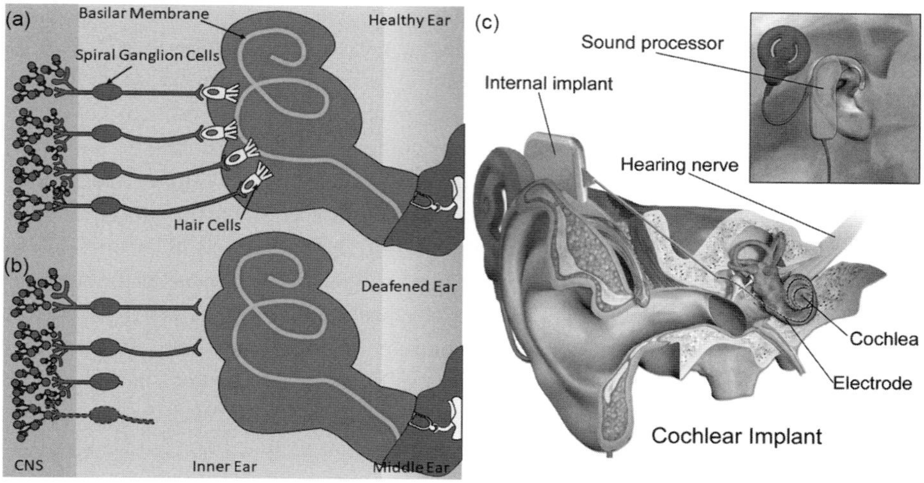

Figure 7.4 Structural differences in the inner ear: (a) Normal and (b) deafened ear. The hair cells are partially or totally lost in the deafened ear, but the remaining path (spiral ganglion cells and peripheral neural parts) is still partially viable. (c) Different components of a cochlear implant. (Medical gallery of Blausen Medical 2014, *Wikijournal of Medicine*.)

7.3.2.1 Cochlear Implants

With the advent of cochlear implants in 1980s, restoring the sense of hearing has strikingly evolved – from a simple sound cadence sensor, aiding lip-reading, to a functional multichannel prosthesis with signal processing capabilities, owing to the numerous advances in biotechnological techniques and biomaterials. An alternative to the dysfunctions of the natural auditory system, cochlear implants are the convertors of mechanical acoustic stimulation to electrical signals, which are designed to circumvent the dysfunctional hair cells, thus avoiding the neuronal disconnection by targeting the remaining neural pathway and spiral ganglion cells (Figure 7.4c). A cochlear implant is composed of an external and an internal implanted component [38]. The external component consists of a wearable sound processor and microphone, which transmits the auditory information to the internal receiver through a transcutaneous module. The interior platform consists of a receiver for decoding the received information, and a biomimetic tonotopic bipolar electrode array, which is inserted into the scala tympani (ST) part of the cochlea via surgery. The electrode array is designed to replicate the mechanism of the BM through a spatially differentiated pattern which, based on the frequency of the stimuli, can stimulate different sets of neurons by a certain subset of electrodes. By directly stimulating the ganglion cells, the electrodes compensate for the dysfunctional ties of the inner ear and trigger the function of the remaining auditory nerve to the brain.

7.3.2.2 Challenges

Known as one of the most successful bionic prostheses, cochlear implants are able to restore levels of hearing. However, several challenges have obstructed these implants from being a fully functional, seamless bionic alternative. For instance, these implants cannot target patients whose auditory nerve is completely destroyed or when implantation inside cochlea is not possible. To address this issue, a new type of alternative prosthesis is proposed, the auditory midbrain implant (AMI) [39], which benefits from electrically stimulating the inferior colliculus of the midbrain.

Furthermore, invasiveness, required surgery, device failure, and infections are some of the major risks of cochlear implants, some of which may necessitate additional revision surgeries after implantation [40]. Moreover, cochlear implants fail to address the physical damage to the auricular cartilage of the ear due to trauma, and the external part of the implant is placed on a physically viable ear auricle. Therefore, the need for a physical bionic ear still remains an ongoing challenge due to the complexity of the ear geometry and anatomy, which are difficult to reconstruct via the conventional tissue engineering methods. However, cartilaginous tissues such as the ear auricle, benefit from a simpler tissue structure and less vasculature [41,42] and thus are easier to be targeted through novel 3D bioprinting strategies to target the feasibility of the bionic ear.

7.3.2.3 Novel Approaches: 3D-Printed Bionic Ears

To address the various challenges of common cochlear implants, a conceptually new approach has been introduced, which interweaves functional electronic components with biological tissue via 3D printing of electronic materials and viable cell-seeded hydrogels in the precise anatomic geometries of the human ear [13,43]. Specifically, the bionic ear consists of a precisely 3D-printed auricle, and an electrically conductive silver nanoparticle (AgNP) infused inductive coil antenna, connecting to cochlear-like electrodes supported on silicone. To print the ear construct, a syringe extrusion-based 3D printer and a CAD drawing of the ear with the anatomical geometry and spatial material heterogeneity is employed (Figure 7.5a). Next, the 3D functional constituents (structural, biological, and electronic) are fed into the 3D printer (Figure 7.5b). To achieve a viable bionic structure, a bioink consisting of an alginate matrix pre-seeded with a viable concentration of chondrocyte cells is used to better mimic the natural ear cartilage structure. The bioprinted ear construct is then cultured *in vitro,* enabling cartilage tissue growth and fusing together to form a "cyborg ear," with expanded auditory senses of radio frequency (RF) reception provided by an inductive coil acting as a receiving antenna (Figure 7.5d and f). The bionic ear showed good structural integrity and shape retention under culture, as can be seen in Figure 7.5g, which illustrates the ear after 10 weeks of *in vitro* culture.

Exhibiting a cell viability of 91.26 $\pm$3.88% with homogeneous chondrocyte distribution, the post-printing viability tests proved that the printing process, including cell encapsulation and deposition, does not negatively impact chondrocyte viability. Moreover, to compare the morphology of chondrocytes in the neocartilage of the bionic ear to that of the native cartilaginous tissue, histological evaluations were performed using hematoxylin and eosin (H&E) and safranin O staining. Consequently, these assays revealed uniform chondrocyte distribution and uniform accumulation of proteoglycans in the cultured ear tissue, which were in line with the development of a new

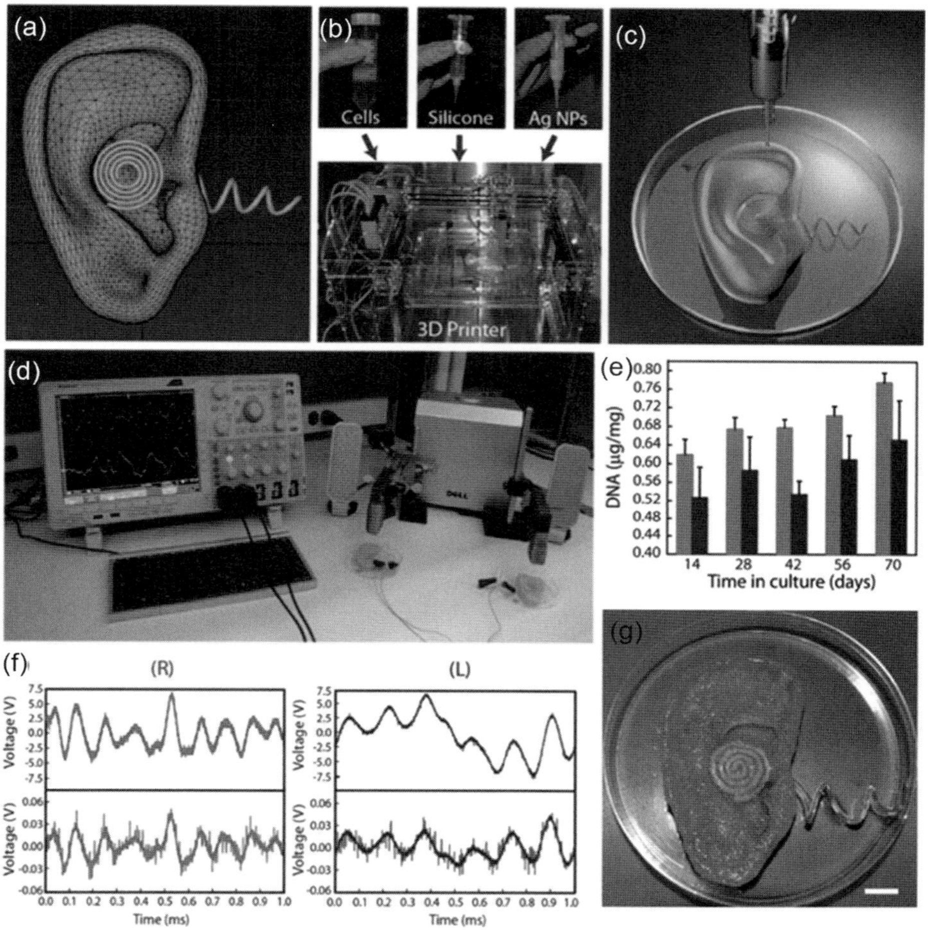

Figure 7.5 The 3D-printed bionic ear, where biological tissue components are interlinked with electronics. (a) CAD drawing of the developed ear. (b) (top) Three materials (chondrocyte cells, silicone, and electronic AgNP-infused silicone) constructing the bioink; (bottom) utilized 3D printer. (c) 3D illustration of the designed ear. (d) The pair of bionic ears exposed to stereophonic audio music. (e) DNA content in the 3D-printed ear at various stages, cultured with 20% FBS (gray) and 10% FBS (black). (Error bars depict standard deviation with $N = 3$). (f) Transmitted (top) and received (bottom) audio signals of the right (R) and left (L) bionic ears. (g) Morphology of the structure after 10 weeks of *in vitro* culture (Scale bar = 1 cm). (Reprinted with permission from Mannoor et al. [43], Copyright 2013 American Chemical Society.)

cartilage [44]. Furthermore, the biomechanical properties of the printed ear were characterized and the Young's modulus and the hardness of the grown cartilaginous tissue of the 3D-printed auricle verified extracellular matrix (ECM) development and improving mechanical properties in the developing tissue.

Current approach of printing a pre-seeded hydrogel matrix eliminates the major problems associated with seeding depth limitations and nonuniform seeding in traditional methods for seeding premolded 3D scaffolds. Moreover, it allows for localization of the cells to a desired geometry for ECM production in defined locations when cultured in nutritive media. Upon tissue development, the polymer scaffold is reabsorbed, and the new tissue retains the shape of the polymer. Another advantage of using the biodegradable scaffold is that it provides each cell with better access to nutrients and more efficient waste removal.

The functionality of the bionic ear was analyzed by performing a series of electrical characterizations and auditory experiments. In particular, a complementary left ear was simply printed by reflecting the original model. Next, left and right channels of stereophonic audio were exposed to the left and right bionic ear via transmitting magnetic loop antennas with ferrite cores. The signals received by the bionic ears were collected from the signal output of the dual cochlear electrodes and fed into a digital oscilloscope and played back by a loudspeaker for auditory and visual monitoring, exhibiting excellent reproduction of the audio signal. Notably, Beethoven's "Für Elise," which was played back from the signal received by the bionic ears, possessed good sound quality. Furthermore, the presented bionic ear enables receiving of electromagnetic signals in a frequency range (up to 5 GHz), which is well beyond the normal perceptible range of human acoustic hearing (20 Hz to 20 kHz).

Ultimately, the fabricated "cyborg ears" showed great potential in combining the versatility of additive manufacturing techniques with novel tissue engineering concepts. Thus, co-3D printing interlaced biological, structural, and electronic components represents a new, general strategy in merging electronics with biological systems. Such hybrids are distinct from either engineered tissue or planar/flexible electronics and offer a unique way of accomplishing a seamless integration of tissue and electronics to generate "off-the-shelf" cyborg organs.

7.3.3 Bionic Nose

With the ability to discriminate thousands of volatile compounds and chemical structures, the olfactory system plays an important role in assisting humans with perception of the outer environment. The olfactory epithelium consists of olfactory sensory neurons (OSN), which are neuron-sensory bipolar cells and consist primarily of olfactory receptors (OR). The chemical structure of the ORs allows the specific binding of each of these proteins to odorant compounds with similar chemical structure. As a type of G protein-coupled receptors (GPCR), ORs interact with G proteins in the plasma membrane of the cilia. After an odorant dissolves in the mucus and is bound to its specific OR in the cilia, the G protein is activated on the cytoplasmic side of the OSN cell, which in turn activates adenyl cyclase. Next, through a chain of chemical reactions inside the cell, cyclic adenosine monophosphate (cAMP) concentration is increased, yielding an action potential by opening cation channels and depolarizing the axon of the olfactory sensory neuron. Eventually, the chemical stimuli and impulse is transmitted to the brain for additional processing.

Olfaction impairment can result from injuries in various sections of the olfactory system from the nasal cavity to the brain [45]. The deficit can appear in the form of partial loss of detection (hyposmia), total loss (anosmia), or loss of odor discrimination or quality (dysosmia) [46]. Olfactory dysfunction has been associated with head trauma, aging [47], and neurodegenerative diseases such as Alzheimer's and Parkinson's [48].

Since 1982, when the first attempts were made by Persaud and Dodd [49] to classify odors using chemical sensors, several approaches have been developed to mimic the mechanism of odor detection and discrimination. Based on the type of elements used for detection, the strategies can be categorized into electronic or bioelectronic (bionic) noses.

The detection mechanism in both categories attempts to mimic the natural biological olfaction mechanism. Specifically, both consist of a sensor part, which acquires and measures the input odor data, and a converter part, which transduces the generated signals. Following a chemical procedure called functionalization, the surface of the sensor substrate is coated with chemical/biological elements specific to particular odorants and targets. When exposed to the target of interest, the contact between the odorant and the substrate leads to the generation of an electric signal. This signal can then be processed and visualized by appropriate methods of data analysis [50]. The detection is based on the type of transducer used; for instance: field-effect transistors (FET), surface plasmon resonance (SPR) sensors, conductive polymers, carbon nanotubes, and graphene.

Although following a similar sensing procedure, the structural differences in electronic and bionic noses causes numerous differences in capabilities and sensitivities. In fact, the core component of the conventional electronic nose is a chemical array sensor, whereas the bionic nose chiefly consists of biological elements. The lack of sufficient biomolecules in the electronic nose causes sensitivity limitations and thus the failure of these devices to precisely mimic the natural olfaction mechanism. This issue is especially problematic when the electronic nose fails to respond to low concentration levels of odor stimuli; even though the same concentration might be easily detected by the natural olfactory system.

7.3.3.1 Bioelectronic Nose

In the bionic nose, efforts are driven toward a more precise simulation of the natural odor detection mechanism to regenerate the signals that are most identical to those produced by native OSN cells. Since isolation and *in vitro* growth of OSNs is challenging, the ORPs are mainly used instead. Indeed, in a large category of bionic nose research studies, the principal strategy is to improve device selectivity and sensitivity by employing ORPs as the active detection element for odor detection. Specifically, the olfactory receptor genes are usually expressed in insect cells, such as *Spodoptera frugiperda* 9 (SF9) [51], or *Cercopithecus aethiops* (COS-7) [52], or in bacteria, such as *Saccharomyces cerevisiae* [53], or *Escherichia coli* [54], using viral vectors. Next, the nonbiological substrate of these devices is functionalized with the produced ORP via chemical procedures. Instead of transducing the sensory signal to a neural signal and transmitting it to the brain, the device transduces the generated signal to an understandable electrical or optical signal.

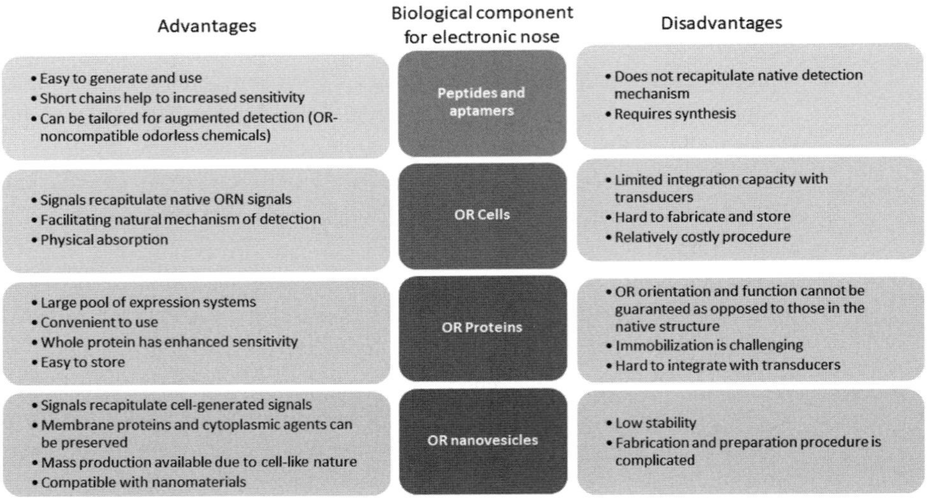

Figure 7.6 Comparison between different sensing elements used in bioelectronic noses. (Wasilewski et al. [50].)

This strategy facilitates noninvasive sample analysis and has high selectivity and relatively low cost. However, there are several limitations to this method. For instance, due to the special seven-helix structure of ORPs, which penetrates the native cell membrane, the correct functioning of ORPs requires a double-layered lipid membrane to imitate the high hydrophobicity in the transmembrane region. Immobilization of the lipid bilayer on the nonbiological substrate to provide the necessary hydrophilic–hydrophobic environment is complicated and difficult to achieve; thus the expression of olfactory receptors is considerably hampered in heterological systems [55]. As a result, the ORP-based bionic noses lack repeatability and reusability. To address this issue, several research groups have used synthesized short peptides as the functional sensing element instead of ORPs [56,57], and to better simulate the cell environment, other studies have reported the use of nanovesicles [58]. The advantages and disadvantages of various types of biological sources used for bioelectronic noses have been depicted in Figure 7.6.

7.3.3.2 Challenges

Although the emergence of the bionic nose has pushed the science forward toward artificial odor perception, there are still several challenges in reaching a full functional and implantable artificial nose. First, although the developed biosensors are capable of distinguishing their target odorants with high sensitivity and selectivity, they still fail to identify complex mixtures and reach a comprehensive odor perception system similar to that of humans [59]. In fact, the natural olfactory–taste system is stimulated by the combinatorial pattern recognition of various ORPs [60–62], and thus, the activation of several ORPs would lead to the perception of a specific odor or taste. Therefore, to reach a higher level of complex odor perception, bionic nose devices require the implementation of a complete set of natural olfactory receptors, which would jointly contribute to differentiation of diverse smells. Recently, a multiplexed biosensor has been proposed by Son et al. [59] that combines human olfactory and taste receptors with a multichannel carbon nanotube FET to analyze four types of taste- and odor-causing compounds produced by food.

The second challenge in reaching a functional implantable bionic nose involves the geometrical complexities of nose anatomy. Currently, two distinct research communities have dedicated efforts to nose organ reconstruction. In the first group, which mostly includes the aforementioned bionic nose sensor devices, the geometry and structural features of a natural nose have been neglected, and most of the research has been dedicated to the simulation of the odor perception instead. On the other hand, the second community of researchers has merely focused on the modification or reconstruction of the anatomy of nose for cosmetic purposes or after the organ has been physically damaged by an accident or other condition. A considerable challenge in this area involves the effective reconstruction of nose anatomy, given the unique geometries of nasal cartilage structures which necessitate a high degree of precision in implant fabrication. Currently, the major method for reconstruction after rhinectomy relies on transposition of autologous mucosal flaps plus cartilage grating and coverage using a skin flap. An alternative approach involves the use of tissue engineering to replace the entire nasal cartilage. However, these approaches do not include retention of odorant perception. Therefore, an optimized and effective bionic nose necessitates targeting both challenges simultaneously and comprehensively.

7.3.3.3 Novel Approaches: 3D-Printed Bionic Nose with Electronic Olfactory Epithelium

As formerly discussed, current bioelectronic noses fail to address the reconstruction issues of nose anatomy. As the 3D bioprinting technology continues to improve, a noninvasive and highly precise approach to repair nasal injuries sounds more promising, and the integration of such technology with biosensor can open new doors to restoring olfaction. A novel proof of concept has been recently suggested to address this problem, as well as retaining the odor perception [63]. Specifically, this method presents 3D-printed nostrils consisting of graphene biosensors that have been biotransferred onto the nose construct, creating a viable human nasal cartilage (Figure 7.7a–c). To reconstruct the biological structure of the nose, a viable density of chondrocyte cells isolated from the articular cartilage of 1-month old calves was cultured and seeded with an alginate hydrogel matrix to construct the bioink for 3D bioprinting (Figure 7.7e). Finally, a viable 3D-bioprinted nose construct was successfully achieved, ready to be integrated with the functionalized graphene biosensor. The developed 3D-bioprinted nose is shown in Figure 7.7f.

7.3.3.3.1 Augmentation of the Olfactory System

When compared to animals and insects, the olfactory system of humans is highly limited and less sensitive due to the lower number of receptors and respiratory tracts [64–66]. Functionalized with a variety of protein–ligand binding forms, including

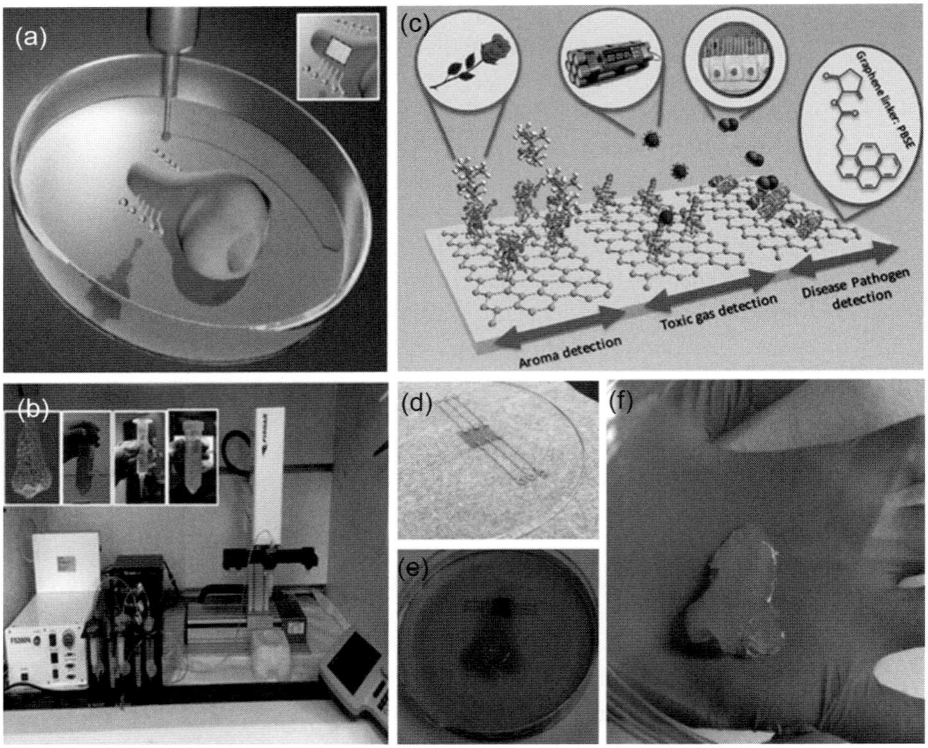

Figure 7.7 The 3D-printed bionic nose. (a) Schematic of 3D-printed bionic nose. (b) 3D printing setup (inset shows CAD drawing and various constituent materials). (c) Illustration of the prospective target analytes, which the augmented bionic nose is capable of detecting (a natural odor, an explosive toxic, and a disease biomarker) (d) Functionalized graphene sensor with different biosensing agents. (e) Integrated printed structure and biosensor in DMEM culture medium. (f) 3D-printed bionic nose construct after *in vitro* culture.

peptide–bacteria, antibody–virus, and peptide–chemicals, the developed bionic nose can be easily "tuned" to "sniff out" a variety of targets, such as odorants, specific disease indicative biomarkers, explosives, and toxics which cannot be detected by the native human nose (Figure 7.7c). As a demonstrative design, three targets of interest were selected for detection: an explosive chemical: 2,4,6-trinitrotoluene (TNT); an airborne virus: influenza virus; and a respiratory tract pathogen: *Staphylococcus aureus* bacteria, encompassing a varied spectrum of targets. To detect each target, chemical vapor deposition (CVD) grown graphene films transferred on a polydimethylsiloxane (PDMS) substrate were functionalized with the corresponding peptides or antibodies and were exposed to the ligand in a custom-made chamber with controllable flow rate (Figure 7.7d). The results from sensing experiments proved the ability of the bionic nose to selectively detect the target molecules.

Another major functionality imparted by the integrated electronic nanosensor nostrils of the bionic nose is the ability to provide an on-body health quality monitoring via "sniffing" breath to provide an early, first-order monitoring and detection of disease conditions. Monitoring dynamic characteristics of respiration, including the presence of biomarkers and volatile organic compounds (VOCs) in exhaled breath, is of growing interest for noninvasive disease diagnosis. For example, by functionalizing these sensors with antibody agents, detection of key asthma-related target analyte cytokines, including TNF-α, IL-4, and IL-5, with very low detection limits could be achieved. By sensing these key cytokines of asthma from exhaled human breath, the bionic nose could eventually promote accurate and noninvasive diagnosis to provide early warning of an impending asthma attack.

In addition to these potentially revolutionary technological applications, the creation of a bionic nose may allow for a more fundamental understanding of the mechanisms employed by the natural olfactory system in achieving selectivity. Eventually, the developed bionic nose could promote an accurate and noninvasive strategy for early diagnosis of an impending disease condition via detection of the appropriate biomarkers. In the long run, *in vitro* culture of printed hybrid architecture would empower the growth of "cyborg organs" that exhibit enhanced functionalities over human biology.

7.3.4 Bionic Skin

Skin is the largest organ in human body, serving a multitude of functions, most notably providing the sense of cutaneous touch, that allows us to receive complex signals from our external environment including temperature, humidity, texture, and a wide range of forces including pressure, shear, twist strain, and vibration. Least notable, yet far more vital, it acts as a barrier, helping us maintain the delicate physiological balance necessary for normal biological function, called *homeostasis*. Moreover, skin plays important roles in protein and vitamin D metabolism.

Skin defects may be caused by chronic ulcers, skin cancer, surgery, or other cutaneous malignancies or injuries such as "road rash" and burns. According to World Health Organization estimates, each year about 11 million burn injuries have been reported worldwide, of which around 265,000 lead to death [67]. Traditionally, one of the most popular treatments for such defects has been autologous grafts (auto-grafts) where the surgeon obtains a graft of skin tissue from a different area of the patient's body to facilitate wound treatment. This strategy reduces the treatment time, avoids complications due to immune rejections, and improves outcome significantly, leading to much-reduced scarring and often preserving, albeit limited, functionality. However, since autografts are obtained from donor sites on the patient's body, their availability is limited, and in case of large wounds or defects, may be insufficient. Thus, much effort has been focused to overcome limitations of traditional treatments by providing alternative engineered skin substitutes [68]. Another compelling reason for pursuit of bionic skin stems from the potential application of such alternatives in replacing or reducing pharmaceutical and skin care industry use of animal models during research and development of cosmetic products. Beyond obvious ethical issues and concerns related to the use of animal models in cosmetics development, using engineered bionic skin as alternative tissue for this purpose has the potential to improve fidelity and translatability of test results, thus reducing the development course and product development costs.

These motivations, along with recent advances in bionic integration and robotics, have triggered the emergence of bionic skins, with the ultimate goal of simulating the comprehensive properties of the tactile sense and multifunctional complexity of skin. Two communities of scientists with different backgrounds have undertaken this challenge, each with a different approach and focus. The tissue engineering community focus has been on restoring the barrier functionality of the skin and addressing challenges in regeneration of defective or damaged skin, as well as reducing scar formation and improving cosmetics. On the other hand, inspired by the multimodal sensory function of the skin and propelled by advances in nanomaterials and bioelectronics, many engineers are devising methods to restore and augment sensory functions of the skin, and to enable myriad new applications in robotics, human-machine interfaces, next generation smart devices, and artificial intelligence via integration of bionic skin with electronics.

7.3.4.1 Tissue Engineering Approach to Biomimetic Skin Substitutes

Tissue engineering approaches to fabricate suitable skin substitutes must satisfy certain requirements to be relevant in clinical and cosmetic applications. For clinical use, major considerations include the ability to prevent dehydration, providing protection against bacterial infections, mechanical properties similar to human skin including flexibility and stretchability, and sufficient resistance against shear and pressure forces, ability to accommodate irregular defect geometries, compatibility to avoid undesirable immune responses, and essential logistical requirements such as affordability, broad accessibility, ease of storage and application, and sufficient shelf life. Satisfying such requirements would enable a suitable skin substitute for wound healing and regenerative applications, yet does not guarantee functionality, because innervation and vascularization remain substantial challenges. In fact, most tissue engineering products to date either include no cellular components at all (acellular), or at best include only components of the connective tissue, most notably fibroblasts and keratinocytes. It is evident that sensory functions of the native tissue may not be replaced unless sensory components, such as mechanoreceptors, thermoreceptors, and pain receptors are integrated or substituted with electromechanical functional devices. Moreover, the absence of melanocytes (i.e., cells that produce melanin, pigments responsible for skin tone) may lead to undesirable aesthetic results.

Two main subgroups of tissue engineering approaches may be identified; First are scaffold-based strategies, where the required 3D structure is designed and realized with degradable biomaterials and then seeded with appropriate cells and growth factors or other chemical agents that promote cell adhesion, growth, and proliferation. Second are bottom-up approaches, where microscale building blocks of biomaterial and biological components are integrated with high spatial resolution to achieve the final structure. Of special interest in bottom-up approaches is 3D bioprinting, the basics of which were discussed earlier in this chapter. A successful example is the use of the laser-assisted bioprinting technique by Michael et al. [69] to create a fully cellularized skin substitute, where layers of fibroblast (cells from the dermal layer of skin that serve a multitude of functions, including collagen production) and keratinocytes (keratin-producing cells of the epidermis) were positioned over a commercial regenerative matrix, Matriderm®, in a manner mimicking the natural structure of skin. Implantation of these skin substitutes on to full-thickness skin wounds on mice was followed by histological and fluorescent analysis. It was shown that, 11 days after implantation, an epidermis layer was formed by keratinocytes, and fibroblasts started to migrate into the Matriderm®. Moreover, deposition of collagen and penetration of blood vessels into the skin construct was detected.

7.3.4.2 Electronic Skins

As was mentioned earlier, properties such as lack of flexibility and stretchability, make current silicon-based electronics unsuitable for applications such as electronic skin, where the goal is to reproduce the multimodal sensory function of the skin via arrays of electronic sensors. One of the early attempts at creating flexible electronics was a pressure sensor matrix with organic field-effect transistors by Someya et al. [70]. The significance of this paper was twofold: (i) field-effect transistors are realized on a bendable substrate that remains functional even when wrapped around a 2-mm cylinder, and (ii) a conductive grid is demonstrated that can be used to send commands to each individual sensor based on where it lies in the grid, solving the complexities involved in wiring large numbers of sensors in an array. Moreover, any kind of sensor could potentially replace some of these pressure sensors allowing the electronic skin to sense temperature, humidity, or pressure. The main shortcoming, however, is that although these substrates are bendable, like a piece of paper, they were not stretchable and could not be used on joints and during movements that require stretching of the e-skin. Many successful designs have since accomplished stretchabilities up to 230%, utilizing materials such as ultrathin polymer sheets [71], rubber and PDMS [72], and silk [73]. and epidermal electronics [81]. These designs may find applications in health monitoring, robotics, restoring lost skin functionality, and bionic augmentation.

7.3.5 Other Bionic Organs

So far, we have discussed a number of bionic sensory organs, along with the challenges and some state-of-the-art approaches to improve the design of these devices. In addition to the aforementioned organs, extensive studies have been performed on confronting other organ impairments in the human body using artificial bionic organs. Developing a bionic heart to maintain blood circulation [74], a bionic pancreas for insulin and glucagon delivery [75], bionic limbs [76], and electronic dura mater for chronic neural biointegration with the spinal cord [77] are a few examples of the ongoing advances in biomimetic implants. In some cases, not only have these devices addressed the main organ issue, but they have also managed to outperform the natural human organ by implementing new functionalities and capabilities. Detailed discussion of these organs is beyond the scope of this chapter.

7.4 Concluding Remarks and Prospects

The mere prospect of reproducing lost function of entire organs via bionic or bioprinted equivalents has elicited unprecedented enthusiasm. This is not surprising at all, considering the potential benefits to society as the direct result of such achievements. Yet, despite tremendous and ever-increasing success stories, bionic organs need to overcome significant and fundamental challenges before their full potential is realized. Foremost among these challenges is the need for seamless integration with the peripheral and central nervous systems. Sustained long-term improvements are required until perception of sensory information from a bionic device is indistinguishable from the natural counterpart.

Moreover, most current approaches to bionic organs are not well suited to mass production and off-the-shelf availability, and they have faced difficulties transitioning from laboratories and animal models into clinical trials. Also, concerns of adverse immune or systematic reactions persist.

We predict a paradigm shift toward neural prosthetics and neurobionics, where current bionic capabilities will be integrated with and controlled by the nervous system. Advancements in intracortical microelectrode arrays [32] that provide unprecedented and selective access to the neurons of the central nervous system, use of engineered materials such as conductive polymers and composites in neural interfaces [78], and improving models to process and decode neural signals [79] promise tremendous enhancement in the next generation of bionic devices. Before then, several challenges should be addressed, including: (i) long-term electrode reliability, (ii) extensive neural tissue damage during electrode implantation, (iii) meningeal responses, and (iv) degradation of electrode and insulation materials [80].

Despite challenges, it is not farfetched to envision future bionics capable of restoring lost and damaged organs using the patient's own cells, derived either from induced pluripotent stem cells (iPSCs) or by direct differentiation in a matter of weeks, to bioprint a replacement organ in minutes. Furthermore, the augmentation potentials are limitless. An example would be a bionic eye with super resolution, both visible and infrared capabilities, and ability to connect to the Cloud to record and recall limitless amounts of data. Through advancements in neural prosthetics, that eye would be seamlessly integrated with the nervous system to the point that it can be considered an extension of the body and would be controlled not through mediation of our bodies the way our current gadgets and devices are, but the same way our own arms and legs and eyes are controlled, with our brain and our thoughts. Such devices belong less to the realm of fiction and more to reality with each passing day.

References

1. Famulari, A., De Simone, P., Verzaro, R., et al. (2003). Artificial organs as a bridge to transplantation. *Artificial Cells, Blood Substitutes, and Biotechnology* 31, 163–168.
2. Anderson, J. M., & McNally, A. K. (2011). Biocompatibility of implants: Lymphocyte/macrophage interactions. *Seminars in Immunopathology*, 33, 221–233.

3. Veiseh, O., Doloff, J. C., Ma, et al. (2015). Size- and shape-dependent foreign body immune response to materials implanted in rodents and non-human primates. *Nature Materials*, 14, 643–651.
4. Zhang, L., Cao, Z., Bai, T., et al. (2013). Zwitterionic hydrogels implanted in mice resist the foreign-body reaction. *Nature Biotechnology*, 31(6), 553–556.
5. Vincent, J. F. V., & Wegst, U. G. K. (2004). Design and mechanical properties of insect cuticle. *Arthropod Structure Development*, 33(3), 187–199.
6. Ji, B., & Gao, H. (2004). Mechanical properties of nanostructure of biological materials. *Journal of the Mechanics and Physics of Solids*, 52(9), 1963–1990.
7. Ratner, B. D., Hoffman, A. S., Schoen, F. J., & Lemons, J. E. (2006). Biomaterials science: An introduction to materials in medicine. *MRS Bulletin*, 31, 59.
8. Vincent, J. (2012). *Structural biomaterials* (3rd ed.) Princeton, NJ: Princeton University Press.
9. Cebon, D., & Ashby, M. F. (1992). Materials selection in mechanical design. In T. Barry & K. Reynard (Eds.), *Computerization and networking of materials databases: Third volume*. West Conshohocken, PA: ASTM International..
10. Agache, P. G., Monneur, C., Leveque, J. L., & De Rigal, J. (1980). Mechanical properties and Young's modulus of human skin in vivo. *Archives of Dermatological Research*, 269(3), 221–232.
11. Kong, Y. L., Gupta, M. K., Johnson, B. N., & McAlpine, M. C. (2016). 3D printed bionic nanodevices. *Nano Today*, 11(3), 330–350.
12. Fratzl, P., & Weinkamer, R. (2007). Nature's hierarchical materials. *Progress in materials Science*, 52(8), 1263–1334.
13. Sebastian Mannoor, M. (2014). *Bionic Nanosystems*.
14. Marro, A., Bandukwala, T., & Mak, W. (2016). Three-dimensional printing and medical imaging: A review of the methods and applications. *Current Problems in Diagnostic Radiology*, 45(1), 2–9.
15. Fenster, A., & Downey, D. B. (1996). 3-D ultrasound imaging: A review. *IEEE Engineering in Medicine and Biology Magazine*, 15(6), 41–51.
16. Murphy, S. V., & Atala, A. (2014). 3D bioprinting of tissues and organs. *Nature Biotechnology*, 32(8), 773–785.
17. Nakamura, M., Iwanaga, S., Henmi, C., Arai, K., & Nishiyama, Y. (2010). Biomatrices and biomaterials for future developments of bioprinting and biofabrication. *Biofabrication*, 2(1), 014110.
18. Rowley, J. A., Madlambayan, G., & Mooney, D. J. (1999). Alginate hydrogels as synthetic extracellular matrix materials. *Biomaterials*, 20(1), 45–53.
19. Re'em, T., Tsur-Gang, O., & Cohen, S. (2010). The effect of immobilized RGD peptide in macroporous alginate scaffolds on TGFβ1-induced chondrogenesis of human mesenchymal stem cells. *Biomaterials*, 31(26), 6746–6755.
20. Xu, T., Jin, J., Gregory, C., Hickman, J. J., & Boland, T. (2005). Inkjet printing of viable mammalian cells. *Biomaterials*, 26(1), 93–99.
21. Roth, E. A., Xu, T., Das, M., Gregory, C., Hickman, J. J., & Boland, T. (2004). Inkjet printing for high-throughput cell patterning. *Biomaterials*, 25(17), 3707–3715.
22. Xu, T., Gregory, C. A., Molnar, P., et al. (2006). Viability and electrophysiology of neural cell structures generated by the inkjet printing method. *Biomaterials*, 27(19), 3580–3588.
23. Boland, T., Xu, T., Damon, B., & Cui, X. (2006). Application of inkjet printing to tissue engineering. *Biotechnology Journal: Healthcare Nutrition Technology*, 1(9), 910–917.

24. Saunders, R. E., Gough, J. E., & Derby, B. (2008). Delivery of human fibroblast cells by piezoelectric drop-on-demand inkjet printing. *Biomaterials*, 29(2), 193–203.
25. Bohandy, J., Kim, B. F., & Adrian, F. J. (1986). Metal deposition from a supported metal film using an excimer laser. *Journal of Applied Physics*, 60(4), 1538–1539.
26. Chuang, A. T., Margo, C. E., & Greenberg, P. B. (2014). 2 Retinal implants: A systematic review. *British Journal of Ophthalmology*, 98(7), 852–856.
27. Stingl, K., Bartz-Schmidt, K. U., Besch, D., et al. (2013). Artificial vision with wirelessly powered subretinal electronic implant alpha-IMS. *Proceedings of the Royal Society B: Biological Sciences*, 280(1757), 20130077.
28. Lewis, P. M., Ayton, L. N., Guymer, R. H., et al. (2016). Advances in implantable bionic devices for blindness: A review. *ANZ Journal of Surgery*, 86(9), 654–659.
29. Brindley, G. S., & Lewin, W. S. (1968). The sensations produced by electrical stimulation of the visual cortex. *The Journal of Physiology*, 196(2), 479–493.
30. Srivastava, N. R., Troyk, P. R., Towle, V. L., et al. (2007). Estimating phosphene maps for psychophysical experiments used in testing a cortical visual prosthesis device. In *2007 3rd International IEEE/EMBS Conference on Neural Engineering*. pp. 130–133. doi:10.1109/CNE.2007.369629
31. Lowery, A. J. (2013). Introducing the Monash vision group's cortical prosthesis. In *2013 IEEE International Conference on Image Processing*. pp. 1536–1539. doi:10.1109/ICIP.2013.6738316
32. Fernández, E., Greger, B., House, P. A., et al. (2014). Acute human brain responses to intracortical microelectrode arrays: Challenges and future prospects. *Frontiers in Neuroengineering*, 7, 24
33. Hochberg, L. R., Bacher, D., Jarosiewicz, B., et al. (2012). Reach and grasp by people with tetraplegia using a neurally controlled robotic arm. *Nature*, 485(7398), 372–375.
34. Collinger, J. L., Wodlinger, B., Downey, J. E., et al. (2013). High-performance neuroprosthetic control by an individual with tetraplegia. *The Lancet*, 381(9866), 557–564.
35. Lorach, H., Marre, O., Sahel, J.-A., Benosman, R., & Picaud, S. (2013). Neural stimulation for visual rehabilitation: Advances and challenges. *Journal of Physiology – Paris*, 107(5), 421–431.
36. Yue, L., Weiland, J. D., Roska, B., & Humayun, M. S. (2016). Retinal stimulation strategies to restore vision: Fundamentals and systems. *Progress in Retinal and Eye Research*, 53, 21–47.
37. Ruiters, S., Sun, Y., Jong, S. de, Politis, C., & Mombaerts, I. (2016). Computer-aided design and three-dimensional printing in the manufacturing of an ocular prosthesis. *British Journal of Ophthalmology*, 100(7), 879–881.
38. Wilson, B. S., & Dorman, M. F. (2008). Cochlear implants: Current designs and future possibilities. *Journal of Rehabilitation Research and Development*, 45(5), 695–730.
39. Lim, H. H., & Lenarz, T. (2015). Auditory midbrain implant: Research and development towards a second clinical trial. *Hearing Research*, 322, 212–223.
40. Tan, F., Walshe, P., Viani, L., & Al-Rubeai, M. (2013). Surface biotechnology for refining cochlear implants. *Trends in Biotechnology*, 31(12), 678–687.
41. Cao, Y., Vacanti, J. P., Paige, K. T., Upton, J., & Vacanti, C. A. (1997). Transplantation of chondrocytes utilizing a polymer-cell construct to produce tissue-engineered cartilage in the shape of a human ear. *Plastic and Reconstructive Surgery*, 100(2), 297–302; discussion 303–304.
42. Bichara, D. A., O'Sullivan, N. A., Pomerantseva, I., et al. (2012). The tissue-engineered auricle: Past, present, and future. *Tissue Engineering Part B: Reviews*, 18(1), 51–61.

43. Mannoor, M. S., Jiang, Z., James, T., et al. (2013). 3D Printed bionic ears. *Nano Letters* 13(6), 2634–2639.
44. Xu, T., Binder, K. W., Albanna, M. Z., et al. (2013). Hybrid printing of mechanically and biologically improved constructs for cartilage tissue engineering applications. *Biofabrication*, 5(1), 015001.
45. Coelho, D. H., & Costanzo, R. M. (2016). Posttraumatic olfactory dysfunction. *Auris Nasus Larynx*, 43(2), 137–143.
46. Hong, S.-C., Holbrook, E. H., Leopold, D. A., & Hummel, T. (2012). Distorted olfactory perception: A systematic review. *Acta Otolaryngologica (Stockh.)*, 132(Suppl. 1), S27–S31.
47. Conley, D. B., Robinson, A. M., Shinners, M. J., & Kern, R. C. (2003). Age-related olfactory dysfunction: Cellular and molecular characterization in the rat. *American Journal of Rhinology*, 17(3), 169–175.
48. Doty, R. L. (2012). Olfactory dysfunction in Parkinson disease. *Nature Reviews Neurology*, 8(6), 329–339.
49. Persaud, K., & Dodd, G. (1982). Analysis of discrimination mechanisms in the mammalian olfactory system using a model nose. *Nature*, 299(5881), 352–355.
50. Wasilewski, T., Gębicki, J., & Kamysz, W. (2017). Bioelectronic nose: Current status and perspectives. *Biosensors and Bioelectronics*, 87, 480–494.
51. Goldsmith, B. R., Mitala Jr., J. J., Josue, J., et al. (2011). Biomimetic chemical sensors using nanoelectronic readout of olfactory receptor proteins. *ACS Nano*, 5(7), 5408–5416.
52. Cook, B. L., Ernberg, K. E., Chung, H., & Zhang, S. (2008). Study of a synthetic human olfactory receptor 17-4: Expression and purification from an inducible mammalian cell line. *PLoS ONE*, 3(8) p.e2920.
53. Sanz, G., & Pajot-Augy, E. (2013). Deciphering activation of olfactory receptors using heterologous expression in *Saccharomyces cerevisiae* and bioluminescence resonance energy transfer. *Methods in Molecular Biology Clifton NJ*, 1003, 149–160.
54. Lee, S. H., Jin, H. J., Song, H. S., Hong, S., & Park, T. H. (2012). Bioelectronic nose with high sensitivity and selectivity using chemically functionalized carbon nanotube combined with human olfactory receptor. *Journal of Biotechnology*, 157(4), 467–472.
55. Zhang, X., De la Cruz, O., Pinto, J. M., et al. (2007). Characterizing the expression of the human olfactory receptor gene family using a novel DNA microarray. *Genome Biology*, 8(5), R86.
56. Mannoor, M. S., Zhang, S., Link, A. J., & McAlpine, M. C. (2010). Electrical detection of pathogenic bacteria via immobilized antimicrobial peptides. *Proceedings of the National Academy of Sciences*, 107(4), 19207–19212.
57. Pavan, S., & Berti, F. (2012). Short peptides as biosensor transducers. *Analytical and Bioanalytical Chemistry*, 402(10), 3055–3070.
58. Jin, H. J., Lee, S. H., Kim, T. H., et al. (2012). Nanovesicle-based bioelectronic nose platform mimicking human olfactory signal transduction. *Biosensors and Bioelectronics*, 35(1), 335–341.
59. Son, M., Kim, D., Ko, H. J., Hong, S., & Park, T. H. (2017). A portable and multiplexed bioelectronic sensor using human olfactory and taste receptors. *Biosensors and Bioelectronics* 87, 901–907.
60. Firestein, S. (2001). How the olfactory system makes sense of scents. *Nature*, 413(6852), 211–218.
61. Malnic, B., Hirono, J., Sato, T., & Buck, L. B. (1999). Combinatorial receptor codes for odors. *Cell*, 96(5), 713–723.

62. Saito, H., Chi, Q., Zhuang, H., Matsunami, H., & Mainland, J. D. (2009). Odor coding by a Mammalian receptor repertoire. *Science Signaling*, 2(60), ra9.
63. Jodat, Y. A., Kiaee, K., Vela Jarquin, D., et al. (2020). A 3D-printed hybrid nasal cartilage with functional electronic olfaction. *Advanced Science*, 7(5), 1901878.
64. Quignon, P., Giraud, M., Rimbault, M., et al. (2005). The dog and rat olfactory receptor repertoires. *Genome Biology*, 6(10), R83.
65. Tan, J., Savigner, A., Ma, M., & Luo, M. (2010). Odor information processing by the olfactory bulb analyzed in gene-targeted mice. *Neuron*, 65(6), 912–926.
66. Zhang, X., De la Cruz, O., Pinto, J. M., Nicolae, D., Firestein, S., & Gilad, Y. (2007). Characterizing the expression of the human olfactory receptor gene family using a novel DNA microarray. *Genome Biology*, 8(5), R86.
67. Peck, M. D. (2011). Epidemiology of burns throughout the world. Part I: Distribution and risk factors. *Burns*, 37(7), 1087–1100.
68. Supp, D. M., & Boyce, S. T. (2005). Engineered skin substitutes: Practices and potentials. *Clinics in Dermatology*, 23(4), 403–412.
69. Michael, S., Sorg, H., Peck, C. T., et al. (2013). Tissue engineered skin substitutes created by laser-assisted bioprinting form skin-like structures in the dorsal skin fold chamber in mice. *PLoS ONE*, 8(3), e57741.
70. Someya, T., Sekitani, T., Iba, S., et al. (2004). A large-area, flexible pressure sensor matrix with organic field-effect transistors for artificial skin applications. *Proceedings of the National Academy of Sciences USA*, 101(27), 9966–9970.
71. Kaltenbrunner, M., Sekitani, T., Reeder, J., et al. (2013). An ultra-lightweight design for imperceptible plastic electronics. *Nature*, 499(7459), 458–463.
72. Rogers, J. A., Someya, T., & Huang, Y. (2010). Materials and mechanics for stretchable electronics. *Science*, 327(5973), 1603–1607.
73. Kim, D.-H., Ghaffari, R., Lu, N., et al. (2012). Electronic sensor and actuator webs for large-area complex geometry cardiac mapping and therapy. *Proceedings of the National Academy of Sciences*, 109(49), 109, 19910–19915.
74. Dowling, R. D., Gray Jr, L. A., Etoch, S. W., et al. (2003). The AbioCor implantable replacement heart. *Annals of Thoracic Surgery*, 75(6), S93–S99.
75. Russell, S. J., El-Khatib, F. H., Sinha, M., et al. (2014). Outpatient glycemic control with a bionic pancreas in type 1 diabetes. *New England Journal of Medicine*, 371(4), 313–325.
76. Dhillon, G. S., Lawrence, S. M., Hutchinson, D. T., & Horch, K. W. (2004). Residual function in peripheral nerve stumps of amputees: Implications for neural control of artificial limbs. *The Journal of Hand Surgery*, 29(4), 605–615; discussion 616–618.
77. Minev, I. R., Musienko, P., Hirsch, A., et al. (2015). Biomaterials. Electronic dura mater for long-term multimodal neural interfaces. *Science*, 347(6218), 159–163.
78. Green, R., & Abidian, M. R. (2015). Conducting polymers for neural prosthetic and neural interface applications. *Advanced Matererials*, 27(46), 7620–7637.
79. Vidal, G. W. V., Rynes, M. L., Kelliher, Z., & Goodwin, S. J. (2016). Review of brain-machine interfaces used in neural prosthetics with new perspective on somatosensory feedback through method of signal breakdown. *Scientifica*, 2016, 8956432, 10 pp.
80. Barrese, J. C., Rao, N., Paroo, K., et al. (2013). Failure mode analysis of silicon-based intracortical microelectrode arrays in non-human primates. *Journal of Neural Engineering*, 10(6), 066014.
81. Kim, D.-H., Lu, N., Ma, R., et al. (2011). Epidermal electronics. *Science*, 333(6044), 838–843.

8 Bioinspired Design for Energy Storage Devices

Loza F. Tadesse and Iwnetim I. Abate

8.1 Introduction

Global warming is a pressing issue for both current and future generations. The various impacts of improper environmental handling have led to drought, famine, flooding, and other natural disasters. In addition, current global energy consumption is growing exponentially, and dependence on foreign oil and gas not only negatively affects the environment but also creates national dependencies that endanger social stability [1]. An alternative, environmentally friendly energy source is therefore required to preserve nature and fulfill this ever-growing need for energy. However, clean energy sources, such as solar radiation, wind, and waves, are intermittent and require energy storage platforms [2]. To this end, high energy density rechargeable batteries have recently attracted tremendous research attention as they enable efficient storage of intermittent clean energy and electrification of transportation vehicles. Similar to fossil fuels, batteries store energy as portable chemical energy, which is the most convenient form of storage.

Lithium (Li) batteries have been used to provide high storage volume and gravimetric energy density for meeting the energy demands of cellular phones and portable computers. Current research on these batteries focuses on improving their cost, safety, stored energy density, charge/discharge rates, and service life in order to translate them to large-scale applications such as electric vehicles [3]. In this chapter, we will cover how researchers are using bioinspired solutions to improve current battery technologies.

8.1.1 Sources of Bioinspiration in Nature

To meet the ever-growing energy demand, efforts are being targeted at producing batteries with high specific capacity, high rate capability, and better cyclability. Nature is one of the main sources of inspiration to achieve these goals as its rich resources provide tools and design inspiration for scientific discovery. Currently, various biological molecules and processes are being studied in advancing battery technologies; however, inspiration from nature to solve such technological challenges is not a recent phenomenon. Leonardo da Vinci's attempt in the fifteenth century to create a "flying machine" by mimicking how birds fly is one of the earliest examples of biomimicry [4]. More recent examples include: a unique design for the front end of a high-speed train inspired by the beak of the kingfisher, which results in 15% more energy efficiency, 10% more speed, and reduced noise levels (transportation); filters inspired by lungs, which result in over 90% carbon dioxide filtration from flue-gas stacks (a type of chimney) (environment); a fog-capturing beetle exoskeleton, which inspired wound healing bandages and antibacterial surface coating implants (medicine); brain neural networks that mimic memory devices (neuromorphic devices), which use a 10 times smaller area and greater than 1 million times lower power consumption per learning cycle (computers) (Figure 8.1).

8.1.2 Recent Advancements to Improve Bioinspiration through Nanotechnology

Technological advances and better ways of understanding biological systems have enabled us to further adopt several techniques and materials from nature to create more efficient, robust, and reliable products. New disciplines such as nanotechnology have enabled us to manipulate materials at nanometer (10^{-9} m) scales. As a result, tailoring the properties of materials to make intricate devices such as semiconductors, display devices, catalysts, and energy conversion/storage device has become possible. Furthermore, the study of complex nanostructures in biological systems has further opened up a new era for fabricating nanomaterials through the emulation of biological processes [11,12]. Large surface-to-volume ratio, geometry, and quantum effects at nanometer scale provide distinctive material properties, which are key to overcome the limitations of conventional bulk materials [12]. With simple nanofabrication techniques, improved material property can be achieved via hybridization of nanostructured biomaterials with functional materials.

Self-assembled biomaterials have a widely demonstrated application as providing structural motifs for nonbiological nanostructured materials [13]. This is because biomaterials are composed of organic materials with various functional groups. Hence, their acidity and polarity at surfaces serve as possible adsorption sites for precursors of the target nanostructured material, duplicating the complex shapes of the biomaterials and generating isostructural target materials. Nanomaterials that cover the surface of the biomaterials form nanoarchitecture similar to that of the templates, and their duplication offers versatility in synthesis of various materials in complex forms beyond conventional bioapplications. Furthermore, control of the nanostructures of biomaterials can be achieved relatively easily by adjusting the synthesis conditions.

In general, there is unprecedented success in biomimicry to advance the performance, quality, strength and other properties of devices. Similar achievements in battery technology are aimed at improving synthesis of battery components for reaching greater capacity and implementing environmentally friendly options by using bioinspiration. In the following sections, bioinspired applications in Li-ion batteries and Li-air batteries will be discussed.

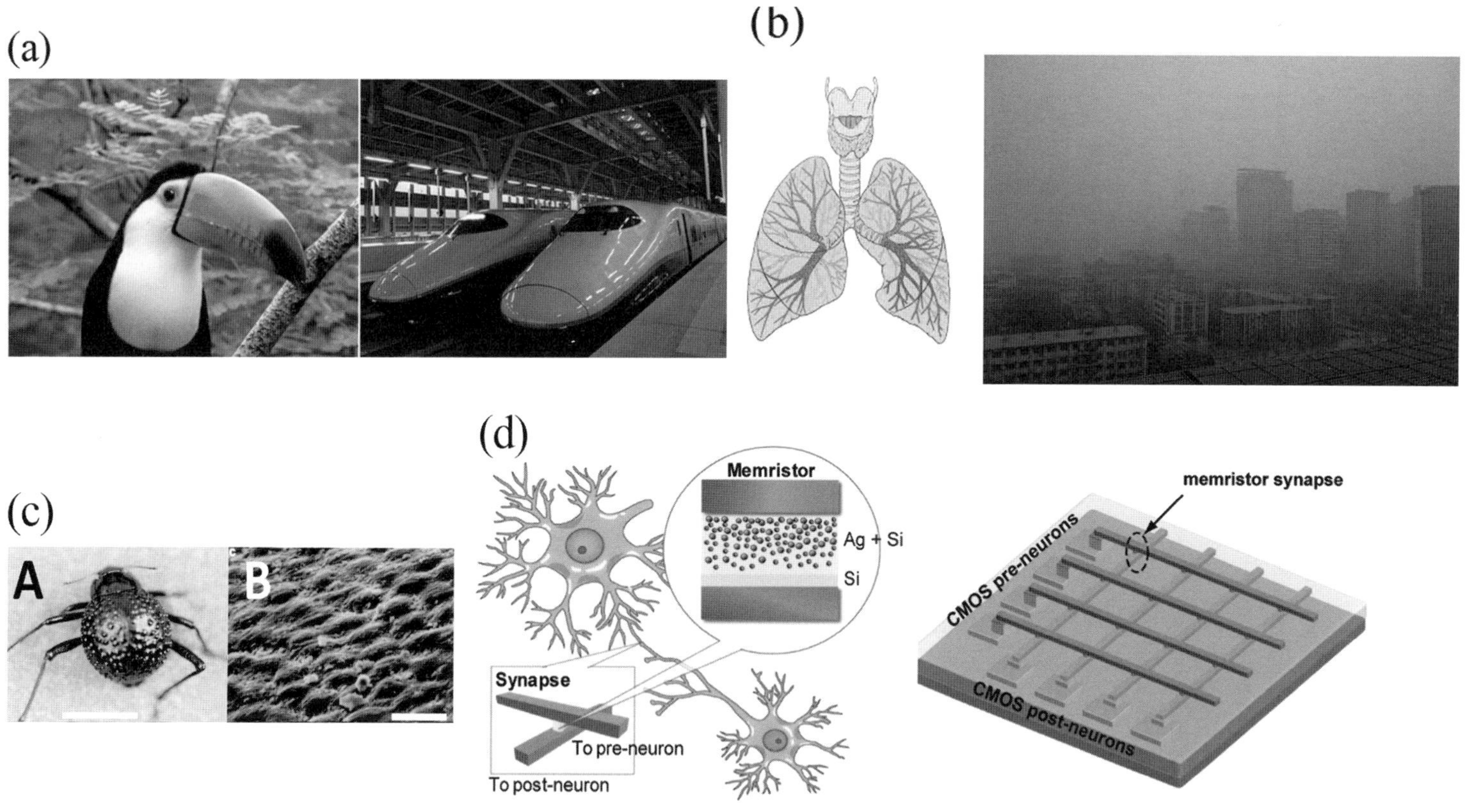

Figure 8.1 Examples of bioinspiration for solutions in various applications. (a) Efficient train inspired from bird beak. (Adapted from Yisris [5] and Sund [6] under Creative Commons attribution license (CC BY 3.0).) (b) Improved carbon sequestration form air pollution inspired from human lungs. (Adapted from Imeto [7] and Lynch [8] under Creative Commons attribution license (CC BY 3.0).) (c) Improving bandages' absorption capability by adapting the cone-like microstructure of a fog-capturing beetle's exoskeleton, which collects water vapor droplets from atmospheric fog. (Reprinted from Green et al. [9], with permission from Springer Nature.) (d) Increasing the energy and storage efficiency of memory devices by adapting the high efficiency of biological systems, which is due to the large connectivity ($\sim 10^4$ in a mammalian cortex) between neurons that offers highly parallel processing power. (Reprinted with permission from Jo et al. [10], copyright 2010, American Chemical Society.)

8.2 Basic Concepts

Batteries are devices that convert chemical energy contained in their active materials into electrical energy via electrochemical oxidation–reduction (redox) reaction. Unlike in heat engines, energy conversion in batteries is not limited by the Carnot cycle, which is dictated by the second law of thermodynamics. As a result, batteries have higher efficiencies compared to engines. Batteries can be broadly categorized into two main types: primary (nonrechargeable) and secondary (rechargeable) batteries. All batteries consist of three major components; namely, anode, cathode, and electrolyte. When a device is connected to a battery – a lightbulb or an electrical circuit – chemical reactions occur on the electrodes that create a flow of electrical energy to the device. The anode is where oxidation takes place; it gives up electrons to the external circuit. The cathode is where reduction takes place; it accepts electrons from the external circuit (Figure 8.2). The electrolyte is a chemical medium that allows the flow of electrical charge (ions) between the cathode and the anode.

For example, if the anodic material is zinc (Zn) and the cathodic material is copper (Cu), the electrochemical reaction at each electrode during discharge is:

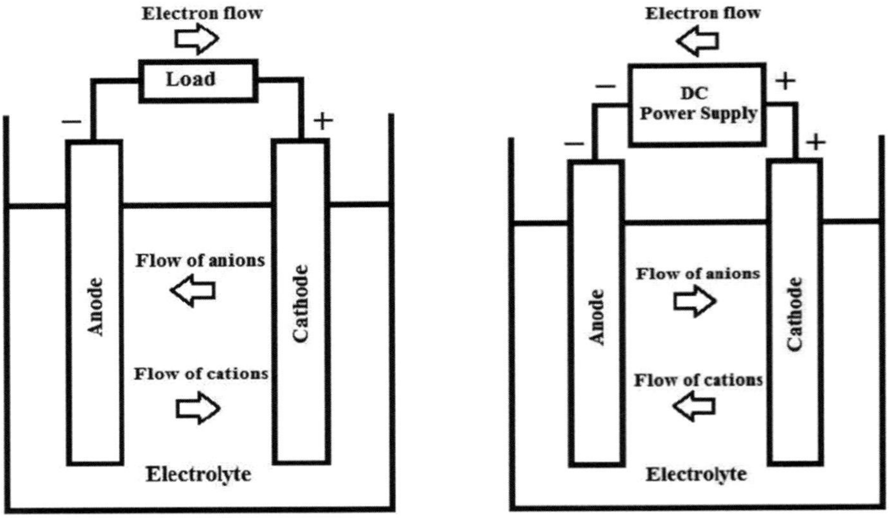

Figure 8.2 Schematic of battery during discharge (left) and charge (right). The direction of flow of electrons during discharging is from the anode to the cathode across the load, which utilizes this current. As the result, anions are attracted toward the mostly positively charged anode and the cations are attracted to the cathode, which is the mostly negatively charged. The process is reversed during charging as electrons from the source flow to the anode, making it negatively charged and attracting cations, with a similar process taking place on the cathode end. (Adapted from Linden [14], with permission from McGraw-Hill Education.)

Negative electrode: anodic reaction (oxidation, loss of electrons)

$$Zn \rightarrow Zn^{2+} + 2e^-$$

Positive electrode: cathodic reaction (reduction, gain of electrons)

$$Cu^{2+} + 2e^- \rightarrow Cu$$

Overall reaction (discharge)

$$Zn + Cu^{2+} \rightarrow Zn^{2+} + Cu$$

During charging, current flow is reversed and oxidation takes place at the positive electrode and reduction occurs at the negative electrode.

Negative electrode: cathodic reaction (reduction, gain of electrons)

$$Zn^{+2} + 2e^- \rightarrow Zn$$

Positive electrode: anodic reaction (oxidation, loss of electrons)

$$Cu \rightarrow Cu^{+2} + 2e^-$$

Overall reaction(charge)

$$Zn^{+2} + Cu \rightarrow Zn + Cu^{+2}$$

The active material in the cell determines the standard potential. This can also be calculated from free-energy data or obtained experimentally. From the standard electrode potentials, the standard potential of a cell can be calculated as follows: (Anode (oxidation potential) + cathode (reduction potential) = standard cell potential), where the oxidation potential is the negative value of the reduction potential. For example, for Zn–Cu cell, the standard cell potential is the sum of 0.76 V ($Zn \rightarrow Zn^{+2} + 2e^-$) and 0.34 V ($Cu^{+2} + 2e^- \rightarrow Cu$), which is equivalent to 1.10 V. The cell voltage also depends on concentration and temperature, which is described by the Nernst relation [14].

8.2.1 Theoretical Capacity (Coulombic)

The theoretical capacity is determined by the amount of active material in the cell and is expressed as the total quantity of electrical charges involved in the electrochemical reaction. It is the change associated with the quantity of electricity obtained from active materials and is defined in units of Coulombs (C). Hence, for example, the theoretical capacity of Zn/Cu is 0.394 ampere-hour (Ah)/g (theoretically 1 g equivalent weight of material delivers 96,486 C) [14].

8.2.2 Theoretical Energy

The theoretical energy describes the capacity of the cell on an energy (watt-hour) basis by taking both voltage and quantity of electricity into consideration, which gives the

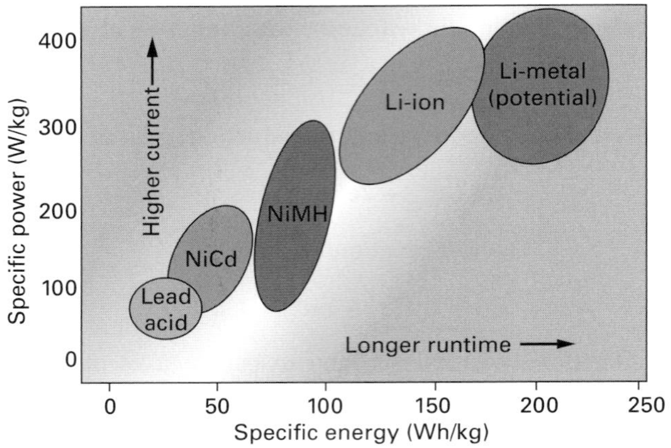

Figure 8.3 Power density versus energy density plot of different battery technologies. Lead–acid battery, the oldest type of rechargeable battery, provides the high current required when starting motor vehicles. Several technologies have been developed after the invention of lead–acid battery that have higher energy density and power. This is achieved by changing the materials used as the electrode and electrolyte. For example, in Li-ion batteries, a Li source such as $LiCoO_2$ is used as cathode and graphite used as anode. As the result, Li-ion batteries have 5–8 times more specific energy and power. (Reprinted with Permission from Tarascon et al. [15], Springer Nature.)

maximum value that can be delivered by a specific electrochemical system (watt-hour = voltage × ampere hour) [14].

8.2.3 Most Frequently Used Terms and Plots in Battery Research

C rate: is the current rate used during discharge or charge. 1 C means a fully charged battery rated at 1 Ah should provide 1 A for 1 hour. The same battery discharging at 0.5 C should provide 500 mA for 2 hours, and at 2 C it delivers 2 A for 30 minutes. The faster the battery charges the sooner it reaches its full capacity and vice versa [15].

Discharge–charge (potential or galvanostatic) curve: shows the amount of the battery's potential as function of time or charge (charge is equal to the product of time and current rate used to discharge or charge the battery) [15].

Regone plot: is a very common chart used to compare the performance of batteries, whereby energy per unit volume describes the specific energy of a battery, while the power density describes how fast energy can be drawn from a battery (Figure 8.3) [15].

8.3 Prior Work

8.3.1 Bioinspired Electrode Synthesis for Battery Application

Electrodes in nanoscale dimensions are favored over bulk materials in electrochemical applications. This is because they provide large surface areas to interact with the electrolytes. They also result in short (Li) ion diffusion lengths and facile strain accommodation induced by volume change [16]. Such electrodes, with the desired dimensions and morphologies, can be easily achieved by using biomaterials with variety of nanostructures, which can be obtained by control of synthesis conditions [17]. Bioinspired methods provide cheaper options compared to commonly used techniques, such as electron beam etching, nanoimprint lithography, or atomic layer deposition. Furthermore, most of these templating techniques generate mesopores [18] or micropores [17], as opposed to nanopores, which enhance ion mobility and hence improve battery capacity.

8.3.2 Virus-Inspired Battery Electrode Synthesis

Viruses are microorganisms that need host organisms for replication. They store their genetic information in either a single- or double-stranded DNA or RNA molecule. The tobacco mosaic virus was the first virus to be observed in 1892, after which multiple others were reported. Among these is M13 bacteriophage, which is one of the most frequently investigated viruses in nanotechnology [19]. It is about 880 nm in length and 6.5 nm in diameter. It contains a circular single-stranded DNA encapsulated by about 2,800 major coat proteins (p8) and minor coat proteins at the end of the virus (p3, p6, p7, and p9) (Figure 8.4). Modifications in the major and minor coat proteins, through genetic engineering, results in hybridized M13 viruses with functional inorganics [20]. These hybridized viruses can serve as biotemplates for nanostructured materials to be used in rechargeable Li battery electrodes.

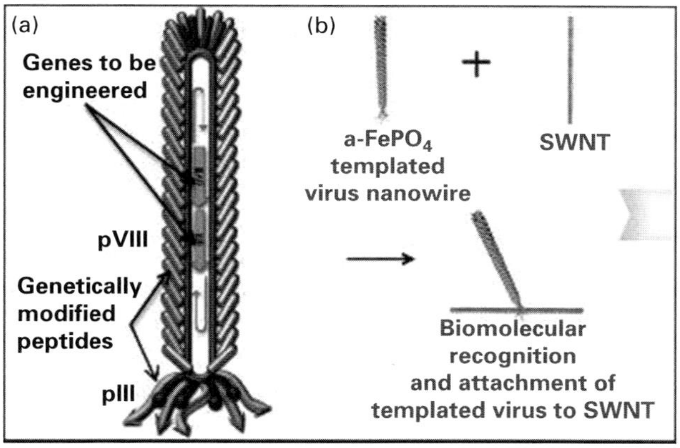

Figure 8.4 M13 macrophage genetic engineering and biomolecular recognition for a high-power Li rechargeable battery. (a) A schematic representation of the multifunctional M13 virus is shown with its genetically engineered surface proteins. When only pVIII of the M13 virus is modified, the virus is called E4 and has increased ionic interactions with cations, which can serve as a template for materials (a-$FePO_4$) growth. With further modification of the pIII virus, affinity for SWNT attachments is increased. (b) The hybridization process with the modified M13 virus to fabricate genetically engineered high-power Li-ion battery cathodes. (Adapted with permission from Lee et al. [22], copyright 2009 American Association for the Advancement of Science.)

8.3.3 Virus-Based Electrodes in Li-Ion Batteries

Li-ion batteries are one of the most widely used secondary batteries. In Li-ion batteries, electrodes store and release electrical energy by the insertion and extraction of Li^+ ions and electrons through the electrode material [15]. The most common anode and cathode materials in Li-ion batteries are lithium intercalated graphite and metal oxides such as $LiCoO_2$, $LiFePO_4$, and LiMnO [15]. During discharge, Li ions move from the anode to the cathode across an electrolyte. The electrolyte is an ionic conductor, which causes electrons produced during discharge to move across an external circuit powering a device. During charge, Li ions move in the reverse direction. These are the most common batteries used in current consumer electronic devices and electric vehicles [15].

Virus-based cathode in Li-ion batteries: Despite their poor electronic/Li-ionic conductivity, amorphous $FePO_4$ (a-$FePO_4$) is a favorable cathodic material in Li-ion batteries. This is because of its high specific capacity and safety, originating from the strong $(PO_4)^{3-}$ covalent bond [21]. By binding $FePO_4$ cathode material with highly conductive carbon nanotubes (CNTs), it is possible to increase the performance of Li-ion batteries [22]. In this application, the gene of VIII protein (pVIII), a major capsid protein of the M13 virus is modified to serve as a template for a-$FePO_4$ growth (Figure 8.4). The gene of III protein (pIII) is also engineered to have a binding affinity for carbon nanotubes. The modified pIII proteins have a high affinity for CNTs; thus, the virus-a-$FePO_4$ nanowire (10–20 nm in diameter) can bind with CNTs (~4–5 nm in diameter) to form the hybrid nanowire [19]. The complex network structure of individual nanowires in the hybrid material has the same morphology as that of the virus template without any hybridization process (Figure 8.5).

The a-$FePO_4$ and CNTs are well connected with each other, indicating that electrons are rapidly supplied to the a-$FePO_4$ cathode through the CNT networks (Figure 8.5). The reduction in penetration distance for Li-ions from shrinkage in a-$FePO_4$ and the continuous supply of electrons from CNTs, improve the performance of the Li-ion battery. Virus-a-$FePO_4$ hybrid nanowires without CNTs (E40), virus-a-$FePO_4$-CNT hybrid nanowires with moderate binding affinity to CNTs (EC#1), and those with strong binding affinity to CNTs (EC#2) were compared in Figure 8.6a. EC#1 and EC#2 have high specific capacity at all current rates compared to E4. This shows that CNT networks in these structures provide sufficient electrons during the electrochemical reactions. Furthermore, EC#2 has outstanding electrochemical performance, with a specific capacity of 170 mAh g^{-1} (comparable with theoretical rate) at a current of C/10 (1 C = 178 mA g^{-1}; Figure 8.6b). Additionally, all of its cycles have excellent cycle retention up to 50 cycles, suggesting the stability of the hybrid nanowires upon cycling (Figure 8.6c). Its high capacity is also maintained at higher current rates, indicating that the hybrid nanowires are promising materials for high-power application [22].

Virus-based anode in Li-ion batteries: Co_3O_4 is one of the most promising anodic materials for Li-ion batteries. This is due to its high specific capacity (~890 mAh g^{-1}) via conversion reactions [23]. However, the product of the conversion reaction of Co_3O_4 is Li_2O phase, which is an electrochemically inactive species. Using a

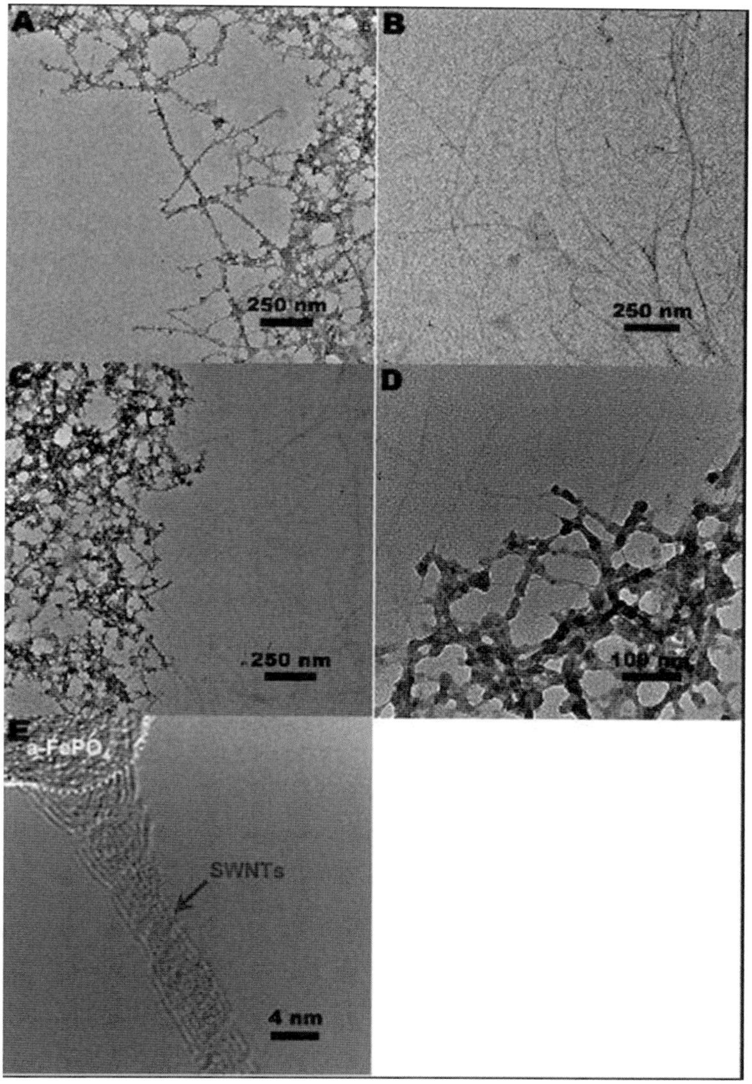

Figure 8.5 Morphology of the a-$FePO_4$ grown on the multifunctional viruses/CNTs hybrid nanostructures. (a) TEM image of a-$FePO_4$ nanowires templated on virus without CNTs, (b) TEM image of pristine CNTs (before interacting with viral a-$FePO_4$, and (c–e) virus-amorphous $FePO_4$-CNT hybrid nanowires at different resolutions. (Adapted with permission from Lee et al. [22], copyright 2009 American Association for the Advancement of Science.)

genetically modified M13 virus, it is possible to reduce the dimension of Li_2O to nanoscale, thus making it more electrochemically active. Nam et al., have genetically altered the M13 virus by modifying p8 proteins into two types: one containing a Co nucleating motif and the other containing a Au binding motif (Figure 8.7a). The Au nanoparticles (~5 nm in diameter) are finely dispersed in virus-Au-Co_3O_4 hybrid

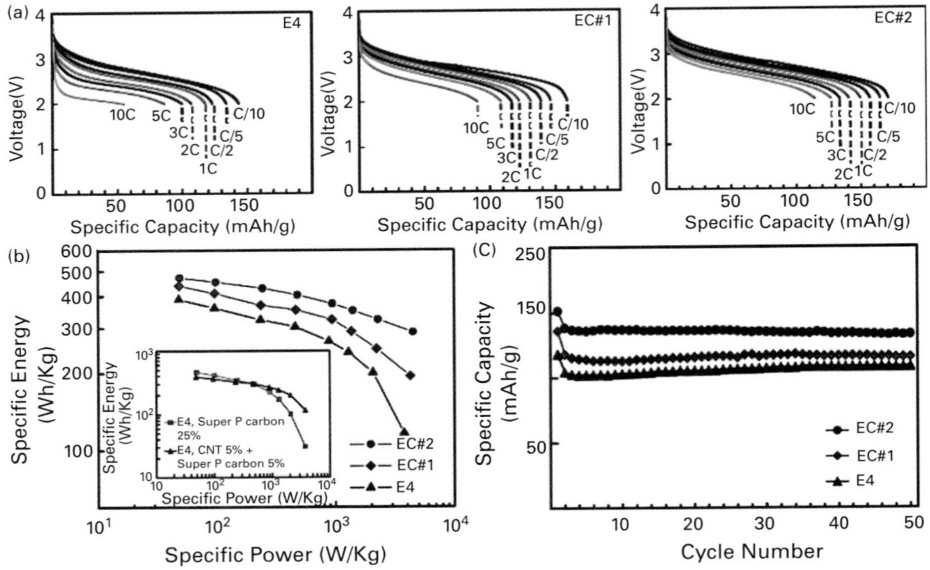

Figure 8.6 Electrochemical performance of hybrid nanowires based on multifunctional M13 virus templates. (a) Discharge profiles of E4 (virus-a-$FePO_4$ nanowires), EC#1 (virus-a-$FePO_4$-CNT nanowires with moderate affinity for CNTs), and EC#2 (virus-a-$FePO_4$-CNT nanowires with strong affinity for CNTs) at current rates from C/10 to 10 C. (b) Ragone plots of the hybrid nanowires (inset: Ragone plot of E4 as a function of carbon content). The highest energy density was achieved with the use of SWNTs, which increase conductivity by providing percolation network. (Only active electrode mass is included in the weight when power and energy capacity is calculated.) (c) Capacity as a function of the number of cycles at a current rate of 1 C for 50 cycles. The capacity does not fade, indicating stability of the electrodes. (Reprinted with permission from Lee et al. [22], copyright 2009 American Association for the Advancement of Science.)

nanowires that are fabricated through Au-binding, followed by Co_3O_4 nucleation and growth (Figure 8.7b and c). Viruses that do not contain the Au binding motif cannot bind Au nanoparticles at all [20].

The hybrid nanowires show excellent electrochemical properties due to the improved electrochemical activity of the conversion reaction at nanoscale. Also, due to the increased electronic conduction, or the catalytic effect of the Au nanoparticles, Au-containing nanowires exhibit better performance (Figure 8.7d). Furthermore, Au nanoparticles increase the reaction rate of the Co_3O_4 anode. This is confirmed by the increase in current density during cyclic voltammetry measurements (Figure 8.7e) [20].

8.3.4 Virus-Based Electrodes in Li-Air Batteries

The next generation secondary batteries are referred to as beyond Li-ion (BLI). Among these groups, the Li-air battery (also known as Li-O_2 battery), discovered two decades ago, has the highest energy density of the metal-air batteries. This has captured much attention due to its higher theoretical specific energy compared to state-of-the-art Li-ion

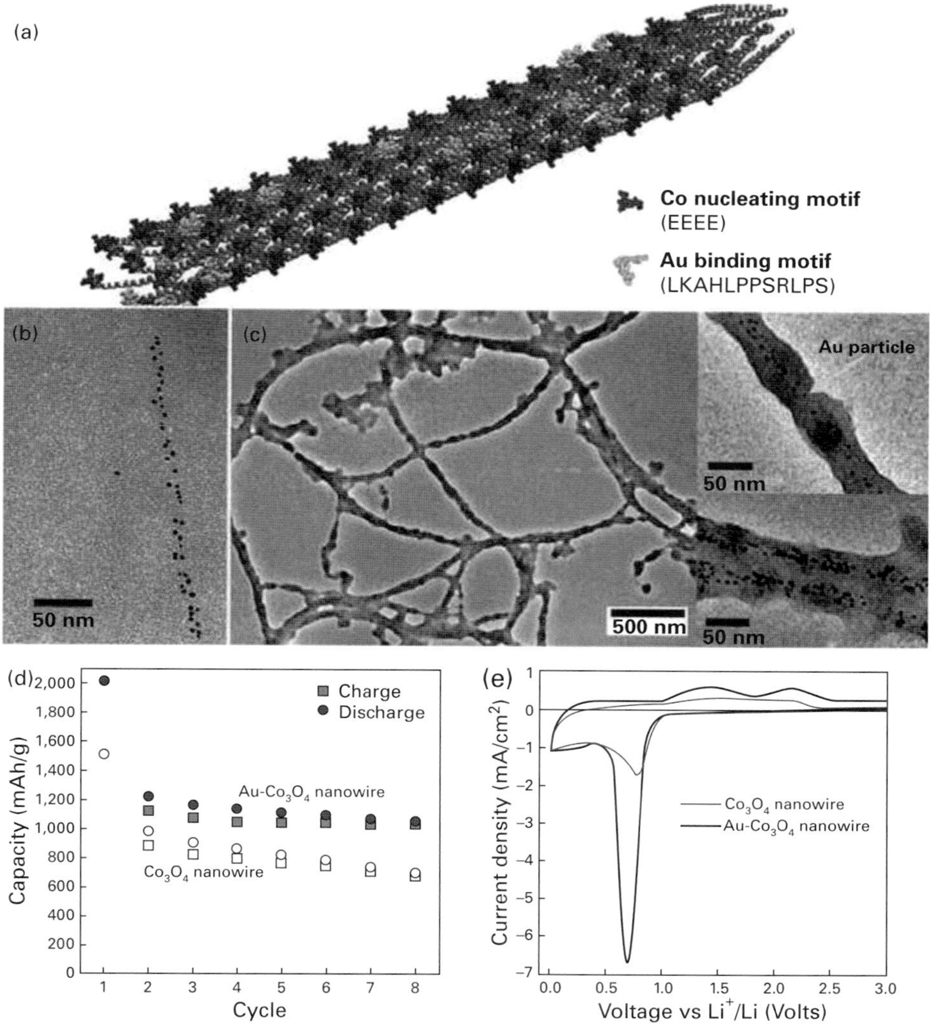

Figure 8.7 Characterization of the hybrid nanostructure of Au nanoparticles incorporated into Co_3O_4 nanowires. (a) Visualization of the modified M13 virus with both a Co-nucleating motif (black) and an Au-binding motif (gray). (b) The virus-Au hybrid nanowire before Co_3O_4 growth. (c) The virus-Au-Co_3O_4 hybrid nanowire. (d) Specific capacity depending on the number of cycles of the hybrid nanowires with and without Au at a current of C/26.5 for 8 cycles. (e) CV curves of the nanowires with and without Au at a scan rate of 0.3 mV s^{-1}. The pick (current density) is higher for hybrid Au-Co_3O_4 than Co_3O_4 nanowires. (Reprinted with permission from Nam et al. [20], American Association for the Advancement of Science.)

battery technology [24]. The reversible reaction, $2Li + O_2 \leftrightarrow Li_2O_2$, governs the electrochemistry of the Li-O_2 cell, which operates at 2.96 V, with a theoretical energy density of 3,458 Wh kg^{-1}. The high energy density is from the use of Li metal as the anode and ambient air as the source of O_2 [25].

However, no practical Li-air battery exists yet. Some of its main drawbacks include: poor practical specific energy, high charge overpotential and poor rechargeability. Thus in order to use Li-oxygen batteries for applications in long-range electric vehicles, there is a need for cathodes with higher porosity and catalytic activity [26]. Noble metals [27], transition metal oxides [18], and carbon-based materials [28] have been used as electrodes. However, these have either high material cost, low conductivities, or side reactions with electrolytes that limit their use.

Virus-based cathode in Li-air batteries: M13 bacteriophages can be assembled via a covalent layer-by-layer process to form a highly nanoporous network capable of organizing nanoparticles and acting as a scaffold for templating metal oxides. Various metals chemically interact with pVIII during the synthesis of nanonetworks for cathode composites. The resulting catalyst structures have high Li-O_2 battery performances that are associated with nanocomposites of biotemplated manganese oxide nanowires (bio MO nanowires), with incorporation of a small weight percent (3 wt%) of Pd nanoparticles (Figure 8.8) [26].

The cathodes were synthesized by first binding Mn^{2+} ions onto the pVIII proteins of the virus. $KMnO_4$ is then allowed to react with it at room temperature. The biotemplating reaction results in a homogenous bio MO nanowires with spherulitic surface morphologies (Figure 8.9). These have higher dimensions, which are favorable for Li-O_2 battery performance. The rough surfaces of the nanowires also provides large catalytic areas and storage space for discharge products, compared to the smooth surfaces of metal oxide nanowires. The wires also have high aspect ratios (~80 nm in diameter and ~1 μm in length) [26].

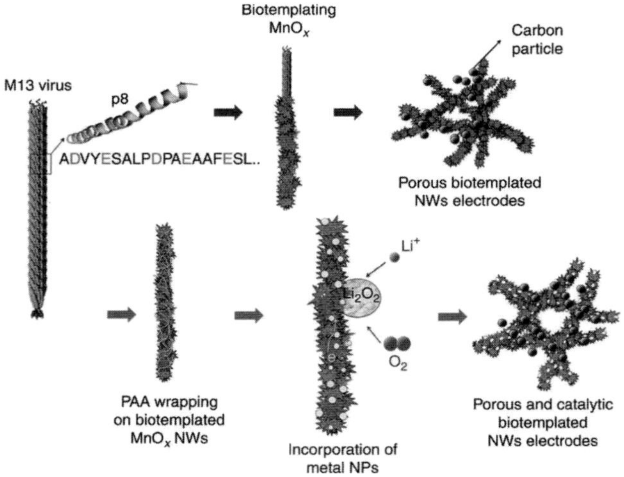

Figure 8.8 Schematic of a nanocomposite structure synthesis. The steps in synthesizing the metal nanoparticle/M13 virus-templated bio MO nanowires and the operational reaction inside Li-O_2 battery cells ($Li^+ + O_2 \rightarrow Li_2O_2$). (Reprinted with permission from Oh et al. [26], Springer Nature.)

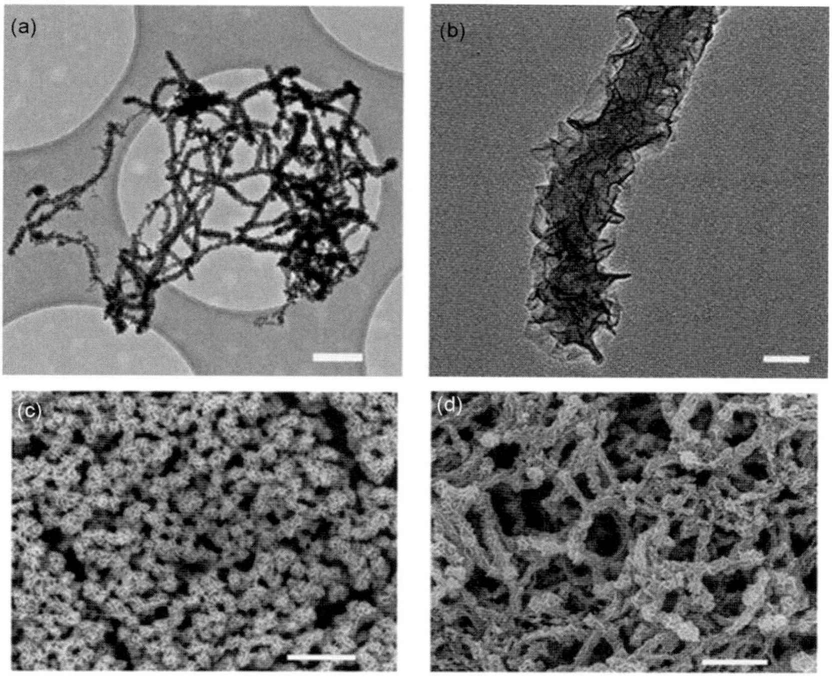

Figure 8.9 TEM and SEM images of M13 virus-templated MO nanowires. (a) TEM images at low magnitude (scale bar, 500 nm). (b) Individual MO nanowire at higher magnitude (scale bar, 50 nm). SEM images of (c) MO nanoparticle showing bulky phase of aggregated particles and (d) homogeneously dispersed bio MO nanowires with larger pores (scale bar, 500 nm). (Reprinted with permission from Oh et al. [26], Springer Nature.)

In this work, MO nanoparticles (60 nm diameter) were also synthesized without viruses. Their morphology was then compared to that of bio MO nanowires in scanning electron microscopy (SEM) images (Figure 8.9c and d). The surface areas that were measured using Brunauer–Emmett–Teller (BET) method, showed that the surface areas of bio MO nanowires (271.7 m^2 g^{-1}) are about three orders of magnitude greater than that of MO nanowires synthesized without viruses (98.5 m^2 g^{-1}). These resulted in an increase of the storage capacity of the Li-O_2 battery by about 40% when bio MO (9,196 mAh g^{-1}c at 0.4 A g^{-1}c compared to 6,545 mAh g^{-1}c when using MO nanoparticles (Figure 8.10). In addition, the discharge voltage of the bio MO nanowires (~2.68 V) is about 80 mV higher, while their charge voltage is about 120 mV lower than that of the MO nanoparticles. This means that bio MO nanowire cathodes for Li-O_2 battery have reduced overpotential [26].

8.3.5 Protein (Peptide)-Inspired Electrode Synthesis for Battery Application

Proteins are continuously being formed in living organisms through the covalent bonding of the carboxyl groups (–COOH) of amino acids with the amino group

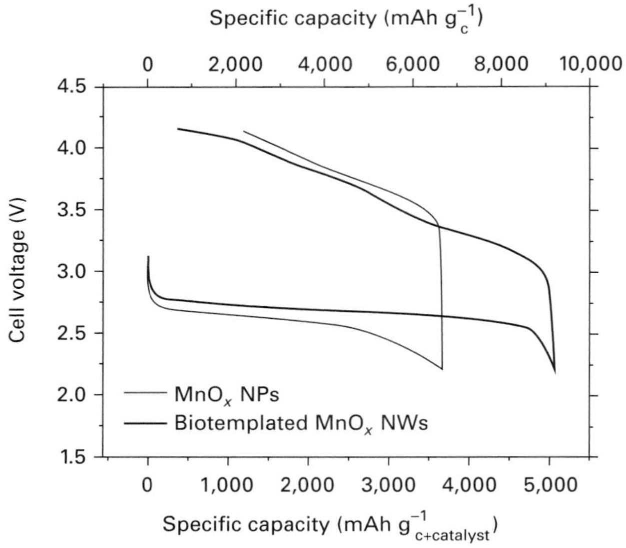

Figure 8.10 Comparison of discharge-specific capacity for MO nanoparticles versus bio MO nanowires. The first galvanostatic cycle comparison, at the current density of 0.4 $Ag^{-1}c$, of MO nanoparticles (with maximum specific capacity ~3,500 mAh g^{-1}) and bio MO nanowires (with maximum specific capacity ~5,000 mAh g^{-1}). Improvement of 40% in capacity is observed with virus (bio) templated MO nanowires. (Reprinted with permission from Oh et al. [26], Springer Nature.)

(–NH_2) of another amino acid. This bonding, (–C(=O)NH–) is called the peptide bond [29]. By controlling the self-assembly conditions, nanostructures of peptides can be easily varied. The structural duplication with the electrode materials is achieved by various surface-coating methods. These include conventional aqueous solution-based synthesis as well as vacuum vapor deposition.

Protein (peptide)-inspired anode in Li-ion battery: The formation of Co_3O_4 nanoparticles on the peptide surface, by a simple reduction and oxidation process, increases the specific capacity of the battery, from 25 to 80 mAh g^{-1}, (Figure 8.11). However, the gravimetric and volumetric energy density of the hybrid nanowires is less than pure Co_3O_4, because the amount of Co_3O_4 nanoparticles in the hybrid is low (Figure 8.12). Removing the peptide by heat treatment not only increases capacity, but also creates a nanotubular structure. The unique property of the nanotube is that the structure has surfaces both inside and outside of the tube. The dual-surface system is beneficial for the electrode material because it allows for bidirectional Li-ion diffusion, doubles the contact area with the electrolyte, and relaxes stress during battery operation via its hollow structure [30].

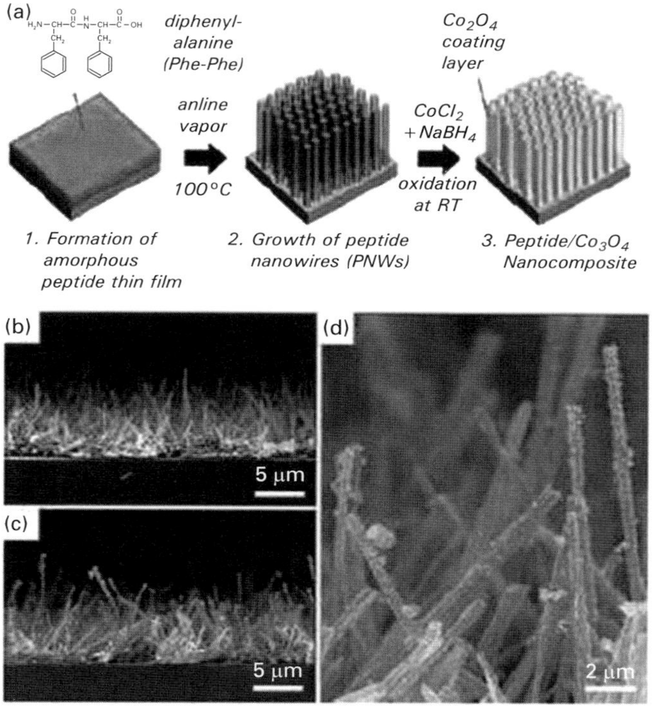

Figure 8.11 Diphenlyalanine-Co_3O_4 hybrid nanowires. (a) Schematic illustration of the fabrication process of hybrid nanowires and SEM images of (b) the pure peptide nanowires and (c, d) the hybrid nanowires. Pure peptide nanowires have smooth surfaces while peptide/Co_3O_4 hybrid nanowires have high surface roughness with a slightly increased diameter. The increased roughness is due to the formation of nanocrystalline inorganic materials (Co_3O_4) on the surface of the peptide nanowires. (Reprinted with permission from Ryu et al. [30], copyright 2010 American Association for the Advancement of Science.)

8.3.6 Bacteria-Inspired Electrode Synthesis for Battery Application

Li-O_2 battery performance greatly depends on the porosity of the cathode, as the O_2 molecules require pores to travel through to reach the other reagents of the electrochemical reaction. Existing synthetic porogens, for instance, pore-forming beads, suffer from process complexity, synthetic challenges, and high cost. Biological alternatives such as nonpathogenic bacteria strains, were studied as porogens for Li-O_2 batteries. This approach provided better porosity control both in terms of pore volume and pore shape, as various strains of bacteria can be used for this purpose (Figure 8.13). A similar approach can also be applied to improve the capacity of Na-O_2 batteries [31]. Furthermore, because bacteria can be easily cultured in media, it has a significant cost reduction ($\approx 10^6$) compared to the cost of synthetic porogens.

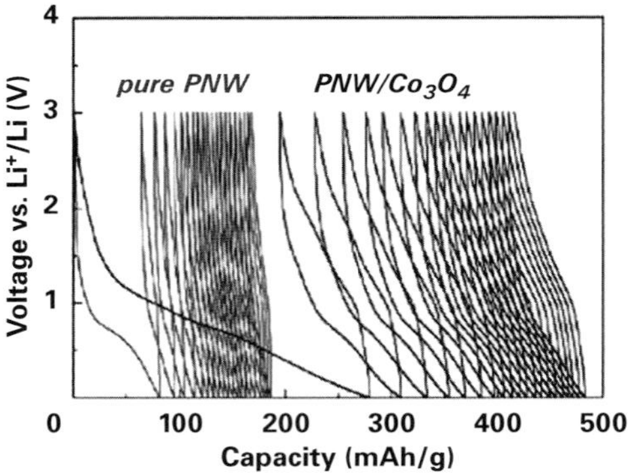

Figure 8.12 Discharge–charge profiles of pure and hybrid Co_3O_4 nanowires. Peptide-Co_3O_4 hybrid nanowires (black) have higher capacity than pure peptide nanowires (gray). (Reprinted with permission from Ryu et al. [30], copyright 2010 American Association for the Advancement of Science.)

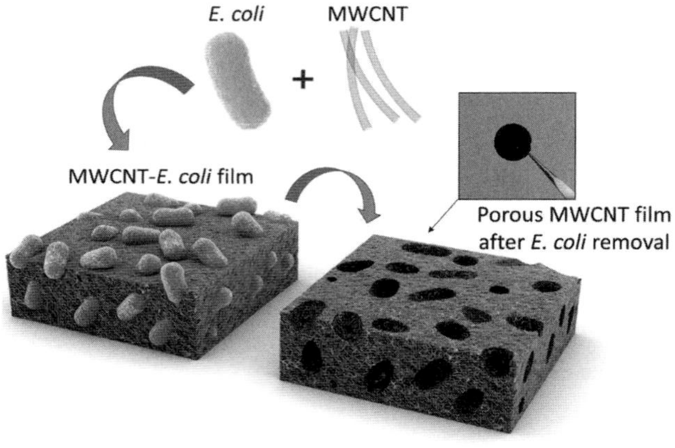

Figure 8.13 Bacteria-inspired cathode synthesis for Li-O_2 batteries. The schematic representation of the porous E-MWCNT film fabrication method, and an image of the final free-standing E-MWCNT film. E-MWCNT films are formed after the removal of biological pore templates, i.e., *E. coli.* (Reprinted with permission from Oh et al. [32], Nature Publishing Group.)

Free-standing porous multi-walled carbon nanotube (MWCNT) films also have $\approx$30% improved oxygen evolution efficiency and double full discharge capacities in cycling studies, compared to compact MWCNT films. In addition, porous MWCNT films offer a specific capacity of 4,942 W/kg (8,649 Wh/kg) at 2 A/ge (1.7 mA/cm^2), which is significantly better than that of compact MWCNTs (Figure 8.14) [32].

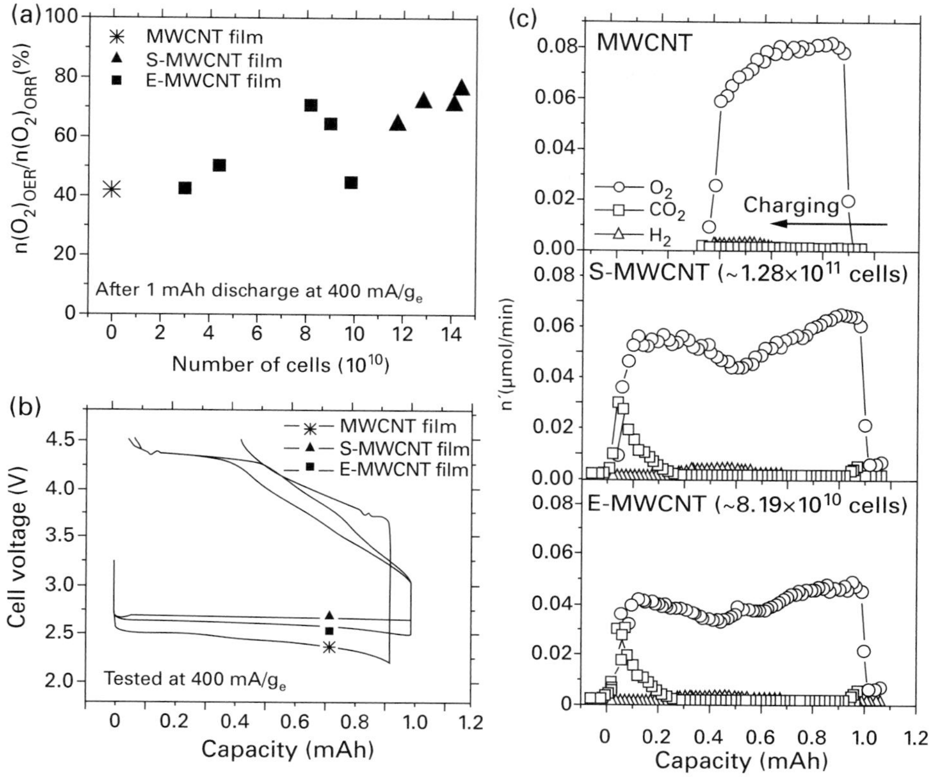

Figure 8.14 Battery performance comparison for compact MWCNT film, spherical porous (S-MWCNT) and cylindrical porous (E-MWCNT). (a) Molar ratio between oxygen gas generated and consumed during the first cycle of the Li-oxygen battery at 400 mA/g at a fixed discharge capacity of 1 mAh (charging cut-off potential was 4.5 V vs. Li/Li^+). (b) The first cycle voltage profiles that correspond to: (a) MWCNT films, E-MWCNT films (~8.19 Å ~10^{10} *E. coli* cells) and S-MWCNT films (1.28 Å ~10^{11} cells). (c) Gas (O_2, CO_2, and H_2) evolution rate profile during the first charging step of Li-oxygen battery corresponding to (b), measured by DEMS. (Reprinted with permission from Oh et al. [32], Nature Publishing Group.)

8.4 Summary and Concluding Remarks

In summary, nature is a great source of inspiration for the engineering of battery technologies. The use of viruses, peptides, bacteria, and other biomolecules enables nanoscale control of synthesis of battery anodes and cathodes that would otherwise be very difficult or very costly to achieve with synthetic materials. Evolution has enabled living organisms in the microcosmos to develop into robust systems that have mastered the design and synthesis processes at smaller scales. As our bottom-up construction principles approach limits in the design of materials at the micro- and nano-scales, inspiration from the expert organisms is becoming increasingly useful. In batteries, specifically, the application of biomaterials and bioinspired synthesis will broaden the material selection scope into diverse areas for the development of electrodes for future energy storage systems.

References

1. Rao, C. S. (2007). *Environmental pollution control engineering*. New Delhi: New Age International.
2. U. S. Energy Information Administration. (2016). Renewable and alternative fuels: Recent data. Retrieved from www.eia.doe.gov/fuelrenewable.html
3. BioAge Group. (2017). Green Car Congress: batteries. Retrieved from www.greencarcongress.com
4. Romei, F. (2008). *Leonardo Da Vinci*. Minneapolis: Oliver Press.
5. Yisris. (n.d.). *Hayabusa* [Photograph]. Flickr, Japan. Retrieved from https://bit.ly/2IUVBxV (Originally photographed 2011, February 20)
6. Sund, S. (n.d.). *Toucan* [Photograph]. Flickr, Honduras. Retrieved from https://bit.ly/2GM4t7h (Originally photographed 2007, November 23)
7. Imeto, K. (n.d.). *Beijing Air Pollution* [Photograph]. Flickr, Beijing, Retrieved from https://bit.ly/2EGasHt (Originally photographed 2014, February 22)
8. Lynch, P. J. (n.d.). Lungs diagram with internal details. *Wikipedia*. Yale University School of Medicine, 23 Dec. 2006, bit.ly/2XLfH1k
9. Green, D. W., Ben-Nissan, B., Yoon, K. S., Milthorpe, B., & Jung, H. S. (2016). Bioinspired materials for regenerative medicine: Going beyond the human archetypes. *Journal of Materials Chemistry B*, 4(14), 2396–2406.
10. Jo, S. H., Chang, T., Ebong, I., Bhadviya, B. B., Mazumder, P., & Lu, W. (2010). Nanoscale memristor device as synapse in neuromorphic systems. *Nano Letters*, 10(4), 1297–1301.
11. Dickerson, M. B., Sandhage, K. H., & Naik, R. R. (2008). Protein-and peptide-directed syntheses of inorganic materials. *Chemical Reviews*, 108(11), 4935–4978.
12. Moriarty, P. (2001). Nanostructured materials. *Reports on Progress in Physics*, 64(3), 297.
13. Sanchez, C., Julián, B., Belleville, P., & Popall, M. (2005). Applications of hybrid organic–inorganic nanocomposites. *Journal of Materials Chemistry*, 15(35–36), 3559–3592.
14. Linden, D. (1984). *Handbook of batteries and fuel cells*. New York: McGraw-Hill.
15. Tarascon, J. M., & Armand, M. (2011). Issues and challenges facing rechargeable lithium batteries. In V. Dusastre (Ed.), *Materials for sustainable energy: A collection of peer-reviewed research and review articles from Nature Publishing Group* (pp. 171–179). Singapore: World Scientific Publishing.
16. Bruce, P. G., Freunberger, S. A., Hardwick, L. J., & Tarascon, J. M. (2012). Li-O2 and Li-S batteries with high energy storage. *Nature Materials*, 11(1), 19–29.
17. Cui, Y., Wen, Z., & Liu, Y. (2011). A free-standing-type design for cathodes of rechargeable Li–O2 batteries. *Energy & Environmental Science*, 4(11), 4727–4734.
18. Débart, A., Paterson, A. J., Bao, J., & Bruce, P. G. (2008). α-MnO_2 Nanowires: A catalyst for the O_2 electrode in rechargeable lithium batteries. *Angewandte Chemie*, 120(24), 4597–4600.

19. Kim, I., Moon J. S., & Oh, J.-W. (2016). Recent advances in M13 bacteriophage-based optical sensing applications. *Nano Convergence*, 3(1), 27.
20. Nam, K. T., Kim, D. W., Yoo, P. J., et al. (2006). Virus-enabled synthesis and assembly of nanowires for lithium ion battery electrodes. *Science*, 312(5775), 885–888.
21. Okada, S., Sawa, S., Egashira, M., et al. (2001). Cathode properties of phospho-olivine $LiMPO_4$ for lithium secondary batteries. *Journal of Power Sources*, 97, 430–432.
22. Lee, Y. J., Yi, H., Kim, W. J., et al. (2009). Fabricating genetically engineered high-power lithium-ion batteries using multiple virus genes. *Science*, 324(5930), 1051–1055.
23. Kang, Y. M., Song, M. S., Kim, J. H., et al. (2005). A study on the charge–discharge mechanism of Co3O4 as an anode for the Li ion secondary battery. *Electrochimica Acta*, 50(18), 3667–3673.
24. Laoire, C. O., Mukerjee, S., Abraham, K. M., Plichta, E. J., & Hendrickson, M. A. (2010). Influence of nonaqueous solvents on the electrochemistry of oxygen in the rechargeable lithium−air battery. *The Journal of Physical Chemistry C*, 114(19), 9178–9186.
25. Girishkumar, G., McCloskey, B., Luntz, A. C., Swanson, S., & Wilcke, W. (2010). Lithium–air battery: Promise and challenges. *The Journal of Physical Chemistry Letters*, 1, 2193–2203.
26. Oh, D., Qi, J., Lu, Y. C., Zhang, Y., Shao-Horn, Y., & Belcher, A. M. (2013). Biologically enhanced cathode design for improved capacity and cycle life for lithium-oxygen batteries. *Nature Communications*, 4, 2756. doi:10.1038/ncomms3756
27. Lu, Y. C., Gallant, B. M., Kwabi, D. G., et al. (2013). Lithium–oxygen batteries: Bridging mechanistic understanding and battery performance. *Energy & Environmental Science*, 6(3), 750–768.
28. Xiao, J., Mei, D., Li, X., et al. (2011). Hierarchically porous graphene as a lithium–air battery electrode. *Nano Letters*, 11(11), 5071–5078.
29. Stoker, H. S. (2010). Proteins. In K. Kennedy (Ed.), *General, Organic, and Biological Chemistry* (pp. 655–697). Monterey, CA: Brooks Cole.
30. Ryu, J., Kim, S. W., Kang, K., & Park, C. B. (2010). Synthesis of diphenylalanine/cobalt oxide hybrid nanowires and their application to energy storage. *ACS Nano*, 4(1), 159–164.
31. Abate, I. I., Thompson, L. E., Kim, H.-C., & Aetukuri, N. B. (2016). Robust NaO2 electrochemistry in aprotic Na–O2 batteries employing ethereal electrolytes with a protic additive. *The Journal of Physical Chemistry Letters*, 7(12), 2164–2169.
32. Oh, D., Ozgit-Akgun, C., Akca, E., et al. (2017). Biotemplating pores with size and shape diversity for Li-oxygen battery cathodes. *Scientific Reports*, 7, 45919. doi:10.1038/srep45919

9 Bioinspired Design of Nanostructures
Inspiration from Viruses for Disease Detection and Treatment

Nicolas Anuku, John David Obayemi, Olushola S. Odusanya,
Karen A. Malatesta, and Wole Soboyejo

9.1 Introduction

A growing number of developments in biology and materials chemistry highlight the notion of bioinspiration – in which biological concepts, mechanisms, functions, and design are the starting points toward new synthetic materials and devices with advanced structures and functions [1]. There is no doubt that emerging and reemerging infectious diseases caused and transmitted by viruses have significantly impacted human health worldwide [2,3]. Clearly, the virus exhibits elegant architectures that could occasionally be cellular macromolecules with structures that are beautifully adapted to the functions of the virion [4]. Virus particles exist in many sizes and shapes, and they vary considerably in the number and nature of the molecules from which they are built. Most viruses show a characteristic size, in the range of tens to hundreds of nanometers [5]. Viruses are intracellular parasites that enter a host cell/body to deliver their genetic material to initiate infection. Usually, the first step in the life cycle of a virus is the attachment to host cells/bodies. These may include their abilities to interact with lipids, proteins, and sugar moieties on the surface of cells and tissue [6].

The design and fabrication of bionanoclusters (BNCs) for the specific detection and treatment of disease via localized chemotherapy or hyperthermia [7] is based on the self-assembled structure of viruses [7] that share identical sizes and functions of entry into the body system. A typical example is organic-coated gold nanoparticles or organic-coated magnetite nanoparticles of 10–50 nm [8]. There are two main kinds of self-assembly: static and dynamic [8,9]. Static self-assembly (S) involves systems that are at global or local equilibrium and do not dissipate energy [8]. Molecular crystals, like most folded, globular proteins are formed by static self-assembly [8]. In static self-assembly, formation of the ordered structure may require energy (for example, in the form of stirring), but once it is formed, it is stable [8]. These BNC sizes allow penetration into the bloodstream by avoiding the reticulo-endothelial system (RES) [9,10] and a series of "biological barriers" [11,12]. The BNCs mimic the attachment of enveloped viruses (HIV and FIV) to receptors by means of a viral glycoprotein membrane [13]. The BNCs then uncoat by disassembling and release the drug to the targeted cancerous cell via localized chemotherapy and hyperthermia.

Generally, virus particles are designed for effective transmission of the nucleic acid genome from one host cell to another within a single animal or among host organisms. A primary function of the virion is, therefore, protection of the genome, which can be damaged irreversibly by a break in the nucleic acid or by mutation during passage through hostile environments [14,15]. During the virus transportation process, usually it may encounter a variety of potentially lethal chemical and physical agents, including nucleolytic and proteolytic enzymes, extremes of temperature or pH, and various types of natural radiation.

Using the inspiration obtained from viruses, nanocluster particles of drug, nucleic acid, or protein attached to organic-coated nanoparticles (Au or magnetite), can encased within a sturdy barrier formed by extensive interactions among the selected proteins comprising the protein coat [16]. Such protein–protein interactions may be hydrogen bonding, electrostatic interactions, hydrophobic bonding, or disulfide bonds tend to maintain stability of the nanoclusters of drugs/particles [16]. The BNC composed of Au or magnetite nanoparticles, multiple proteins (CD47, self-marker protein that can be used to trick the immune system to consider a foreign nanoparticle as its own) [17], and the anticancer drug (paclitaxel), and the nucleic acid (DNA or RNA) must, therefore, survive large variations in temperature, pH, or the chemical composition of the cell environment.

During transport in the body, the BNCs, like viruses, are subjected to hydrodynamic and shear forces that can cause them to fragment into smaller pieces. To prevent this, they must be bound by significant adhesive forces that cannot be overcome by the forces encountered during their transport in the body. There is a need to quantify the adhesion forces between the self-assembled constituents of BNCs that are relevant to disease detection and treatment, as well as nucleic acid delivery and protein delivery. This can be performed using force microscopy during atomic force microscopy (AFM) to measure adhesion [18]. Adhesion measurement is a force microscopy technique that can be used to study bimaterial pairs [18] that simulate pair-wise interactions between interacting nanoparticles [18]. This is the basis for force spectroscopy and the measurement of nano/molecular scale adhesion [18].

It is important to note that the results suggest that force microscopy can be used to measure the adhesive forces between the constituents of BNCs [19]. So far, there have been no prior efforts to use force microscopy to measure the adhesion forces between nanoparticles and the self-assembled constituents of BNCs that are being designed for the specific disease detection and treatment via localized chemotherapy and hyperthermia. Thus, this chapter presents the design and fabrication of BNCs, or bionanocarriers, for the specific detection and treatment of cancer via localized chemotherapy and hyperthermia.

The design and formation of the BNCs are based on viral structure; viral attachment to host cells and the mechanism of viral entry into cells are the source of bioinspiration. The BNC is designed to mimic viral action through specific targeting of cancerous cells. The BNC is based on the self-assembly of biosynthesized magnetic nanoparticles (BMNPs). Self-assembly is an efficient and a preferred process to assemble nanostructures. The self-assembly of magnetic nanoparticles uses the bottom-up fabrication. This offers the advantages of a cheaper fabrication and higher fabrication efficiency with particle sizes ranging between 1.5 and 20 nm [20,21].

Adhesion between the viral proteins and the coated BMNPs of the BNCs shall also be examined for the robustness of the . The adhesion of the ligands upon the reactive surface of the BNC shall also be observed. Adhesion is measured using a dip coating/atomic force microscopy (DC/AFM) technique. The adhesion forces are obtained for dip-coated BMNPs that interact with the peptide or antibody-based molecular recognition units (MRUs) that attach specifically to receptors on the surfaces of breast cancer cells. Adhesion forces between anticancer drugs, such as paclitaxel, and the constituents of MRU-conjugated Au or iron BNCs, are measured using force microscopy techniques. This type of BNCs can be used for nucleic acid delivery and protein delivery. The implications of the results are then discussed for the design of robust Au and/or iron BNCs and for potential applications in localized drug delivery and disease detection and treatment.

9.2 Bioinspirations from Viruses

9.2.1 Structures of Viruses and Protein Folding

Viral architecture [22] is described in terms of structural units of increasing complexity, from the smallest biochemical unit (the polypeptide chain) to the infectious particle (or virion). The entire infectious virus particle, called a virion, is usually is made up of nucleic acid and an outer shell of protein [23]. The protein subunit or protein spike is a single, folded polypeptide chain that helps the virus to attach itself to the host cell. The structural unit, or protomer, is the unit from which capsids or nucleocapsids are formed. This unit may comprise one protein subunit or multiple different protein subunits.

The morphological unit, or capsomere, represents surface structures (e.g., knobs, projections, clusters). The capsid is the coat protein shell surrounding the nucleic acid genome/virion. The nucleocapsid is the core, which is the nucleic acid–protein assembly packaged within the virion (see Figure 9.1) [24]. The envelope, or viral membrane, is a lipid bilayer carrying viral glycoproteins, and it surrounds the nucleocapsid, whereas the virion is the complete infectious virus particle (Figure 9.1).

Viral envelopes are impermeable and block entry of chemicals or enzymes in aqueous solutions into the virus particle. The protective function of virus particles demands that these particles be stable structures. However, virions also mediate attachment of the virus to an appropriate host cell and delivery of the genome to the interior of that cell, where the particle is at least partially disassembled. The functions of virus particles depend on very stable intermolecular interactions among their components during assembly, egress from the virus-producing cell, and transmission, but also ready reversal of these interactions during entry and uncoating in a new cell. It is clear that contact of a virus with the appropriate cell-surface receptor or exposure to a specific intracellular environment can trigger substantial conformational changes in the virion.

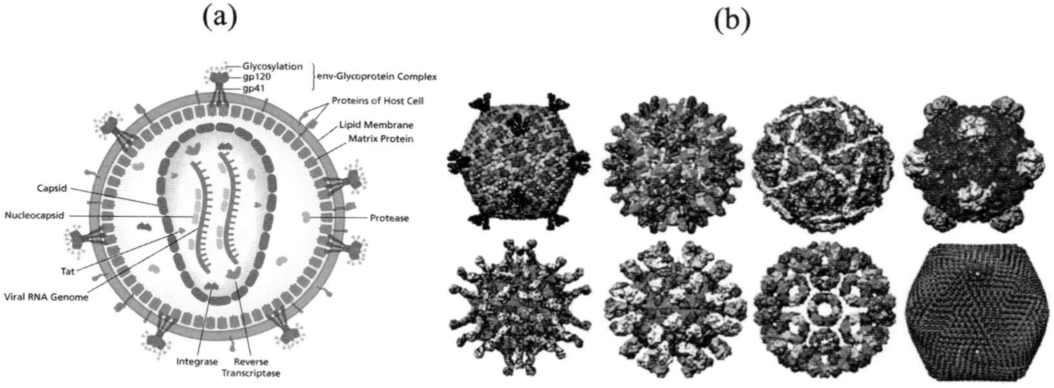

Figure 9.1 (a) A typical structure of HIV virus showing the basic biochemical and morphological units. (Reprinted from www.scistyle.com under CC BY-SA 4.0.) (b) Different virus structures – an inspiration for the nanoarchitectures of BNCs. (The figures were obtained courtesy of VIPERdb (http://viperdb.scripps.edu) [24].)

Generally, virus particles are classified as metastable since they have not yet attained the minimum free energy conformation. The free energy conformation state can be gained only when an unfavorable activation energy state is overcome following induction of irrevocable conformational transitions associated with attachment and entry. Thus, virions are not simply inert structures but are molecular machines that play an active role in delivery of the nucleic acid genome to the appropriate host cell and initiation of the reproductive cycle [25].

9.2.2 Virus Adhesion

A particularly intriguing feature of the nanoscale virus is the scale on which it builds and interacts with biological system's structural components, such as microtubes [26], microfilaments [27], and chromatin [28]. The associations maintaining these and the associations of other cellular components seem relatively simple when examined by high-resolution structural methods, such as crystallography or Nuclear magnetic resonance (NMR) – shape complementarity, charge neutralization, hydrogen bonding, and hydrophobic interactions. A key property of biological nanostructure is molecular recognition, leading to self-assembly. To date, the most successful biomimetic component used for self-assembly has been DNA itself [29]. Branched DNA molecules can be combined by "sticky-ended" cohesion [30].

In synthetic systems, sticky ends may be programmed with a large diversity. A sticky end of sufficient length coheres by base pairing alone, but they can be ligated to covalency. Sticky ends form classic B-DNA when they cohere [30]; in addition to the affinity inherent in complementarity, sticky ends also lead to structural predictability. Virus adhesion (Figure 9.2) to host cell is mediated by virion protein(s) binding to specific host surface molecule(s) (membrane proteins, lipids, or the carbohydrate

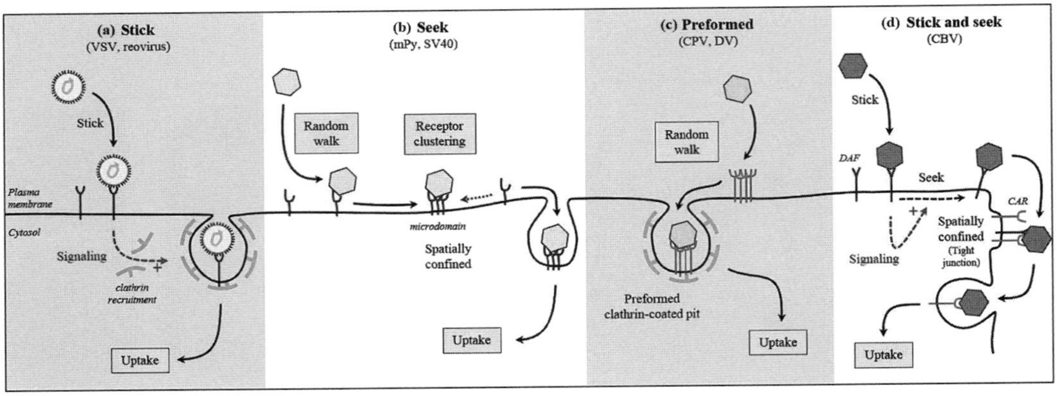

Figure 9.2 Schematics on how viruses find their receptors on the plasma membrane of biological systems and adhere/interact to their host cells. (Reprinted from Boulant et al. [25] under Creative Commons attribution license (CC BY 4.0).)

moieties) that are present either on glycoproteins or glycolipids [31]. In the case of HIV, first the virus attaches to the host cell and then it binds to CD4. This process causes conformational changes in the viral envelop, allowing coreceptor binding that is mediated in part by the V3 loop of viral envelop [32]. This entire process (Figure 9.2) initiates the membrane fusion process as the fusion peptide of gp41 inserts into the target membrane, followed by six-helix bundle formation and complete membrane fusion [32].

Usually, when viruses approach cell surface and binding to receptor(s), some viruses remain stuck at a confined location, from where they will be endocytosed in a passive or signal-induced manner. Other viruses, after binding to their receptor, will diffuse at the cell surface seeking additional receptor molecules [25]. The attachment or adhesion of a virus to a cell plasma membrane results in the interaction of receptor, and endocytosis make virus particles remain spatially confined immediately following interaction, as shown in the surface membrane (see Figure 9.2) [25].

Viruses can form extracellular assemblies that, in composition, resemble the organization and dissemination of bacterial biofilms [33]. Viral biofilms form in tight equilibrium with the infected cell, which provides the machinery for the synthesis and assembly of virus components, and for matrix components that will hold the biofilm together.

Finally, the infected cells provide the initial surface to which the biofilm will adhere. Expression of the viral genome would drive the expression and posttranslational modifications of matrix components, while cell-surface rearrangements (e.g., ruffles and filopodia) [34] would help to maintain viral biofilms on the cell surface. A tightly regulated balance between adhesion, cohesiveness, and dispersion would ensure the equilibrium between viral biofilm generation and dissemination to other cells. An attractive hypothesis is that the production of the viral biofilm could be an antiviral mechanism developed by the infected cell to encase infectious viral particles but which has been hijacked by some viruses to allow them to spread efficiently. The production of the ECM (extracellular matrix) [35] modulated by viral-genome expression in infected cells would favor both viral particles trapped on the surface of the infected cells and virus spread through cell contacts. In the case of influenza viruses, the hemagglutinin (HA) protein binds/adheres to sialic acid residues that are expressed on the airway or alveolar epithelium, triggering endocytosis of the virion [36]. In most cases, influenza A virus infection results in the sequential activation of beneficial and detrimental host-immune pathways in the lung [37].

The finding of virus biofilms suggests many fundamental questions. Furthermore, as extracellular infectious entities involved in virus dissemination, could virus biofilms be potential therapeutic targets? This could be envisaged in two ways: first, virus biofilm formation could be inhibited as a means to reduce virus transmission; second, and probably more interesting, viral biofilms could be modified as a way to allow or enhance specific host immune responses that would reduce chronic infection [38].

9.2.3 Receptor-Mediated Endocytosis and Exocytosis of Viruses

Receptor-mediated endocytosis is a process that transports viruses and bionanoparticles into a target cell host. Equipped with a limited amount of nucleic acid, viruses propagate by parasitizing host cells and multiplying their own viral nucleic acid and protein capsid via the biochemical machinery of the host [39]. The life cycle of a virus follows a sequential route through various sections of the host. It takes about 20–40 minutes for many bacterial phages to finish one life cycle from injection to lysis. For most animal viruses, entering and leaving a host cell are mediated by specific binding of outer coat proteins to specific mobile receptors on the host cell surface. The endocytic pathway is also of interest for understanding possible mechanisms by which nanoclusters might enter human or animal cells, a significant issue for the development of gene and drug delivery [40,41].

Virus entry into the cytoplasm of host cells can be via endocytosis and escape from endosomal vesicles in a process referred to as receptor-mediated endocytosis or by direct penetration from the plasma membrane, known as endocytosis-independent receptor-mediated entry (see Figure 9.3) [25]. In the case of influenza A virus, initially there is binding of HA that is expressed on the surface of the influenza virion, with sialic acid residues linked to cell surface glycans, which induces binding and fusion of the virion with the plasma membrane of the target cell [37]. Usually, the HA in human viruses interacts with sialic acid residues linked to surface glycans via an α-2,6 linkage then enters the host cell via endocytosis where it is trafficked to the lysosome [37].

Studies on targeted drug delivery in cells have identified particle size as an important factor in cellular uptake of bionanoparticles. It has been shown that particle with radii <50 nm exhibit significantly greater uptake compared with particles > 50 nm [36,37]. Size effects and receptor contributions in glycoviral gene delivery have been investigated [38–40]. Receptor-mediated endocytosis is strongly size-dependent with an optimal size of approximately 25 nm.

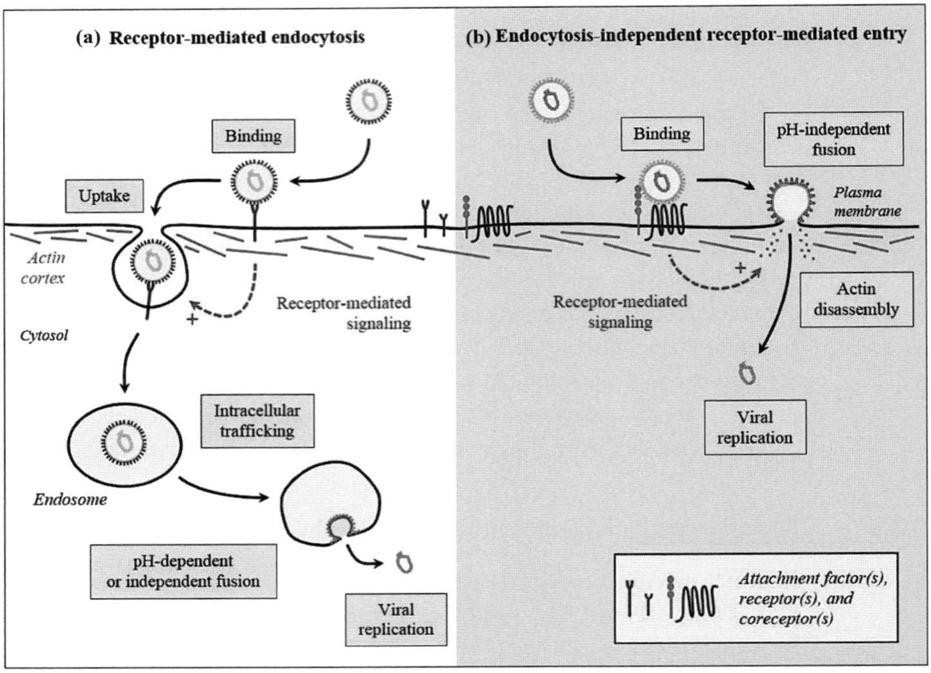

Figure 9.3 (I) Strategy of virus entry into cytoplasm via receptor-mediated endocytosis or endocytosis-independent receptor-mediated entry. (Reprinted from Boulant et al. [25] under Creative Commons attribution license (CC BY 4.0).)

However, in the case of nanoparticles, controlling the intracellular fate under *in vivo* conditions is the great challenge. Most nanoparticles are internalized by cells via endocytosis – the process where nanoparticles bind to receptors and are internalized in membrane-vesicles and are trapped in lysosomes. It has been shown that during this process, BNC is attached to CD47, a self-marker protein that can be used to trick the immune system to consider a foreign nanoparticle as its own [42]. Several nanoparticles have been designed to promote endosomal escape; however, their success remains limited. Macrophages of the liver and spleen are very effective in the removal of nanoparticles from the body. In fact, a majority of the injected nanoparticle dose often ends up sequestered in the liver and spleen [43–45].

9.2.4 Viral-Mediated Gene Delivery

Generally, genomes in viral can consist of DNA or RNA only and depending on the type of virus, their DNA and RNA molecules can be linear or circular as well as double stranded or single stranded [23] (see Figure 9.4). The simplest viruses contain only enough RNA or DNA to encode four proteins [23]. All viruses utilize normal cellular ribosomes, tRNAs, and translation factors for synthesis of their proteins; but in most

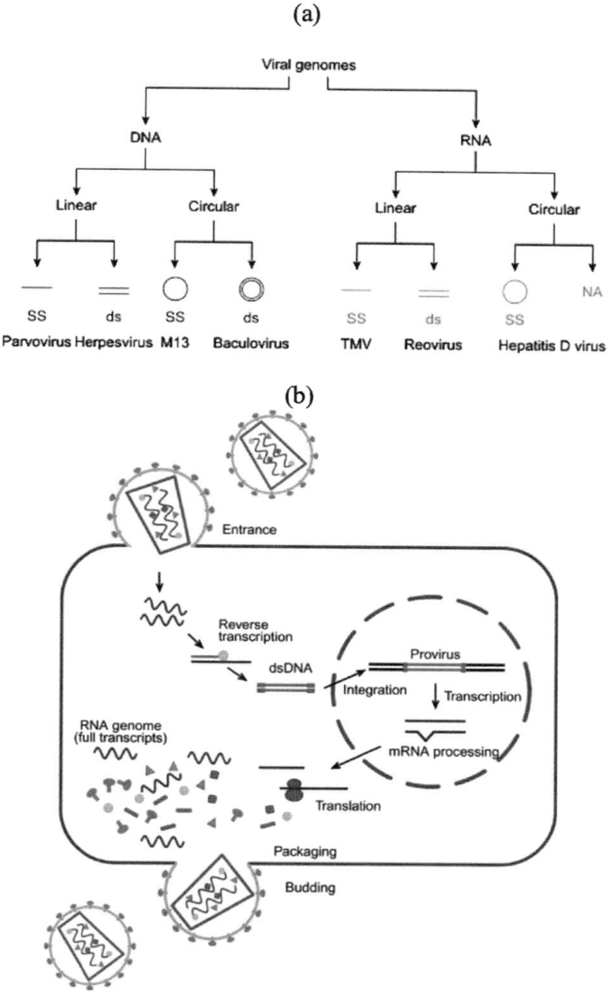

Figure 9.4 (a) Variety of viral genome. (Reprinted from Marintcheva [48], copyright 2018 with permission from Elsevier.) (b) Schematic showing life cycle of retrovirus showing how virus-mediate gene delivery/transcription of DNA/RNA. (Reprinted from Koo et al. [49], copyright 2014 with permission from Elsevier.)

cases, DNA viruses utilize cellular enzymes for synthesis of their DNA genomes and mRNAs [23].

Viral-mediated gene delivery is the best option for gene delivery vehicles [46]. This is why when certain genes carried by cancer-causing viruses integrate into chromosomes of a normal animal cell, the normal cell can be converted to a cancer cell [23]. It has been shown that human immunodeficiency virus (HIV) that have genomes comprises of RNA tend to supply their own polymerase as cells they infect only produce enzymes capable of copying DNA templates [47]. Therefore, targeting of the viral polymerase in HIV with azidothymidine (AZT) tends to enable selective inhibition and minimizes host cell toxicity [47]. Similarly, in the case of a replication-deficient recombinant virus, it is able to efficiently infect quiescent cells, integrate into the host genome, mediate long-term and stable expression, and remain controllable with minimal nonspecific effects [46].

Several types of viral-mediated gene delivery systems exist, each having its own unique qualities. The major classes of viral vectors include the retrovirus (including the lentiviral vector), adenovirus, and adeno-associated virus (AAV) [46]. Retroviruses are RNA viruses that carry a gene for reverse transcriptase, which transcribes the viral genetic material into a double-stranded DNA intermediate that is incorporated into the host DNA [50] (Figure 9.4). This resulting double-stranded DNA can integrate into a mammalian genome system using a virally encoded integrase. In the case of adenoviruses, they are large and are referred to as linear double-stranded DNA viruses [46], which are hairpin-like, with inverted terminal repeats (ITRs). The ITR thus acts as a self-primer to promote primase-independent DNA synthesis, making them important elements in DNA multiplication [51]. Finally, AAV is a single-stranded DNA virus that is helper-dependent and small but is capable of transducing both dividing and nondividing cells by delivering a predominantly episomal transgene product [52,53].

9.3 Implications for Bioinspired Design Nanoparticles/Drug Nanoclusters

Even with the most advanced targeting strategy, intravascularly injected nanoparticles primarily accumulate in the liver and spleen [43–45]. Clearly, better targeting strategies are desired. It is also increasingly recognized that therapeutic targets, especially tumors, are heterogeneous in their biology. Accordingly, targeting strategies must take this heterogeneity into account. Viruses are clearly a source of inspiration to in the design of BNCs. These qualities range from nanoparticles/clusters structure, robustness, attachment capability, endocytosis, systemic drug delivery, and exocytosis. Benefits and limitations of various features of nanoparticles must be taken into account while designing the future generation of therapeutic nanoparticles. These include the use of particle morphology and flexibility to control their *in vivo* behavior, the development of compartmentalized nanoparticles for encapsulating multiple agents, and the use of biomimicry and strategies for enhanced endosomal escape [54].

For example, modification of the nanoparticle surface "stealth" polymers is often used to reduce their clearance by the immune system [55]. Polyethylene glycol (PEG) is often used for this purpose [56]. Other materials such as dextran [57], di-block copolymers [58], and albumin [59] are also used as stealth polymers. Elongated particles exhibit reduced macrophage uptake and enhanced circulation time [60]. The flexibility of nanoparticles has also been shown to enhance circulation times and target accumulation [61,62]. In another strategy, nanoparticles are placed on the surface of red blood cells to reduce their immune clearance [63]. Nanoparticles have also been encapsulated in autologous red blood cells to prolong the circulation [64]. In this section, we explore lessons/inspirations learned and implications from viral assembly by nanoparticle synthesis, viral adhesion to membrane host cells by understanding nanoparticle adhesion, viral-mediated endocytosis and exocytosis by expediting nanoparticle entry/uptake into mammalian cells, and viral gene delivery by exploring nanocluster/nanoparticle drug delivery.

9.3.1 Nanocluster/Nanoparticle Synthesis

9.3.1.1 Biosynthesis and Conjugation of Magnetite Using *Magnetospirillum magneticum*

Obayemi et al. [65] successfully biosynthesized magnetite nanoparticles, which has resulted in the formation of particle with clinical sizes between 10 nm and 60 nm. These biocompatible magnetic nanoparticles are produced in the presence of *Magnetospirillum magneticum* (MM) bacteria that respond to magnetic fields. MM bacteria were cultured in an enriched magnetic spirillum growth medium (EMSGM) at different pH levels. The dependence of the synthesized nanoparticle structure, shape, and size on pH and time were characterized using a combination of transmission electron microscopy (TEM), environmental scanning electron microscopy (ESEM), UV-visible spectrophotometry (UV-Vis) and x-ray diffraction (XRD).

The particles formed were also functionalized via carbodiimide reduction with luteinizing hormone releasing hormone (LHRH), a ligand whose receptors are over-expressed in most breast cancer cells [65] (Figure 9.5). The resulting particles were examined with Fourier transform infrared (FTIR) and quantitative image analysis. The implications of the results are then discussed for the potential development of magnetic nanoparticles for the targeting and treatment of breast cancer. This suggests that these particles may be used for specific targeting and to enrich the magnetic resonance

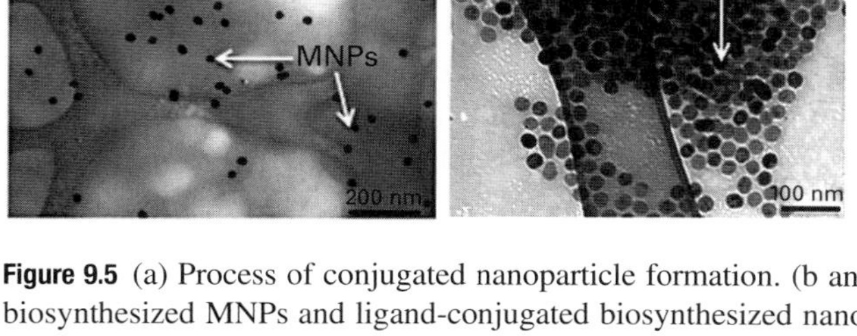

Figure 9.5 (a) Process of conjugated nanoparticle formation. (b and c) TEM micrograph of biosynthesized MNPs and ligand-conjugated biosynthesized nanoparticles, respectively. (Reprinted from Obayemi et al. [65], copyright 2015 with permission from Elsevier.)

imaging (MRI) images associated with the early detection and localized treatment of cancer and arteriosclerosis.

9.3.1.2 Biosynthesis and Conjugation of Gold Nanoparticles/Drug from Bacteria and Plant

In a separate study, Hampp et al. [66] showed that gold nanoparticles (AuNPs) were synthesized using the common soil bacterium, *Bacillus megaterium* (BM). These nanoparticles were conjugated to breast-specific antibody. Confirmation of the AuNPs formation was demonstrated with the aid of UV–Vis spectrophotometry and TEM [66] (Figure 9.6). Also, the TEM images show that well-developed mostly spherical and homogenous nanoparticles develop from the reaction of culture media with aqueous chloroaurate ions at pH 4.

They also showed that the AuNPs were synthesized extracellularly. These results suggest that biosynthesized AuNPs and ligand-conjugated AuNP may be used for the rapid screening of potential ligands and design novel types of AuNPs for the specific targeting of diseased cells such as cancer cells [66].

Furthermore, Dozie Nwachukwu et al. [67] showed in their earlier studies that *Serratia marcescens* can be used for the synthesis of AuNPs both intra- and extracellularly (Figure 9.6a). However, the use of cell-free extract (extracellular) proved to be more effective than the biomass, as it requires lower reaction times. Also, the study indicates that pH 4 is the optimum pH for the synthesis. The SEM image revealed AuNPs within the bacterial cells (Figure 9.6d). This is a strong indication that the factor responsible for the reduction of gold chloride is located within the cell [67]. Using the same extract, a model anticancer drug was synthesized for the effective treatment of cancer.

Strategies learned from these kinds of nanoparticle synthesis studies and assembly can lead to comprehensive understanding and insights in the design of nanoparticles/BNCs with theranostic capability in the specific targeting and treatment of disease cells.

9.3.2 Nanoparticles Adhesion

In this section, the need to understand the role of adhesive interactions between functionalized nanoparticles and cancer cells or diseased cells is relevant in developing effective nanocarriers/BNCs for targeted therapy. In 2018, Hu et al. [68] used the combination of experimental and computational approaches to study the adhesion between triptorelin/LHRH agonist-conjugated PEG-coated magnetite nanoparticles and breast cells. Under *in vitro* conditions, atomic force microscopy technique (Figure 9.7) and molecular dynamics simulations were combined to evaluate the adhesive interaction of the ligand-conjugated nanoparticles to cells [68]. Their results also showed that triptorelin-magnetic nanoparticles (MNPs) have higher adhesion to breast cancer cells than to normal breast cells [68]. Therefore, the presence of triptorelin molecules leads to increased specificity during the targeting of breast cancer cells. This increase in specificity was attributed to the increased incidence of overexpressed LHRH receptors on the surfaces of breast cancer cells [68].

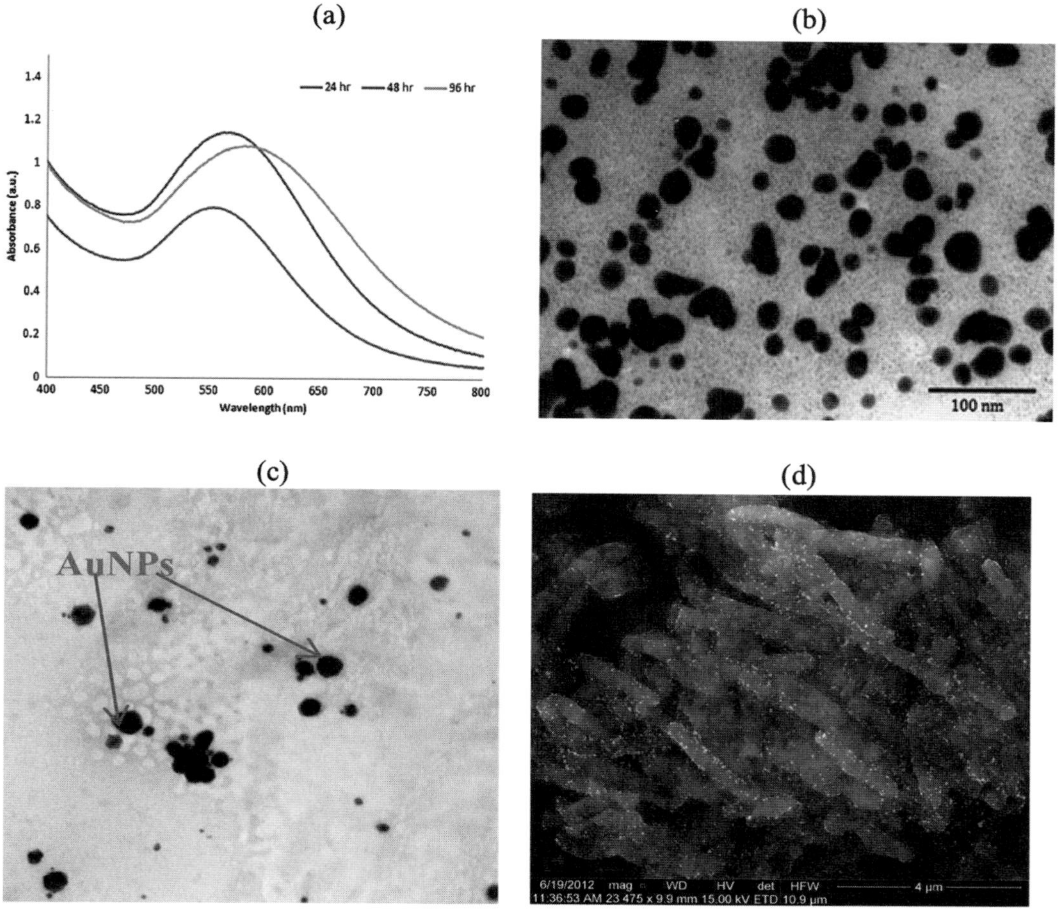

Figure 9.6 (a) TEM micrograph of AuNPs from reactions with culture media and 1 mM HAuCl4 at pH 4 after. (b) UV–Vis absorption spectra of AuNPs. ((a–b) reprinted from Hampp et al. [66], copyright 2012 with permission from Cambridge University Press.) (c) TEM images of the biosynthesized AuNPs obtained from cell-free extract of Serratia marcescens at a pH 4. (d) SEM micrograph showing rod-shaped Serratia marcescens with intracellular AuNPs after 24 h. ((c–d) reprinted by permission from Springer Nature, Dozie-Nwachukwu et al. [67], Copyright 2017.)

These were the specific conclusions made from the study: The work of adhesion between triptorelin-MNP-coated AFM tips and breast cancer cells is 14 times more than the work of adhesion between triptorelin-MNP-coated AFM tips and normal breast cells. Also, the molecular dynamics simulations revealed lower interaction energies in PEG/cell membrane and triptorelin/cell membrane bimaterial pairs (Figure 9.7I) than those in PEG/LHRH and triptorelin/LHRH bimaterial pairs [68]. It was deduced that the

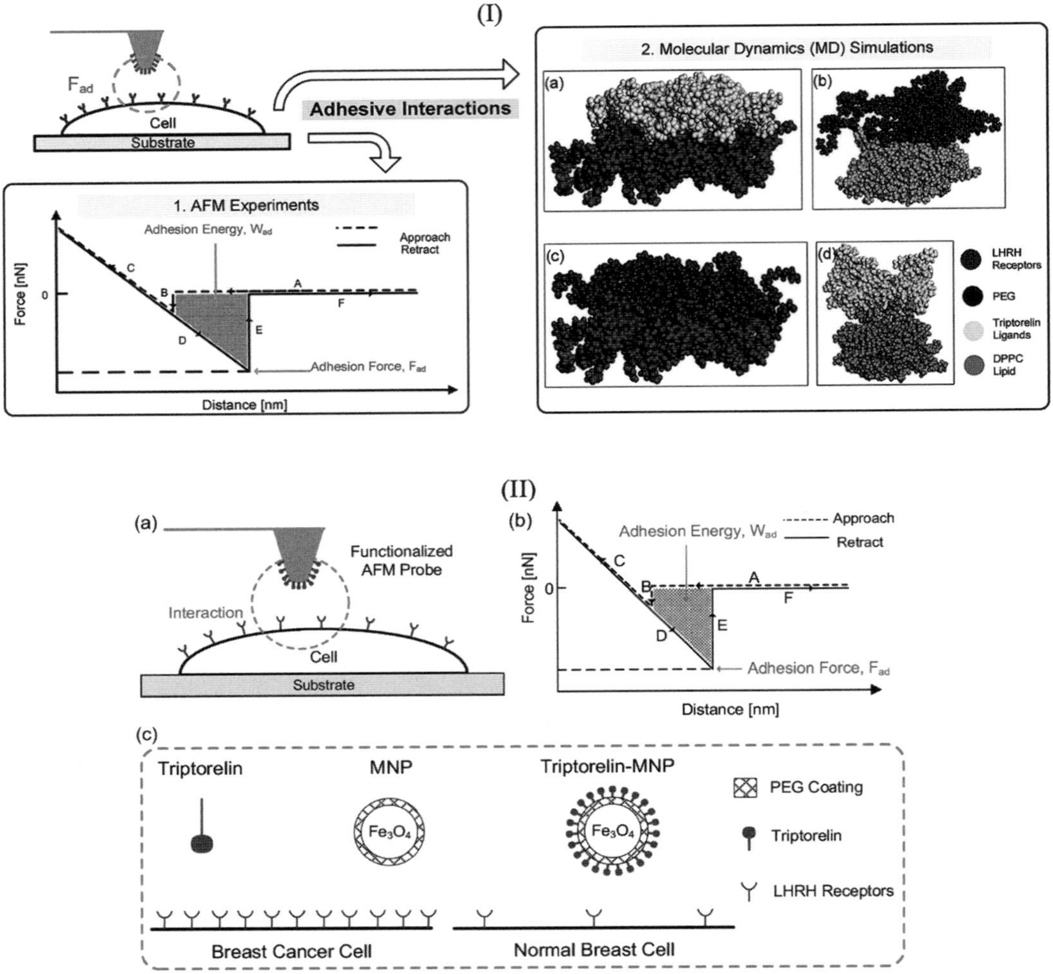

Figure 9.7 (I) Schematic of experimental setup with cell-nanoparticle interaction as well as molecular dynamic simulation of receptor-ligand interactions. (II) (a) Cell-ligand-conjugated nanoparticle interaction, (b) typical force–displacement plot with corresponding stages of force displacement, and (c) schematics of interaction subsystems: AFM tips are functionalized with triptorelin, MNP, or triptorelin-MNPs that interact with breast cancer cells or normal breast cells, respectively. (Reprinted from Hu et al. [68], copyright 2018 with permission from Elsevier.)

former two interactions were dominated by electrostatic interactions while the latter two have overall larger interactions due to the presence of van der Waals interactions [68]. They also showed that the effects of hydrogen bonding are minimal in all of the cases examined in this study [68]. The study concluded that adhesive interactions of LHRH and PEG are additive and thus promote the attachment of triptorelin-MNPs to breast cancer cells [68].

In a similar fashion, Obayemi et al. [69] in 2017 evaluated adhesion forces between components of model nanoparticle systems that were conjugated to LHRH or breast specific antibody (BSA) for the respective targeting of LHRH receptors and BSA receptors on the surfaces of breast cancer cells (MDA-MB-231 cell line) and normal breast cells (MCF 10A). They found that average adhesion forces between ligand-conjugated BMNPs-LHRH coated-AFM tips and MDA-MB-231 triple negative breast cancer (TNBC) cells are over six times greater than those to normal breast cells [69]. They concluded that the increased adhesion of the LHRH-conjugated or EphA2-conjugated MNPs to TNBC cells is attributed to the increased incidence of LHRH and EphA2 receptors on the surfaces of breast cancer cells. The understanding and insights from these adhesion forces and energies from these kinds of studies are critical in the development of BNCs/nanoparticles for specific targeting of disease cells.

9.3.3 Endocytosis and Exocytosis of Nanoparticles

There is no doubt that receptor-mediated endocytosis and exocytosis is one of the crucial phenomena a virus can adopt to enter or leave an animal cell. Using this kind of inspiration from viruses, engineered ligand-conjugated nanoparticles with receptors that are expressed on disease cells when injected into the body can find their way specifically to target the corresponding disease cells. To enable cell-type specific targeting, nanoparticles are often surface-coated with biopolymers or macromolecules and ligand-conjugated nanoparticles to enable them bind specifically to their respective complementary receptors on the cell membrane of disease cells.

It was established that endocytosis of nanoparticles or nanoparticle entry into cells are categorized in four pathways namely: clathrin/caveolar-mediated endocytosis, phagocytosis, macropinocytosis, and pinocytosis [70], whereas exocytosis of nanoparticles, or nanoparticle exit from the cell, uses the following pathways: lysosome secretion, vesicle-related secretion, and nonvesicle-related secretion (Figure 9.8).

Hu et al. [71], explored the entry and endocytosis of LHRH-conjugated PEG-coated magnetite nanoparticles into TNBC cells and normal breast cells (Figure 9.8a). The entry of the ligand-conjugated nanoparticles was studied using thermodynamics and kinetics models under *in vitro* conditions [71]. Their results revealed that LHRH-MNPs enter preferentially into breast cancer cells via the receptor-mediated endocytosis pathway and that the interactions between LHRH receptors on breast cancer and ligands results in significant LHRH-MNP uptake after 3 hours [71]. The cluster size distribution and uptake efficacy show that the entry of LHRH-MNPs into breast cancer cells depend on the highly efficient receptors [71]. Both the thermodynamic and kinetic model predictions are in good agreement with experimental observations.

By critically understanding endocytosis and exocytosis of viruses in relation to nanoparticles [72], it paves way for new design of safe nanoparticles that can efficiently attach, enter, deliver a drug, and leave human cells and tissues. It can also help us to develop more efficient intracellular drug-delivery nanosystems [70].

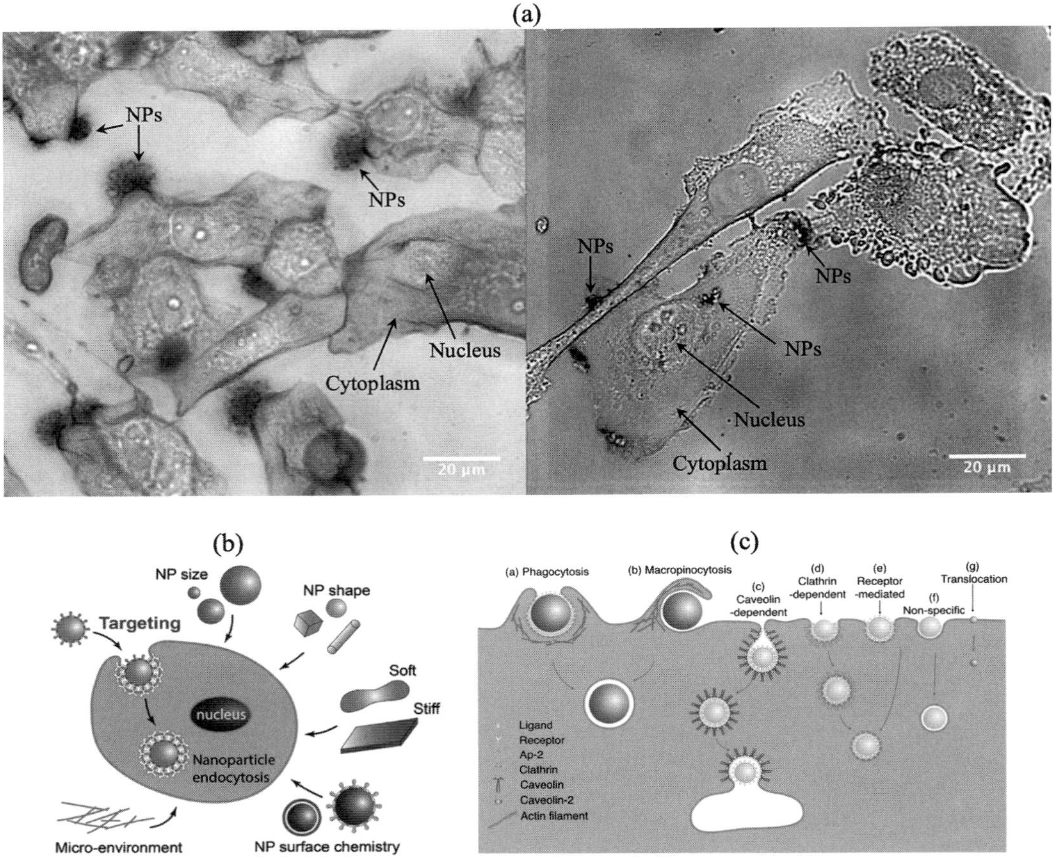

Figure 9.8 (a) LHRH-MNPs internalization into MDA-MB-231 breast cancer cells. (Reprinted from Hu et al. [71], copyright 2018 with permission from Elsevier.) (b–c) Schematic of possible endocytosis and exocytosis pathways of nanoparticles. (Adapted with permission from Zhang et al. [72]. copyright 2015 American Chemical Society.

9.3.4 Nanoparticle-Mediated Drug Delivery System

In a similar fashion to viral-mediated gene delivery, we will now discuss drug-loaded nanocarriers in the delivery of drug to sites of interest (Figure 9.9a). Recently, nanoparticles, or nanosystems, have gained significant attention in various biomedical applications. The role of nanoparticles in drug delivery is crucial because of the small dimensions that enable them cross the blood–brain barrier and operate at a cellular level (Figure 9.9b). Thus incorporating drug molecules in nanosystems provides greater advantages, like biodistribution of active compounds, protection against degradation, and improved drug attachment, passage, targeting, expulsion, and communication with biological barriers [72].

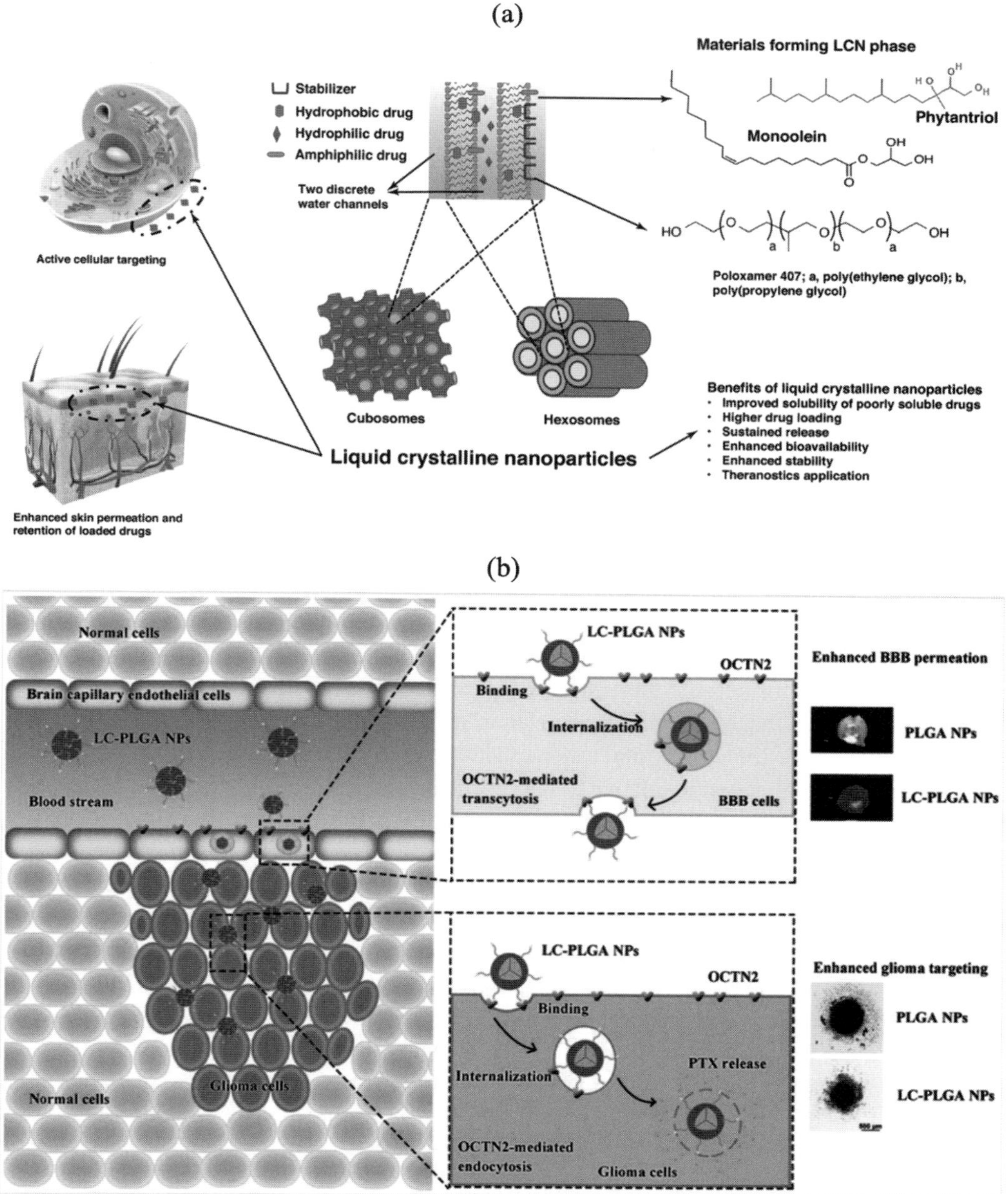

Figure 9.9 (a) Perspective of nanodrug delivery systems (DDS)-mediated treatment for acute atherosclerosis. (Reprinted from Madheswaran et al. [73], copyright 2019 with permission from Elsevier.) (b) Schematic of ligand-based targeted therapy drug delivery system. (Reprinted with permission from Kou et al. [74], © 2017 Taylor & Francis Group.)

Studies have also explored the effectiveness of using nanosystems like magnetic nanoparticles, micelles, dendrimers, polymers, nanospheres, and liposomes as carriers in drug delivery. In most cases, drug molecules and/or diagnostic reporters are loaded in these nanosystems and are transported in the bloodstream, adhere to the endothelium, diffuse through the intercellular space, specifically adhere to the diseased cells, enter the cells via different mechanisms and release the drug molecules effectively [72].

The idea that nanosystems have unique physical and biological properties that might be used to overcome the obstacles of gene and drug delivery has gained interest in recent years. Thus, the development of drug-loaded and multifunctional nanosystems will help to promote the future medical applications of nanotechnology. Again, nanosystems can be designed with different compositions and biological properties in a way that they can be made to have characteristic self-assembled properties that include coatings for stability, conjugated-ligands with targeting biomolecules [18,20,21]. Usually, for a well-designed and robust nanosized drug delivery system, when they are injected into the body system, rapid and nonspecific interactions with albumin occur through ionic and hydrophobic interactions [75].

9.4 Summary and Concluding Remarks

This chapter highlights progress and inspiration from viruses for the design of nanosystems for specific targeting and treatment of disease cells. These include the self-assembly nature of viruses in relation to nanoparticle synthesis with case studies on biosynthesized magnetite nanoparticles and gold nanoparticles. Also, bioinspiration from viral adhesion relative to nanoparticles adhesion to disease cells was emphasized under *in vitro* conditions. The implication of this were discussed in the development of robust nanosystems/nanocarriers in the specific targeting of breast cancer. Also, this chapter presents the concept of endocytosis and exocytosis using the inspiration gained from viruses to discuss their implications in the context of nanoparticle entry/uptake and exit from host cells. Finally, nanoparticle-mediated drug delivery systems as inspired/motivated by virus-mediated gene delivery were expatiated with their implications in specific drug delivery process. All of these boinspired phenomena from viruses, when integrated into a nanostructure, can effectively open up great potentials in biomedical sciences in the area of disease targeting and treatment. For instance, the bioinspiration from the development of nanoparticles creates some novel opportunities for the specific detection and treatment of diseases such as cancer.

References

1. Sanchez, C., Arribart, H., & Giraud Guille, M. M. (2005). Biomimetism and bioinspiration as tools for the design of innovative materials and systems. *Nature Materials*, 4, 277–288.
2. Fauci, A. S. (2006). Emerging and re-emerging infectious diseases: Influenza as a prototype of the host-pathogen balancing act. *Cell*, 124, 665–670.

3. Morens, D. M., Folkers, G. K., & Fauci, A. S. (2008). Emerging infections: A perpetual challenge. *Lancet Infectious Diseases*, 8, 710–719.
4. Flint, S. J., Enquist, L. W., Krug, R. M., Racaniello, V. R., & Skalka, A. M. (2000). *Principles of virology, molecular biology, pathogenesis, and control.* ASM Press.
5. Daniel, M. -C., & Astruc, D. (2004). Gold nanoparticles: Assembly, supramolecular chemistry, quantum-size-related properties, and applications toward biology, catalysis, and nanotechnology. *Chemical Reviews*, 104(1), 293–346.
6. Cossart, P., & Helenius, A. (2014). Endocytosis of viruses and bacteria. *Cold Spring Harbor Perspectives in Biology*, 6, 1–28.
7. Wang, S., Chen, K.-J., Wu, T.-H., et al. (2010). Photothermal effects of supramolecularly assembled gold nanoparticles for the targeted treatment of cancer cells. *Angewandte Chemie International Edition*, 49, 3777–3781.
8. Whitesides, G. M., & Grzybowski, B. (2002) Self-assembly at all scales. *Science*, 295(5564), 2418–2421.
9. Guo, S., & Huang, L. (2011). Nanoparticles escaping RES and endosome: Challenges for siRNA delivery for cancer therapy. *Journal of Nanomaterials*, 2011, 742895.
10. Jain, R. K., & Stylianopoulos, T. (2010). Delivering nanomedicine to solid tumors. *Nature Reviews Clinical Oncology*, 7, 653–664.
11. Jain, R. K. (1999). Transport of molecules, particles, and cells in solid tumors. *Annual Review of Biomedical Engineering*, 1, 241–263.
12. Galdiero, S., Falanga, A., Vitiello, M., Cantisani, M., Marra, V., & Galdiero, M. (2011). Silver nanoparticles as potential antiviral agents. *Molecules*, 16(10), 8894–8918.
13. Jain, R. K. (1990). Vascular and interstitial barriers to delivery of therapeutic agents in tumors. *Cancer and Metastasis Review*, 9, 253–266.
14. Mariampillai, A., Leung, M. K. K., Jarvi, M., et al. (2010). Optimized speckle variance OCT imaging of microvasculature. *Optics Letters*, 35(8), 1257–1259.
15. Krissinel, E., & Henrick, K., (2007). Inference of macromolecular assemblies from crystalline state. *Journal of Molecular Biology*, 372(3), 774–797.
16. Mitragotri, S., & Stayton, P. (2014). Organic nanoparticles for drug delivery and imaging. *MRS Bulletin*, 39(03), 219–223.
17. Oni, Y., Hao, K., & Dozie-Nwachukwu, S., et al. (2014). Gold nanoparticles for cancer detection and treatment: The role of adhesion. *Journal of Applied Physics*, 115(8), 084305–084305-8.
18. Russell, J. T., Lin, Y., Böker, A., et al. (2005). Self-assembly of cross-linking of bionanoparticles at liquid–liquid interfaces. *Angewandte Chemie International Edition*, 44, 2420–2426.
19. Burnett, R. M. (1985). The structure of the adenovirus capsid: II. The packing symmetry of hexon and its implications for viral architecture. *Journal of Molecular Biology*, 185(1), 125–143.
20. Chen, S.-H., Wang, D.-C., Chen, G.-Y., & Jan, C.-L. (2006). Self-control of the self-assembly gold nanoparticles. *Journal of Medical and Biological Engineering*, 26(3), 137–142.
21. Bishop, K. J. M., Wilmer, C. E., Sohand, S., & Grzybowski, B. A. (2009). Nanoscale forces and their uses in self-assembly. *Small*, 5(14), 1600–1630.
22. Remskar, M., Skraba, Z., Cléton, F., Sanjinés R., & Lévy, F. (1996). MoS_2 as microtubes. *Applied Physics Letters*, 69, 351.
23. Lodish H, Berk A, Zipursky SL, et al. *Molecular Cell Biology*. 4th edition. New York: W. H. Freeman; 2000. Section 6.3, Viruses: Structure, Function, and Uses. Available from: https://www.ncbi.nlm.nih.gov/books/NBK21523/

24. Carrillo-Tripp, M., Shepherd, C. M., Borelli, I.A., et al. (2009). VIPERdb2: An enhanced and web API enabled relational database for structural virology. *Nucleic Acids Research*, 37, D436–D442.
25. Boulant, S., Stanifer, M., & Lozach, P. Y. (2015). Dynamics of virus-receptor interactions in virus binding, signaling, and endocytosis. *Viruses*, 7, 2794–2815.
26. Wessells, N. K., Spooner, B. S., Ash, J. F., et al. (1971). Microfilaments in cellular and developmental processes. *Science*, 171(3967), 135–143.
27. Kronberg, R. (1974). Chromatin structure: A repeating unit of histones and DNA. *Science*, 184(4139), 868–871.
28. Douglas, S. M., Dietz, H., Liedl, T., et al. (2009). Self-assembly of DNA into nanoscale three-dimensional shapes. *Nature*, 459, 414–418.
29. Seeman, N. C. (2003). DNA in a material world. *Nature*, 421, 427–431.
30. Costerton, J. W., Stewart, P. S., & Greenberg, E. P. (1999). Bacterial biofilms: A common cause of persistent infections. *Science*, 284(5418), 1318–1322.
31. Grove, J., & Marsh, M. (2011). The cell biology of receptor-mediated virus entry. *Journal of Cell Biology*, 195, 1071–1082.
32. Wilen, C. B., Tilton, J.C., & Doms, R.W. (2012). HIV: cell binding and entry. *Cold Spring Harbor Perspectives in Medicine*, 2, a006866.
33. Chhabra, E. S., & Higgs, H. N. (2007). The many faces of actin: Matching assembly factors with cellular structures. *Nature Cell Biology*, 9, 1110–1121.
34. Bryant, S. J., & Anseth, K. S. (2002). Hydrogel properties influence ECM production by chondrocytes photoencapsulated in poly(ethylene glycol) hydrogels. *Journal of Biomedical Materials Research*, 59(1), 63–72.
35. Gao, H., Shi, W., & Freund, L. B. (2005). Mechanics of receptor-mediated endocytosis. *Proceedings of the National Academy of Sciences of the United States of America*, 102(7), 9469–9474.
36. Rust, M. J., Lakadamyali, M., & Zhang, F., et al. (2004). Assembly of endocytic machinery around individual influenza viruses during viral entry. *Nature Structural & Molecular Biology*, 11, 567–573.
37. Herold, S., Becker, C., Ridge, K. M., & Budinger, G. R. (2015). Influenza virus-induced lung injury: Pathogenesis and implications for treatment. *European Respiratory Society*, 45(5), 1463–1478.
38. Thoulouze, M. I., & Alcover, A. (2011). Can viruses form biofilms? *Trends in Microbiology*, 19, 257–262.
39. Lodish, H., Berk, A., Zipursky, S. L., et al. (2000). Viruses: Structure, function, and uses. In *Molecular cell biology* (4th ed.). W. H. Freeman.
40. Gao, H., Shi, W., & Freund, L. B. (2005). Mechanics of receptor-mediated endocytosis. *Proceedings of the National Academy of Sciences of the United States of America*, 102(27), 9469–9474.
41. Husseiny, M. I., Abd El-Aziz, M., Badr, Y., & Mahmoud, M. A. (2007). Biosynthesis of gold nanoparticles using *Pseudomonas aeruginosa*. *Spectrochimica Acta Part A: Molecular and Biomolecular Spectroscopy*, 67(3–4), 1003–1006.
42. Luo, Y. H., Chang, L. W., & Lin, P. (2015). Metal-based nanoparticles and the immune system: Activation, inflammation, and potential applications. *BioMed Research International*, 2015, 143720.
43. Bae, Y. H., & Park, K. (2011). Targeted drug delivery to tumors: Myths, reality and possibility. *Journal of Controlled Release*, 153(3), 198–205.

44. Brannon-Peppas, L., & Blanchette, J. O. (2012). Nanoparticle and targeted systems for cancer therapy. *Advanced Drug Delivery Reviews*, 64, 206–212.
45. Moein Moghimi, S., Christy Hunter, A., & Clifford Murray, J. (2001). Long-circulating and target-specific nanoparticles: Theory to practice. *Pharmacological Reviews*, 53(2), 283–318.
46. Sarkissian, S. D & Raizada, M. K. (2007). Therapeutic Potential of Systemic Gene Transfer Strategy for Hypertension and Cardiovascular Disease. In G. Y.H. Lip and J. E. Hall (Eds). Comprehensive Hypertension, (pp 429–445). Oxford, UK: Elsevier Publishing. doi: https://doi.org/10.1016/B978-0-323-03961-1.50040-4.
47. Söderström A, Norkrans G, Lindh M. Hepatitis B virus DNA during pregnancy and post partum: aspects on vertical transmission. Scand J Infect Dis 2003; 35: 814–9.
48. Marintcheva, B. (2018). Introduction to viral structure, diversity and biology. In *Harnessing the power of viruses* (p. 8). Academic Press.
49. Koo, B. C., Kwon, M. S., & Kim, T. (2014). Retrovirus-mediated gene transfer. In C. A. Pinkert (Ed.), *Transgenic animal technology: A laboratory handbook* (3rd ed.). Elsevier; pp. 167–194.
50. Mali, S. (2013). Delivery systems for gene therapy. *Indian Journal of Human Genetics*, 19(1), 3–8.
51. Lee, C. S., Bishop, E. S., & Zhang, R., et al. (2017). Adenovirus-mediated gene delivery: Potential applications for gene and cell-based therapies in the new era of personalized medicine. *Genes and Diseases*, 4, 43–63.
52. Samulski, R. J., & Muzyczka, N. (2014). AAV-mediated gene therapy for research and therapeutic purposes. *Annual Review of Virology*, 1(1), 427–451.
53. Balakrishnan, B., & Jayandharan, G. R. (2014). Basic biology of adeno-associated virus (AAV) vectors used in gene therapy. *Current Gene Therapy*, 14(2), 86–100.
54. Doshi, N., & Mitragotri, S. (2009). Designer biomaterials for nanomedicine. *Advanced Functional Materials*, 19(24), 3843–3854.
55. Luis Elechiguerra, J., Burt, J. L., Morones, J. R., et al. (2005). Interaction of silver nanoparticles with HIV-1. *Journal of Nanobiotechnology*, 3, 6.
56. Alexis, F., Pridgen, E., Molnar, L. K., & Farokhzad, O. C. (2008). Factors affecting the clearance and biodistribution of polymeric nanoparticles. *Molecular Pharmaceutics*, 5(4), 505–515.
57. Simpson, C. A., Agrawal, A. C., Balinski, A., Harkness, K. M., & Cliffel, D. E. (2011). Short-chain PEG mixed monolayer protected gold clusters increase clearance and red blood cell counts. *ACS Nano*, 5(5), 3577–3584.
58. Amoozgar, Z., & Yeo, Y. (2012). Recent advances in stealth coating of nanoparticle drug delivery systems. *Wiley Interdisciplinary Reviews: Nanomedicine and Nanobiotechnology*, 4(2), 219–233.
59. Farokhzad, O. C., & Langer, R. (2006). Nanomedicine: Developing smarter therapeutic and diagnostic modalities. *Advanced Drug Delivery Reviews*, 58(14), 1456–1459.
60. Jin-Wook, Y., Chambers, E., & Samir, M. (2010). Factors that control the circulation time of nanoparticles in blood: Challenges, solutions and future prospects. *Current Pharmaceutical Design*, 16(21), 2298–2307.
61. Berry, C. C., & Curtis, A. S. G. (2003). Functionalisation of magnetic nanoparticles for applications in biomedicine. *Journal of Physics D: Applied Physics*, 36, R198.
62. Wang, M., & Thanou. M. (2010). Targeting nanoparticles to cancer. *Pharmacological Research*, 62(2), 90–99.
63. Gupta, A. K., & Gupta, M. (2005). Synthesis and surface engineering of iron oxide nanoparticles for biomedical applications. *Biomaterials*, 26(18), 3995–4021.

64. Yoo, J. -W., Irvine, D. J., Discher, D. E., & Mitragotri, S. (2011). Bio-inspired, bioengineered and biomimetic drug delivery carriers. *Nature Reviews Drug Discovery*, 10, 521–535.
65. Obayemi, J. D., Dozie-Nwachukwu, S., Danyuo, Y., et al. (2015). Biosynthesis and the conjugation of magnetite nanoparticles with luteinizing hormone releasing hormone (LHRH). *Materials Science and Engineering C*, 46, 482–496.
66. Hampp, E., Botah, R., Odusanya, S., Anuku, N., Malatesta, K., & Soboyejo, W. O. (2012). Biosynthesis and adhesion of gold nanoparticles for breast cancer detection and treatment. *Journal of Materials Research*, 27(22), 2891.
67. Dozie-Nwachukwu, S., Obayemi, J. D., Danyuo, Y., et al. (2017). Biosynthesis of gold nanoparticles and gold/prodigiosin nanoparticles with *Serratia marcescens* bacteria. *Waste and Biomass Valorization*, 8(6), 2045–2059. doi 10.1007/s12649-016-9734-7
68. Hu, J., Youssefian, S., Obayemi, J., Malatesta, K., Rahbar, N., & Soboyejo, W. (2018). Investigation of adhesive interactions in the specific targeting of triptorelin-conjugated PEG-coated magnetite nanoparticles to breast cancer cells. *Acta Biomaterialia*, 71, 363–378.
69. Obayemi, J. D., Hu, J., Uzonwanne, V. O., et al. (2017). Adhesion of ligand-conjugated biosynthesized magnetite nanoparticles to triple negative breast cancer cells. *Journal of the Mechanical Behavior of Biomedical Materials*, 30(8), 2141.
70. Oh, N., & Park, J. H. (2014). Endocytosis and exocytosis of nanoparticles in mammalian cells. *International Journal of Nanomedicine*, 9(Suppl. 1), 51–63.
71. Hu, J., Obayemi, J. D., Malatesta, K., Košmrlj, A., & Soboyejo, W. O. (2018). Enhanced cellular uptake of LHRH-conjugated PEG-coated magnetite nanoparticles for specific targeting of triple negative breast cancer cells. *Materials Science and Engineering C*, 88, 32–45.
72. Zhang, S., Gao, H., & Bao, G. (2015). Physical principles of nanoparticle cellular endocytosis. *ACS Nano*, 9, 8655–8671.
73. Madheswaran, T., Kandasamy, M., Bose, R. J., & Karuppagounder, V. (2019). Current potential and challenges in the advances of liquid crystalline nanoparticles as drug delivery systems. *Drug Discovery Today*, 24(7), 1405–1412.
74. Kou, L., Hou, Y., Yao, Q., et al. (2017). L-Carnitine-conjugated nanoparticles to promote permeation across bloodbrain barrier and to target glioma cells for drug delivery via the novel organic cation/carnitine transporter OCTN2. *Artificial Cells, Nanomedicine and Biotechnology*, 3, 1–12.
75. Kumar, A., Mansour, H. M., Friedman, A., & Blough, E. R. (2013). *Nanomedicine in drug delivery*. CRC Press.

Part III

Natural Phenomena

10 Aquatic Animals Operating at High Reynolds Numbers

Biomimetic Opportunities for AUV Applications*

Frank E. Fish

10.1 Introduction

Size does matter. Whether small or large in body size, all organisms obey the laws of physics and thus are subjected to forces imposed by the physical environment. These forces place constraints on the level of performance in regard to physiology (e.g., metabolic rate, heat transfer), morphological design (e.g., skeletal framework, muscle mechanics), and behavior (e.g., predator–prey interactions, flight, locomotor speed). The structural and functional consequences of a change in size are referred to as scaling [1].

Evolutionarily, there has been a trend to push the limits of size as organisms became more complex [2–5]. This trend over millions of years was initiated with microscopic single-celled life forms and culminated in giant sequoias, dinosaurs, and whales. However, the transition from small to large has not always been a smooth regular change, as the forces that impinge on any particular body size may radically vary according to scale.

When size is changed, there can be a shift in the dominant forces on a body with regard to its function. This shift can place constraints on the biological system and require a change in design, materials, or performance [1]. The force due to gravity (i.e., weight) varies with the mass of an animal, although the gravitational acceleration remains relatively constant on Earth. In this case, there is a smooth transition as weight increases regularly with increasing size. However, if a heavy animal attempts to walk on water, the surface tension of the fluid will not support the increased weight as it would for a smaller species [6]. A water strider is able to remain on the water surface, as the surface tension from cohesion of the water molecules is greater than the weight of the insect [7,8]. A change in design is necessary by increasing surface area and use of water

* I would like to thank John Dabiri for recommending me to the editors of the book to write this review. I appreciate the comments and assistance of Danielle Adams, Hilary Bart-Smith, Haibo Dong, Janet Fontanella, William Gough, Keith Moored, Alexander Smits, Kelsey Tennett, Paul Webb, Timothy Wei, Daniel Weihs, and Terrie Williams in support of much of the fundamental research reported in this review. This work was supported by grants from the Multidisciplinary University Research Initiative program from the Office of Naval Research (Robert Brizzolara, program officer: N000140810642, N000141410533).

displacement to provide buoyancy to remain afloat. In this case, an elephant can float in water but is unable to prevent full immersion and walk on the surface of the water [9,10].

Bioinspired design and biomimetic solutions are currently being applied to the creation of biological autonomous underwater vehicles (BAUVs) based on animals [11,12]. From an engineering perspective, an aquatic animal is an autonomous underwater vehicle (AUV) and can be described as a mobile vehicle that possesses a multimodal, high-band-width interface to its environment [13]. The propulsive and maneuvering systems of aquatic animals hold the promise for improved performance for AUVs and vessels operating throughout the ocean environment [12,14–18]. Naval interests are concerned with improvements to propulsive systems for speed, maneuverability, efficiency/economy, and stealth, including silence and wakelessness [15,18–20]. In attempting to use nature as the inspiration for BAUVs and marine vessels, scale becomes of paramount importance. Despite the overwhelming number of animals that exist in the oceans, the vast majority are small – often composing the plankton. Nekton are animals capable of swimming independent of water currents, and a number of these can travel long distances in the open ocean. However, many of these species are still too small to be scaled to dimensions and actions that can be effectively transitioned to engineered maritime systems. Such discrepancies of scale limit the effective transition of natural designs to engineered technologies. The problem of scale for use in biomimicry may be reduced by finding areas of overlap in size and performance between a biological organisms and the engineered application.

The focus of this chapter is to examine why large fast-swimming animals that use lift-based propulsion represent the ideal systems to be emulated for biomimetic application of marine vehicles. Identification of the hydrodynamic scaling relationships demonstrate the practicality of emulating high Reynolds number, inertial-based swimming for large-scale engineered systems that are associated with current technologies and tasks. An understanding of the advantages of modeling high Reynolds number swimmers will help to develop models for control systems and alternative methods of propulsion.

10.2 Hydrodynamic Scaling

In the field of fluid mechanics, the nonlinear dynamics of fluid flow define particular regions of operation that affect the forces experienced by swimming animals. The forces experienced by a small animal swimming at a slow speed are radically different from the forces experienced by a large animal swimming at a high speed. Although the medium is the same, moving through water for a very small organism versus a large animal would, from our own perspective, be like moving through highly viscous molasses [21].

The hydrodynamic forces determine the pattern of fluid movement around a swimming animal. Depending on the size of the animal and its swimming speed, the dominant forces are viscous, inertial, or a combination of the two. Viscous forces are due to friction from the cohesive attraction of adjacent water molecules and adhesion between the water and a solid surface. The friction from deformation causes the fluid to shear. The resistance to deformation is expressed as the viscosity (μ) of the fluid [22,23]. Inertial forces resist changes in velocity and are related to the density (ρ) of the fluid [22–24].

The ratio of the inertial and viscous forces is related to the characteristics of the water (density, viscosity), size, and velocity, and it is expressed as the dimensionless Reynolds number [22,23]. Reynolds number (*Re*) is the velocity (*U*) times a characteristic length (*L*; typically body length) divided by the kinematic viscosity (ν):

$$Re = UL/\nu,$$

where ν is μ/ρ. When Reynolds numbers are equal between two situations, the physical characteristics of the flow will be equivalent [23]. Equal *Re* displays geometric similarity in flow patterns. The implication is that regardless of phylogeny or medium, when operating at equivalent *Re*, the shape of the body and propulsor will be similar if the same propulsion mechanism is used [23]. While operating at similar *Re*, tuna and dolphins generate thrust from lift by oscillating a relatively stiff semilunate caudal fin [22,24,25]. The caudal fin has a fusiform cross-sectional design, high aspect ratio planform with tapering tips, and is attached to a rigid, streamlined body with a narrow, streamlined caudal peduncle [22,24,25]. The phylogenetic diversity of thunniform swimmers includes not only tuna and cetaceans (e.g., whales and dolphins), but also lamnid sharks (e.g., great white, mako), and extinct ichthyosaurs. The similarity of operational *Re* and swimming mechanics has mandated morphologies that represent the quintessential example of evolutionary convergence [26].

At low *Re*, viscous forces dominate. Small swimmers (e.g., bacteria, sperm) that swim at relatively slow speeds fall within a low *Re* range of <1 [27]. At this *Re*, inertial forces are insignificant. The flow adjacent to the body stays attached to the surface and does not separate until the trailing edge. The drag (i.e., resistance to forward movement) is derived from frictional forces and decreases with increasing *Re* [28]. Most planktonic animals (e.g., copepods) are within an intermediate range of *Re* ($1 < Re < 1{,}000$) [29–31], where both viscous and inertial forces are important (Figure 10.1). Within this range of intermediate *Re*, pressure and friction are important in determining the magnitude of the drag, as the adjacent fluid tends to separate upstream of the trailing edge of a body, producing a broad wake [28]. However, drag continues to decrease. Nektonic animals (e.g., fish, whales) swim at high *Re* of $>1{,}000$ and generally at $10^4 < Re < 10^8$ [27]. In the high *Re* range, work is performed to accelerate a mass of water for propulsion. Viscous forces are small and confined to a thin boundary layer that is attached to the surface of the body. The distribution of pressure around the body further affects the drag. In nektonic animals, adverse pressure gradients that can greatly increase drag are minimized by designing bodies with a highly streamlined fusiform body shape [22,23,32,33].

Above *Re* of 1,000, the drag remains effectively constant [28,34]. However as the *Re* increases in the inertial dominated region, flow in the boundary layer becomes unstable and susceptible to surface texture and disturbance that can cause turbulence. The transition to turbulent flow does not correspond to a single *Re* but may range from

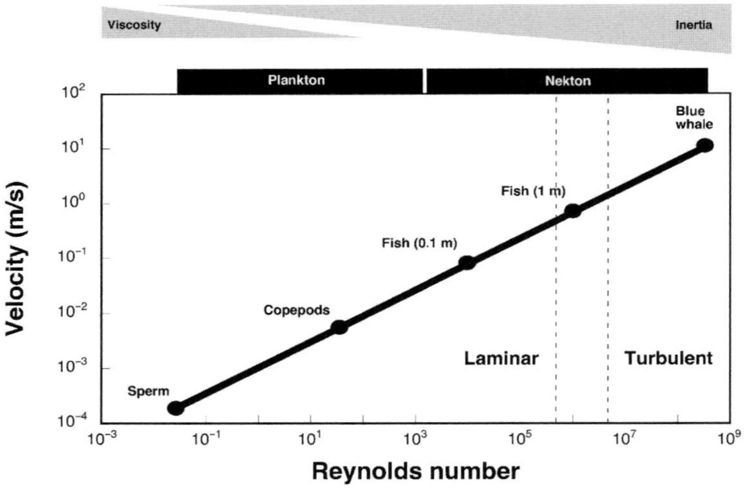

Figure 10.1 Relationship of swimming velocity and Reynolds number over a size range from sperm to whale. Gray triangles show the relative proportion of the viscous and inertial forces with respect to the Reynolds number. Black boxes indicate the range of Reynolds numbers for plankton and nekton. Vertical dashed lines indicate the Reynolds number boundaries for laminar and turbulent flows based on a flat plate. (Adapted from Videler [40].)

5×10^5 to 5×10^6 (Figure 10.1). Below 5×10^5 under stable conditions, the boundary layer flow will be laminar over the surface of a flat plate or streamlined body [22]. Transition to turbulent flow above 5×10^5 will depend on environmental conditions, body shape, and surface roughness [22,32,35]. However, at very high *Re* ($>10^6$), even minute disturbances will result in turbulent flow. The *Re* of a fish 0.1 m in length swimming at one body length per second (0.1 m/s) has a *Re* of 9.58×10^3, so flow is laminar over most of the fish's body. However, the *Re* of fish 1.0 m in length swimming at one body length per second (1.0 m/s) is 9.58×10^5, respectively, so flow is transitional and may be either laminar or turbulent over part or all of the body. The *Re* for large, fast swimming fishes and all marine mammals is even higher ($\geq 10^6$) due to their large size, so flow is turbulent over most of the body at routine swim speeds (Figure 10.1). A 30-m blue whale swimming at 10.2 m/s [36] would have a *Re* of 2.93×10^8, indicating turbulence. The flow around high *Re* swimmers would be consistent with engineered marine vehicles, as *Re* is within a similar range. Although a Seawolf-class submarine at top speed (20.1 m/s) would have a *Re* of 2.2×10^9, which is an order of magnitude greater than the *Re* for the blue whale, the flow dynamics of the submarine and the whale are expected to be similar [37]. The REMUS (Remote Environmental Monitoring UnitS) 600 AUV has a *Re* of 8.5×10^6 at 2.6 m/s (http://auvac.org/uploads/platform_pdf/remus600web.pdf). In addition, both propeller-driven and gliding AUVs have a range of *Re* (4.5×10^5–1.8×10^7) that is equivalent to most fast swimming animals [22,38–41].

The transition from laminar to turbulent flow is produced when perturbations in the flow about the body or generated by the body amplify the inertial forces and cannot be

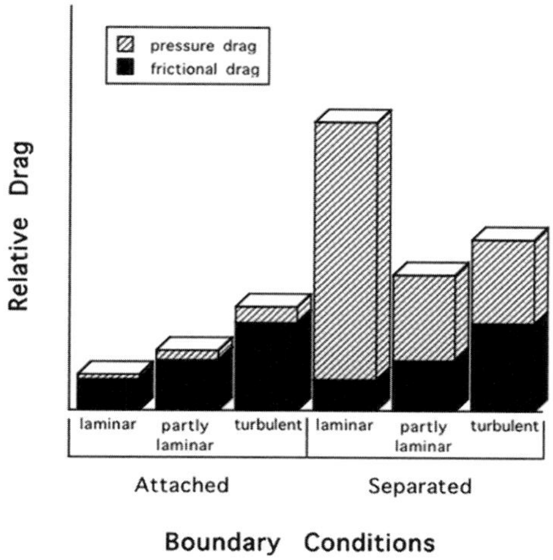

Figure 10.2. Relative drag associated with boundary conditions. (Redrawn from Webb [22] with permission.)

dampened by the viscosity (24). This transition results in a nonlinear change in the trajectory of increasing drag with increasing swimming speed and/or body size [42]. With the transition from laminar to turbulent flow, the drag on body with respect to *Re* is further reduced [22,28,32,34,35,42]. This reduction in drag with turbulent boundary layer is due to maintenance of an attached boundary layer with added energy to overcome the adverse pressure gradient in the rear of the body; whereas at high *Re*, a laminar boundary layer is inclined to separate from the body surface [22,23]. The premature separation of a laminar boundary layer disrupts streamlining and affects the pressure differential around the body by increasing the pressure component of drag as manifested by a broad wake with substantial energy loss.

Overall, total drag (frictional drag + pressure drag) is lower with an attached laminar boundary layer than a turbulent boundary layer [22]. However for large animals moving at high speed, it is more advantageous to maintain a turbulent boundary layer to minimize the pressure drag and thus the total drag compared to a separated laminar boundary layer (Figure 10.2).

Cetaceans (e.g., whales, dolphins, porpoises) and large fish have been the subject of considerable attention as animals that achieve high speed and high efficiency (see reviews by Fish and Hui [43], Fish and Rohr [16], Fish [44]). In particular, the so-called "Gray's Paradox" postulated that the dolphin was able to achieve speeds that are considerably higher than its muscle mass and shape would allow [45]. Much speculation was generated regarding the possible mechanisms of drag reduction [46,47]. This speculation led to large research efforts focusing on compliant surfaces and surface textures for turbulent flow control [38,46–52] and non-Newtonian effects due to surfactant or polymer excretion as a source of drag reduction [53–56]. It is now recognized that the original estimates of power output for dolphins were incorrect, in that they failed to account for influences such as drafting, and were probably based on flawed estimates of speed and incorrect assumptions regarding the muscular physiology of dolphins [48,49,57–59]. Special mechanisms in animals that have been proposed to maintain laminar flow and reduce drag have been mostly unsubstantiated [44,59,60].

10.3 Lift-Based Swimming Mode

For inertial, high *Re* swimmers, three primary mechanisms of propulsion are recognized: drag-based, acceleration reaction, and lift-based [22,26]. Each mechanism operates most effectively within a particular range of *Re* (Figure 10.3). All of these mechanisms result in the transfer of momentum for the motion of the animal's propulsor to the water, but the forces that they produce are different [24]. In drag-based propulsion, the propulsor is an appendage or paddle that has its maximum planar area oriented perpendicular to its motion through the water. Pressure drag on the propulsor is directed opposite to the motion of the animal and by action–reaction generates the thrust. The use of pressure drag for propulsion is most effective at *Re* of 10^2 to 10^3 that swim by paddling or undulation of the body [24]. However at higher *Re*, paddling animals the size of ducks (*Anas platyrhynchos*) and muskrats (*Ondatra zibethicus*) rely on this drag-based mode for propulsion [61,62]. At *Re* between 10^3 and 10^5 (Figure 10.3), acceleration reaction can be used effectively for thrust production. Acceleration reaction is an unsteady force that is generated by accelerating a mass of water, an additional inertial mass called the 'added mass' [24,31,63,64]. The acceleration reaction is dominant in some types of undulatory swimming [31]. From Re $>10^5$, lift is the dominant force for propulsion.

Lift-based swimming generates thrust by oscillating an appendage with the design of a wing-like hydrofoil [22,32]. The oscillations of the appendage move with a sinusoidal trajectory as the appendage is canted at a slight angle to the pathway (angle of attack) to the relative flow [22]. The hydrofoil produces a lift perpendicular to the pathway that is resolved into an anterior thrust force [32,65]. The lift is the result of pressure asymmetry that is normal to the direction of motion of the propulsor [64]. Lift can only be produced at high *Re* where viscous forces are unimportant, so that a flow around the propulsor can develop [66]. The oscillating movements of a hydrofoil result in unsteady shedding of vorticity from the trailing edge [67]. The vorticity in thrust production is necessary to

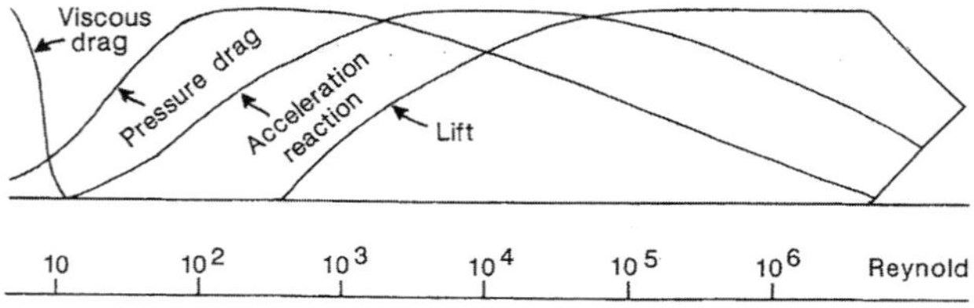

Figure 10.3. Diagram showing the relative importance of the propulsive forces in relation to the Reynolds number. (From Webb [24] with permission.)

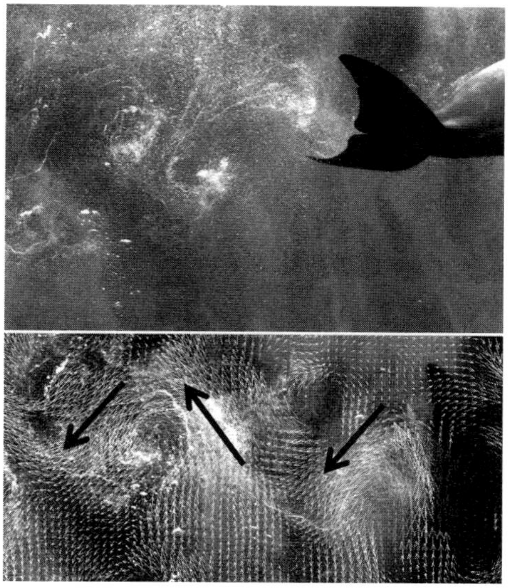

Figure 10.4 Dolphin swimming through a microbubble curtain to visualize the vortices and thrust jet in the wake produced by the vertically oscillating caudal flukes. The image on top is from raw video footage and shows the interconnecting vortices. The bottom image shows the vector field in the wake from digital particle image velocimetry analysis [60]. The vortices are indicated by colors, where dark gray corresponds to positive or counterclockwise rotation and light gray indicates negative or clockwise rotation. The black arrows indicate the jet flow between the vortices.

transport momentum from the hydrofoil into the water. The vorticity shed into the wake is manifest as a staggered array of vortices with alternating spins. Each vortex is formed from a reversal in the direction of the hydrofoil at the end of each half stroke requiring a reversal in circulation [68]. A new reverse circulation is formed as the hydrofoil is accelerated in the opposite direction [23]. The oscillating motion of the hydrofoil produces two parallel trails of staggered vortices that are perpendicular to the plane of oscillation and with opposite spins. In three dimensions, the alternating vortices are linked by tip vortices forming a folded chain of vortex rings. The pattern and spin of the vortices generate a rearward jet flow, which produces thrust to overcome the drag of the body (Figure 10.4). Maintenance of a positive angle of attack ensures thrust generation throughout most of the stroke cycle [25]. Unsteady lift, through control and production of vorticity, is one of the key factors contributing to the large lift coefficients generated by biological systems [69]. To maximize lift and thrust, the propulsor has a high aspect ratio (propulsor span^2/propulsor planar area) compared to the lower aspect ratio of undulatory swimmers (Figure 10.5) [25]. In addition, the planform of the oscillating propulsor has a crescent shape with tapering tips (Figure 10.6) [70].

Lift-based swimming has advantages over drag-based swimming. Propulsion using lift-based modes is faster than drag-based modes (Figure 10.7) [23]. Drag-based propulsion operates most effectively at low speeds, whereas lift-based hydrofoils perform best at higher speeds. A flow field needs to be established for a lifting

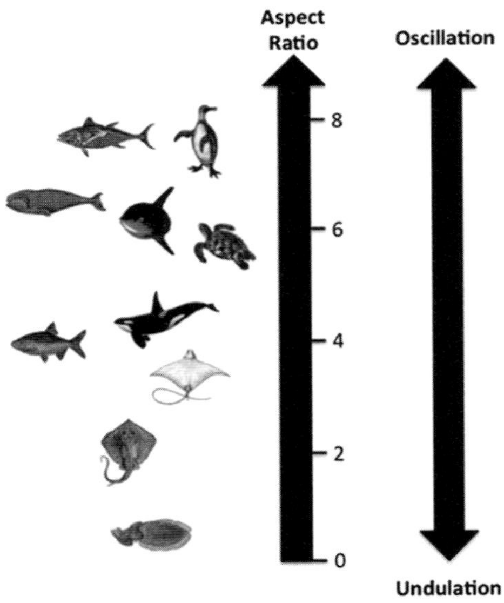

Figure 10.5 Aspect ratio of propulsive fins and flippers in relation to the oscillatory–undulatory gradient. (Adapted from Daniel et al. [66].)

surface to function, thus hydrofoils are limited in use to conditions where the body of an animal is already in motion. As lift is directly dependent on the square of the velocity, the greater the velocity, the more lift and thus thrust can be produced. The velocity of the hydrofoil is a combination of the velocity of a whole organism plus the velocity due to the movements of the reciprocating propulsor, which further enhances thrust. The limitation of drag-based swimming is that thrust is dependent on the speed differential of the forward velocity of the animal and rearward velocity of the propulsor [71–73]. At a speed where the body and paddle speeds are equivalent, thrust can no longer be produced [23]. Another advantage for lift-based oscillatory systems is that thrust is generated throughout most of the stroke cycle [22,26,60], whereas thrust is produced through only half the stroke cycle for a drag-based paddler [26,61]. However, lift-based propulsion is not optimal at low speeds (Figure 10.7) or for accelerations from an initial velocity of zero, as circulation takes time to develop and the changes in circulation are not synchronized with changes in direction of oscillatory propulsors [66].

Lift-based swimming can produce large propulsive power outputs by unsteady effects of the oscillating propulsors [22,25,32,74]. Whereas stall (i.e., dramatic loss of lift) on an airfoil occurs near 12° of angle of attack, oscillating an airfoil sinusoidally about its quarter-chord point can increase the stall angle to more than 30° while doubling the lift [14]. Unsteady aerodynamics has demonstrated various flow control mechanisms for enhanced lift production [75,76]. In addition, the propulsors of lift-based swimmers show various degrees of flexibility [77–79]. Flexibility can further enhance the thrust production and efficiency of lift-based propulsors [80,81]. Flexibility can be present across the

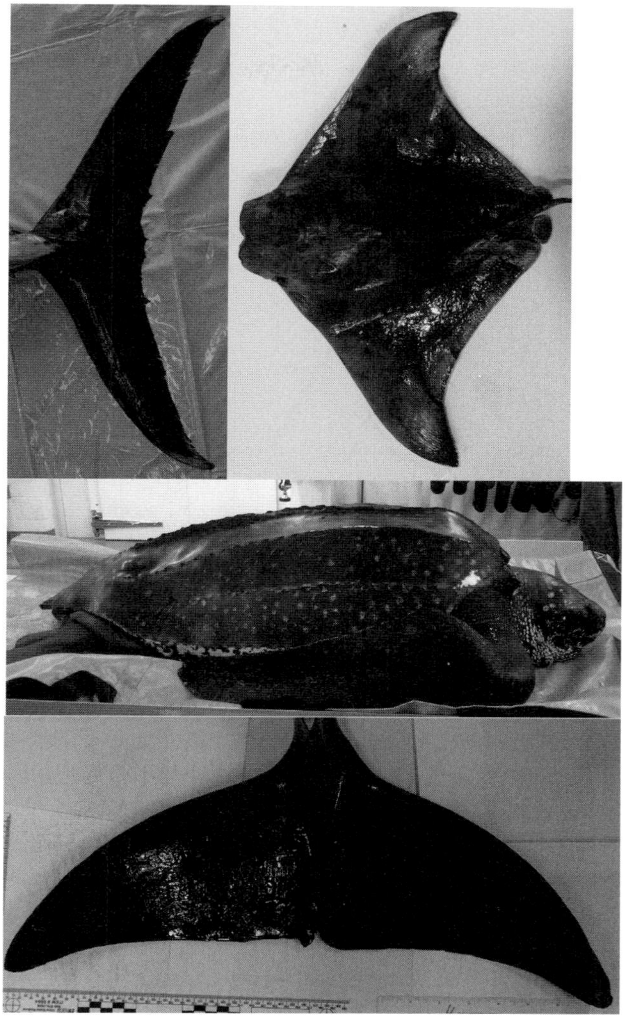

Figure 10.6 Planforms of wing-like, lift-based propulsors, including tuna caudal fin (top left), cownose ray pectoral fins (top right), pectoral flippers of leatherback sea turtle (center), and dolphin caudal flukes (bottom).

span and the chord of an oscillating hydrofoil. Spanwise flexibility prevents the total loss of thrust at the reversal of an oscillatory stroke [82]. On the other hand, chordwise flexibility at the trailing edge of a fin could increase the efficiency by up to 20% with only a moderate decrease in the overall thrust [80,81,83–85]. A computational study by Moored et al. [86] found a 133% increase in efficiency for a flexible panel compared to an equivalent rigid panel. A combination of spanwise and chordwise flexibility produces a narrow thrust jet in the wake to improve efficient thrust generation [87].

Lift-based swimming is associated with the radiation of large, highly derived animals into pelagic habitats, where high speeds and steady swimming are required [88]. The

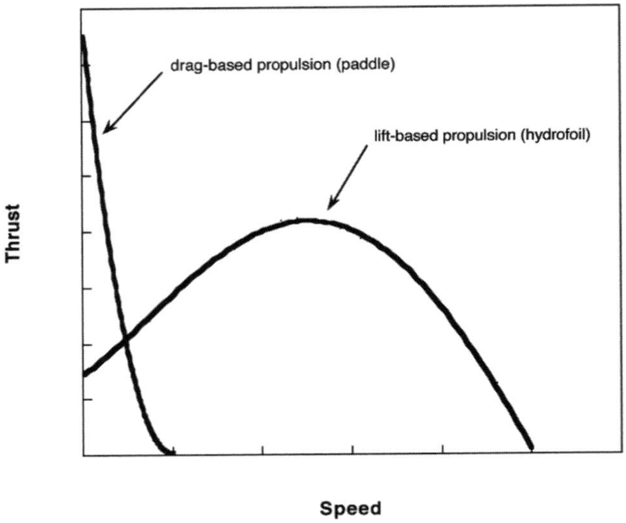

Figure 10.7 Comparison of lift-based and drag-based thrust production in relation to swimming speed. Drag-based paddling can result in substantially greater thrust production than lift-based swimming when a body is stationary. As speed increases, drag-based paddling becomes less effective compared to lift-based thrust production. (Redrawn from Vogel [23].)

diversity of lift-based swimmers is divided into two different modes of swimming, based on the propulsive structures and kinematics, thunniform and subaqueous flight [22,26,89,90]. Thunniform swimmers are propelled by an oscillating caudal fin that is moved either side-to-side or dorsoventrally. The caudal fin is relatively stiff and has a semilunar or sickle-shaped design, with a cross-section similar to an engineered hydrofoil [25,91]. Movement of the caudal fin is generally restricted to the narrow and streamlined caudal peduncle. Subaquaeous flight uses oscillatory motions of the paired pectoral appendages in the dorsoventral direction.

For fishes, the thunniform mode is used by tuna and mackerels (Scombridae), swordfish (Xiphiidae), sailfish and marlins (Istiophoridae), and lamnid sharks (Lamnidae). An extinct group of reptiles, the ichthyosaurs (Ichthyosauridae), swam in the thunniform mode and had fins and a fusiform body shape analogous to present day sharks and dolphins [52,92,93]. The main mammalian groups that use the thunniform mode are the cetaceans [26,94], which swim with caudal flukes as the propulsor and include the whales (Balaenopteridae, Eschrichtiidae, Balaenidae, Physeteridae, Ziphiidae, Monodontidae), dolphins (Delphinidae, Platanisidae), and porpoises (Phocoenidae). Thunniform swimming is also used by manatees and dugongs (Sirenia), and the true seals (Phocidae) and walrus (Odobenidae) of the Pinnipedia. The phocid seals and walrus are different from other mammalian thunniform swimmers in laterally oscillating the hind flippers, which are alternately spread [65,95].

The subaqueous flight swimmers flap their forelimbs in a manner similar to the flight of birds. The manta, eagle, and cownose rays (Myliobatidae) have greatly expanded

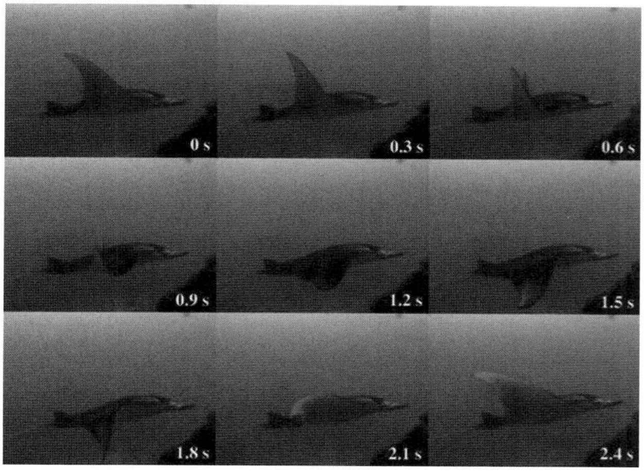

Figure 10.8 Sequential frames of video of propulsive movements of the pectoral fins of a swimming manta. The time elapsed between each frame is indicated in the bottom right corner of each frame. (Video courtesy of Julie Morton, Yap Divers, Yap, Micronesia.)

pectoral fins with a triangular shape (Figure 10.6). The oscillatory pattern of swimming by the myliobatoids has been referred to as the mobuliform mode (Figure 10.8) [96–100]. Flapping of elongate pectoral flippers is performed by penguins (Spheniscidae), sea lions and fur seals (Otariidae), sea turtles (Testudines), and the extinct plesiosaurs (Plesiosauria) [93,101–105].

10.4 Speed

The application of the mechanisms of high *Re* swimming has been motivated by accounts of extraordinary speeds, particularly for dolphins, associated with unresolved mechanisms to reduce hydrodynamic drag [44,45,59]. Accurate swimming speed measurements of free-ranging marine mammals have been rare, and few data are available concerning the duration of swimming effort [16,22]. Reports on swimming speeds of large pelagic swimmers (e.g., swordfish, tuna, whales, dolphins) have in many instances been anecdotal and often unreliable [16,39,43,44]. The record speeds reported for swordfish and marlin were 36.1 m/s – without any information on the method of speed determination [106]. Estimates of swimming speeds for swordfish of 17.9 and 25 m/s were based on the penetration of wood by the rostrum [107] and comparing the penetration of wood by the rostrum with a projectile fired from a small-caliber cannon [108], respectively. Even porpoises were exaggerated to swim speeds as high as 31.3 m/s with no supporting information [109]. The reason for questioning these and other extraordinary reports is that estimates of swimming speeds based on observations from ships, airplanes, or shorelines were often made without fixed reference points, information on currents, or accurate timing instruments. In addition, maximum speeds of fish and cetaceans were often based on distressed and wounded animals [16,108,110–113]. Indeed, marine animals most often cruise at routine speeds that are much lower than their maximal speeds [16].

Swimming speeds of animals performing routine behaviors are collected by several means, including boat measurements, timing of captive animals, theodolite tracking, observations correlated with map locations, sonar tracking, and radio-tagged animals [16]. These methods still have limitations with respect to accurate speed measurements. For example, the resolution and sampling rates using satellite tracking systems are too low to accurately measure short-duration swimming efforts, such as occur for sprints. The diving pattern and high probability that animals do not swim in straight lines between consecutive locations measured by satellites tend to skew the data toward minimum transit speeds, which are well below routine and maximum speeds [16].

Swimming speeds are categorized as (1) sustained and (2) sprint swimming [22]. Sustained swimming is characterized by an activity level maintained for long time durations. These routine speeds are associated with steady low-level activities. Sustained speeds are the preferred swimming speeds and are associated with behaviors, including foraging, migrating, exploration, sleep swimming, schooling, and diving. In addition, these sustained speeds are related to a level of energy expenditure that is the most efficient for travel [26,40,94,114–120]. In contrast, sprints are high speed, maximum effort movements. These are short-duration activities. Sprints are used during chases and rapid accelerations when energetic efficiency is not critical.

Thunniform and subaqueous flight swimmers generally have the fastest sustained and sprint swimming speeds due to their propulsive mechanics (lift-based) and large body size compared to other swimming animals. These swimmers operate within a range of *Re* from 1.1×10^6 to 3.0×10^8 [38,39]. Although regarded as slow swimmers, even the green sea turtle (*Chelonia mydas*; 3.0×10^6) and manatee (*Trichechus manatus*; 9.0×10^6) operate within the range of *Re* with other lift-based swimmers [38].

Animals using the lift-based thunniform and subaqueous flight modes are generally faster than paddling and undulatory swimmers [22,26,38,94]. Tuna can swim steadily from 0.2 m/s to 2.7 m/s and sprint from 3.5 to 20.7 m/s (110,121,122). Sustained swimming by swordfish has been measured at 0.3 to 0.9 m/s and sprinting while chasing prey was reported as 36.1 m/s [123–125]. Sailfish have a sprint speed that is equivalent to the maximum swimming speed of swordfish [113]. Swimming speeds for the manta (*Manta birostris*) range from 0.3 to 0.5 m/s while foraging when measured by satellite tags [126]. For cetaceans, sustained swimming speeds are 0.2–8.4 m/s and sprint speeds are 2.9–13.4 m/s [16]. The low sprint speed was recorded for a bowhead whale (*Balaena mysticetus*) that was fleeing an approaching ship [127], whereas the high value was for a sei whale (*Balaenoptera borealis*) that had been wounded during a hunt [36]. Pinnipeds exhibit sustained swimming speeds of 0.6–2.6 m/s. Maximum swimming speeds are higher for otariids than phocid seals. The otariids can maintain a speed of 5.1 m/s for 3–5 minutes, although bursts of 6.7–8.0 m/s have been reported [128,129]. Maximum speed for phocids is 3.6–4.8 m/s. Penguins have average swimming species of 1.6–2.4 m/s and maximum speeds up to 3.35 m/s [129–131]. Leatherback sea turtles (*Dermochelys coriacea*) typically swim at 0.56–0.84 m/s, but have a maximum speed of 1.9–2.8 m/s [132]. Although comparatively slow, these animals, like other lift-based swimmers, are capable of migrations of thousands of kilometers over several years at sea [133,134].

Swimming speed is related to duration of activity. A trained bottlenose dolphin (*Tursiops truncatus*) was found to swim at 3.1 m/s indefinitely, 6.1 m/s for 50 s, 7.0 m/s for 10 s, and 8.3 m/s for 7.5 s [135]. Based on such data, one record for sustained swimming by a killer whale (*Orcinus orca*), although widespread in the literature is probably incorrect. The killer whale was reported to swim for 20 min at 12.5–15.4 m/s [136]. The data were obtained from questionnaires placed aboard a ship. Because the killer whale was "playing" around the ship, the animal was probably bow riding and thereby swimming at a higher speed with less effort.

There is a marked dependence of specific speed on size [22]. Small dolphins have high length-specific swimming speeds compared to larger whales. For example, the striped dolphin (*Stenella attenuata*), can sprint at 6.0 body lengths/s; whereas, a 38 times more massive killer whale (*Orcinus orca*) can sprint at only 1.5 body lengths/s. Furthermore, a 27.4-m blue whale (*Balaenoptera musculus*) sprinting at 10.2 m/s would have a length-specific speed of 0.37 body lengths/s. This trend is explained as a matter of scaling, whereby animals maintain the same proportion of muscle mass in the body independent of size, but smaller animals have relatively greater muscular power outputs than larger animals [22].

10.5 Efficiency

Efficiency is a critical parameter associated with locomotion as it can determine the distance that can be traveled for a given quantity of fuel or energy. Animals have an advantage over autonomous engineered systems in that the living organism can consume food found in the water to replenish its energy reserves and not just rely on stored energy reserves. However as food resources may be widely distributed temporally and spatially, aquatic animals should be adapted to swim in a manner that minimizes metabolic and mechanical energy expenditure [26]. Efficiency as a performance measure is defined as the useful rate of energy expenditure divided by the total rate of energy available [137]. The useful rate of energy expenditure is associated with the production of thrust power for aquatic propulsion, but the total rate of energy consumption can be the total mechanical power that is expended to swim (i.e., thrust power plus the rate of all energy loss) or the total metabolic energy consumed that is used by the animal to perform work. The ratio of thrust power to total power is the propulsive or Froude efficiency (η), which is comparable between animals and engineered systems.

The propulsive efficiency for high *Re*, lift-based swimmers exceeds that for conventional marine propellers. A fixed-pitch propeller has a maximum η of 0.7 [138,139], whereas, η is 0.76–0.90 for tuna (*Euthynuus affinis*), 0.89 for the manta ray (*Manta birostris*), and 0.75–0.90 for cetaceans [22,74,100,140]. In addition, the maximum η is limited to a narrow range of speeds, but cetaceans maintain high η over an extended range of operation (Figure 10.9).

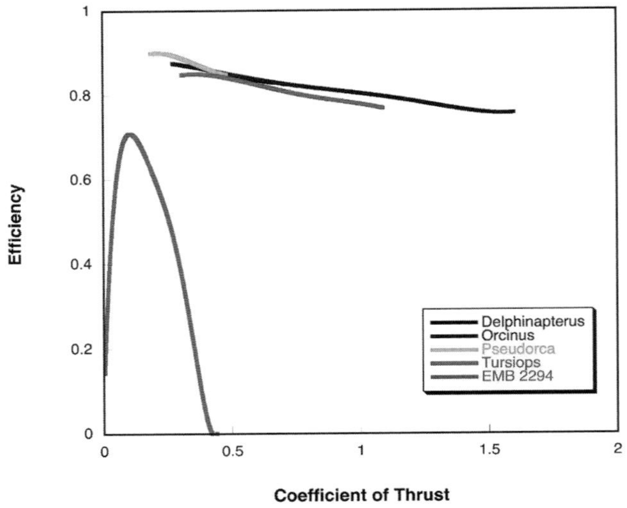

Figure 10.9 Comparison of relationships of propulsive efficiency and thrust coefficient for four species of small whales and a typical marine propeller. (Data for the whales were obtained from Fish [74] and data for the propeller (EMB 2294) from Saunders [138].)

As structural and kinematic parameters determine locomotor performance, the flexible components of the lift-based propulsors, in conjunction with their propulsive oscillations, are expected to enhance efficiency and energy economy [141,142]. Compared to rigid propellers, the higher η for high *Re*, lift-based swimmers may be a function of the control of flexibility of the propulsor [78,79,100]. A dolphin robot with a rigid fluke achieved a maximum η of 0.7, which was 7–22% below the η obtained for live cetaceans [74,143]. The white-sided dolphin (*Lagenorhynchus acutus*) shows 35% and 13% deflections across the fluke chord and span, respectively [144]. Flexible pitching panels demonstrate increases in thrust and efficiency when the frequency of motion was tuned to the structural resonant frequency [86,145]. Flexibility across the chord of an oscillating, flexible hydrofoil can increase propulsive efficiency compared to a rigid propulsor executing similar movements [80,85].

Bending across the chord throughout the stroke could provide camber to the symmetric flukes and increase lift and thrust production. Active cambering of the blades of an experimental propeller led to a 15% increase in thrust [17]. Cambering changes the flow over the wing surface to increase the forces generated around the wing. Cambering would be beneficial to maintain lift production at the end of each stroke as an oscillating propulsor changes direction, when there is a period of feathering (i.e., the foils are parallel to the incident flow, producing no thrust and reducing efficiency).

Flexibility at the tips of an oscillating propulsor, such as dolphin flukes would bend down slightly from the plane of the fluke during the up-stroke and lag behind the center, whereas bending in the opposite direction would occur during the down-stroke. Bose et al. [146] suggested that the phase difference due to such spanwise flexibility would

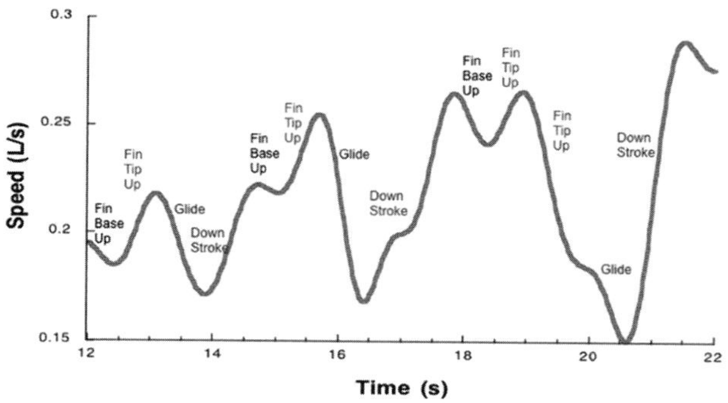

Figure 10.10 Plot of the length-specific velocity against time over multiple fin strokes for a manta swimming in the field. Labels indicate the movement of the oscillating pectoral fin. The manta accelerates during the downstroke of fin. As the base of the pectoral fin starts to initiate and upstroke, the manta decelerates, indicating an increase in drag. At the end of the upstroke, the manta accelerates due to the distal fin tip generating thrust. The manta glides with raised pectoral fins and decelerates before initiating another downstroke.

prevent the total loss of thrust at the end of the stroke. The flexibility at the tips could act passively as winglets, helping to reduce energy losses due to wing tip vortices and direct a more coherent thrust jet into the wake. Active movements of the flexible spanwise component of the propulsors could also contribute to thrust generation and increase efficiency. Movements of the flexible tips of pectoral fins of manta rays were shown to be out of phase with the oscillating fins (Figures 10.8 and 10.10). Although the large portion of the propulsion was produced by downstroke of the fins, the upstroke of the base of the fins increased the drag and decelerated the body. However, the movement of the tips of the fins, which lagged behind the base of the fin, accelerated the manta (Figure 10.10).

Another factor that is associated with high efficiency for propulsion by lift-based swimmers is the Strouhal number (St) [14,147]. St is an indicator of the association of efficiency and flow pattern in the wake. The maximum spatial amplification and optimum creation of thrust-producing jet vortices lays in a narrow range of St [14,147,148]. The highest propulsive efficiencies are found for St between 0.2 and 0.4. St is equal to $A f / U$, where A is the peak-to-peak or heave amplitude and f is the stroke frequency. Animals that use oscillating propulsive systems for swimming and flying have St within the optimal range over an extended range of locomotor speeds [14,148–150]. Except at low swimming speeds, lift-based swimmers (cetaceans, mobuliform rays, phocid seals, and scombrids) have St values concentrated in the optimal range (Figure 10.11).

As large inertial masses, high Re swimmers can use certain behavioral strategies to increase propulsive efficiency. One strategy is burst-and-coast swimming, which is a two-phase periodic behavior of alternating accelerations (burst phase) with glides (coast phase) [151–154]. The burst-and-coast strategy was estimated to reduce the energy expenditure of swimming by over 50% compared to steady swimming [155]. The inertia of large, well-streamlined swimmers would allow an extended period of coasting

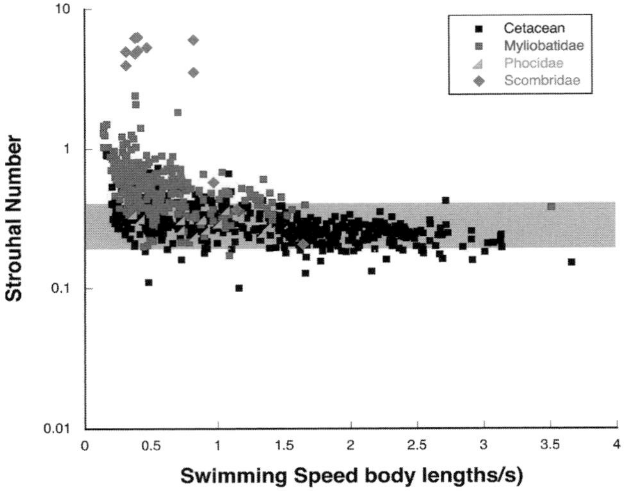

Figure 10.11 Strouhal number (*St*) as a function of length-specific swimming speed. The light gray shaded area indicates the region of $St = 0.2$–0.4, where maximum propulsive efficiency occurs. The lift-based swimmers that are plotted include various species of cetaceans, mobuliform rays, phocid seals, and tuna. (Data are from Fish [74,253], Fish et al. [65,248], Magnuson [140], and unpublished data.)

without excessive deceleration due to viscous drag and limit excess energy loss in reaccelerating the body. Small, low *Re* swimmers would be at a disadvantage, as the relatively large viscous forces would rapidly decelerate the body once swimming motion had abated [156].

The ability to reduce the energy cost of swimming to obtain a free ride is dependent on the flow conditions at high *Re*. A high-pressure field in front of a large moving body can be used to generate the force to move a smaller body anteriorly. Dolphins are renown for bow riding on the bows of ships and whales [43] and the pilotfish (*Naucrates* sp.) swims in the flow field in front and along the sides of larger fish like sharks [157]. Trout (*Oncorhynchus mykiss*) are able to hold station in the upstream side of a cylinder in a water current [158].

Whereas bow riding uses the high pressure region of a large body to "push" a smaller animal forward without physical contact, drafting or slipstreaming of another body uses a hydrodynamically favorable region along the sides or the wake of one animal to "pull" another along. The close proximity of the two bodies induces an interaction of their pressure and flow fields. The presence of the neighboring bodies causes the flow between them to increase, which produces a pressure drop (Bernoulli effect). The pressure drop creates a suction force resulting in a mutual attraction that exerts an equal and opposite force on each animal [159,160]. Neonate dolphins are able to maintain speed close to their mothers and gain 90% of the thrust required [160]. Large bowhead whales similarly gain an energetic advantage by swimming in a close echelon formation [161].

Certain locomotor aggregates are arranged as highly structured formations, whereby individuals are oriented in the same direction, maintain a defined spacing, and are

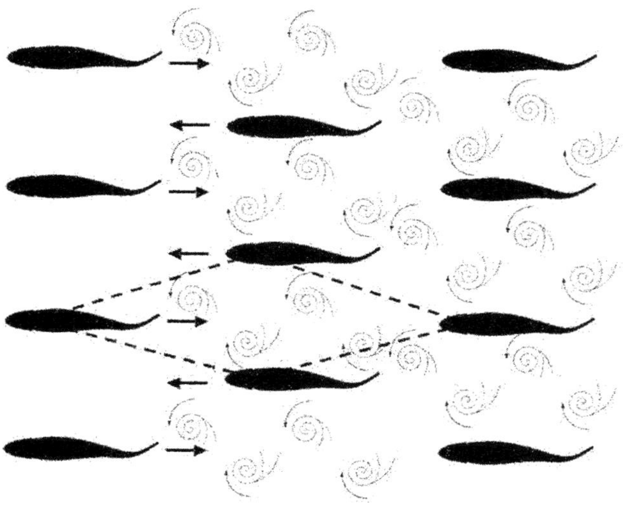

Figure 10.12 Optimal arrangement for maximum energy savings of a fish school swimming in a horizontal layer. Vortex patterns in the wake of the fish are illustrated. Arrows show direction of induced flow relative to vortices. The dashed lines show the diamond configuration of fish. (Based on an illustration from Weihs [162].)

organized in discrete patterns (i.e., schools). For example, tuna swimming together in a school can be arranged in an elongate diamond formation to take advantage of hydrodynamic effects [162]. The position of leading fish in a school in the flow is maintained by the oscillatory sideways propulsive motions of the fish from its body and caudal fin. The propulsive motions produce the vortex wake, which is experienced by the fish swimming behind and to the side of the leading fish (Figure 10.12). The tangential velocity of each vortex is in the same direction as the velocity of the trailing fish, reducing its relative velocity [162,163]. As drag is proportional to the square of the velocity, the fish will experience a reduction in energy. Weihs [162] estimated that the trailing fish could experience a relative velocity of 40–50%.

A variant of using shed vorticity to increase propulsive efficiency and reduce the cost of swimming is the use by animals of the wake from obstructions in a flow stream [164]. By swimming behind an object shedding a vortex wake, an animal could extract energy from the flow. The flow behind a rigid, bluff (nonstreamlined) body is nonsteady. The pressure on the downstream side of the bluff body is low, forming a suction and drawing fluid back toward the body [68]. This reversal in flow direction leads to instabilities in the boundary flow adjacent to the body and causes flow separation from its surface. Flow separates alternately from each side of the body, producing two staggered rows of vortices, which are shed into the wake. All of the vortices in one row rotate in the same direction, but opposite to that of the other row. This flow pattern is a drag-type of vortex street or Kármán vortex street and is opposite the thrust-type vortex street [23,68,162]. The vortex pattern is stable for a long distance downstream if the distance between successive vortices on the same side is 3.56 times the distance between the two rows [23]. Due to the direction of rotation in the vortex street, the flow in the center of the street is directed anteriorly. By swimming between the parallel vortex rows of the Kármán vortex street [158,165,166], the cost of locomotion can be reduced [163]. The trailing fish will experience a reduced relative velocity compared to the free stream velocity. Trout (*Oncorhynchus mykiss*) swimming in a Kármán street of a stationary cylinder alter the body kinematics. Tail-beat frequency is lower than when swimming in the free stream and matches the frequency of vortex shedding by the cylinder [158].

10.6 Maneuverability and Agility at High *Re*

The ability to maneuver is an important aspect of the locomotor performance of aquatic animals and engineered vehicles [167]. Maneuvering is particularly important in steering, course correction, negotiating obstacles in complex surroundings, maintaining stability in high energy environments, catching prey, acquiring targets, avoidance behaviors, swimming through turbulent energetic waters, and ritualistic display and mating [167–174]. Aquatic animals display high maneuvering performance with little loss of stability [167].

Animals can move freely about three translational planes and about three rotational axes that are orthogonally arranged and intersect at the center of gravity (CG) when suspended within the water column (Figure 10.13). CG is the point at which all the weight of the animal is considered to be concentrated and acts as the balance point. Movement about CG permits three translational and three rotational movements for a total of six degree of freedom. The degrees of freedom for the translational planes are surge (anterior–posterior movement), heave (vertical displacement), and slip (lateral displacement); whereas rotation about the rotational axes is termed roll for the

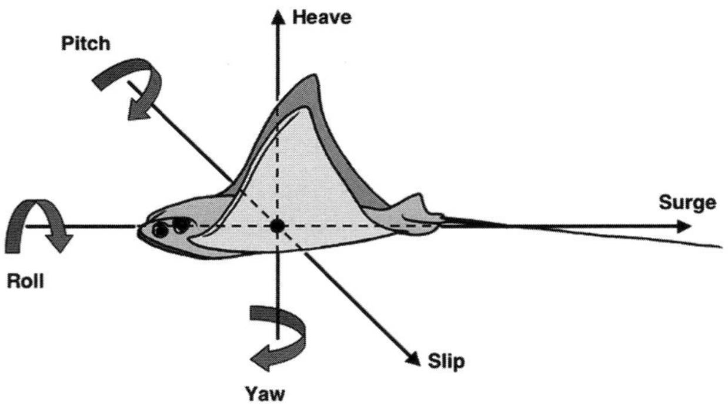

Figure 10.13 Translational and rotational movements associated with a three-dimensional axis system superimposed on cownose ray. Rotational movements include roll (rotation around the longitudinal axis), pitch (rotation around the transverse axis), and yaw (rotation around the vertical axis). Translational movements along the three axes include surge (longitudinal axis), heave (vertical axis) and slip (transverse axis).

longitudinal axis, pitch for the lateral axis, and yaw for the vertical axis [167,175]. Stability about the roll axis is lateral stability, about the yaw axis is directional stability, and about the pitch axis is longitudinal stability. Longitudinal stability is related to "trim," which relates to the alignment of the centers of gravity and buoyancy in the vertical axis, and the longitudinal axis of the horizontally oriented body [176].

The design and position of control surfaces determine how maneuverable a body will be in water [177,178]. For high *Re* swimmers, these control surfaces include fins, flippers, flukes, and caudal peduncle. Lift-based propulsion and maneuvers work best with high aspect ratio appendages used as control surfaces [179,180]. Lift-based maneuvering systems have the advantage of generating a centripetal force to effect turning without acquiring a large decelerating drag [23,177,179]. The position of the lift-based control surface is typically far from CG with the concentration of the control surface area posterior of CG [178]. These positions effectively make the body similar to an arrow, which has self-stabilizing characteristics [177,181]. In a stable body, as opposed to a maneuvering body, the sum of all forces and all turning moments is zero. A stable body reduces the energetic cost of locomotion by reducing corrective forces and minimizing the distance traveled. However, destabilizing torques can be produced by the control surfaces for turning by changing the angle of attack to generate unbalanced lift and drag forces. These changes can regulate turning performance by changing the shape of the control surface with internal muscles, reorienting the control surface into the flow directly by muscular insertion, and bending the body [52,78,171,180,182].

Maneuverability for aquatic animals is typically defined as turning with a minimum radius, typically with respect to yaw [167,171,183]. In the case of the boxfish (*Ostracion meleagris*), which possesses multiple mobile fins, the fish is able to spin around with a turning radius near zero [171]. Although having fewer fins to maneuver, the bat ray is also capable of a pure rotational yawing turn [174]. However, the inflexibility of the body means that as the fish turns, it still occupies a space of the diameter equal to the body length (171). More flexible-bodied animals can fit into smaller spaces. Coral-reef fishes were shown to have high maneuverability with turning radii of less than 0.06 body length (*L*) [171,184]. These fish must operate in a habitat that is confining due to the three-dimensional complexity of the corals.

Predatory behavior also necessitates high maneuverability due to the scaling effects between the predator and its prey [168,173]. The turning radius of a large aquatic predator will generally be larger than smaller prey, because turn radius is directly related to body mass. Although a large predator can swim at higher absolute speeds, the prey has superior turning performance for escape [168,180]. Large, fast-swimming predators have small *L*-specific turn radii, which are produced as gliding unpowered turns using lift from the control surfaces and flexible bodies. Turning speeds for sea lions (*Zalophus californianus*) executing hairpin loops range from 3.6 to 4.5 m/s with minimum turn radii of 0.09–0.16 *L* [180]. Similarly, cetaceans are capable of turn radii of 0.11 to 0.17 *L* [177]. The flexible-bodied sandbar (*Carcharhinus plumbeus*) and scalloped hammerhead (*Sphyrna lewini*) sharks have turn radii of 0.19 *L* and 0.18 *L*, respectively [185], whereas the turn radii for stiffer-bodied bluefish (*Pomatomus saltatrix*), Atlantic

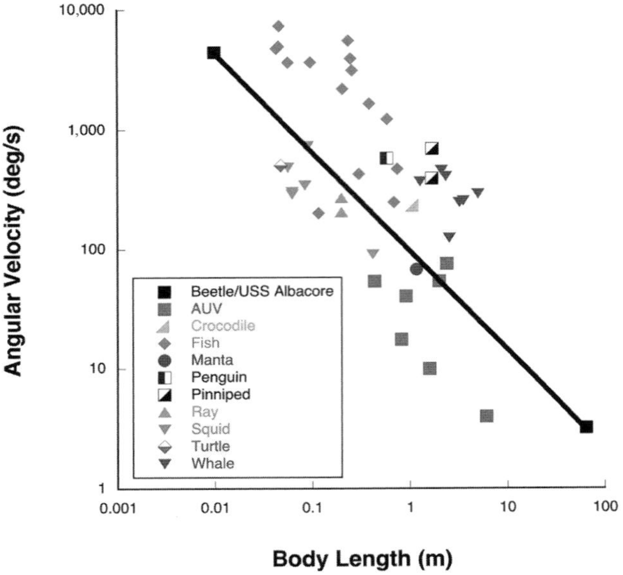

Figure 10.14 Comparison of turning rate with respect to size. The line connects the beetle and submarine, which both have inflexible bodies. Animals with flexible bodies are located above the line, whereas animals with stiff bodies are below the line. AUVs have generally poorer agility than animals of equivalent size. (Data for animals from Webb [169,183], Hui [191], Foyle and O'Dor [192], Blake et al. [188], Gerstner [185], Frey and Salisbury [193], Walker [171], Fish [194], [177], Fish et al. [180,182,201], Rivera et al. [195], Helmer et al. [196], Geurten et al. [197], and Jastrebsky et al. [198]; data for AUVs and submarine from Miller [199], Kumph [240], Anderson and Chhabra [231], Menozzi et al. [246], Stanway [241], Liang et al. [247], Zhan et al. [244], and Fish et al. [248].)

mackerel (*Scomber scombrus*), and yellowfin tuna (*Thunnus albacares*) are 0.19 *L*, 0.21 *L*, and 0.47 *L*, respectively [186,187].

Agility is defined as the rapidity in which direction can be changed and is measured as change in angular velocity and thus the rate of turn [171,180,188,189]. Yawing rates are a function of body size, with small animals turning faster than large swimmers (Figure 10.14). Small fish (body length of 0.1–0.6 m) can turn at rates of 1,221 to 7,301°/s [169,183,184]. Rigid-bodied fish are less agile, such as the boxfish and tuna, turning at 147°/s and 426°/s, respectively [171,187]. For cetaceans, turn rates of 124–453°/s have been reported [177], although turn rates of up to 1,372°/s have been observed when bottlenose dolphins (*Tursiops truncatus*) are pursuing fish [190]. Large fast-swimming animals that are propelled by wing-like motions of the flippers have high turn rates, such as penguins (*Spheniscus humboldti*) at 576°/s, cownose rays at 267°/s, and sea lions at 690°/s [191,180]. During maximum turning, sea lions were capable of a centripetal acceleration of 5.1 *g*.

Turning around the yaw axis is not the only means of maneuvering for agile lift-based swimmers. Manta rays are able to use their large pectoral fins to pitch the body and execute somersaults at 41.5°/s in a looping circular trajectory [200,201]. This somersault maneuver is used by mantas for feeding on concentrations of krill. During lunge feeding, rorqual whales (*Balaenoptera* sp.) will perform rolling maneuvers of 33.8–39.3°/s (Goldbogen et al. [202,203]. Sea lions can roll at a rate as high as 415°/s (Leftwich, personal communication). Rolling is used in banking to support yawing turns [174,180,188,201,204].

10.7 Stealth

The propulsive mechanisms used by high *Re* swimmers have oscillatory frequencies that are lower blade loadings than standard marine propellers [205]. A potential advantage of this means of propulsion is that there can be a significant reduction in noise generation. To achieved sufficient thrust to propel a vessel, marine propellers spin at a high rate, which can generate noise by displacement of water by the propeller blade profile, induce high flow over the propeller blades, produce large pressure differences between suction and pressure surfaces of the propeller blades, produce periodic fluctuations while operating in a variable wake field, and cavitate [205].

Information on hydrodynamic noise generated by biological swimmers is extremely limited. Hydrodynamic noise is generally near-field displacement of low frequency, ranging down to subsonic levels [206,207]. Moulton [208] found that the swimming sounds produced by four species of schooling fish (5.1–30.5 cm) were nonharmonic with the dominant frequencies below 100 Hz. In one laboratory study, *Anchoviella choerostoma* in a group of 500 to 1,000 individuals did not produce sound above extraneous noise [208]. The maximum sound intensities can reach 0.7–2 kHz and are associated with rapid accelerations and turning maneuvers [207–209]. Body shape can also affect the amplitude and frequency characteristics of hydrodynamic sound [209].

Cavitation is caused by excessive propeller speed or loading that results in a reduced pressure below ambient on the suction side (rear) of the blades, creating bubbles [210]. The implosion of the cavitation bubbles generates a shock wave, which is the source of noise of a frequency up to 1 MHz [205]. Iosilevskii and Weihs [211] calculated that 10–15 m/s was the maximum cavitation-free velocity at a shallow depth for fast-swimming tuna, lamnid sharks, and dolphins. The cavitation-free velocity coincides with the maximum swimming speeds reported for lamnid sharks and dolphins [16,211]. Tuna are known to swim at higher speeds, potentially risking cavitation. The collapse of cavitation bubbles near the surface of a propulsor can cause damage [205], as has been observed in the caudal fin of tuna [211]. However as the onset of cavitation is dependent on the pressure differential between the propeller and ambient pressure, tuna can swim continuously at depth with high pressure, negating the cavitation risk. The risk for shallow-swimming dolphins may increase as the flexible flukes will increase leading edge suction, inducing cavitation at speeds lower than the maximum cavitation-free velocity [211].

The leading edge tubercles on the pectoral flippers of the humpback whales modify the hydrodynamic flow over the flipper [204,212,213]. The main effects are a slight increase in lift, delay of stall, and no drag penalty pre-stall [212]. The tubercles have been used to improve the aero/hydrodynamic performance of various wing-like structures [213,214]. In addition, the flow pattern induced by the tubercles has been found to reduce noise generation [215–220]. Noise reduction may be due to a reduction in the tip vortex as the flow pattern over a wing-like structure reduces spanwise flow, which would feed the tip vortex [213,214]. The flow experienced by and modified by the tubercles is within the Reynolds regime coinciding with typical engineered applications. The Reynolds number of a flipper is 1.6×10^6 when the whale is lunge feeding at 2.6 m/s [214].

10.8 Limitations and Applications

Nature presents various solutions for improving mechanical devices and operations, as well as developing new technologies [221–225]. It is viewed that evolution, through natural selection, has provided innovations in design that have fostered enhanced functionality and efficiency. As novel morphologies and physiological operations are investigated by biologists, they are serving as the inspiration for engineers to advance technological development. Bioinspiration and biomimetics attempt to produce engineered systems that possess characteristics that resemble living systems or function like them [221,224,226]. The goal of this approach is to engineer systems that emulate the performance of living systems, in order to improve current human-engineered technologies or utilize the distinct features of organisms that would offer novel solutions for particular functions or missions [44,59,227–229].

In recent years, engineers have looked to natural solutions through biomimetics for application to marine systems [11]. Of specific interest is the development of bioinspired autonomous underwater vehicles (BAUV) based on swimming animals [14,17,18,59,67,100,143,201,228,230–239]. Such BAUVs would have applications for the military, resource exploration, ocean mapping, search and rescue, diver support, pollution localization, marine infrastructure inspection, and biological and oceanographic research. However, compared to the current operational and research AUVs, animals demonstrate superior swimming performance with regard to major parameters of speed, efficiency, maneuverability, agility, and stealth that are required for future BAUVs.

The swimming speeds of high *Re* swimming animals outpaces standard and BAUVs for which there is data. Experimental BAUVs demonstrate speeds of 0.09 m/s [240]–1.4 m/s [241]. This is below the routine cruising speeds of animals of about 0.2–8.4 m/s and far below the maximum sprint speeds that can top off at 36.1 m/s. REMUS is currently the main operational AUV and has a maximum speed of 2.6 m/s. Only torpedoes surpass animals [37], with speeds of over 22–36 m/s for standard torpedoes and supercavitating torpedoes reaching 100 m/s [242,243].

In comparison to the maneuvering ability of animals, the turn radius and rate for engineered marine craft are poor (Figure 10.14). In order to turn and reverse its direction a ship must reduce speed by 50%, whereas a fish can perform the same maneuver without braking with a length-specific turn radius that is 10 times lower than the ship [12]. With baseline turn rates of 17.5°/s for torpedoes and 9.0°/s for submarines for simulations of antitorpedo tactics [244], these values are considerably lower than the capability of animal designs. In addition, a 1950s torpedo has a turn radius of 5.0 *L* [186]. REMUS has turn radius of 2.9 *L* and turn rate of 9.9°/s [241], and a conventional AUV was reported to have a turn radius and rate of 2.7 *L* and 4.0°/s, respectively [231]. The addition of a swivel tail can reduce the turn radius on a standard AUV body to 0.99 *L* [245]. Although minimum turn radius and maximum turn rate are improved with many BAUVs, these performance measures are not equivalent to the maneuverability of animals (Figure 10.14) [186,231,240,241,246,247]. The closest to turning performance of the animals was the Vorticity Control Unmanned Undersea Vehicle (VCUUV) that was based on the propulsive mode, body shape, and flexibility of a tuna. This robotic tuna was capable of swimming at 1.25 m/s and could perform a yawing turn of 0.5 *L* at 75°/s [231].

Recently, a BAUV based on myliobatoid rays (MantaBot) has been designed and built (Figure 10.15) [100,248]. The speed, efficiency, and maneuverability of myliobatoid rays are all characteristics that are desirable in an AUV. The main actuation mechanism to provide the lift-based propulsion of the pectoral fins used active tensegrity structures by multiple cable-routed actuation [237]. This approach has the benefits of being both weight efficient and reducing the number of actuators. Preliminary results demonstrated the potential for this system to utilize the key mechanisms employed by biology to achieve its high performance targets. Data collected on the MantaBot indicate swimming speeds of approximately 1 *L*/s (0.43 m/s) with a flapping frequency of 1.1 Hz and heave amplitude of approximately 0.4 *L*. The Strouhal number was calculated to be about 0.44, which was close to the optimal

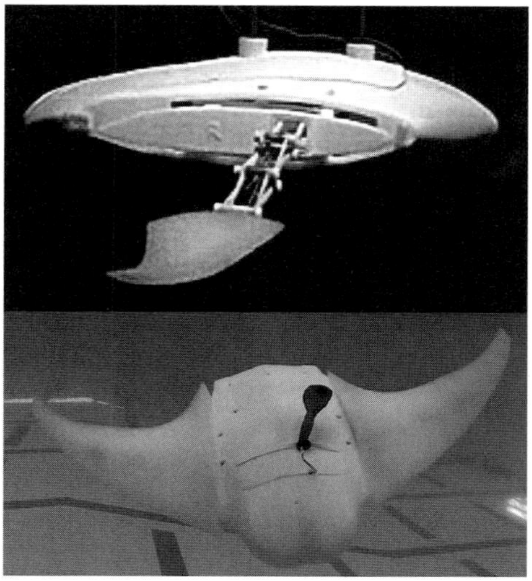

Figure 10.15 Images of MantaBot showing tensegrity structures to actuate fins (top) and while free-swimming (bottom).

Strouhal number range for maximal propulsive efficiency [100]. The minimum turn radius was approximately 0.3 *L*.

The information gained from the present analysis has demonstrated that high *Re*, lift-based swimmers outperform conventional AUVs. There is the potential to learn from these animals for application in the production of BAUVs [18]. The body shape and mechanism of propulsion can be used to reduce drag and increase thrust production to operate at high speeds with high efficiency or low speeds to extend battery life and operational duration. Mechanisms to actuate various control surfaces [171,178,250] can aid in enhanced maneuverability and agility. Such enhanced capabilities can be applied to AUV functions in stability, docking, terrain hugging, station-holding, trajectory planning, obstacle avoidance, and operation in the high-energy littoral zone [250–252]. The mode of swimming used by high *Re*, lift-based swimmers has distinct advantages in reducing the amount of noise generated and changing the noise signature compared to the distinct sound of a standard propeller [18]. Finally, copying the morphology and inertial-dominated propulsive systems of large, fast-swimming animals mitigates scaling problems in applications to AUV and marine vehicle technology [254].

Acknowledgments

I would like to thank John Dabiri for recommending me to the editors of the book to write this review. I appreciate the comments and assistance of Danielle Adams, Hilary Bart-Smith, Haibo Dong, Janet Fontanella, William Gough, Keith Moored, Alexander Smits, Kelsey Tennett, Paul Webb, Timothy Wei, Daniel Weihs, and Terrie Williams in support of much of the fundamental research reported in this review. This work was supported by grants from the Multidisciplinary University Research Initiative program from the Office of Naval Research (Robert Brizzolara, program officer: N000140810642, N000141410533).

References

1. Schmidt-Nielsen, K. (1977). Problems of scaling: Locomotion and physiological correlates. In T. J. Pedley (Ed.), *Scale effects in animal locomotion*. London: Academic Press; p. 545.
2. Newell, N. D. (1949). Phyletic size increase – An important trend illustrated by fossil invertebrates. *Evolution*, 3, 103–124.
3. Pianka, E. R. (1970). On r-and K-selection. *American Naturalist*, 104(940), 592–597.
4. Hildebrand, M. (1975). *Analysis of vertebrate structure*. New York: Wiley; p. 657.
5. Bonner, J. T. (2006). *Why size matters*. Princeton, NJ: Princeton University Press; p. 161.
6. Kardong, K. V. (2012). *Vertebrates: Comparative anatomy, function, evolution*. New York: McGraw-Hill; p. 794.
7. Milne, L. J., & Milne, M. (1978). Insects of the water surface. *Scientific American*, 238, 134–142.

8. Bush, J. W. M., & Hu, D. L. (2006). Walking on water: Biolocomotion at the interface. *Annual Review of Fluid Mechanics*, 38, 339–369.
9. Johnson, D. L. (1980). Problems in the land vertebrate zoogeography of certain islands and the swimming powers of elephants. *Journal of Biogeography*, 7, 383–398.
10. Wes, J. B. (2002). Why doesn't the elephant have a pleural space? *News in Physiological Sciences*, 17, 47–50.
11. Fish, F. E., & Kocak, D. M. (2011). Biomimetics and marine technology: An introduction. *Marine Technology Society Journal*, 45, 8–13.
12. Scaradozzi, D., Palmieri, G., Costa, D., & Pinelli, A. (2017). BCF swimming locomotion for autonomous underwater robots: A review and novel solution to improve control and efficiency. *Ocean Engineering*, 130, 437–453.
13. Webb, B., & Consi, T. R. (2001). *Biorobotics: Methods and applications*. Menlo Park, CA: American Association for Artificial Intelligence; p. 208.
14. Triantafyllou, G. S., & Triantafyllou, M. S. (1995). An efficient swimming machine. *Scientific American*, 272, 64–70.
15. Bushnell, D. M. (1998). Drag reduction "designer fluid mechanics" – Aeronautical status and associated hydrodynamic possibilities (an "embarrassment of technical riches"). In J. C. S. Meng (Ed.), *Proceedings of the International Symposium on Seawater Drag Reduction*. Newport, RI: Naval Undersea Warfare Center.
16. Fish, F. E., & Rohr, J. (1999). *Review of dolphin hydrodynamics and swimming performance. SPAWARS Technical Report 1801*. San Diego, CA: SPAWARS.
17. Bandyopadhyay, P. R. (2005). Trends in biorobotic autonomous undersea vehicles. *Journal of Oceanic Engineering*, 30, 109–139.
18. Fish, F. E. (2013). Advantages of natural propulsive systems. *Marine Technology Society Journal*, 47, 37–44.
19. Shaw, W. C. (1959). Sea animals and torpedoes. In *U. S. Naval Ordinance Test Station. NOTS TP 2299, NAVORD Report 6573*. China Lake, CA: Naval Ordinance Test Station.
20. McKenna, T. M. (2011). Developing bioinspired autonomous systems. *Marine Technology Society Journal*, 45, 19–23.
21. Purcell, E. M. (1977). Life at low Reynolds number. *American Journal of Physics*, 45, 3–11.
22. Webb, P. W. (1975). Hydrodynamics and energetics of fish propulsion. *Bulletin of the Fisheries Research Board of Canada*, 190, 1–159.
23. Vogel, S. (1994). *Life in moving fluids*. Princeton, NJ: Princeton University Press; p. 467.
24. Webb, P. W. (1988). Simple physical principles and vertebrate aquatic locomotion. *American Zoologist*, 28, 709–725.
25. Lighthill, J. (1975). *Mathematical biofluid dynamics*. Philadelphia: Society for Industrial and Applied Mathematics.
26. Fish, F. E. (1996). Transitions from drag-based to lift-based propulsion in mammalian swimming. *American Zoologist*, 36, 628–641.
27. Wu, T. Y. (1977). Introduction to the scaling of aquatic animal locomotion. In T. J. Pedley (Ed.), *Scale effects in animal locomotion*. London: Academic Press; p. 545.
28. Potter, M. C., & Foss, J. F. (1975). *Fluid mechanics*. New York: Ronald Press; p. 588.
29. Jordan, C. E. (1992). A model of rapid-start swimming at intermediate Reynolds number: Undulatory locomotion in the chaetognath *Sagitta elegans*. *Journal of Experimental Biology*, 163, 119–137.
30. Yen, J. (2000). Life in transition: Balancing inertial and viscous forces by planktonic copepods. *Biological Bulletin*, 198, 213–224.

31. McHenry, M. J., Azizi, E., & Strother, J. A. (2003). The hydrodynamics of locomotion in intermediate Reynolds numbers: Undulatory swimming in ascidian larvae (*Botrylloides* sp.). *Journal of Experimental Biology*, 206, 327–343.
32. Fish, F. E. (1993). Influence of hydrodynamic design and propulsive mode on mammalian swimming energetics. *Australian Journal of Zoology*, 42, 79–101.
33. Vogel, S. (2008). Modes and scaling in aquatic locomotion. *Integrative and Comparative Biology*, 48, 702–712.
34. Fox, R. W., Pritchard, P. J., & McDonald, A. T. (2009). *Introduction to fluid mechanics*. Hoboken, NJ: Wiley; p. 752.
35. Blake R. W. (1983). *Fish locomotion*. Cambridge: Cambridge University Press; p. 208.
36. Tomilin, A. G. (1957). *Mammals of the U.S.S.R. and adjacent countries* (Vol. IX, *Cetacea*). Moskva: Izdatel'stvo Akademi Nauk SSSR; p. 717 (translated from Russian).
37. Hutchinson, R. (2001). *Submarines: War beneath the waves from 1776 to the present day*. New York: HarperCollins; p. 223.
38. Aleyev, Yu. G. (1977). *Nekton*. The Hague: Junk; p. 433.
39. Kooyman, G. L. (1989). *Diverse divers*. Berlin: Springer-Verlag; p. 200.
40. Videler, J. J. (1993). *Fish swimming*. London: Chapman & Hall; p. 260.
41. Gafurov, S. A., & Klochkov E. V. (2015). Autonomous unmanned underwater vehicles development tendencies. *Procedia Engineering*, 106, 141–148.
42. Hoerner, S. F. (1965). *Fluid-dynamic drag*. Brick Town, NJ: Author.
43. Fish, F. E., & Hui C. A. (1991). Dolphin swimming: A review. *Mammal Review*, 21, 181–196.
44. Fish, F. E. (2006). The myth and reality of Gray's paradox: Implication of dolphin drag reduction for technology. *Bioinspiration and Biomimetics*, 1, R17–R25.
45. Gray, J. (1936). Studies in animal locomotion VI. The propulsive powers of the dolphin. *Journal of Experimental Biology*, 13, 192–199.
46. Kramer, M. O. (1960). Boundary layer stabilization by distributed damping. *Journal of the American Society for Naval Engineering*, 72, 25–33.
47. Kramer, M. O. (1960). The dolphins' secret. *New Scientist*, 7, 1118–1120.
48. Bechert, D. W., Bruse, M., & Hage, W. (2000). Experiments with three-dimensional riblets as an idealized model of shark skin. *Experiments in Fluids*, 28, 403–412.
49. Bechert, D. W., Bruse, M., Hage, W., & Meyer, R. (2000). Fluid mechanics of biological surfaces and their technological application. *Naturwissenschaffen*, 87, 157–171.
50. Carpenter, P. W., Davies, C., & Lucey, A. D. (2000). Hydrodynamics and compliant walls: Does the dolphin have a secret? *Current Science*, 79, 758–765.
51. Romanenko, E. V. (2002). *Fish and dolphin swimming*. Sofia: Pensoft; p. 429.
52. Fish, F. E. (2004). Structure and mechanics of nonpiscine control surfaces. *IEEE Journal of Oceanic Engineering*, 28, 605–621.
53. Rosen, M. W., & Cornford, N. E. (1971). Fluid friction of fish slimes. *Nature*, 234, 49–51.
54. Hoyt, J. W. (1975). Hydrodynamic drag reduction due to fish slimes. In T. Y. Wu, C. J. Brokaw, & C. Brennen (Eds.), *Swimming and flying in nature* (Vol. 2). New York: Plenum Press; p. 1005.
55. Hoyt, J. W. (1990). Drag reduction by polymers and surfactants. In D. M. Bushnell & J. N. Hefner (Eds.), *Viscous drag reduction in boundary layers*. Washington, DC: American Institute of Aeronautics and Astronautics, Inc.; p. 510.
56. Videler, J. J., Haydar, D., Snoek, R., Hoving, H. J. T., & Szabo, B. G. (2016). Lubricating the swordfish head. *Journal of Experimental Biology*, 219, 1953–1956.

57. Lang, T. G., & Daybell, D. A. (1963). Porpoise performance tests in a seawater tank. In *Naval Ordinance Test Station Technical Report* 3063 China Lake, CA: Naval Ordinance Test Station.
58. Fish, F. E. (2005). A porpoise for power. *Journal of Experimental Biology*, 208, 977–978.
59. Fish, F. E. (2006). Limits of nature and advances of technology in marine systems: What does biomimetics have to offer to aquatic robots? *Applied Bionics and Biomechanics*, 3, 49–60.
60. Fish, F. E., Legas, P., Williams, T. M., & Wei, T. (2014). Measurement of hydrodynamic force generation by swimming dolphins using bubble DPIV. *Journal of Experimental Biology*, 217, 252–260.
61. Fish, F. E. (1984). Mechanics, power output, and efficiency of the swimming muskrat (*Ondatra zibethicus*). *Journal of Experimental Biology*, 110, 183–201.
62. Fish, F. E. (1995). Kinematics of ducklings swimming in formation: Energetic consequences of position. *Journal of Experimental Zoology*, 272, 1–11.
63. Daniel, T. L. (1984). Unsteady aspects of aquatic locomotion. *American Zoologist*, 24, 121–134.
64. Daniel, T. L., & Webb, P. W. (1987). Physical determinants of locomotion. In P. Dejours, L. Bolis, C. R. Taylor, & E. R. Weibel (Eds.), *Comparative physiology: Life in water and on land*. New York: Liviana Press, Springer-Verlag; p. 556.
65. Fish, F. E., Innes, S., & Ronald, K. (1988). Kinematics and estimated thrust production of swimming harp and ringed seals. *Journal of Experimental Biology*, 137, 157–173.
66. Daniel, T., Jordan, C., & Grunbaum, D. (1992). Hydromechanics of swimming. In R. N. Alexander (Ed.), *Advances in comparative and environmental physiology* (Vol. 11). London: Springer-Verlag; p. 304.
67. Anderson, J. M. (1998). The vorticity control unmanned undersea vehicle: A biologically inspired autonomous vehicle. In J. C. S. Meng (Ed.), *Proceedings of the international symposium on seawater drag reduction*. Newport, RI: Naval Undersea Warfare Center; pp. 479–483.
68. Fish, F. E. (2010). Swimming strategies for energy economy. In P. Domenici & B. G. Kapoor (Eds.), *Fish swimming: An etho-ecological perspective*. Enfield, NH: Science Publishers; p. 534.
69. Wang, Z. J. (2000). Vortex shedding and frequency selection in flapping flight. *Journal of Fluid Mechanics*, 410, 323–341.
70. Van Dam, C. P. (1987). Efficiency characteristics of crescent-shaped wings and caudal fins. *Nature*, 325, 435–437.
71. Alexander R. N. (1983). *Animal mechanics*. Oxford: Blackwell; p. 301.
72. Blake, R. W. (1979). The mechanics of labriform locomotion. I. Labriform locomotion in the angelfish (*Pterophyllum eimekei*): An analysis of the power stroke. *Journal of Experimental Biology*, 82, 255–271.
73. Blake, R. W. (1980). The mechanics of labriform locomotion. II. An analysis of the recovery stroke and the overall fin-beat cycle propulsive efficiency in the angelfish. *Journal of Experimental Biology*, 85, 337–342.
74. Fish, F. E. (1998). Comparative kinematics and hydrodynamics of odontocete cetaceans: Morphological and ecological correlates with swimming performance. *Journal of Experimental Biology*, 201, 2867–2877.
75. Ellington, C. P., van den Berg, C., Willmott, A. P., & Thomas, A. L. R. (1996). Leading-edge vortices in insect flight. *Nature*, 384, 626–630.

76. Dickinson, M. H., Lehmann, F.-O., & Sane, S. P. (1999). Wing rotation and the aerodynamic basis of insect flight. *Science*, 284, 1954–1960.
77. Fierstine, H. L., & Walters, V. (1968). Studies of locomotion and anatomy of scombrid fishes. *Memoirs of the Southern California Academy of Sciences*, 6, 1–31.
78. Fish, F. E., & Lauder, G. V. (2006). Passive and active flow control by swimming fishes and mammals. *Annual Review of Fluid Mechanics*, 38, 193–224.
79. Fish, F. E., Nusbaum, M. K., Beneski, J. T., & Ketten, D. R. (2006). Passive cambering and flexible propulsors: Cetacean flukes. *Bioinspiration and Biomimetics*, 1, S42–S48.
80. Katz, J., & Weihs, D. (1978). Hydrodynamic propulsion by large amplitude oscillation of an airfoil with chordwise flexibility. *Journal of Fluid Mechanics*, 88, 485–497.
81. Prempraneerach, P., Hover, F. S., & Triantafyllou, M. S. (2003). The effect of chordwise flexibility on the thrust and efficiency of a flapping foil. In *Proceedings of the Thirteenth International Symposium on Unmanned Untethered Submersible Technology: Proceedings of the Special Session on Bio-Engineering Research Related to Autonomous Underwater Vehicles*. Lee, NH: Autonomous Undersea Systems Institute.
82. Liu, P. & Bose, N. (1997). Propulsive performance from oscillating propulsors with spanwise flexibility. *Proceedings of the Royal Society of London A*, 453, 1763–1770.
83. Katz, J., & Weihs, D. (1979). Large amplitude unsteady motion of a flexible slender propulsor. *Journal of Fluid Mechanics*, 90, 713–723.
84. Bose, N., & Lien, J. (1989). Propulsion of a fin whale (*Balaenoptera physalus*): Why the fin whale is a fast swimmer. *Proceedings of the Royal Society of London B*, 237, 175–200.
85. Bose, N. (1995). Performance of chordwise flexible oscillating propulsors using a time-domain panel method. *International Shipbuilding Progress*, 42, 281–294.
86. Moored, K. W., Dewey, P. A., Boschitsch, B. M., Smits, A. J., & Haj-Hariri, H. (2014). Linear instability mechanisms leading to optimally efficient locomotion with flexible propulsors. *Physics of Fluids*, 26, 041905.
87. Ren, Y., Liu, G., & Dong, H. (2015). Effect of surface morphing on the wake structure and performance of pitching-rolling plates. *53rd AIAA Aerospace Sciences Meeting, AIAA-2015-1490*, Kissimmee, FL.
88. Webb, P. W., & de Buffrénil, V. (1990). Locomotion in the biology of large aquatic vertebrates. *Transactions of the American Fisheries Society*, 119, 629–641.
89. Breder, C. M. (1926). The locomotion of fishes. *Zoologica*, 4, 159–297.
90. Robinson, J. A. (1975). The locomotion of plesiosaurs. *Neues Jahrbuch für Geologie und Paläontologie*, 149, 286–332.
91. Lang, T. G. (1966). Hydrodynamic analysis of dolphin fin profiles. *Nature*, 209, 1110–1111.
92. Howell, A. B. (1930). *Aquatic mammals*. Springfield, IL: Charles C. Thomas; p. 338.
93. Massare, J. A. (1988). Swimming capabilities of Mesozoic marine reptiles: Implications for method of predation. *Paleobiology*, 14, 187–205.
94. Fish, F. E. (2000). Biomechanics and energetics in aquatic and semiaquatic mammals: Platypus to whale. *Physiological and Biochemical Zoology*, 73, 683–698.
95. Gordon, K. R. (1981). Locomotor behaviour of the walrus (*Odobenus*). *Journal of Zoology, London*, 195, 349–367.
96. Klausewitz, W. (1964). Der lokomotionsmodus der flugelrochen (Myliobatoidei). *Zoologischer Anzeiger*, 173, 111–120.
97. Lindsey, (1978). Form, function, and locomotory habits in fish. In W. S. Hoar & D. J. Randall (Eds.), *Fish physiology: Locomotion* (Vol. 7). New York: Academic Press; p. 576.

98. Heine, C. (1992). *Mechanics of flapping fin locomotion in the cownose ray, Rhinoptera bonasus (Elasmobranchii: Myliobatidae).* [PhD Dissertation]. Durham, NC: Duke University.
99. Rosenberger, L. J. (2001). Pectoral fin locomotion in batoid fishes – undulation versus oscillation. *Journal of Experimental Biology*, 204, 379–394.
100. Fish, F. E., Schreiber, C. M., Moored, K. M., Liu, G., Dong, H., & Bart-Smith, H. (2016). Hydrodynamic performance of aquatic flapping: Efficiency of underwater flight in the manta. *Aerospace*, 3, 20. doi:10.3390/aerospace3030020
101. Godfrey, S. J. (1984). Additional observations of subaquaeous locomotion in the California sea lion (*Zalophus californianus*). *Aquatic Mammals*, 11(2), 53–57.
102. Feldkamp, S. D. (1987). Foreflipper propulsion in the California sea lion, *Zalophus californianus*. *Journal of Zoology, London*, 212, 43–57.
103. Wyneken, J. (1997). Sea turtle locomotion: Mechanisms, behavior, and energetics. In PL Lutz & JA Musick (Eds.), *The biology of sea turtles*. Boca Raton, FL: CRC Press; p. 432.
104. Motani, R. (2002). Scaling effects in caudal fin propulsion and the speed of ichthyosaurs. *Nature*, 415, 309–312.
105. Rivera, A. R. V., Wyneken, J., & Blob, R. W. (2011). Forelimb kinematics and motor patterns of swimming loggerhead sea turtles (*Caretta caretta*): Are motor patterns conserved in the evolution of new locomotor strategies? *Journal of Experimental Biology*, 214, 3314–3323.
106. Barsukov, V. V. (1960). The speed of movement of fishes. *Priroda*, 3, 103–104.
107. Gudger, E. W. (1940). The alleged pugnacity of the swordfish and the spearfishes as shown by their attacks on vessels: A study of their behavior and the structures which make possible these attacks. *Memoirs of the Royal Asiatic Society of Bengal*, 12, 215–315.
108. Shuleykin, V. V. (1949). *Essays on physics of the sea*. Moskva: Akademi Nauk, SSSR; p. 334.
109. Lane, F. W. (1941). How fast do fish swim? *Country Life*, 90, 534–535.
110. Walters, V., & Fiersteine, H. L. (1964). Measurements of swimming speeds of yellowfin tuna and wahoo. *Nature*, 202, 208–209.
111. Berzin, A. A. (1972). *The sperm whale*. Jerusalem: Israel Program for Scientific Translation; p. 394.
112. Williamson, G. R. (1972). The true body shape of rorqual whales. *Journal of Zoology, London*, 167, 277–286.
113. Tinsley, J. B. (1984). *The sailfish: Swashbuckler of the open seas*. Gainesville, FL: University of Florida Press; p. 216.
114. Tucker, V. A. (1970). Energetic cost of locomotion in animals. *Comparative Biochemistry and Physiology*, 34, 841–846.
115. Tucker, V. A. (1975). The energetic cost of moving about. *American Scientist*, 63, 413–419.
116. Schmidt-Nielsen, K. (1972). Locomotion: Energy cost of swimming, flying, and running. *Science*, 177, 222–228.
117. Prange, H. D. (1976). Energetics of swimming of a sea turtle. *Journal of Experimental Biology*, 64, 1–12.
118. Williams, T. M., & Kooyman, G. L. (1985). Swimming performance and hydrodynamic characteristics of harbor seals *Phoca vitulina*. *Physiological Zoology*, 58, 576–589.
119. Williams, T. M. (1987). Approaches for the study of exercise physiology and hydrodynamics in marine mammals. A. C. Huntley, D. Costa, G. A. J. Worthy, & M. A. Castellini (Eds.), *Approaches to marine mammal energetics*. Yarmouth Port, MA: Society for Marine Mammalogy; p. 253.

120. Williams, T. M. (1999). The evolution of cost efficient swimming in marine mammals: Limits to energetic optimization. *Philosophical Transactions Royal Society of London B Biological Sciences*, 353, 1–9
121. Sepulveda, C. H., & Dickson, K. A. (2000). Maximum sustainable speeds and cost of swimming in juvenile kawakawa tuna (*Euthynnus affinis*) and chub mackerel (*Scomber japonicus*). *Journal of Experimental Biology*, 203, 3089–3101.
122. Guinet, C., Domenici, P., De Stephanis, R., Barrett-Lennard, L., Ford, J. K. B., & Verborgh, P. (2007). Killer whale predation on bluefin tuna: Exploring the hypothesis of the endurance-exhaustion technique. *Marine Ecology Progress Series*, 347, 111–119.
123. Golenchenko, A. P. (1960). The swordfish. *Priroda*, 4, 115.
124. Carey, F. G., & Robinson, B. H. (1981). Daily patterns in the activities of swordfish, *Xiphias gladius*, observed by acoustic telemetry. *Fisheries Bulletin*, 79, 277–292.
125. Sedberry, G., & Loefer, J. (2001). Satellite telemetry tracking of swordfish, *Xiphias gladius*, off the eastern United States. *Marine Biology*, 139, 355–360.
126. Graham, R. T., Witt, M. J., Castellanos, D. W., et al.(2012). Satellite tracking of manta rays highlights challenges to their conservation. *PLOS ONE*, 7, e36834.
127. Richardson, W. J., & Finley, K. J. (1989). *Comparison of behavior of bowhead whales of the Davis Strait and Bering/Beaufort stocks. NTIS No. PB89–195556*. King City, ON: LGL Ltd.
128. Ponganis, P. J., Ponganis, E. P., Ponganis, K. V., Kooyman, G. L., Gentry, R. L., & Trillmich, F. (1990). Swimming velocities in otariids. *Canadian Journal of Zoology*, 68, 2105–2112.
129. Sato, K., Watanuki, Y., Takahashi, A., et al. (2007). Stroke frequency, but not swimming speed, is related to body size in free-ranging seabirds, pinnipeds and cetaceans. *Proceedings of the Royal Society of London B: Biological Sciences*, 274, 471–477.
130. Clark, B. D., & Bemis, W. (1979). Kinematics of swimming of penguins at the Detroit Zoo. *Journal of Zoology*, 188, 411–428.
131. Culik, B., Wilson R., & Bannasch, R. U. (1994). Underwater swimming at low energetic cost by pygoscelid penguins. *Journal of Experimental Biology*, 197, 65–78.
132. Eckert, S. A. (2002). Swim speed and movement patterns of gravid leatherback sea turtles (*Dermochelys coriacea*) at St Croix, US Virgin Islands. *Journal of Experimental Biology*, 205, 3689–3697.
133. Luschi, P., Hays, G. C., & Papi, F. (2003). A review of long-distance movements by marine turtles, and the possible role of ocean currents. *Oikos*, 103, 293–302.
134. James, M. C., Myers, R. A., & Ottensmeyer, C. A. (2005). Behaviour of leatherback sea turtles, *Dermochelys coriacea*, during the migratory cycle. *Proceedings of the Royal Society of London B: Biological Sciences*, 272, 1547–1555.
135. Lang, T. G. (1975). Speed, power, and drag measurements of dolphins and porpoises. In T. Y. Wu, C J. Brokaw, & C. Brennen (Eds.), *Swimming and flying in nature*. New York: Plenum Press; p. 1005.
136. Johannessen, C. L., & Harder, J. A. (1960). Sustained swimming speeds of dolphins. *Science*, 132, 1550–1551.
137. Daniel, T. L. (1991). Efficiency in aquatic locomotion: Limitations from single cells to animals. In R. W. Blake (Ed.), *Efficiency and economy in animal physiology*. Cambridge: Cambridge University Press; p. 187.
138. Saunders, H. E. (1957). *Hydrodynamics in ship design* (Vol. II). New York: Society for Naval Architects and Marine Engineers; p. 980.
139. Larrabee, E. E. (1980). The screw propeller. *Scientific American*, 243, 134–148.

140. Magnuson, J. J. (1978). Locomotion by scombrid fishes: Hydrodynamics, morphology and behaviour. In W. S. Hoar & J. D. Randall (Eds.), *Fish physiology* (Vol. 7). London: Academic Press; p. 576.
141. Alexander, R. N. (1988). *Elastic mechanisms in animal movement*. Cambridge: Cambridge University Press; p. 141.
142. Pabst, D. A. (1996). Springs in swimming animals. *American Zoologist*, 36, 723–735.
143. Nakashima, M., & Ono, K. (1999). Experimental study of two-joint dolphin robot. In *Proceedings of the 11th International Symposium on Unmanned Untethered Submersible Technology, 99-8-01*. Lee, NH: Autonomous Undersea Systems Institute.
144. Curren, K. C., Bose, N., & Lien J. (1994). Swimming kinematics of a harbor porpoise (*Phocoena phocoena*) and an Atlantic white-sided dolphin (*Lagenorhynchus acutus*). *Marine Mammal Science*, 10, 485–492.
145. Dewey, P. A., Boschitsch, B. M., Moored, K. W., Stone, H. A., & Smits, A. J. (2013). Scaling laws for the thrust production of flexible pitching panels. *Journal of Fluid Mechanics*, 732, 29–46.
146. Bose, N., Lien, J., & Ahia J. (1990). Measurements of the bodies and flukes of several cetacean species. *Proceeding of the Royal Society of London B*, 242, 163–173.
147. Triantafyllou, G. S., Triantafyllou, M. S., & Grosenbaugh, M. A. (1993). Optimal thrust development in oscillating foils with application to fish propulsion. *Journal of Fluids and Structures*, 7, 205–224.
148. Triantafyllou, M. S. Triantafyllou, G. S., & Yue, D. K. (2000). Hydrodynamics of fishlike swimming. *Annual Review of Fluid Mechanics*, 32, 33–53.
149. Taylor, G. K., Nudds, R. L., & Thomas, A. L. R. (2003). Flying and swimming animals cruise at a Strouhal number tuned for high power efficiency. *Nature*, 425, 707–711.
150. Rohr, J. J., & Fish, F. E. (2004). Strouhal numbers and optimization of swimming by odontocete cetaceans. *Journal of Experimental Biology*, 207, 1633–1642.
151. Videler, J. J., & Weihs, D. (1982). Energetic advantages of burst-and-coast swimming of fish at high speeds. *Journal of Experimental Biology*, 97, 169–178.
152. Blake, R. W. (1983). Functional design and burst-and-coast swimming in fishes. *Canadian Journal of Zoology*, 61, 2491–2494.
153. Fish, F. E., Fegely, J. F., & Xanthopoulos, C. J. (1991). Burst-and-coast swimming in schooling fish (*Notemigonus crysoleucas*) with implications for energy economy. *Comparative Biochemistry and Physiology Part A: Physiology*, 100, 633–637.
154. Wu, G., Yang, Y., & Zeng, L. (2007). Kinematics, hydrodynamics and energetic advantages of burst-and-coast swimming of koi carps (*Cyprinus carpio koi*). *Journal of Experimental Biology*, 210, 2181–2191.
155. Weihs, D. (1974). Energetic advantages of burst swimming of fish. *Journal of Theoretical Biology*, 48, 215–229.
156. Chung, M. H. (2009). On burst-and-coast swimming performance in fish-like locomotion. *Bioinspiration & Biomimetics*, 4, 036001.
157. Magnuson, J. J., & Gooding, R. M. (1971). Color patterns of pilotfish (*Naucrates ductor*) and their possible significance. *Copeia*, 1971, 314–316.
158. Liao, J., Beal, D. N., Lauder, G. V., & Triantafyllou, M. S. (2003). The Kármán gait: Novel body kinematics of rainbow trout swimming in a vortex street. *Journal of Experimental Biology*, 206, 1059–1073.
159. Kelly, H. R. (1959). A two-body problem in the echelon-formation swimming of porpoise. In *U. S. Naval Ordinance Test Station, Technical Note 40606-1*. China Lake, CA: Naval Ordinance Test Station.

160. Weihs, D. (2004). The hydrodynamics of dolphin drafting. *Journal of Biology*, 3, 8.1–8.16.
161. Fish, F. E., Goetz, K. T., Rugh, D. J., & Brattström, L. V. (2013). Hydrodynamic patterns associated with echelon formation swimming by feeding bowhead whales (*Balaena mysticetus*). *Marine Mammal Science*, 29, E498–E507.
162. Weihs, D. (1973). Hydromechanics of fish schooling. *Nature*, 241, 290–291.
163. Fish, F. E. (1999). Energetics of swimming and flying in formation. *Comments on Theoretical Biology*, 5, 283–304.
164. Webb, P. W. (1998). Entrainment by river chub *Nocomis micropogon* and smallmouth bass *Micropterus dolomieu* on cylinders. *Journal of Experimental Biology*, 201, 2403–2412.
165. Liao, J., Beal, D. N., Lauder, G. V., & Triantafyllou, M. S. (2003a). Fish exploiting vortices decrease muscle activity. *Science*, 302, 1566–1569.
166. Liao, J. (2004). Neuromuscular control of trout swimming in a vortex street: Implications for energy economy during the Kármán gait. *Journal of Experimental Biology*, 207, 3495–3506.
167. Webb, P. W. (2004). Maneuverability: General issues. *IEEE Journal of Oceanic Engineering*, 29, 547–555.
168. Howland, H. C. (1974). Optimal strategies for predator avoidance: The relative importance of speed and manoeuvrability. *Journal of Theoretical Biology*, 47, 333–350.
169. Webb, P. W. (1976). The effect of size on the fast-start performance of rainbow trout, *Salmo gairdneri*, and a consideration of piscivorous predator-prey interactions. *Journal of Experimental Biology*, 65, 157–177.
170. Domenici, P., & Blake, R. W. (1997). The kinematics and performance of fish fast-start swimming. *Journal of Experimental Biology*, 200, 1165–1178.
171. Walker, J. A. (2000). Does a rigid body limit maneuverability? *Journal of Experimental Biology*, 203, 3391–3396.
172. Walker, J. A. (2004).Kinematics and performance of maneuvering control surfaces in teleost fishes. *Journal of Oceanic Engineering*, 29, 572–584.
173. Domenici, P. (2001). The scaling of locomotor performance in predator-prey encounters: From fish to killer whales. *Comparative Biochemistry and Physiology Part A*, 131, 169–182.
174. Parson, J., Fish, F. E., & Nicastro, A. J. (2011). Turning performance in batoid rays: Limitations of a rigid body. *Journal of Experimental Marine Biology and Ecology*, 402, 12–18.
175. Webb, P. W. (2006). Stability and maneuverability. In R. E. Shadwick & G. V. Lauder (Eds.), *Fish physiology; fish biomechanics* (Vol. 23). Amsterdam: Academic Press; p. 542.
176. Burcher, R., & Rydill, L. (1994). *Concepts in submarine design*. Cambridge: Cambridge University Press; p. 300.
177. Fish, F. E. (2002). Balancing requirements for stability and maneuverability in cetaceans. *Integrative and Comparative Biology*, 42, 85–93.
178. Triantafyllou, M. S. (2017). Tuna fin hydraulics inspire aquatic robotics. *Science*, 357, 251–252.
179. Marchaj, C. A. (1988) *Aero-hydrodynamics of sailing*. Camden, ME: International Marine Publishing; p. 743.
180. Fish, F. E., & Nicastro, A. J. (2003). Aquatic turning performance by the whirligig beetle: Constraints on maneuverability by a rigid biological system. *Journal of Experimental Biology*, 206, 1649–1656.

181. Harris, J. E. (1936). The role of the fins in the equilibrium of the swimming fish. I. Wind-tunnel tests on a model of *Mustelus canis* (Mitchill). *Journal of Experimental Biology*, 13, 476–493.
182. Fish, F. E., Hurley, J., & Costa, D. P. (2003). Maneuverability by the sea lion *Zalophus californianus*: Turning performance of an unstable body design. *Journal of Experimental Biology*, 206, 667–674.
183. Webb, P. W. (1983). Speed, acceleration and manoeuverability of two teleost fishes. *Journal of Experimental Biology*, 102, 115–122.
184. Gerstner, C. L. (1999). Maneuverability of four species of coral-reef fish that differ in body and pectoral-fin morphology. *Canadian Journal of Zoology*, 77, 1102–1110.
185. Kajiura, S. M., Forni, J. B., & Summers, A. P. (2003). Maneuvering in juvenile carcharhinid and sphyrnid sharks: The role of the hammerhead shark cephalofoil. *Zoology*, 106, 19–28.
186. Bandyopadhyay, P. R., Rice, J. Q., Corriveau, P. J., & Macy, W. K. (1995). Maneuvering hydrodynamics of fish and small underwater vehicles, including the concept of an agile underwater vehicle. *NUWC-NCT Technical Report*, 10, 494.
187. Blake, R. W., Chatters, L. M., & Domenici, P. (1995). Turning radius of yellowfin tuna (*Thunnus albacares*) in unsteady swimming manoeuvres. *Journal of Fish Biology*, 46, 536–538.
188. Norberg, U. M. (1990). *Vertebrate flight: Mechanics, physiology, morphology, ecology and evolution*. Berlin: Springer-Verlag; p. 290.
189. Webb, P. W. (1994). The biology of fish swimming. In L. Maddock, Q. Bone, & J. M. V. Rayner (Eds.), *Mechanics and physiology of animal swimming*. Cambridge: Cambridge University Press; p. 250.
190. Maresh, J. L., Fish, F. E., Nowacek, D. P., Nowacek, S. M., & Wells, R. S. (2004). High performance turning capabilities during foraging by bottlenose dolphins (*Tursiops truncatus*). *Marine Mammal Science*, 20, 498–509.
191. Hui, C. A. (1985). Maneuverability of the Humboldt penguin (*Spheniscus humboldti*) during swimming. *Canadian Journal of Zoology*, 63, 2165–2167.
192. Foyle, T. P. & O'Dor, R. K. (1988). Predatory strategies of squid (*Illex illecebrosus*) attacking small and large fish. *Marine Behavior and Physiology*, 13, 155–168.
193. Frey, E., & Salisbury, S. W. (2000). The kinematics of aquatic locomotion in *Osteolaemus tetraspis*. In G. C. Grigg, F. Seebacher, & C. E. Franklin (Eds.), *Cope crocodilian biology and evolution*. Chipping Norton, UK: Surry Beatty & Sons; p. 446.
194. Fish, F. E. (1997). Biological designs for enhanced maneuverability: Analysis of marine mammal performance. In *Proceedings of the 10th International Symposium on Unmanned Untethered Submersible Technology, pp. 109–117*. Lee, NH: Autonomous Undersea Systems Institute.
195. Rivera, G., Rivera, A. R., Dougherty, E. E., & Blob, R. W. (2006). Aquatic turning performance of painted turtles (*Chrysemys picta*) and functional consequences of a rigid body design. *Journal of Experimental Biology*, 209, 4203–4213.
196. Helmer, D., Geurten, B. R., Dehnhardt, G., & Hanke, F. D. (2016). Saccadic movement strategy in common cuttlefish (*Sepia officinalis*). *Frontiers in Physiology*, 7, 660. doi:10.3389/fphys.2016.00660
197. Geurten, B. R., Niesterok, B., Dehnhardt, G., & Hanke, F. D. (2017). Saccadic movement strategy in a semiaquatic species – The harbour seal (*Phoca vitulina*). *Journal of Experimental Biology*, 220, 1503–1508.

198. Jastrebsky, R. A., Bartol, I. K., & Krueger, P. S. (2017). Turning performance of brief squid *Lolliguncula brevis* during attacks on shrimp and fish. *Journal of Experimental Biology*, 220, 908–919.
199. Miller (1991).
200. Duffy, C. A. J., & Abbott, D. (2003). Sightings of mobulid rays from northern New Zealand, with confirmation of the occurrence of *Manta birostris* in New Zealand waters. *New Zealand Journal of Marine and Freshwater Research*, 37, 715–721.
201. Fish, F. E., Smits, A. J., Haj-Hariri, H., Bart-Smith, H., & Iwasaki, T. (2012). Biomimetic swimmer inspired by the manta ray. In Y. Bar-Cohen (Ed.), *Biomimetics: Nature-based innovation*. Boca Raton, FL: CRC Press; p. 735.
202. Goldbogen, J. A., Calambokidis, J., Friedlaender, A. S., et al. (2013). Underwater acrobatics by the world's largest predator: 360° rolling manoeuvres by lunge-feeding blue whales. *Biology Letters*, 9, 20120986.
203. Segre, P. S., Cade, D. E., Fish, F. E., et al. (2016). Hydrodynamic properties of fin whale flippers predict maximum rolling performance. *Journal of Experimental Biology*, 219, 3315–3320.
204. Fish, F. E., & Battle, J. M. (1995). Hydrodynamic design of the humpback whale flipper. *Journal of Morphology*, 225, 51–60.
205. Carlton, J. S. (2012). *Marine propellers and propulsion*. Amsterdam: Elsevier; p. 516.
206. Tavolga, W. N. (1967). Underwater sound in marine biology. *Underwater Acoustics*, 2, 35–41.
207. Tavolga, W. N. (1971). Sound production and detection. In W. S. Hoar & D. J. Randall (Eds.), *Fish physiology* (Vol. V). New York: Academic Press; p. 600.
208. Moulton, J. M. (1960). Swimming sounds and the schooling of fishes. *Biological Bulletin*, 119, 210–223.
209. Kasumyan, A. O. (2008). Sounds and sound production in fishes. *Journal of Ichthyology*, 48, 981–1030.
210. Geer, D. (2001). *Propeller handbook*. Camden, ME: International Marine Publishing; p. 152.
211. Iosilevskii, G., & Weihs, D. (2008). Speed limits on swimming of fishes and cetaceans. *Journal of the Royal Society Interface*, 5, 329–338.
212. Miklosovic, D. S., Murray, M. M., Howle, L. E. & Fish, F. E. (2004) Leading edge tubercles delay stall on humpback whale (*Megaptera novaeangliae*) flippers. *Physics of Fluids*, 16, L39–L42.
213. Fish, F. E., Weber, P. W., Murray, M. M., & Howle, L. E. (2011). The humpback whale's flipper: Application of bio-inspired tubercle technology. *Integrative and Comparative Biology*, 51, 203–213.
214. Fish, F. E., Weber, P. W., Murray, M. M., & Howle, L. E. (2011). Marine applications of the biomimetic humpback whale flipper. *Marine Technology Society Journal*, 45, 198–207.
215. Hansen, K. L., Kelso, R. M., & Doolan, C. J. (2010). Reduction of flow induced tonal noise through leading edge tubercle modifications. In *16th AIAA/CEAS Aeroacoustics Conference, 7-9 June 2010*, Stockholm, Sweden: AIAA/CEAS.
216. Lau, A. S. H., & Kim, J. W. (2013). The effect of wavy leading edges on aerofoil-gust interaction noise. *Journal of Sound and Vibration*, 332, 6234–6253.
217. Polacsek, C., Reboul, G., Clair, V., Le Garrec, T., & Deniau, H. (2011). Turbulence-airfoil interaction noise reduction using wavy leading edge: An experimental and numerical study. In *INTER-NOISE and NOISE-CON Congress and Conference Proceedings, 2011*. pp. 4464–4474. Osaka, Japan: Institute of Noise Control Engineering.

218. Kim, J. W., Haeri, S., & Joseph, P. (2016). On the reduction of aerofoil–turbulence interaction noise associated with wavy leading edges. *Journal of Fluid Mechanics*, 792, 526–552.
219. Shi, W., Atlar, M., Rosli, R., Aktas, B., & Norman, R. (2016). Cavitation observations and noise measurements of horizontal axis tidal turbines with biomimetic blade leading-edge designs. *Ocean Engineering*, 121, 143–155.
220. Wang, J., Zhang, C., Wu, Z., Wharton, J., & Ren, L. (2017). Numerical study on reduction of aerodynamic noise around an airfoil with biomimetic structures. *Journal of Sound and Vibration*, 394, 46–58.
221. Vogel, S. (1998). *Cat's paws and catapults*. New York: W. W. Norton; p. 382.
222. Forbes, P. (2005). *The gecko's foot*. New York: Norton; p. 272.
223. Dabiri, J. O. (2007). Renewable fluid dynamic energy derived from aquatic animal locomotion. *Bioinspiration & Biomimetics*, 2, L1.
224. Bar-Cohen, Y. (2006). Biomimetics –Using nature to inspire human innovation. *Bioinspiration & Biomimetics*, 1, P1–P12.
225. Allen, R. (2010). *Bulletproof feathers*. Chicago: University of Chicago Press; p. 192.
226. Ralston, E., & ,Swain G. (2009). Bioinspiration – The solution for biofouling control? *Bioinspiration & Biomimetics*, 4, 015007.
227. Taubes, G. (2000). Biologists and engineers create a new generation of robots that imitate life. *Science*, 288, 80–83.
228. Fish, F. E. (2009). Biomimetics: Determining engineering opportunities from nature. *Proceedings SPIE Conference*, 7401, 740109.
229. Mohseni, K., Mitta, R., & Fish, F. E. (2006). Preface: Special issue featuring selected papers from the mini-symposium on biomimetic & bio-inspired propulsion. *Bioinspiration and Biomimetics*, 1, E01.
230. Nakashima, M., & Ono, K. (2002). Development of a two-joint dolphin robot In J. Ayers, J. L. Davis, & A. Rudolph (Eds.), *Neurotechnology for biomimetric robots*. Cambridge, MA: MIT Press; p. 650.
231. Anderson, J. M., & Chhabra, N. K. (2002). Maneuvering and stability performance of a robotic tuna. *Integrative and Comparative Biology*, 42, 118–126.
232. Kato, N. (2005). Median and paired fin controllers for biomimetic marine vehicles. *Applied Mechanics Reviews*, 58, 238–252.
233. Lauder, G. V., Anderson, E. J., Tangorra, J., & Madden, P. G. (2007). Fish biorobotics: Kinematics and hydrodynamics of self-propulsion. *Journal of Experimental Biology*, 210, 2767–2780.
234. Bozkurttas, M., Tangorra, J., Lauder, G., & Mittal, R. (2008). Understanding the hydrodynamics of swimming: From fish fins to flexible propulsors for autonomous underwater vehicles. *Advances in Science and Technology*, 58, 193–202.
235. Low, K. H., & Chong, C. W. (2010). Parametric study of the swimming performance of a fish robot propelled by a flexible caudal fin. *Bioinspiration & Biomimetics*, 5, 046002.
236. Low, K. H. (2011). Current and future trends of biologically inspired underwater vehicles. *IEEE Defense Science Research Conference and Expo (DSR)*, 2011, 1–8.
237. Moored, K. W., Fish, F. E., Kemp, T. H., & Bart-Smith, H. (2011). Batoid fishes: Inspiration for the next generation of underwater robots. *Marine Technology Society Journal*, 45, 99–109.
238. Low, K. H., Hu, T., Mohammed, S., Tangorra, J., & Kovac M. (2015). Perspectives on biologically inspired hybrid and multi-modal locomotion. *Bioinspiration & Biomimetics*, 10, 020301.

239. Raj, A., & Thakur, A. (2016). Fish-inspired robots: Design, sensing, actuation, and autonomy – A review of research. *Bioinspiration & Biomimetics*, 11, 031001.
240. Kumph, J. M. (2000). *Maneuvering of a robotic pike* [Master Thesis]. Cambridge, MA: Massachusetts Institute of Technology; p. 76.
241. Stanway, J. (2008). The turtle and the robot. *Oceanus*, 47, 22–25.
242. Kirk, J., & Klein, A. (1987). *Ships of the US Navy*. New York: Exeter Books; p. 192.
243. Ashley, S. (2001). Warp drive underwater. *Scientific American*, 284, 70–79.
244. Zhan, K., Yu, B., & Wang, J. (2011). Simulations of the anti-torpedo tactic of the conventional submarine using decoys and jammers. *Applied Mechanics and Materials*, 65, 165–168.
245. Basic, J., & Blagojevic, B. (2015). Hydrodynamic performance of an autonomous underwater vehicle with a swivel tail. In C. Guedes Soares, R. Dejhalla & D. Pavletic (Eds.), *Towards green marine technology and transport*. Boca Raton, FL: CRC Press; p. 922.
246. Menozzi, A., Leinhos, H. A., Beal, D. N., & Bandyopadhyay, P. (2008). Open-loop control of a multifin biorobotic rigid underwater vehicle. *IEEE Journal of Oceanic Engineering*, 33, 59–68.
247. Liang, J., Wang, T., & Wen, L. (2010). Development of a two-joint robotic fish for real-world exploration. *Journal of Field Robotics*, 28, 70–79.
248. Fish, F. E., Dong, H., Zhu, J., & Bart-Smith, H. (2017). Swimming kinematics of mobuliform rays: Oscillatory winged propulsion by large pelagic batoids. *Marine Technology Society Journal*, 51(5), 35–47.
249. Pavlov, V., Rosental, B., Hansen, N. F., et al. (2017). Hydraulic control of tuna fins: A role for the lymphatic system in vertebrate locomotion. *Science*, 357, 310–314.
250. Singh H., Bellingham, J. G., Hover, F., et al. (2001). Docking for an autonomous ocean sampling network. *IEEE Journal of Oceanic Engineering*, 26, 498–514.
251. Colgate, J. E., & Lynch, K. M. (2004). Mechanics and control of swimming: A preview. *IEEE Journal of Oceanic Engineering*, 29, 660–673.
252. Murthy, K., & Rock, S. (2010). Spline-based trajectory planning techniques for benthic AUV operations. In *Proceedings of IEEE Autonomous Underwater Vehicles (AUV) Conference*. Stanford, CA: IEEE/OES; pp. 1–9.
253. Fish, F. E. (1993). Power output and propulsive efficiency of swimming bottlenose dolphins (*Tursiops truncatus*). *Journal of Experimental Biology*, 185, 179–193.
254. Fish, F. E. (2020). Advantages of aquatic animals as models for bio-inspired drones over present AUV technology. *Bioinspiration & Biomimetics*, 15, 025001.

11 Flying of Insects

Bo Cheng

11.1 Introduction

The abundance of flying insects in nature may make them seem ordinary to most of us. However, for approximately 350 million years [1], flying insects have been experimenting successfully with various aspects of flight, including aerodynamics [2,3], wing design [4], sensors [5,6], and flight control [7–9]. As a result, they have developed miniaturized flight apparatus and efficient computation architectures for executing aerobatic feats that are not yet emulated in engineering flight (Figure 11.1). This makes flying insects truly extraordinary small-scale aircraft from nature, and their design and working principles have received wide interest in both engineering and biology communities.

This review describes various aspects of insect flight and the existing engineering designs inspired by insect flight. Without being exhaustive, it aims to present to readers the sophistications of insect flight systems that enable their extraordinary flight capability, from which meaningful bioinspiration for engineered flight can be generated. It is critical to bear in mind that the evolutionary adaptations of insect flight rest upon their aerial survival and sexual selection, which are shaped by a number of competing objectives under a variety of constraints other than those related to flight alone. From an engineering perspective, it can be argued that the diverse forms of insect flight [10] are solutions of multi-objective optimizations to various ecological demands. Also, nature is limited in the raw material available to insects for building flight apparatus and computational circuits as compared with those available to modern engineers; this may constrain or at least play a significant role in determining the available design solutions [11]. In the meanwhile, the evolution of flying insects, or of any biological systems, is also phylogenetically constrained [12], i.e., new adaptations can only evolve from modifications of older ones [11]; engineering designs, on the other hand, can be completely developed from scratch and engineers have often invented solutions that were never part of the natural repertoire, but regardless may offer a superior solution for a particular function. Therefore, as the main purpose of this review, it is critical for engineers to gain understanding of various aspects of insect flight and the underlying factors driving its forms and functions, thereby avoiding uninformed biomimicry of insect flight.

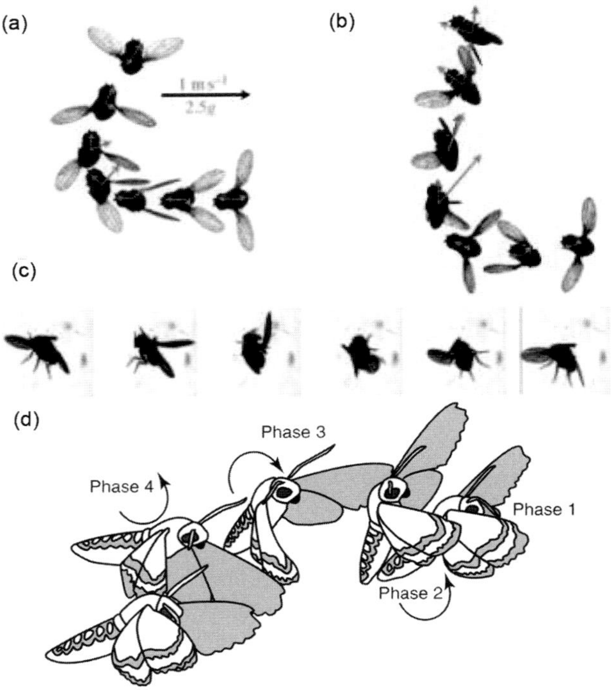

Figure 11.1 Examples of flight maneuvers of insects. Fruit flies performing (a) a saccadic turn (adapted from Dickinson and Muijres [7] and originally from Muijres et al. [221]) and (b) an evasive maneuver elicited by looming visual stimulus (adapted from [7] and originally from Muijres et al. [167]). (c) A fruit fly responding to extreme perturbation generated by a pulse train of magnetic torques, which has rotated the fly at approximately 60,000°s^{-1} (adapted from Beatus et al. [128]). (d) A hawkmoth performing an escape maneuver after receiving a frontal startling stimuli (adapted from Cheng and Hedrick [19]).

11.2 Basic Concepts

Like any human-engineered aircraft, flying insects need all the essential apparatus to achieve powered flight (Figure 11.2), these include wings for creating aerodynamic forces, sensors for measuring the flight states, motors for actuating the wings, and neural circuits for processing sensory information and flight control. During flight, these apparatuses work together to collect information from the environment and then act upon the surrounding fluid through the motion of wings. The interaction between the wings and fluid gives rise to highly complex aerodynamics, which determines the forces acting on the insect and subsequently its movements. Note, that the working principles of each and every one of these flight apparatuses are fundamentally different from those in engineered aircraft, while they are not only miniaturized for the body size of insects but also operate in faster time scales from flight control perspectives, despite the corresponding neural firing rates being significantly lower than the clock-speed of modern processors.

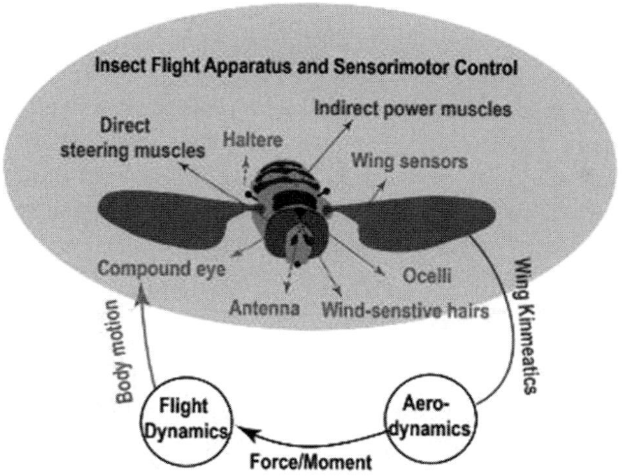

Figure 11.2 A schematic diagram illustrating the major components in a general dipteran insect's flight system, and its interactions, in a simplistic closed-loop fashion.

Unlike the fixed or rotary wings used in human-engineered aircraft, insects use flapping wings that exploit highly unsteady aerodynamics of vortex-dominated flow. They also fly in a flow region where neither viscous nor inertial forces acting on the fluid particles dominates (the Reynolds number (*Re*) range in insect flight is about $10–10^4$). Their wings' structural properties, resulting from the intricate architecture of the wing design (venation, flexion lines, relief, etc.) and the corresponding materials, also play an indispensable role when the surface of the wing interacts with surrounding fluids. In particular, wing deformations, such as camber and twist [13], resulting from the fluid–structure interaction, or, although less likely, from direct thoracic muscle control [14], are found to be favorable to the aerodynamic efficiency [15,16]. However, other factors, such as the need to reduce wing mass, due to constraints in the muscle mechanical power or in the elastic storage capacity, may also contribute to the distribution of wing flexural stiffness [17].

Flapping wings create periodic aerodynamic forces with fast varying magnitude and direction [18–20]. For most flying insects, the frequency of wing flapping is sufficiently fast so that only cycle-averaged, not instantaneous, forces affect the body movements and corresponding flight stability and maneuverability. The flight dynamics of insects, as shown by extensive computational [21–23], analytical [24–27], and experimental [28] studies, are inherently unstable, or weakly stable, without feedback control [29]. However, the instability is found to be insignificant for most insects (i.e., the time scale of the unstable modes is slow) as the unstable modes of flight are relatively slow compared with the wingbeat frequency and the processing speed of an insect's sensorimotor system.

To perform voluntary maneuvers or to maintain equilibrium states using compensatory reflexes, insects collect a large amount of information through a suite of sensory modalities. These include vision sensors, such as compound eyes [30] and ocelli [31], and mechano-sensors such as antennae [32], haltere (in dipteran insects) [33–36], and wing sensors [37]. These sensory modalities differ in the physical quantities they measure and their bandwidth and associated gain for flight control [5,38–40]. Using a suite of behavioral modules driven by various sensory cues, insects are able to solve flight-control problems efficiently at fast rates, and their flight-control systems are sometimes modeled by a set of coupled feedback/feedforward loops [7,41–43]. Notably, specific patterns of optical flow derived from the moving images projected on the retina in the insects' compound eyes are undoubtedly the most critical factors shaping the basic behavioral modules of insects during flight [8]. More complex flight behaviors, such as escape, landing, and avoidance maneuvers, might emerge by coordinating the behavioral modules in the central brain that process complex temporal and spatial information.

The neural control circuits generate signals that excite a small set of highly constrained flight-control (or steering) muscles, the precise actions of which are used to "perturb" the wing motion for flight control [44–46]. For example, it has been identified that there are a dozen of these flight-control muscles in fruit flies (*Drosophila*) [47]. These muscles are also called synchronous flight muscles because their activation is synced with the firings of the motoneurons. On the other hand, dipteran insects and bees are equipped asynchronous power muscles, which are used to generate the baseline wing flapping motion using slower neural signals, however, relatively primitive insects, such as *Lepidoptera* and *Odonata*, only use synchronous muscles [48].

In the following, basic concepts relevant to the key design aspects of insect flight systems are introduced in more detail. Due to the vast of amount of literature available, the review is certainly not exhaustive, as it only aims to present to readers, from the perspective of engineering, the key design aspects that enable the unparalleled fight capability of insects.

11.2.1 Basics of Insect-Wing Design and Aerodynamics

Without question, the design of insect wings is primarily shaped by their aerodynamic functions. Early studies of insect wings focused mainly on using their features, such as vein pattern, flexion lines, and three-dimensional relief as a collection of traits for classifying and identifying insects [4]. In recent decades, the aerodynamic function of insect wings has received increasing amounts of attention [2,3], although studies that directly relate the insect-wing design features and the corresponding structural properties to aerodynamic functions are still scarce [15,49].

Insect wings have diverse shapes, and the relationship between wing shape and flight performance has been discussed by Wootton, to a certain extent based on empirical observations [4]; however, quantitative analysis is still lacking. The most critical shape parameter is probably the aspect ratio (AR), which is relatively small compared with most fixed or rotary wings of engineered aircraft and has direct aerodynamic significance. As pointed by Lentink [50], AR is equivalent to Rossby number (Ro), which is a dimensionless fluid number that measures the relative magnitude of convective force and Coriolis force in the fluid flow. The majority of animal wings have AR within 3–4

[51], and this property is believed to be essential for the existence of stably attached leading-edge vortex (LEV), or delayed stall phenomenon, which results in a low-pressure region on the top of an insect wing that augments aerodynamic lift.

To understand the design of insect wings and its aerodynamic consequences, one should understand both the kinematics of the gross wing motion and its concomitant deformation; the former is a result of muscular control at the wing base coupled to certain extent with passive wing structural properties, while the latter mostly results from the wing structural properties, unlike vertebrate fliers such as birds [52] and bats [53]. Leaving the wing deformation aside, the aerodynamics of insect flight is often studied assuming a rigid wing undergoing three-degrees-of-freedom (DoF) motion, characterized by back-and-forth wing strokes accompanied by fast wing rotation during the end of each half stroke [54] and sometimes out-of-plane wing motion (deviation). Therefore, the wing kinematics is generally described by three Euler angles: wing-stroke position; rotation; and deviation angles [19]. Note, that in many cases, the gross wing motion cannot be strictly separated from the wing deformation, for example, the change of the wing rotation angle, and therefore angle of attack (AoA) is tightly related to the passive torsion along the wing span [13].

Because of the multi-DoF and periodic wing motion, flows created by the flapping wings have strong unsteady and three-dimensional (3D) effects, which have been the subject of a large number of studies, again often assuming rigid wings [55–61]. These studies have revealed a variety of aerodynamic mechanisms possible to insect flight, such as delayed stall [61], rotational lift, added mass [62], wake capture [63], and clap-and-fling [64]. For comprehensive reviews of the aerodynamics of insect flight, one can refer to Sane [3], Shyy et al. [65], and Chin and Lentink [2]. It is worth noting that these unconventional aerodynamic mechanisms have made flapping flight an attractive alternative to fixed or rotary flight for designing micro air vehicles flying at low Reynolds number (*Re*) [66]. In the literature, although debatable, flapping flight is often considered superior to fixed or rotary flight, as a more stable, efficient, and agile form of flight at low *Re* regime.

In addition to the aerodynamic mechanisms revealed by studying rigid wings, it is widely accepted that wing deformation under the action of fluid and/or wing inertial forces [67,68] is an indispensable factor driving the design of insect wings for higher aerodynamic performance, as wing design features such as vein patterns, flexion lines (Figure 11.3), and relief (e.g., corrugation) may all have their own aerodynamic purposes [14,69]. Note that, unlike vertebrate fliers, it is commonly believed that insect wings are passive structures that transmit muscular forces from base to the tip under the influence of aerodynamic and/or inertial forces. These forces together determine both the gross wing motion and the wing-surface deformation. Wing deformation is commonly observed and quantified as chordwise camber, spanwise twist, and sometimes spanwise bending [14]. Figure 11.4 shows an example of camber and twist observed in hoverflies. Recent studies have shown that both camber and twist improve the power economy and the force vectoring in locusts during tethered forward flight [15], while twist, but not camber, improves aerodynamic efficiency [16] and flight stability [70] in butterflies. Despite these efforts, studies that directly quantify the effects of wing deformation on

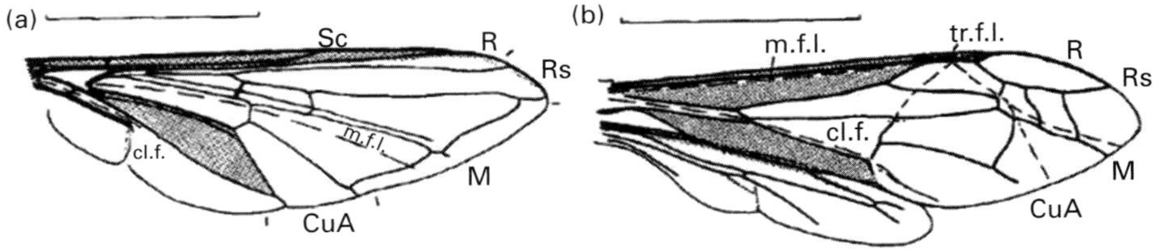

Figure 11.3 Examples of insect wings showing the distribution of veins, relatively rigid supporting areas (shaded), deformable areas (unshaded), and flexion lines (shaded). (a) Hoverfly (*Syrphus ribesii*); (b) European wasp (*Vespulu germunic*). m.f.l., median flexion line; cl.f., claval furrow; tr. f.l., transverse flexion line. Scale lines = 5 mm. (Adapted from Wootton [222].)

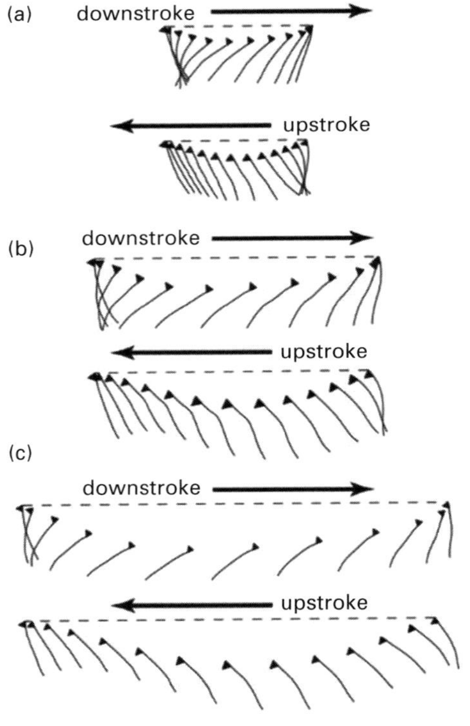

Figure 11.4 Illustration of wing camber and twist of a hoverfly (*Eristalis tenax*). Diagram showing the wing motion and the instantaneous wing profile at 25% (a), 50% (b), and 75% (c) of the wing length. (Adapted from Walker et al. [223].)

flight performance in free-flying insects are still scarce, as it is difficult to measure and reproduce the 3D surface deformation in either simulations or experiments.

Another line of research focuses on understanding how wing deformation arises from the action of fluid and inertial forces due to wing flexibility, i.e., a type fluid–structure interaction (FSI) problem [71–75]. As these problems are widely studied in aerospace engineering communities for pitching and plunging wings at relatively higher *Re* compared with those of insect wings [68], investigations on insect wings are again relatively scarce. This is because the difficulties in reproducing the detailed wing structural design using either experimental or numerical models, as well as in solving the Navier-Stokes equations coupled with the wing structural dynamics involving large 3D body deformations. For example, recent studies on flexible insect wings with FSI typically assume isotropic and homogenous wings [71,76], which can produce spanwise twist but not chordwise camber under FSI. Note, that the wing structural properties and the deformation pattern depend critically on the distribution of resilin patches on veins and resilin stripes on connections of veins and membranes [77,78], where resilin is a rubber-like structure that permits flexibility and provides elastic energy storage for the wing. Again, due to the complexities in wing venation, relief, and deflection lines, it is daunting to model insects to full complexity using experimental or numerical models for FSI studies. As pointed out by Combes and Daniel [79], the FSI problem can potentially be simplified if the wing deformation is mainly dominated by the inertial force, so that the structural dynamics can be decoupled from and solved prior to solving the Navier-Stokes equation. Whether aerodynamic or inertial force dominates the wing deformation depends on the density ratio of wing and the air [80], which varies among insects. For example, it was found the wing deformation in the hawkmoth is dominated by inertial forces [79], while that of dragonflies is dominated by aerodynamic force [81].

Without directly studying the FSI problems, and with simplifying assumptions on the aerodynamic loads, early studies have used conceptual analogy and physical, analytical, and numerical wing models to explain a variety of mechanical behaviors of the wing deformation, which is well summarized in Wootton et al. [14]. Notably, wing camber and twist can be generated by inertial forces during stroke reversals in the form of torsion waves, as described in [82]; however, it was also sometimes observed that they can be maintained in the rest of the stroke before the next stroke reversal, possibly due to aerodynamic loading. Ennos have used a mechanical wing model with posterodistally arranged veins to explain such automatic generation of wing camber and twist due to aerodynamic loading [13].

11.2.2 Basics of Insect-Wing Hinge Design and Motor System

The 3-DoF wing movements of insect wings are produced by a highly articulated motor system embedded in the meso- and meta-thoracic segments of the insect body [10]. In dipteran insects, power muscles (Figure 11.5a) create baseline wing movements by transmitting their forces through mesothorax oscillations to a sophisticated wing-hinge system (Figure 11.5c) [44,83]. As mentioned previously, the insect wing is mostly a passive structure, so that its deformation cannot be directly regulated for flight control, which is however common in vertebrate fliers. Instead, the flight control is mainly realized through the alteration of 3-DoF wing movements using a set of steering muscles attached to the wing-hinge articulation (Figure 11.5b). In this regard, there are two physiologically and functionally different classes of flight muscles: indirect

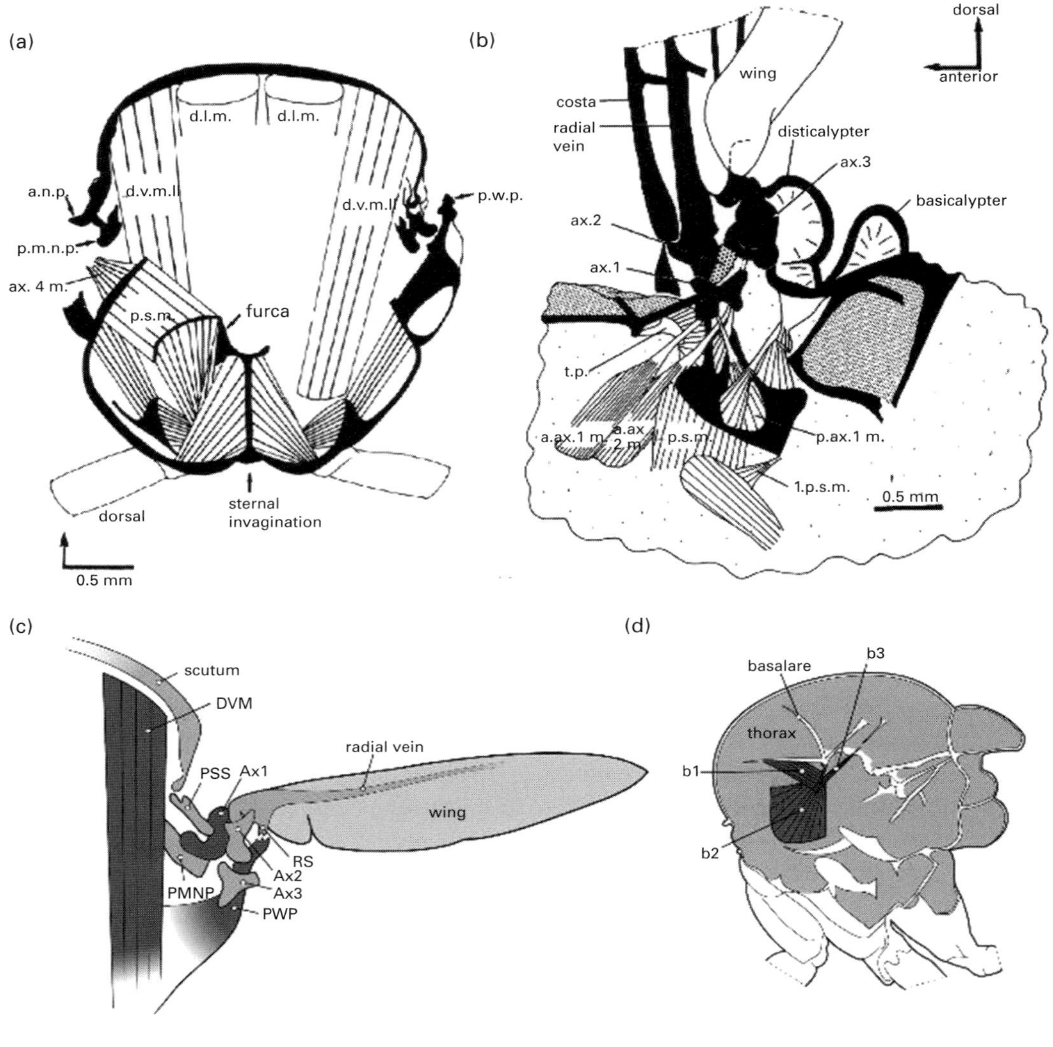

Figure 11.5 Examples of insect thorax and wing-hinge designs showing both sclerites and musculature. (a) A transverse view through the thorax of savannah' flies (*Glossina m. morsitans*), a species of dipteran insect. Indirect power muscles (e.g. dorsal–ventral muscles (d.v.m.) and dorsal longitudinal muscles (d.l.m.)), a steering muscle attached to the fourth axillary sclerite (ax.4), and a tension muscle, the pleurosternal muscle (p.s.m) are shown. (b) The axillary sclerites (ax.1, ax.2, and ax.3) and some of the attached steering muscles (a.ax.1 m., p.ax.1 m. and a.ax.3 m.) of *G. m. morsitans*. (c) A rear transverse view of the thorax of a *G. m. morsitans*, showing the axillary sclerites. Note, the first axillary is connected with post-medial notal process (PMNP) and parascutal shelf (PSS), and the radial stop (RS) involved in the wing-hinge gearing mechanism. (d) Basalar steering muscles b1–b3 inserted to the nail-shaped basalare sclerite. (Parts (a–b) are adapted from Miyan and Ewing [84]; Part (c) is adapted from Hedenström [224] and originally from [84]; Part (d) is adapted from [224] and originally from Dickinson and Tu [44].)

power muscles and direct steering muscles. The indirect muscles, mainly the antagonistically arranged dorsal–ventral muscles (d.v.m.) and dorsal longitudinal muscles (d.l.m.) [84], are stretch-activated and asynchronous in dipteran insects, meaning their contractions are not triggered by motor-neuron spikes in one-to-one fashion [44]. Instead, their oscillation is activated and maintained mechanically, thanks to their antagonistic arrangements and the elasticity of the thorax. On the other hand, the direct steering muscles are synchronous since they respond to motor spikes in one-to-one fashion [47]. These relatively tiny muscles insert directly on a group of sclerites within the wing hinge (Figure 11.5b). Sclerites are hardened skeletal elements constituting the wing articulation and their movements are directly translated to the 3-DoF movements of the wing [84]. In flies, there are four axillary sclerites (ax. 1, ax. 2, ax. 3, and ax. 4), three of them (i.e. not ax. 2) have steering muscle inserted (i1 and i2 on ax. 1; iii1, iii3, and iii4 on ax. 3, hg1, hg2, hg3, and hg4 on ax. 4). Ax. 1, ax. 2, ax. 3 are shown in Figure 11.5b and c, and part of the attached steering muscles are shown in Figure 11.5b. In addition, there are two basalar sclerites that are directly involved in wing articulation (Figure 11.5d), which are inserted by three steering muscles (b1, b2, and b3). The total of twelve steering muscles identified in fruit flies are summarized in Figure 11.6a.

Subtle changes in wing movement for flight maneuver can be produced by regulating the contractions of steering muscles through changing the firing rate and timings of the corresponding motor neurons [45,46,85]. With minimal mechanical power, steering muscles tune the movements of the sclerites and in turn control the wing movements (Figure 11.6 shows a hypothetical model for the influence of the twelve steering muscles of flies on their wing movements) and also determines how energy from power muscles is transferred into wing motion [44]. In addition, the wing-hinge system also bears gearing properties [86,87] that yield different modes of operation [46,88] (Figure 11.5c shows the RS involved in the wing-hinge gearing mechanism).

In addition, there are tension muscles that put the thorax under tension [89], such as pleurosternal muscles (Figure 11.5a). These muscles have elastic properties that conserve the kinetic energy of the wing, and therefore are considered to increase the overall efficiency of the system [48,90–92]. The flapping frequency of the wing is determined by the mechanical properties of the thorax, indirect muscles (including the tension muscles) and the wings, instead of by the prescribed neuromuscular control signals [44,93].

The sophistication of the insect-wing hinge design is unquestionably one of the marvelous adaptations that enable insect flight at small scale, and it is probably the most complicated joint design in animal kingdom [94]. Such complexity is needed for providing graded and continuous alteration of 3-DoF wing motion to a sufficient magnitude and precision for flight control. However, whether such complexity needs to be emulated in an engineered flapping-wing system is questionable, for the reasons mentioned in the very beginning of this review. For example, one has to consider the fact that an insect muscle, unlike those of vertebrate fliers such as hummingbirds, is innervated only by a single motor neuron [47] and thus insects cannot rely on the graded recruitment of motor units to regulate muscle tension. With this constraint, steering muscles can be divided into two groups, "tonic" and "phasic," according to the activities

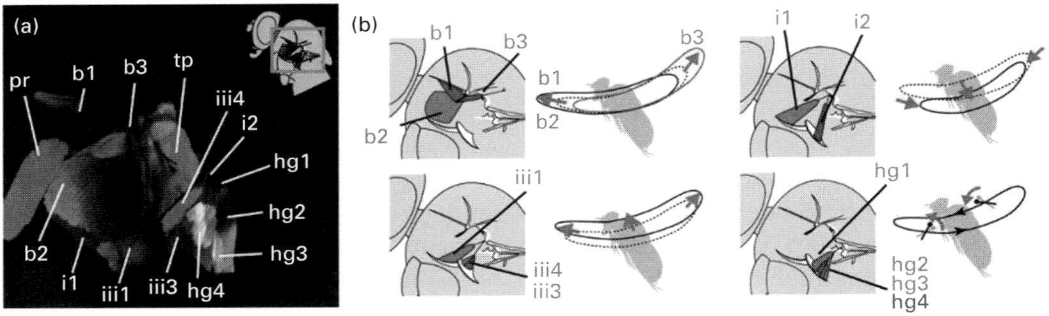

Figure 11.6 Steering muscles and their hypothesized influence on wing motion. (a) Anatomical arrangement of the twelve steering muscles of fruit flies (*Drosophila*). b1, b2, and b3; i1 and i2; iii1, iii3, and iii4; and hg1, hg2, hg3, and hg4. (b) Hypothesized influence of the twelve steering muscles on wing motion. (Adapted from Lindsay et al. [47].)

in their motor neurons during flight, and therefore the way they regulate tension [44,47]. Tonic steering muscles (e.g., b1, b3, i2, iii3, and hg4 in flies) have sustained activities during flight and can provide graded continuous regulation through variations in motorneuron firing timings and frequencies, and therefore they are able to generate feedback-based graded compensatory flight control. On the other hand, phasic muscles (e.g., b2, i1, iii1, hg1, and hg3 in flies) show only scattered activities and are capable of producing large transient changes in wing motion, therefore are believed to realize spontaneous feedforward steering control. For example, when a fly performs a saccadic maneuver, phasic muscles are likely used to initiate the maneuver by generating a large turning moment, while the tonic muscles are used to regulate the magnitude of the maneuver based on the feedback from haltere [47], the gyroscopic sensor of dipteran insects. This is supported by the discovery that the firings of the tonic b1 muscle are phase-locked with the wingbeat cycle, while its firing phase can be altered by the haltere afferents [35].

Using high-speed videography techniques and electromyograms (EMGs), researchers were also able to correlate the muscle activities to the output wing kinematics by recording them simultaneously. For example, it was found that the b1 and b2 muscles of blowflies control the flapping amplitude and produce turning moments [45]. These experiments were first performed in tethered conditions [e.g., 45,46], but recently achieved in free flight [e.g., 85,95]. Nonetheless, investigations on how neuromuscular control affects the resultant aerodynamic force are still quite rare, with the exception of Balint and Dickinson [96], showing that modulation of wing motion through the control of a distinct group of steering muscles results in a distinct manipulation of the aerodynamic force vector.

11.2.3 Basics of Insect Sensory System and Sensorimotor Control

Flying insects are equipped with a variety of sensory apparatus, collectively providing a rich information set for flight control. Sensory measurements are multimodal or even redundant, as different sensory apparatuses may measure the same quantities with different bandwidth, the outputs of which can be fused by descending interneurons via convergent sensorimotor pathways toward specific flight-control and stabilization tasks. Based on the physical quantities measured, there are mainly two types of sensors used for flight control, visual and mechanosensors. Visual sensors include compound eye and ocelli (simple eye), which provide optical information produced by the surrounding environment and the movements of the insects themselves. For reviews of the basic physiological elements and organization of vision sensors, readers can refer to [5,97]. Mechanosensors measure flight states and environment conditions through various mechanical strains induced. For examples, body angular rates are measured by the strains induced by Coriolis force in haltere, air speeds and body linear acceleration can be measured by the strains induced by the wind in wind-sensitive hairs and antenna, and wing loadings can be measured by the strains induced by wing deformations. For the classifications and physiological structures of mechanosensors, readers can refer to [5,98].

Insects rely heavily on motion vision for flight control [99–104], especially for generating feedforward signals that initiate flight maneuvers. For example, visual expansion was found to trigger either rapid turns [105] or landing [106,107]. A remarkable aspect of insect flight is that insects use relatively simple, yet robust neural mechanisms to extract motion information from noisy and complex visual environments. For flies such as *Calliphora*, approximately two-thirds of the neurons in the brain contribute to visual information processing [5,108]. Therefore, vision-elicited behaviors and the underlying neural mechanisms are at the center stage of insect neuroscience research [5,109–113]. In particular, three types of visual responses of flies and their neural control circuits have been studied on both functional and cellular levels [8,109]: optomotor responses, escape-and-landing responses, and fixation responses. Based on optomotor responses, the mechanism for the fundamental elementary motion detection (EMD, e.g., using "Hassenstein–Reichardt" model) [114] (Figure 11.7c), the subsequent spatial processes performed by the tangential cells (Figure 11.7a and b) in the lobula plate through selectively pooling the outputs of the motion detectors and the subsequent temporal integration processing have been widely studied [5,8,115].

Visually guided turns (both rapid and smooth turns) were extensively studied in both tethered [116], semi-tethered [117], and free-flying [118] insects. An important observation in visual-guided turning responses in fruit flies is that after the turning is initiated and when there is a turning moment created, the fly does not respond to visual stimulation to the same direction of the turning, but vigorously responds to the displacement in the opposite direction [119,120]. This phenomenon was explained using an efference copy (reafferent principle [120,121]), which estimates the expected sensory feedbacks and eliminates the self-orientated disturbances. Several classic control circuits have been proposed to model this principle [121]. Recently, it was shown that the control strategy, including the reafference principle, can be represented by a PID control circuit [122]. In addition, it has been shown recently that dragonflies, when pursing prey, use a certain level of planning and prediction for head saccades, which can be explained by the existence of internal models and efference copies [121,123].

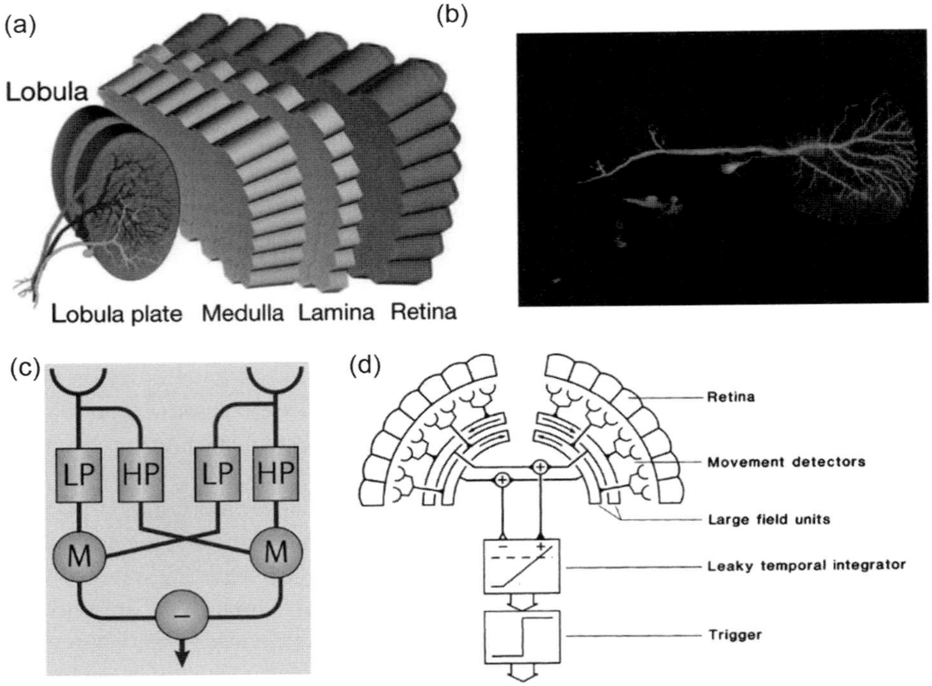

Figure 11.7 Insect visual processing system. (a) Schematics of the fly optic lobe, showing three motion-sensitive tangential cells in the Lobula plate (adapted from Maisak et al. [225]). (b) A picture showing one of the large tangential cells filled with a fluorescent dye (adapted from Borst [125]). (c) Schematics of an EMD consisting of a low-pass filter (LP), a high-pass filter (HP) and a multiplier (M) (adapted from Borst [8]). (d) Spatio-temporal integration of motion (STIM) model based on EMD, which is proposed to model the insect leg-extension landing response (adapted from [125]).

Compared with visually guided turns, neither the behaviors nor the underlying neural mechanisms of landing responses received the same amount of attention. This is partly due to the fact that landing responses are more complex in terms of the related behavioral modules, which are separated in space and time. For the same reason, they also pose more difficulties in experimental investigations. Nonetheless, progress has been made in the past, for example, functional models such as time-to-contact model (or relative retinal expansion velocity model, RREV) [124] and STIM model (Figure 11.7d) [125,126] have been proposed and tested, as the former predicts the triggering time and the latter predicts the latency of the leg-extension responses. In addition, the behavioral modules involved in landing on a vertical post have been studied, together with their neural mechanisms [107]. Finally, note that although landing behaviors are triggered by vision, the final landing maneuver (moment before touchdown) could be too fast to be guided by vision, therefore it is likely that the insects use mechanosensory inputs or simply perform the landing open-loop.

In addition to vision, mechanosensors are critical for producing fast compensatory responses for flight stabilization. A unique adaptation of such is probably the haltere in dipteran insects. These tiny sensory organs have small dumb-bell structure and are evolved from the hind wings; they oscillate exactly at the wingbeat frequency with precise phase differences from the wing motion [127]. The rotations of the insect body cause the halteres to deviate from its nominal plane of oscillation and the corresponding Coriolis force is measured at its base by a group of mechanoreceptors [33]. The ability of haltere to measure body rotations was first recognized in 1948 by Pringle [34], and subsequent studies have further quantified its function [88]. It is worth noting that the synchrony between the haltere and the wing motion is likely facilitated by the mechanical oscillation of the thorax instead of by rhythmic neural stimulation [87] (which is slower than wingbeat frequency). Therefore, the haltere is able to provide fast and phased measurement of body rates (less than 5 ms delay [39]) which are transformed directly into motor commands of the wing steering muscle without going through any interneuron. In this way, the halteres and steering muscles form a fast reflexive control loop that ensures wingbeat-by-wingbeat flight control. The fast reflexes from this unique sensorimotor pathway provide active damping that improves local flight stability [42,128].

For insects without halteres, antennae were proposed to offer similar functions [32,129,130], as an active structure actuated by its own group of muscles; however, its ability to measure Coriolis force was later questioned [5]. Wing-load sensing is achieved by mechanoreceptors located on the wing surfaces; these sensors are sensitive to wing deformations at wingbeat frequency [131–134]. Recently, it has been proposed that mechanosensors distributed on the wing surface of hawkmoth can potentially sense Coriolis forces in similar fashion to haltere [135].

With a large amount of information collected from different sensory modalities, how they are fused and used for flight control is also a paramount question investigated over the past decades [38,40,136,137]. Sherman and Dickinson [138] showed that the sensory signals from haltere and visual systems are combined as a weighted sum, with haltere dominant. They proposed that the dominance of the haltere feedback may be explained by visual feedback inaccuracy due to the complex spatial composition of the visual world. Another study shows that the motoneurons at muscles driving the haltere receive strong input from visual signals, suggesting the visual control may be mediated in part by equilibrium reflexes from haltere [139]. Dickinson and co-workers also indicated that vision is more sensitive to slow motions while the haltere is more sensitive to fast motion [38,140]. It is also found that the rapid turning maneuvers in fruit fly is initiated by visual cues but requires subsequent haltere-mediated stabilization control [141], the muscular control of which is briefly discussed in Section 11.2.2. In addition, Krapp and co-workers [142] proposed an interesting hypothesis suggesting that the measurements from various sensory organs are fused and then represented in the non-orthogonal coordinates representing the natural modes of the dynamic system.

Another critical aspect of sensorimotor control in insect flight is that neural circuits have limited computation and transmission speed, which gives rise to time delays during the sensorimotor transduction. Such delays have been previously identified for fruit flies [39,42,128], hawkmoths [143,144], and hummingbirds [145] through model fitting to free-flight data or by measuring their response time. Recent studies have shown that neural time delays could be one of the limiting factors for both the stability and the maneuverability of flying insects [43,146]. For example, it is shown that the rotational maneuverability of fruit flies is limited by neural delays but not by muscle mechanical power [43].

11.3 Prior Work

With the basic concepts established regarding the design and principles underlying various biological processes that give rise to such high control authority, here prior work that aimed at emulating insect flight as a novel form of micro air vehicles is summarized. Primary efforts to date have focused on the development of actuation mechanisms at scales of insects or hummingbirds, mainly for achieving high frequency flapping with certain control authority, while there are also efforts made toward emulating the sensing apparatus and optical-flow-based control algorithms used by insects.

11.3.1 Emulating Insect Flight: Flapping-Wing Micro Air Vehicles

Developing an actuation mechanism that not only achieves high wingbeat frequency for lift generation but also with high authority for wing trajectory control is a challenging task, especially if it is to match the complexity and performance of an insect's motor system. This difficulty arises fundamentally because of the scales required for mimicking insect or hummingbird flight, which is between microelectromechanical systems (MEMS) and "macro" devices [147], and therefore often demands novel fabrication method such as smart composite microstructures(SCMs) [148]. To date, electromagnetic [149–155] and piezoelectric motors [156–159] combined with custom-designed transmission mechanisms have proven successful for generating reciprocating wing motion with stroke amplitude and frequency required for sustained hovering. Most of these actuators have also achieved a certain degree of wing trajectory control authority [151,160–166], however still relatively limited compared with those that achieved by insects [47,167] or hummingbirds [145,168]. Overall, although with significant progresses made in the past decade, these actuators still significantly underperform the insect motor system in terms of efficiency, endurance and control authority.

Piezoelectric actuators have been used extensively in the past for creating flapping flight at insect scale [148,158,159,169,170]. When coupled with transmission mechanisms made of flexure joints, they offer high displacements, fast response, and moderate efficiency levels at high wingbeat frequencies. Recently, an insect-scale robot utilizing dual piezoelectric actuators has demonstrated stable hover, trajectory tracking [162], and landing [171] capabilities, with improved control authority derived from the bilaterally asymmetric control of wing-stroke trajectories. Attempts have been made in the past to improve both the efficiency and the control authority of the actuator. For example, using a compliant thoracic mechanism mimicking those of insects at a functional level [172], Lau and his co-workers have achieved 20–30% saving of power expenditure for a flapping-wing MAV. A differential AoA mechanism driven by a single piezoelectric actuator was also used to provide yaw-and-roll attitude control authority [173]. Other attempts at improving the control authority include the out-of-stroke-plane wing deviation generation [174] and bilateral independent wing amplitude modulation using two control actuators [175]. Additional efforts have also been made to reproduce the complexity of insect-wing structure [176,177], as the effects of wing flexural and torsional flexibility have also been produced and tested [178]. Despite progress made in the past, the high voltage requirement of piezoelectric actuators still limits autonomy due to the availability of adequately sized power circuitry [179], although recent work on optimized drive circuit topologies has shown significant progress [180].

Recently, custom-designed electromagnetic actuators operating at resonant states have proven successful as alternative actuation mechanisms [150,181]. For example, direct-drive electromagnetic actuators coupled with magnetic springs were developed and optimized [182] (Figure 11.8a). They operate as forced nonlinear oscillators and are able to achieve lift-to-weight ratio over one according to theoretical modeling [183] and later validated in experiments [184].

Commercial DC motors paired with gear/linkage transmissions [151,185–187] (e.g., Figure 11.8b) or string-and-pulley systems (Jupiter 4 and Nano Hummingbird, Aero-Vironment [152]) have also demonstrated sustained stable hover, however at the scale of hummingbirds which is larger than most insects. These types of transmission generally have fixed wing-stroke kinematics and the potential for high mechanical friction and wear, resulting in an increased likelihood of failure. With the introduction of torsional stiffness, direct-drive DC motors [153,163,188] have also become a feasible wing-actuation method (e.g., Figure 11.8c), recently achieving lift-off and a certain degree of flight-control authority. With these forms of actuation, operating the system near resonance can be used to conserve energy, thereby recovering work invested in overcoming inertial mass, functioning in a similar manner to the elastic elements in the flight systems of insects [84]. Other transmission designs have also been attempted in the past based on DC motors [189–191].

11.3.2 Biomimetic Sensors and Flight Control

Another key aspect leading to the success of insect flight is their highly specialized sensors and sensory architecture for flight control, which have also inspired novel designs for sensing devices [192–195] and convergent flight sensing and control strategies [196–200]. In fact, the MEMS gyroscopic sensors used widely nowadays [201] share the same working principle as the halteres of dipteran insects in terms of using Coriolis force to measure body angular rates [33].

A suite of biomimetic sensors and the corresponding algorithms were developed and tested in the Micromechanical Flying Insect (MFI) project more than a decade ago [193]. Recently, an angular-rate sensor that is morphologically similar to halteres was

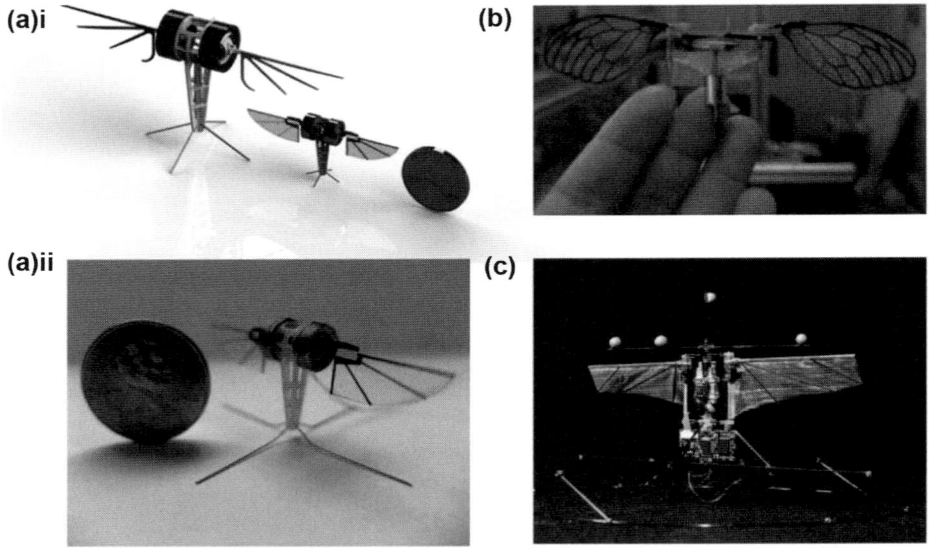

Figure 11.8 Examples of flapping-wing MAVs designed at Purdue University. (a) Electromagnetic flapping-wing robots (courtesy of Jesse Roll and Xinyan Deng). (b) A motor-driven flapping-wing robot [226]. (c) Hummingbird-mimicking flapping-wing robot with direct drive (courtesy of Fei Fan, Zhan Tu, Jian Zhang, and Xinyan Deng)

also developed [159,194]. Artificial compound eyes of insects for wide-fielding sensing have been investigated extensively in past decades [202] and some designs have successfully achieved similar characteristics to natural compound eyes [192,203]. In addition, optical-flow processing algorithms inspired by insect flight have been investigated extensively for navigation [204–207] and egomotion sensing [208,209], and many of these bioinspired algorithms were originally generated from insect behavioral experiments [210–212]. Notably, an efficient flight-control algorithm based on wide-field integration of optic flow was applied successfully to autonomous navigation [213], and the algorithm for gazing landing of bees has been applied successfully to a robotic system [214].

In addition, luminance or optical-flow sensors using analog or digital Very-Large-Scale Integration (VLSI) technology mimicking the neural computation principle of insects have been investigated extensively in the past for their potential application in MAVs [215,216]. Compared with existing MAV guidance systems and other distance sensors such as laser rangefinders and ultrasonic sensors, these sensors have the potential to provide high bandwidth and wide field of view while satisfying low-weight requirement [198,216]. The well-known visual processing circuits of the insect compound eye, i.e., Reichardt-type EMDs were implemented on a Field Programmable Gate Array (FPGA) platform with high sensitivity to image motion and low computational cost [217]. Subsequent study by the same group also successfully implemented the lobula plate tangential cell network of the fly on the FPGA for rotation sensing of

MAVs [195]. In addition, a fully analog, light-sensing ocellar sensor was developed for estimating angular rate during flight [218], and more recently, a micro ocellar sensor and an artificial airspeed-sensing antenna were successfully developed and applied together to control a flapping-wing MAV [219,220].

11.4 Future Directions

Understanding the design sophistications and the principles of insect flight will continue to be a challenging task, as it requires multidisciplinary research coalescing tools and knowledge from different fields in engineering, biology, and physics. In the future, it is expected that new experimental approaches, alongside existing tools, in high-speed videography, electrophysiology, neuroscience, and fabrication will be needed for studying insect flight. For example, the principles behind the wing structural design can only be understood with a more detailed engineering modeling of insect-wing structure combined with experimental or numerical tools to investigate its aerodynamic functions. Understanding the muscular design and function of insect-wing hinges also requires simultaneous investigations of aerodynamics and flight dynamics. In addition, since neural computations are ultimately integrated with the biomechanics of flight as well as with the flight-control problems that flies need to solve, traditionally separated bodies of work on insect flight mechanics, behaviors, physiology, and neuroscience will need to be better coalesced for a holistic understanding of insect flight.

11.5 Concluding Remarks

From efficient neural processing and intricate biomechanical flight apparatus to the complex unsteady fluid flow and fast aerodynamic maneuvers, insect flight encompasses many complex processes in physics, biology, and engineering, packaged into a miniature flying machine designed by nature. Therefore, as ordinary as they might appear, flying insects embody extraordinary biological designs for flight, which will continue to inspire engineering design in the future; the complex physical and biological processes involved will also continue to spawn a treasure trove of scientific problems. It is also critical that bioinspiration from insect flight should be deeply grounded in the scientific understanding of insect flight, in particular the different factors driving biological and engineering design.

References

1. Wootton, R. (1981). Palaeozoic insects. *Annual Review of Entomology*, 26, 319–344.
2. Chin, D. D., & Lentink, D. (2016). Flapping wing aerodynamics: From insects to vertebrates. *Journal of Experimental Biology*, 219, 920–932.
3. Sane, S. P. (2001). The aerodynamics of flapping wings. *Doctor of Philosophy, Integrative Biology*. University of California.

4. Wootton, R. J. (1990). The mechanical design of insect wings. *Scientific American*, 263(5), 114–120.
5. Taylor, G. K., Krapp, H. G., & Simpson, S. J. (2007). Sensory systems and flight stability: What do insects measure and why? In J. Casas and S. J. Simpson (Eds.), *Advances in insect physiology*, vol. 34, pp. 231–316. Academic Press.
6. Daniel, T., Aldworth, Z., Hinterwirth, A., & Fox, J. (2012). *Insect inertial measurement units: Gyroscopic sensing of body rotation*. Springer.
7. Dickinson, M. H., & Muijres, F. T. (2016). The aerodynamics and control of free flight manoeuvres in drosophila. *Philosophical Transactions of the Royal Society B*, 371, 20150388.
8. Borst, A. (2014). Fly visual course control: Behaviour, algorithms and circuits. *Nature Reviews Neuroscience*, 15, 590–599.
9. Srinivasan M. V., & Zhang, S. (2004). Visual motor computations in insects. *Annual Review of Neuroscience*, 27, 679–696.
10. Dudley, R. (2000). *The biomechanics of insect flight*. Princeton University Press.
11. Sane, S. P. (2016). Bioinspiration and biomimicry: What can engineers learn from biologists? *Journal of Applied Science and Engineering*, 1–6.
12. Hennig, W. (1981). *Insect phylogeny*. John Wiley & Sons Ltd.
13. Ennos, A. R. (1988). The importance of torsion in the design of insect wings. *Journal of Experimental Biology*, 140, 137–160.
14. Wootton, R., Herbert, R., Young, P., & Evans, K. (2003). Approaches to the structural modelling of insect wings. *Philosophical Transactions of the Royal Society of London B: Biological Sciences*, 358, 1577–1587.
15. Young, J., Walker, S. M., Bomphrey, R. J., Taylor, G. K., & Thomas, A. L. (2009). Details of insect wing design and deformation enhance aerodynamic function and flight efficiency. *Science*, 325, 1549–1552.
16. Zheng, L., Hedrick, T. L., & Mittal, R. (2013). Time-varying wing-twist improves aerodynamic efficiency of forward flight in butterflies. *PLOS ONE*, 8, e53060.
17. Combes, S. A., & Daniel, T. L. (September 2003). Flexural stiffness in insect wings. II. Spatial distribution and dynamic wing bending. *Journal of Experimental Biology*, 206, 2989–2997.
18. Fry, S. N., Sayaman, R., & Dickinson, M. H. (April 2003). The aerodynamics of free-flight maneuvers in Drosophila. *Science*, 300, 495–498.
19. Cheng, B. X., & Hedrick, T. L. (December 15, 2011). The mechanics and control of pitching manoeuvres in a freely flying hawkmoth (Manduca sexta). *Journal of Experimental Biology*, 214, 4092–4106.
20. Sun, M., & Tang, H. (January 2002). Unsteady aerodynamic force generation by a model fruit fly wing in flapping motion. *Journal of Experimental Biology*, 205, 55–70.
21. Sun, M., & Xiong, Y. (February 2005). Dynamic flight stability of a hovering bumblebee. *Journal of Experimental Biology*, 208, 447–459.
22. Gao, N., Aono, H., & Liu, H. (2011). Perturbation analysis of 6DoF flight dynamics and passive dynamic stability of hoveringfruitfly Drosophila melanogaster. *Journal of Theoretical Biology*, 270, 98–111.
23. Wu, J. H., & Sun, M. (2012). Floquet stability analysis of the longitudinal dynamics of two hovering model insects. *Journal of the Royal Society Interface*, 9, 2033–2046.
24. Cheng, B., & Deng, X. (2011). Translational and rotational damping of flapping flight and its dynamics and stability at hovering. *IEEE Transactions on Robotics*, 27, 849–864.

25. Taha, H. E., Hajj, M. R., & Nayfeh, A. H. (2014). Longitudinal flight dynamics of hovering MAVs/insects. *Journal of Guidance, Control, and Dynamics*, 37, 970–979.
26. Taha, H. E., Nayfeh, A. H., & Hajj, M. R. (2014). Effect of the aerodynamic-induced parametric excitation on the longitudinal stability of hovering MAVs/insects. *Nonlinear Dynamics*, 78, 2399–2408.
27. Karásek, M., & Preumont, A. (2012). Flapping flight stability in hover: A comparison of various aerodynamic models. *International Journal of Micro Air Vehicles*, 4, 203–226.
28. Taylor, G. K., & Thomas, A. L. R. (August 2003). Dynamic flight stability in the desert locust *Schistocerca gregaria*. *Journal of Experimental Biology*, 206, 2803–2829.
29. Sun, M. (2014). Insect flight dynamics: Stability and control. *Reviews of Modern Physics*, 86, 615–646.
30. Krapp, H. G., Hengstenberg, B., & Hengstenberg, R. (1998). Dendritic structure and receptive-field organization of optic flow processing interneurons in the fly. *Journal of Neurophysiology*, 79, 1902–1917.
31. Taylor, Charles P. (1981). Contribution of compound eyes and ocelli to steering of locusts in flight: I. Behavioural analysis. *Journal of Experimental Biology*, 93, 1–18.
32. Mamiya, A., Straw, A. D., To´masson, E., & Dickinson, M. H. (2011). Active and passive antennal movements during visually guided steering in flying Drosophila. *The Journal of Neuroscience*, 31(18), 6900–6914.
33. Nalbach, G. (1993). The halteres of the blowfly Calliphora. *Journal of Comparative Physiology A*, 173, 293–300.
34. Pringle, J. W. S. (1948). The gyroscopic mechanism of the halteres of Diptera. *Philosophical Transactions of the Royal Society of London B: Biological Sciences*, 233, 347–384.
35. Fayyazuddin, A., & Dickinson, M. H. (1996). Haltere afferents provide direct, electrotonic input to a steering motor neuron in the blowfly, Calliphora. *The Journal of Neuroscience*, 16, 5225–5232.
36. Fox, J. L., Fairhall, A. L., & Daniel, T. L. (2010). Encoding properties of haltere neurons enable motion feature detection in a biological gyroscope. *Proceedings of the National Academy of Sciences*, 107, 3840–3845.
37. Dickerson, B. H., Aldworth, Z. N., & Daniel, T. L. (2014). Control of moth flight posture is mediated by wing mechanosensory feedback. *Journal of Experimental Biology*, 217, 2301–2308.
38. Bender, J. A., & Dickinson, M. H.(December 2006). A comparison of visual and haltere-mediated feedback in the control of body saccades in Drosophila melanogaster. *Journal of Experimental Biology*, 209, 4597–4606.
39. Fuller, S. B., Straw, A. D., Peek, M. Y., Murray, R. M., & Dickinson, M. H. (2014). Flying Drosophila stabilize their vision-based velocity controller by sensing wind with their antennae. *Proceedings of the National Academy of Sciences*, 111, E1182–E1191.
40. Roth, E., Hall, R. W. Daniel, T. L., & Sponberg, S. (2016). Integration of parallel mechanosensory and visual pathways resolved through sensory conflict. *Proceedings of the National Academy of Sciences*, 201522419.
41. Cowan, N. J., Ankarali, M. M., Dyhr, J. P., et al. (2014). Feedback control as a framework for understanding tradeoffs in biology. *Integrative and Comparative Biology*, icu050.
42. Ristroph, L., Bergou, A. J., Ristroph, G., Coumes, K. Berman, G. J., & Guckenheimer, J. (March 16, 2010). Discovering the flight autostabilizer of fruit flies by inducing aerial stumbles. *Proceedings of the National Academy of Sciences*, 107, 4820–4824.

43. Liu P., & Cheng, B. (2017). Limitations of rotational manoeuvrability in insects and hummingbirds: Evaluating the effects of neuro-biomechanical delays and muscle mechanical power. *Journal of the Royal Society Interface*, 14, 20170068.
44. Dickinson, M., & Tu, M. S. (1997). The function of Dipteran flight muscle. *Comparative Biochemistry and Physiology*, 116A, 223–238.
45. Tu, M. S., & Dickinson, M. H. (1995). The control of wing kinematics by two steering muscles of the blowfly (*Calliphora vicina*). *Journal of Comparative Physiology*, 178, 813–830.
46. Balint, C. N., & Dickinson, M. H. (2001). The correlation between wing kinematics and steering muscle activity in the blowfly *Calliphora vicina*. *Journal of Experimental Biology*, 204, 4213–4226.
47. Lindsay, T., Sustar, A., & Dickinson, M. (2017). The function and organization of the motor system controlling flight maneuvers in flies. *Current Biology*, 27, 345–358.
48. Ellington, C. (1985). Power and efficiency of insect flight muscle. *Journal of Experimental Biology*, 115, 293–304.
49. Mountcastle, A. M., & Combes, S. A. (2013). Wing flexibility enhances load-lifting capacity in bumblebees. In *Proceedings of the Royal Society B*, 20130531.
50. Lentink, D., & Dickinson, M. H. (August 15, 2009). Biofluiddynamic scaling of flapping, spinning and translating fins and wings. *Journal of Experimental Biology*, 212, 2691–2704.
51. Lentink, D., & Dickinson, M. H. (August 15, 2009). Rotational accelerations stabilize leading edge vortices on revolving fly wings. *Journal of Experimental Biology*, 212, 2705–2719.
52. Chin, D. D., Matloff, L. Y., Stowers, A. K., Tucci, E. R., & Lentink, D. (2017). Inspiration for wing design: How forelimb specialization enables active flight in modern vertebrates. *Journal of the Royal Society Interface*, 14, 20170240.
53. Swartz, S. M., Iriarte-Diaz, J., Riskin, D. K., et al. (2007). Wing structure and the aerodynamic basis of flight in bats. *AIAA Journal*, 1.
54. Ellington, C. P. (1984). The aerodynamics of hovering insect flight. 3. Kinematics. *Philosophical Transactions of the Royal Society of London B: Biological Sciences*, 305, 41–78.
55. Ozen, C. A., & Rockwell, D. (2012). Three-dimensional vortex structure on a rotating wing. *Journal of Fluid Mechanics*, 707, 541.
56. Garmann, D., & Visbal, M. (2014). Dynamics of revolving wings for various aspect ratios. *Journal of Fluid Mechanics*, 748, 932–956.
57. Cheng, B., Roll, J., Liu, Y., Troolin, D. R., & Deng, X. (2014). Three-dimensional vortex wake structure of flapping wings in hovering flight. *Journal of the Royal Society Interface*, 11, 20130984.
58. Wojcik, C. J., & Buchholz, J. H. (2014). Vorticity transport in the leading-edge vortex on a rotating blade. *Journal of Fluid Mechanics*, 743, 249–261,
59. Birch, J. M., Dickson, W. B., & Dickinson, M. H. (March 1, 2004). Force production and flow structure of the leading edge vortex on flapping wings at high and low Reynolds numbers. *Journal of Experimental Biology*, 207, 1063–1072.
60. Sane, S. P., & Dickinson, M. H. (October 2001). The control of flight force by a flapping wing: Lift and drag production. *Journal of Experimental Biology*, 204, 2607–2626.
61. Dickinson, M. H., Lehmann, F.-O., & Sane, S. P. (1999). Wing rotation and the aerodynamic basis of insect flight. *Science*, 284, 1881–2044.
62. Sane, S. P., & Dickinson, M. H. (April 2002). The aerodynamic effects of wing rotation and a revised quasi-steady model of flapping flight. *Journal of Experimental Biology*, 205, 1087–1096.

63. Birch. J. M., & Dickinson, M. H. (July 1, 2003). The influence of wing–wake interactions on the production of aerodynamic forces in flapping flight. *Journal of Experimental Biology*, 206, 2257–2272.
64. Weisfogh, T. (1973). Quick estimates of flight fitness in hovering animals, including novel mechanisms for lift production. *Journal of Experimental Biology*, 59, 169–230.
65. Shyy, W., Lian, Y., Tang, J., Viieru, D., & Liu, H. (2008). *Aerodynamics of low Reynolds number flyers*. Cambridge University Press.
66. Pines, D. J., & Bohorquez, F. (2006). Challenges facing future micro-air-vehicle development. *Journal of Aircraft*, 43, 290–305,
67. Shyy, W., Berg, M., & Ljungqvist, D. (1999). Flapping and flexible wings for biological and micro air vehicles. *Progress in Aerospace Sciences*, 35, 455–505,
68. Shyy, W., Aono, H., Chimakurthi, S. K., Trizila, P., Kang, C.-K., & Cesnik, C. E. (2010). Recent progress in flapping wing aerodynamics and aeroelasticity. *Progress in Aerospace Sciences*, 46, 284–327.
69. Wootton, R. J. (1992). Functional morphology of insect wings. *Annual Review of Entomology*, 37, 113–140,
70. Senda, K., Obara, T., Kitamura, M., Yokoyama, N., Hirai, N., & Iima, M. (2012). Effects of structural flexibility of wings in flapping flight of butterfly. *Bioinspiration & Biomimetics*, 7, 025002.
71. Tian, F.-B., Dai, H., Luo, H., Doyle, J. F., & Rousseau, B. (2014). Fluid–structure interaction involving large deformations: 3D simulations and applications to biological systems. *Journal of Computational Physics*, 258, 451–469.
72. Dai, H., Luo, H., & Doyle, J. F. (2012). Dynamic pitching of an elastic rectangular wing in hovering motion. *Journal of Fluid Mechanics*, 693, 473–499.
73. Nakata, T., & Liu, H. (2012). A fluid–structure interaction model of insect flight with flexible wings. *Journal of Computational Physics*, 231, 1822–1847.
74. Ishihara, D., Horie, T., & Denda, M. (2009). A two-dimensional computational study on the fluid–structure interaction cause of wing pitch changes in dipteran flapping flight. *Journal of Experimental Biology*, 212, 1–10.
75. Sotiropoulos, F., & Yang, X. (2014). Immersed boundary methods for simulating fluid–structure interaction. *Progress in Aerospace Sciences*, 65, 1–21.
76. Wang, Q., Goosen, J., & van Keulen, F. (2017). An efficient fluid–structure interaction model for optimizing twistable flapping wings. *Journal of Fluids and Structures*, 73, 82–99.
77. Ma, Y., Ning, J. G., Ren, H. L., Zhang, P. F., & Zhao, H. Y. (2015). The function of resilin in honeybee wings. *Journal of Experimental Biology*, 218, 2136–2142.
78. Haas, F., Gorb, S., & Blickhan, R. (2000). The function of resilin in beetle wings. *Proceedings of the Royal Society of London B: Biological Sciences*, 267, 1375–1381.
79. Daniel T. L., & Combes, S. A. (2002). Flexible wings and fins: Bending by inertial or fluid-dynamic forces?. *Integrative and Comparative Biology*, 42, 1044–1049.
80. Yin, B., & Luo, H. (2010). Effect of wing inertia on hovering performance of flexible flapping wings. *Physics of Fluids*, 22, 11902.
81. Chen, J.-S., Chen, J.-Y., & Chou, Y.-F. (2008). On the natural frequencies and mode shapes of dragonfly wings. *Journal of Sound and Vibration*, 313, 643–654.
82. Ennos, A. R. (1988). The inertial cause of wing rotation in Diptera. *Journal of Experimental Biology*, 140, 161–169.
83. Ennos, A. R. (1986). A comparative study of the flight mechanism of diptera. *Journal of Experimental Biology*, 127, 355–372.

84. Miyan, J. A., & Ewing, A. W. (1985). How Diptera move their wings: A re-examination of the wing base articulation and muscle systems concerned with flight. *Philosophical Transactions of the Royal Society of London B: Biological Sciences*, 311, 271–302.
85. Wang, H., Ando, N., & Kanzaki, R. (2008). Active control of free flight manoeuvres in a hawkmoth, *Agrius convolvuli*. *Journal of Experimental Biology*, 211, 423–432.
86. Walker, S. M., Thomas, A. L., & Taylor, G. K. (2012). Operation of the alula as an indicator of gear change in hoverflies. *Journal of the Royal Society Interface*, 9, 1194–1207.
87. Deora, T., Singh, A. K., & Sane, S. P. (2015). Biomechanical basis of wing and haltere coordination in flies. *Proceedings of the National Academy of Sciences*, 201412279.
88. Nalbach, G. (1989). The gear change mechanism of the blowfly (*Calliphora erythrocephala*) in tethered flight. *Journal of Comparative Physiology*, 165, 321–331.
89. Wisser, A., & Nachtigall, W. (1984). Functional-morphological investigations on the flight muscles and their insertion points in the blowfly *Calliphora erythrocephala* (Insecta, Diptera). *Zoomorphology*, 104, 188–195.
90. Tu, M. S., & Dickinson, M. H. (1994). Modulation of negative work output from a steering muscle of the blowfly *Calliphora vicina*. *Journal of Experimental Biology*, 192, 207–224.
91. Dickinson, M. H., & Lighton, J. R. B. (1995). Muscle efficiency and elastic storage in the flight motor of Drosophila. *Science*, 268, 87–90.
92. Lehmann, F.-O., & Dickinson, M. (1997). The changes in power requirements and muscle efficiency during elevated force production in the fruit fly Drosophila. *Journal of Experimental Biology*, 200, 1133–1143.
93. Pringle, J. W. S. (1949). The excitation and contraction of the flight muscles of insects. *Journal of Physiology*, 108, 226–232.
94. Pringle, J. W. S. (2003). *Insect flight*, vol. 9. Cambridge University Press.
95. Springthorpe, D., Fernández, M. J., & Hedrick, T. L. (2012). Neuromuscular control of free-flight yaw turns in the hawkmoth Manduca sexta. *Journal of Experimental Biology*, 215, 1766–1774.
96. Balint, C. N., & Dickinson, M. H. (2004). Neuromuscular control of aerodynamic forces and moments in the blowfly, *Calliphora vicina*. *Journal of Experimental Biology*, 207, 3813–3838.
97. Paulk, A., Millard, S. S., & Swinderen, B. v. (2012). Vision in Drosophila: Seeing the world through a model's eyes. *Annual Review of Entomology*, 58, 313–332.
98. Keil, T. A. (1997). Functional morphology of insect mechanoreceptors. *Microscopy Research and Technique*, 39, 506–531.
99. Maimon, G., Straw, A. D., & Dickinson, M. H. (2008). A simple vision-based algorithm for decision making in flying drosophila. *Current Biology*, 18, 464–470.
100. Tammero, L. F., Frye, M. A., & Dickinson, M. (2004). Spatial organization of visuomotor reflexes in Drosophila. *Journal of Experimental Biology*, 207, 113–122.
101. Collett, T., Nalbach, H., & Wagner, H. (1993). Visual stabilization in arthropods. *Reviews of Oculomotor Research*, 5, 239.
102. Krapp, Holger G. (2000). Neuronal matched filters for optic flow processing in flying insects. *International Review of Neurobiology*, 44, 93–120.
103. Srinivasan, M. V., & Zhang, S.-W. (2000). Visual navigation in flying insects. *International Review of Neurobiology*, 44, 67.
104. Heisenberg, M., & Wolf, R. (1992). The sensory-motor link in motion-dependent flight control of flies. *Reviews of Oculomotor Research*, 5, 265–283.

105. Bender, J. A., & Dickinson, M. H. (August 2006). Visual stimulation of saccades in magnetically tethered Drosophila. *Journal of Experimental Biology*, 209, 3170–3182.
106. Tammero, L. F., & Dickinson, M. (2001). The influence of visual landscape on the free flight behavior of the fruit fly Drosophila melanogaster. *Journal of Experimental Biology*, 205, 327–343.
107. Van Breugel, F., & Dickinson, M. H. (2012). The visual control of landing and obstacle avoidance in the fruit fly Drosophila melanogaster. *Journal of Experimental Biology*, 215, 1783–1798.
108. Strausfeld, N. J. (2012). *Atlas of an insect brain*. Springer Science & Business Media.
109. Borst, A. (2009). Drosophila's view on insect vision. *Current Biology*, 19, R36–R47.
110. Tammero, L. F., Frye, M. A., & Dickinson, M. H. (2004). Spatial organization of visuomotor reflexes in Drosophila. *Journal of Experimental Biology*, 207, 113–122.
111. Srinivasan, M. V. (2011). Honeybees as a model for the study of visually guided flight, navigation, and biologically inspired robotics. *Physiological Reviews*, 91, 413–460.
112. Srinivasan, M. V., Zhang, S., Altwein, M., & Tautz, J. (2000). Honeybee navigation: Nature and calibration of the "odometer." *Science*, 287, 851–853.
113. Barth, F. G., Humphrey, J. A., & Srinivasan, M. V. (2012). *Frontiers in sensing: From biology to engineering*: Springer Science & Business Media.
114. Reichardt, W. (1961). Autocorrelation, a principle for the evaluation of sensory information by the central nervous system. *Sensory Communication*, 303–317.
115. Borst, A., Haag, J., & Reiff, D. F. (2010). Fly motion vision. *Annual Review of Neuroscience*, 33, 49–70.
116. Tammero, L. F., & Dickinson, M. H. (September 2002). Collision-avoidance and landing responses are mediated by separate pathways in the fruit fly, Drosophila melanogaster. *Journal of Experimental Biology*, 205, 2785–2798.
117. Mayer, M., Vogtmann, K., Bausenwein, B., Wolf, R., & Heisenberg, M. (1988). Flight control during free yaw turns in Drosophila-melanogaster. *Journal of Comparative Physiology A – Sensory Neural and Behavioral Physiology*, 163, 389–399.
118. Boeddeker, N., & Egelhaaf, M. (2005). A single control system for smooth and saccade-like pursuit in blowflies. *Journal of Experimental Biology*, 208, 1563–1572.
119. Heisenberg, M., & Wolf, R. (1979). On the fine structure of yaw torque in visual flight orientation of Drosophila melanogaster. *Journal of Physiology*, 130, 113–130.
120. Heisenberg, M., & Wolf, R. (1988). Reafferent control of optomotor yaw torque in Drosophila-melanogaster. *Journal of Comparative Physiology A – Sensory Neural and Behavioral Physiology*, 163, 373–388.
121. Varju, D. (1990). A note on the reafference principle. *Biological Cybernetics*, 63, 315–323,
122. Roth, E., Reiser, M. B., Dickinson, M. H., & Cowan, N. J. (2012). A task-level model for optomotor yaw regulation in Drosophila melanogaster: A frequency-domain system identification approach. *Proceedings of the Conference on Decisions and Controls (CDC)*, Maui, Hawaii.
123. Mischiati, M., Lin, H.-T., Herold, P., Imler, E., Olberg, R., & Leonardo, A. (2015). Internal models direct dragonfly interception steering. *Nature*, 517, 333–338.
124. Wagner, H. (1982). Flow-field variables trigger landing in flies. *Nature*, 297 (5862), 147–148.
125. Borst, A. (1990). How do flies land? *Bioscience*, 40, 292–299.
126. Borst, A., & Bahde, S. (1986). What kind of movement detector is triggering the landing response of the housefly? *Biological Cybernetics*, 55, 59–69.

127. Hall, J. M., McLoughlin, D. P., Kathman, N. D., Yarger, A. M., Mureli, S., & Fox, J. L. (2015). Kinematic diversity suggests expanded roles for fly halteres. *Biology Letters*, 11, 20150845.
128. Beatus, T., Guckenheimer, J. M., & Cohen, I. (2015). Controlling roll perturbations in fruit flies. *Journal of the Royal Society Interface*, 12, 20150075.
129. Sane, S. P., Dieudonné, A., Willis, M. A., & Daniel, T. L. (2007). Antennal mechanosensors mediate flight control in moths. *Science*, 315, 863–866.
130. Sane, S. P., & Jacobson, N. P. (January 2006). Induced airflow in flying insects II. Measurement of induced flow. *Journal of Experimental Biology*, 209, 43–56.
131. Dickinson, M. (1990). Comparison of encoding properties of campaniform sensilla on the fly wing. *Journal of Experimental Biology*, 151, 245–261.
132. Dickinson, M. (1992). Directional sensitivity and mechanical coupling dynamics of campaniform sensilla during chordwise deformations of the fly wing. *Journal of Experimental Biology*, 169, 221–233.
133. Dickinson, M. (1990). Linear and nonlinear encoding properties of an identified mechanoreceptor on the fly wing measured with mechanical noise stimuli. *Journal of Experimental Biology*, 151, 219–244.
134. Dickinson, M. H., & Palka, J. (1987). Physiological properties, time of development, and central projection are correlated in the wing mechanoreceptors of Drosophila. *The Journal of Neuroscience*, 7, 4201–4208.
135. Dickerson, B. H., Aldworth, Z. N., & Daniel, T. L. (2014). Control of moth flight posture is mediated by wing mechanosensory feedback. *Journal of Experimental Biology*, 217, 2301–2308.
136. Parsons, M. M., Krapp, H. G., & Laughlin, S. B. (2010). Sensor fusion in identified visual interneurons. *Current Biology*, 20, 624–628.
137. Fayyazuddin, A., & Dickinson, M. H. (October 1999). Convergent mechanosensory input structures the firing phase of a steering motor neuron in the blowfly, Calliphora. *Journal of Neurophysiology*, 82, 1916–1926.
138. Sherman A., & Dickinson, M. (2004). Summation of visual and mechanosensory feedback in Drosophila flight control. *Journal of Experimental Biology*, 207, 133–142,
139. Chan, W. P., Prete, F., & Dickinson, M. (1998). Visual input to the efferent control system of a fly's gyroscope. *Science*, 280, 289–292.
140. Sherman A., & Dickinson, M. H. (January 2003). A comparison of visual and haltere-mediated equilibrium reflexes in the fruit fly Drosophila melanogaster. *Journal of Experimental Biology*, 206, 295–302.
141. Dickinson, M. H. (April 2005). The initiation and control of rapid flight maneuvers in fruit flies. *Integrative and Comparative Biology*, 45, 74–81.
142. Krapp, H. G., Taylor, G. K., & Humbert, J. S. (2011). The mode-sensing hypothesis: Matching sensors, actuators and flight dynamics. In Friedrich G. Barth, Joseph A. C. Humphrey, and Mandyam V. Srinivasan (Eds.), *Frontiers in sensing – biology and engineering*. Springer Verlag.
143. Sponberg, S., Dyhr, J. P., Hall, R. W., & Daniel, T. L. (2015). Luminance-dependent visual processing enables moth flight in low light. *Science*, 348, 1245–1248.
144. S. P. Windsor, R. J. Bomphrey, & Taylor, G. K. (2014). Vision-based flight control in the hawkmoth Hyles lineata. *Journal of the Royal Society Interface*, 11, 20130921.
145. Cheng, B., Tobalske, B. W., Powers, G. T., et al. (2016). Flight mechanics and control of escape manoeuvres in hummingbirds. I. Flight kinematics. *Journal of Experimental Biology*, 219, 3518–3531.

146. Ristroph, L., Ristroph, G., Morozova, S., Bergou, A. J., Chang, S., & Guckenheimer, J. (2013). Active and passive stabilization of body pitch in insect flight. *Journal of the Royal Society Interface*, 10, 20130237.
147. Trimmer, W. S. (1989). Microrobots and micromechanical systems. *Sensors and Actuators*, 19, 267–287.
148. Wood, R. J., Avadhanula, S., Sahai, R., Steltz, E., & Fearing, R. S. (2008). Microrobot design using fiber reinforced composites. *Journal of Mechanical Design*, 130, 052304.
149. Zou, Y., Zhang, W., & Zhang, Z. (2016). Liftoff of an electromagnetically driven insect-inspired flapping-wing robot. *IEEE Transactions on Robotics*, 32, 1285–1289.
150. Roll, J., Cheng, B., & Deng, X. (2015). An electromagnetic actuator for high-frequency flapping-wing micro air vehicles. *IEEE Transactions on Robotics*, *31*, 400–414.
151. Coleman, David, Moble Benedict, Vikram Hrishikeshavan, and Inderjit, Chopra. "Design, development and flight-testing of a robotic hummingbird." In *American Helicopter Society 71st Annual Forum*, pp. 5–7. 2015.
152. Keennon, M., Klingebiel, K., Won, H., & Andriukov, A. (2012). Development of the nano hummingbird: A tailless flapping wing micro air vehicle. In *50th AIAA Aerospace Sciences Meeting Including the New Horizons Forum and Aerospace Exposition, Nashville, TN*. January, 9–12.
153. Hines, L., Campolo, D., & Sitti, M. (2014). Liftoff of a motor-driven, flapping-wing microaerial vehicle capable of resonance. *IEEE Transactions on Robotics*, 30, 220–232.
154. Lentink, D., Jongerius, S. R., & Bradshaw, N. L. (2009). The scalable design of flapping micro-air vehicles inspired by insect flight. In Dario Floreano, Jean-Christophe Zufferey, Mandyam V. Srinivasan, Charlie Ellington (Eds.), *Flying insects and robots*. Springer, 185–205.
155. Baek, S. S., Ma, K. Y., & Fearing, R. S. (2009). Efficient resonant drive of flapping-wing robots. In Intelligent Robots and Systems, 2009. *IROS 2009. IEEE/RSJ International Conference,* 2854–2860.
156. Mateti, K., Byrne-Dugan, R. A., Tadigadapa, S. A., & Rahn, C. D. (2012). Wing rotation and lift in SUEX flapping wing mechanisms. *Smart Materials and Structures*, 22, 014006.
157. Deng, X. Y., Schenato, L., Wu, W. C., & Sastry, S. S. (August 2006). Flapping flight for biomimetic robotic insects: Part I – System modeling. *IEEE Transactions on Robotics*, 22, 776–788.
158. Wood, R. J. (2008). The first takeoff of a biologically inspired at-scale robotic insect. *IEEE Transactions on Robotics*, 24, 341–347,
159. Polcawich, R. G., Pulskamp, J. S., Bedair, S., et al. (2010). Integrated PiezoMEMS actuators and sensors. In *Sensors, 2010 IEEE*, 2193–2196.
160. Perez-Arancibia, N. O., Chirarattananon, P., Finio, B. M., & Wood, R. J. (2011). Pitch-angle feedback control of a biologically inspired flapping-wing microrobot. In *Robotics and Biomimetics (ROBIO), 2011 IEEE International Conference*, 1495–1502.
161. Pérez-Arancibia, N. O., Ma, K. Y., Galloway, K. C., Greenberg, J. D., & Wood, R. J. (2011). First controlled vertical flight of a biologically inspired microrobot. *Bioinspiration & Biomimetics*, 6, 036009.
162. Ma, K. Y., Chirarattananon, P., Fuller, S. B., & Wood, R. J. (2013). Controlled flight of a biologically inspired, insect-scale robot. *Science*, 340, 603–607.
163. Zhang, J., Cheng, B., & Deng, X. (2016). Instantaneous wing kinematics tracking and force control of a high-frequency flapping wing insect MAV. *Journal of Micro-Bio Robotics*, 11, 67–84.

164. Zhang, J., Tu, Z., Fei, F., & Deng, X. (2017). Geometric flight control of a hovering robotic hummingbird. In *Robotics and Automation (ICRA), 2017 IEEE International Conference*, 5415–5421.
165. Karásek, M., Hua, A., Nan, Y., Lalami, M., & Preumont, A. (2014). Pitch and roll control mechanism for a hovering flapping wing MAV. *International Journal of Micro Air Vehicles*, 6, 253–264.
166. Yan, J., Wood, R. J., Avadhanula, S., Sitti, M., & Fearing, R. S. (2001). Towards flapping wing control for a micromechanical flying insect. In *Robotics and Automation, 2001. Proceedings 2001 ICRA. IEEE International Conference*, 3901–3908.
167. Muijres, F. T., Elzinga, M. J., Melis, J. M., & Dickinson, M. H. (2014). Flies evade looming targets by executing rapid visually directed banked turns. *Science*, 344, 172–177.
168. Altshuler, D. L., Quicazán-Rubio, E. M., Segre, P. S., & Middleton, K. M. (2012). Wingbeat kinematics and motor control of yaw turns in Anna's hummingbirds (Calypte anna). *Journal of Experimental Biology*, 215, 4070–4084.
169. Bronson, J., Pulskamp, J., Polcawich, R., Kroninger, C., & Wetzel, E. (2009). PZT MEMS actuated flapping wings for insect-inspired robotics. In *Micro Electro Mechanical Systems, 2009. MEMS 2009. IEEE 22nd International Conference*, 1047–1050.
170. Sitti, M. (2003). Piezoelectrically actuated four-bar mechanism with two flexible links for micromechanical flying insect thorax. *IEEE/ASME Transactions on Mechatronics*, 8, 26–36.
171. Graule, M., Chirarattananon, P., Fuller, S., et al. (2016). Perching and takeoff of a robotic insect on overhangs using switchable electrostatic adhesion. *Science*, 352, 978–982.
172. Lau, G.-K., Chin, Y.-W., Goh, J. T.-W., & Wood, R. J. (2014). Dipteran-insect-inspired thoracic mechanism with nonlinear stiffness to save inertial power of flapping-wing flight. *IEEE Transactions on Robotics*, 30, 1187–1197.
173. Teoh, Z. E., & Wood, R. J. (2013). A flapping-wing microrobot with a differential angle-of-attack mechanism. In *Robotics and Automation (ICRA), 2013 IEEE International Conference*, 1381–1388.
174. Finio, B. M., Whitney, J. P., & Wood, R. J. (2010). Stroke plane deviation for a microrobotic fly. In *Intelligent Robots and Systems (IROS), 2010 IEEE/RSJ International Conference*, 3378–3385.
175. Finio, B. M., Shang, J. K., & Wood, R. J. (2009). Body torque modulation for a microrobotic fly. In *Robotics and Automation, 2009. ICRA'09. IEEE International Conference*, 3449–3456.
176. Shang, J., Combes, S. A., Finio, B., & Wood, R. J. (2009). Artificial insect wings of diverse morphology for flapping-wing micro air vehicles. *Bioinspiration & Biomimetics*, 4, 036002.
177. Tanaka, H., & Wood, R. J. (2010). Fabrication of corrugated artificial insect wings using laser micromachined molds. *Journal of Micromechanics and Microengineering*, 20, 075008.
178. Tanaka, H., Whitney, J. P., & Wood, R. J. (2011). Effect of flexural and torsional wing flexibility on lift generation in hoverfly flight. Oxford University Press.
179. Steltz, E., Seeman, M., Avadhanula, S., & Fearing, R. S. (2006). Power electronics design choice for piezoelectric microrobots. In *Intelligent Robots and Systems, 2006 IEEE/RSJ International Conference*, 1322–1328.
180. Karpelson, M., Wei, G.-Y., & Wood, R. J. (2012). Driving high voltage piezoelectric actuators in microrobotic applications. *Sensors and Actuators A: Physical*, 176, 78–89.

181. Meng, K., Zhang, W., Chen, W., et al. (2012). The design and micromachining of an electromagnetic MEMS flapping-wing micro air vehicle. *Microsystem Technologies*, 18, 127–136.
182. Roll, J. A., Bardroff, D. T., & Deng, X. (2016). Mechanics of a scalable high frequency flapping wing robotic platform capable of lift-off. In *Robotics and Automation (ICRA), 2016 IEEE International Conference, 4664–4671.*
183. Cheng, B., Roll, J., & Deng, X. (2013). Modeling and optimization of an electromagnetic actuators for flapping-wing micro air vehicle. In *IEEE International Conference on Robotics and Automation (ICRA)*, Karlsruhe, Germany, 4035–4041.
184. Roll, J., Cheng, B., & Deng, X. (2013). Design, fabrication and testing of an electromagnetic actuator for flapping wing micro air vehicles. *Submitted to Proceedings of IEEE ICRA*, Karlsruhe, Germany.
185. Croon, G. C. H. E. d., Groen, M. A., Wagter, C. D., Remes, B., Ruijsink, R., & Oudheusden, B. W. v. (2012). Design, aerodynamics and autonomy of the DelFly. *Bioinspiration & Biomimetics*, 7, 025003.
186. Perseghetti, B. M., Roll, J. A., & Gallagher, J. C. (2014). Design constraints of a minimally actuated four bar linkage flapping-wing micro air vehicle. In *Robot intelligence technology and applications 2*. Springer, 545–555.
187. Hu, Z., Cheng, B., & Deng, X. (2010). Lift generation and flow measurements of a robotic insect. In *49th AIAA Aerospace Sciences Meeting*, Orlando, FL.
188. Zhang, J., Fei, F., Tu, Z., & Deng, X. (2017). Design optimization and system integration of robotic hummingbird. In *Robotics and Automation (ICRA), 2017 IEEE International Conference*, 5422–5428.
189. Conn, A., Burgess, S., & Ling, C. (2007). Design of a parallel crank-rocker flapping mechanism for insect-inspired micro air vehicles. *Proceedings of the Institution of Mechanical Engineers, Part C: Journal of Mechanical Engineering Science*, 221, 1211–1222.
190. Galiński, C., & Żbikowski, R. (2005). Insect-like flapping wing mechanism based on a double spherical Scotch yoke. *Journal of the Royal Society Interface*, 2, 223–235.
191. Seshadri, P., Benedict, M., & Chopra, I. (2012). A novel mechanism for emulating insect wing kinematics. *Bioinspiration & Biomimetics*, 7, 036017.
192. Floreano, D., Pericet-Camara, R., Viollet, S., et al. (2013). Miniature curved artificial compound eyes. *Proceedings of the National Academy of Sciences*, 110, 9267–9272.
193. Wu, W.-C., Schenato, L., Wood, R. J., & Fearing, R. S. (2003). Biomimetic sensor suite for flight control of a micromechanical flying insect: Design and experimental results. In *Robotics and Automation, 2003. Proceedings. ICRA'03. IEEE International Conference*, 1146–1151.
194. Smith, G., Bedair, S., Schuster, B., et al. (2012). Biologically inspired, haltere, angular-rate sensors for micro-autonomous systems. In *SPIE Defense, Security, and Sensing*, 83731K–83731K-13.
195. Plett, J., Bahl, A., Buss, M., Kühnlenz, K., & Borst, A. (2012). Bio-inspired visual ego-rotation sensor for MAVs. *Biological Cybernetics*, 106, 51–63,
196. Floreano, D., Zufferey, J.-C., Srinivasan, M. V., & Ellington, C. (2009). *Flying insects and robots*. Springer.
197. Srinivasan, M. V., Chahl, J. S., Weber, K., Venkatesh, S., Nagle, M. G., & Zhang, S.-W. (1999). Robot navigation inspired by principles of insect vision. *Robotics and Autonomous Systems*, 26, 203–216.

198. Hyslop, A. M., & Humbert, J. S. (2010). Autonomous navigation in three-dimensional urban environments using wide-field integration of optic flow. *Journal of Guidance, Control, and Dynamics*, 33, 147–159.
199. Humbert, J. S., & Hyslop, A. M. (2010). Bioinspired visuomotor convergence. *IEEE Transactions on Robotics*, 26, 121–130.
200. Neumann, T., & Bülthoff, H. (2001). Insect inspired visual control of translatory flight. *Advances in Artificial Life*, 627–636.
201. Barbour, N., & Schmidt, G. (2001). Inertial sensor technology trends. *IEEE Sensors Journal*, 1, 332–339.
202. Duparré, J., & Wippermann, F. (2006). Micro-optical artificial compound eyes. *Bioinspiration & Biomimetics*, 1, R1.
203. Jeong, K.-H., Kim, J., & Lee, L. P. (2006). Biologically inspired artificial compound eyes. *Science*, 312, 557–561.
204. Muratet, L., Doncieux, S., & Meyer, J.-A. (2004). A biomimetic reactive navigation system using the optical flow for a rotary-wing UAV in urban environment. *Proceedings of the International Session on Robotics*.
205. Chahl, J., & Srinivasan, M. V. (2000). A complete panoramic vision system, incorporating imaging, ranging, and three dimensional navigation. In *Omnidirectional Vision, 2000. Proceedings. IEEE Workshop*, 104–111.
206. Franceschini, N., Ruffier, F., & Serres, J. (2007). A bio-inspired flying robot sheds light on insect piloting abilities. *Current Biology*, 17, 329–335.
207. Zufferey, J.-C., Beyeler, A., & Floreano, D. (2009). Optic flow to steer and avoid collisions in 3D. In *Flying insects and robots*. Springer, 73–86.
208. Kehoe, J., Watkins, A., Causey, R., & Lind, R. (2006). State estimation using optical flow from parallax-weighted feature tracking. In *AIAA Guidance, Navigation, and Control Conference and Exhibit*, 6721.
209. Franz, M. O., Chahl, J. S., & Krapp, H. G. (2004). Insect-inspired estimation of egomotion. *Neural Computation*, 16, 2245–2260.
210. Srinivasan, M., Zhang, S., Lehrer, M., & Collett, T. (1996). Honeybee navigation en route to the goal: Visual flight control and odometry. *Journal of Experimental Biology*, 199, 237–244.
211. Srinivasan, M. V., Zhang, S., & Chahl, J. S. (2001). Landing strategies in honeybees, and possible applications to autonomous airborne vehicles. *The Biological Bulletin*, 200, 216–221.
212. Srinivasan, M. V., Zhang, S.-W., Chahl, J. S., Barth, E., & Venkatesh, S. (2000). How honeybees make grazing landings on flat surfaces. *Biological Cybernetics*, 83, 171–183.
213. Conroy, J., Gremillion, G., Ranganathan, B., & Humbert, J. S. (2009). Implementation of wide-field integration of optic flow for autonomous quadrotor navigation. *Autonomous Robots*, 27, 189–198.
214. Chahl, J. S., Srinivasan, M. V., & Zhang, S.-W. (2004). Landing strategies in honeybees and applications to uninhabited airborne vehicles. *The International Journal of Robotics Research*, 23, 101–110.
215. Barrows, G., & Neely, C. (2000). Mixed-mode VLSI optic flow sensors for in-flight control of a micro air vehicle. In *Proceedings of SPIE*, 52–63.
216. Xu, P., Humbert, J. S., & Abshire, P. (2011). Analog VLSI implementation of wide-field integration methods. *Journal of Intelligent & Robotic Systems*, 64, 465–487.

217. Zhang, T., Wu, H., Borst, A., Kuhnlenz, K., & Buss, M. (2008). An FPGA implementation of insect-inspired motion detector for high-speed vision systems. In *Robotics and Automation, 2008. ICRA 2008. IEEE International Conference*, 335–340.
218. Gremillion, G., Galfond, M., Krapp, H. G., & Humbert, J. S. (2012). Biomimetic sensing and modeling of the ocelli visual system of flying insects. In *Intelligent Robots and Systems (IROS), 2012 IEEE/RSJ International Conference*, 1454–1459.
219. Fuller, S. B., Karpelson, M., Censi, A., Ma, K. Y., & Wood, R. J. (2014). Controlling free flight of a robotic fly using an onboard vision sensor inspired by insect ocelli. *Journal of the Royal Society Interface*, 11, 20140281.
220. Fuller, S. B., Sands, A., Haggerty, A., Karpelson, M., & Wood, R. J. (2013). Estimating attitude and wind velocity using biomimetic sensors on a microrobotic bee. In *Robotics and Automation (ICRA), 2013 IEEE International Conference*, 1374–1380.
221. Muijres, F. T., Elzinga, M. J., Iwasaki, N. A., & Dickinson, M. H. (2015). Body saccades of Drosophila consist of stereotyped banked turns. *Journal of Experimental Biology*, 218, 864–875.
222. Wootton, R. J. (1981). Support and deformability in insect wings. *Journal of Zoology*, 193, 447–468.
223. Walker, S. M., Thomas, A. L., & Taylor, G. K. (2009). Deformable wing kinematics in free-flying hoverflies. *Journal of the Royal Society Interface*, 7, 131–142.
224. Hedenström, A. (2014). How insect flight steering muscles work. *PLoS Biology*, 12, e1001822.
225. Maisak, M. S., Haag, J., Ammer, G., et al. (2013). A directional tuning map of Drosophila elementary motion detectors. *Nature*, 500, 212.
226. Parks, P., Cheng, B., Hu, Z., & Deng, X. (2011). Translational damping on flapping cicada wings. In *IEEE/RSJ International Conference on Intelligent Robots and Systems (IROS)*, 574–579.

12 Designing Nature-Inspired Liquid-Repellent Surfaces

Birgitt Boschitsch Stogin, Lin Wang, and Tak-Sing Wong

12.1 Introduction

12.1.1 Wetting

Wetting refers to the interactions between a liquid and a solid in a given environment [1–3]. In particular, it refers to the study of how liquids spread on solids. This field of science involves principles found in fluid mechanics and materials science and is relevant to various natural phenomena and industrial applications.

12.1.2 Wetting in Nature

In nature, organisms at various length scales rely on micro- to nano-scale surface designs to achieve macroscopically observable interactions with liquids [4] (Figure 12.1). For example, the intrinsically hydrophobic material and hierarchical micro- and nano-scale texturing of lotus leaves lend these natural surfaces the ability to repel water rapidly [5,6]. Springtails are arthropods living in soil, which have overhanging nanostructured skin patterns that help prevent soiling [7]. Wings exhibit similar hydrophobicity, with additional microscale structure arrangements that facilitate directional/anisotropic droplet motion (i.e., water droplets on butterfly wings travel outward, away from the body) [8]. The peristome of the carnivorous pitcher plant promotes the imbibition and spread of liquid nectar or rain water; in the imbibed state, insects are unable to attach mechanically to the peristome and instead slide off the surface into the plant digestive system [9]. Other organisms that rely on wetting or anti-wetting strategies to survive in their respective environments include Namib Desert beetles [10], water striders [11,12], diving bell spiders [13], geckos [14], and more!

12.1.3 Wetting Fundamentals

To understand how natural organisms or materials interact with liquids, or to design materials with desired wettability, we must first understand the basics of wetting. One important parameter in understanding the wetting behavior of a liquid on a solid is the **contact angle** (Figure 12.2), which is the angle the liquid forms with the

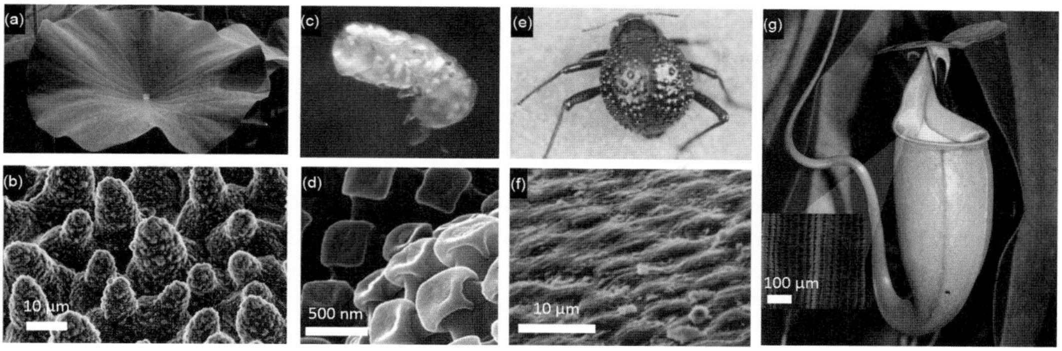

Figure 12.1 Example of liquid-repellent surfaces in nature. (a) A lotus leaf and (b) its hierarchical surface textures. (c) A springtail and (d) its overhanging nanoscale surface texture. (e) A Namib Desert beetle and (f) its surface chemical patterns. (g) A *Nepenthes* pitcher plant and its surface textures at the rim (inset). (Images from panel (a–d) are reproduced under Creative Commons License from Ensikat et al. [6] and Helbig et al. [7]. Images from (e, f) are reproduced from Parker and Lawrence [10]. The pitcher plant image is courtesy of W. Federle and H. Bohn.)

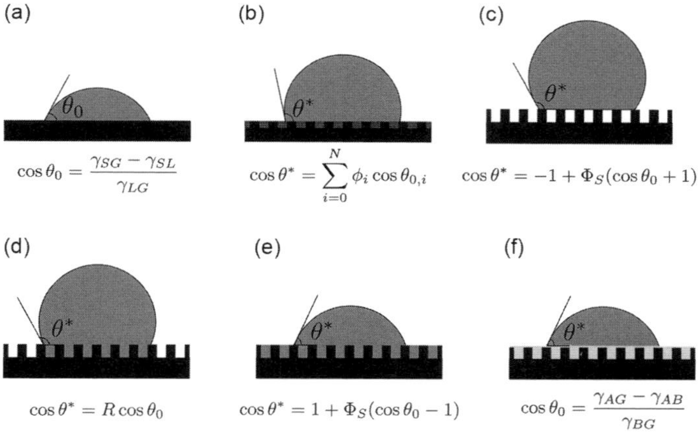

Figure 12.2 Wetting configurations and their respective governing equations. (a) Wetting on a smooth solid surface. (b) Wetting on a chemically heterogeneous smooth solid surface. (c) Cassie–Baxter state wetting. (d) Wenzel state wetting. (e) Hemiwicking. (f) Wetting on a liquid-infused porous surface.

solid in a given environment (liquid or gas). The contact angle of a droplet on a *smooth* surface is referred to as the **intrinsic contact angle** (θ_0) and the contact angle of a droplet on a rough surface is referred to as the **apparent contact angle** (θ^*). All contact angles depend on the liquid, solid, and gas present in the system of interest [1,15].

12.2 Fundamental Equations of Wetting

12.2.1 Wetting on Smooth Surfaces

12.2.1.1 Chemically Homogeneous Surfaces

Young's equation predicts θ_0 of a liquid droplet on a given smooth, chemically homogeneous surface and can be derived using an energy minimization approach [15]. Consider a two-dimensional droplet resting on a smooth, solid surface with contact angle θ_0 (Figure 12.2a). Now suppose we perturb the contact line slightly, moving it outward by dx. Then the liquid–solid interface increases by dx, the solid–gas interface decreases by dx, and the liquid–gas interface increases by $dx \cos\theta_0$. The change in interfacial energy per unit length associated with this perturbation can be represented as

$$dE = (\gamma_{SL} - \gamma_{SG})dx - \gamma_{LG}\cos\theta_0\, dx,$$

where γ_{SG} is the interfacial tension between the solid and gas phases (e.g., surface energy), γ_{SL} is the interfacial tension of solid–liquid interface, and γ_{LG} is the interfacial tension of the liquid–gas interface (e.g., the surface tension). Conceptually, the $(\gamma_{SL} - \gamma_{SG})$ term accounts for replacing a solid–gas interface with a solid–liquid one due to the perturbation dx. The $\gamma_{LG}\cos\theta_0\, dx$ term accounts for stretching of the liquid–gas interface due to perturbation dx. Letting dE go to zero (i.e., finding the minimal interfacial tension) and solving for $\cos\theta_0$ leads to **Young's equation**:

$$\cos\theta_0 = \frac{\gamma_{SG} - \gamma_{SL}}{\gamma_{LG}}.$$

12.2.1.2 Chemically Heterogeneous Surfaces

Surfaces may also exhibit chemical heterogeneity [3,16]. Let us now consider a smooth surface comprising N different materials, each with a different intrinsic contact angle $\theta_{0,\,i}$ where $i = 1, 2, \ldots, N$ (Figure 12.2b). Let ϕ_i denote the fraction of the surface comprising material i (i.e., $\sum_{i=1}^{N}\phi_i = 1$). Assume the different areas comprising different materials are much smaller than the droplet and randomly distributed about the surface [17]. The change in interfacial energy per unit length associated with perturbing the contact line by dx can be represented as

$$\begin{aligned} dE = {} & \phi_1(\gamma_{S_1L} - \gamma_{S_1G})dx + \phi_2(\gamma_{S_2L} - \gamma_{S_2G})dx + \cdots + \phi_N(\gamma_{S_NL} - \gamma_{S_NG})dx \\ & + \gamma_{LG}\cos\theta^*\, dx, \end{aligned}$$

where γ_{S_iL} represents the interfacial energy between the material i and the foreign liquid, and γ_{S_iG} represents the interfacial energy between the material i and the gas phase environment. This can be represented as

$$dE = \left(\sum_{i=1}^{N}\phi_i\left(\gamma_{S_iL} - \gamma_{S_iG}\right) + \gamma_{LG}\cos\theta^*\right)dx.$$

Allowing $dE = 0$, applying Young's equation for a chemically homogeneous surface to each material, and simplifying, we arrive at the generalized **Cassie equation**, which predicts the apparent contact angle on a chemically heterogeneous surface [16]:

$$\cos\theta^* = \sum_{i=1}^{N} \phi_i \cos\theta_{0,i}.$$

12.2.2 Wetting on Rough Surfaces

12.2.2.1 Cassie–Baxter State

Let us now consider chemically homogenous rough surfaces. When a droplet rests partially on the solid and partially on air pockets, the droplet is said to be in the **Cassie–Baxter state** (Figure 12.2c) [3]. This state is typically associated with high droplet mobility. Similar to the way a hockey puck can glide over an air hockey table when the air is turned on, a droplet in the Cassie–Baxter state can glide over the rough surface.

Changing the roughness of a surface changes the contact angle of a droplet on that surface. As mentioned previously, the contact angle on a rough surface is called the apparent contact angle. To analytically determine the thermodynamically most stable apparent contact angle of a droplet in the Cassie–Baxter state, we perform an energy minimization [18]. To do this, we look at a three-phase contact line of a droplet in the Cassie state and perturb it by moving it dx outward. The interfacial energy change per unit length associated with this perturbation is

$$dE_C = \Phi_S(\gamma_{SL} - \gamma_{SG})dx + (1 - \Phi_S)\gamma_{LG}dx + \gamma_{LG}\cos\theta^* dx,$$

where Φ_S is the **solid fraction**, which is the ratio of the area of the liquid–solid interface to the total projected area of the substrate [19]. Note that the last term in the equation above captures the stretching of the liquid–gas interface due to this perturbation. The interfacial energy per unit length E is minimized when dE_C is zero, so the energy minimization and application of Young's Equation yields

$$0 = \Phi_S(\gamma_{SL} - \gamma_{SG}) + (1 - \Phi_S)\gamma_{LG} + \gamma_{LG}\cos\theta^*,$$

$$\cos\theta^* = -1 + \Phi_S(\cos\theta_0 + 1).$$

The second equation above is called **Cassie–Baxter's equation** [3] and predicts the *thermodynamically most stable* apparent contact angle of a Cassie–Baxter state droplet [20]. An important physical insight from this governing equation is that in order to maintain high apparent contact angle (θ^*), one must minimize the solid fraction (Φ_s).

The Cassie–Baxter equation can be used to explain the large apparent contact angle of water on a broad range of natural water-repellent surfaces found in plants, insects, and other animals [4]. However, the classical Cassie–Baxter equation does not provide theoretical guidance as to why nanoscale textures are essential for liquid repellency, as there is no length scale term in the equation. Furthermore, it does not explain why some natural organisms do not minimize Φ_s and yet display large apparent contact angle of water. For example, springtails are able to maintain high liquid repellency, despite their relatively high Φ_s [7]. Recent theories have suggested that **line tension** [21,22] may be the key to explaining how texture size affects wetting properties [23–26]. That is, the change in interfacial energy dE_{total}, corresponding to a perturbation in droplet contact area dA would be the change in interfacial energy dE_{surface} plus the change in line energy dE_{line}. Specifically,

$$dE_{\text{surface}} = \Phi_S(\gamma_{SL} - \gamma_{SG})dA + (1 - \Phi_S)\gamma_{LG}dA + \gamma_{LG}\cos\theta^* dA,$$

$$dE_{\text{line}} = \frac{1}{k}\tau\Phi_s dA,$$

$$dE_{\text{total}} = dE_{\text{surface}} + dE_{\text{line}},$$

where τ is the three-phase contact line tension and k is the geometry factor (with length units) indicating the ratio of the characteristic texture cross-sectional area to pillar cross-section perimeter length. By letting $dE_{\text{total}} = 0$ and solving for $\cos\theta^*$, one can obtain a new governing equation as [23]:

$$\cos\theta^* = -1 + \Phi_S(\cos\theta_0 + 1) - \frac{1}{k}\frac{\tau\Phi_S}{\gamma_{LG}}.$$

12.2.2.2 Wenzel State

When a droplet impregnates the surface roughness, it is said to be in the Wenzel state (Figure 12.2d) [2]. This state is typically associated with droplet pinning and immobility [27].

To derive Wenzel's equation, we perform an energy minimization similar to that in Cassie's equation [18],

$$dE_W = R(\gamma_{SL} - \gamma_{SG})dx + \gamma_{LG}\cos\theta^* dx,$$

where R is the surface **roughness**, defined here as the ratio of total surface area to projected surface area (such that $R \geq 1$). Performing the energy minimization yields

$$0 = R(\gamma_{SL} - \gamma_{SG})dx + \gamma_{LG}\cos\theta^* dx,$$

$$\cos\theta^* = \frac{-R(\gamma_{SL} - \gamma_{SG})}{\gamma_{LG}},$$

so according to Young's Equation,

$$\cos\theta^* = R\cos\theta_0.$$

This equation is called **Wenzel's equation** [2], and predicts the *thermodynamically most stable* apparent contact angle of a droplet in the Wenzel state [2,20]. An important insight from the Wenzel equation is that roughness amplifies surface wettability.

Specifically, the presence of roughness can render an intrinsically hydrophobic material more hydrophobic and an intrinsically hydrophilic material more hydrophilic.

12.2.2.3 Hemiwicking

When a droplet impregnates a surface and continues to spread within the surface roughness, the droplet is said to be in the **hemiwicking state** (Figure 12.2e) [18,28]. To predict analytically the apparent contact angle of a droplet in the hemiwicking state, we again perform an energy minimization analysis. Again moving the contact line of the droplet dx, we get an interfacial energy change per unit length of

$$dE_H = (\gamma_{SL} - \gamma_{SG})\Phi_S dx + (\gamma_{LL} - \gamma_{LG})(1 - \Phi_S)dx + \gamma_{LG} \cos \theta^* dx.$$

Letting $dE_H = 0$ for energy minimization and applying Young's equation leads to,

$$\cos \theta^* = \Phi_S \cos \theta_0 + 1 - \Phi_S.$$

This equation is referred to as the **hemiwicking equation.**

12.2.2.4 Superhydrophobic Surfaces

A smooth surface is conventionally considered hydrophobic if $\theta_0 > 90°$ and hydrophilic if $\theta_0 < 90°$ [15] (though there is some debate regarding this definition in the wetting community [29,30]). The maximum contact angle on a smooth surface can rarely exceed 120° [31]. Considering the Cassie and Wenzel equations, roughening an intrinsically hydrophobic surface will make it more hydrophobic (i.e., $\theta^* > \theta_0$) and roughening an intrinsically hydrophilic surface will make it more hydrophilic (i.e., $\theta^* < \theta_0$). A surface is considered to be a **superhydrophobic surface** (SHS) if $\theta^* > 150°$ [27]. However, as the apparent contact angle does not necessarily capture droplet mobility, it is often important to report both the apparent contact angle and the contact angle hysteresis (see next section) to characterize the wetting properties of a given surface.

12.2.2.5 Contact Angle Hysteresis

Another important parameter in wetting science is the **contact angle hysteresis** ($\Delta\theta^*$). When a droplet of given volume rests on a given surface, that surface must tilt to a certain angle before the droplet begins to slide (this angle is referred to as the **tilting angle** or the **sliding angle**) [32]. Specifically, the maximum possible contact angle at the bottom-most region of the droplet is known as the **advancing angle**, θ_A^*, and the minimum possible angle at the top-most region of the droplet is known as the **receding angle**, θ_R^* (Figure 12.3a). The contact angle hysteresis is defined as the difference between the advancing angle and the receding angle, that is,

$$\Delta\theta^* = \theta_A^* - \theta_R^*.$$

The contact angle hysteresis captures the mobility of a droplet on a given surface [32]. For example, a highly mobile droplet will not deform significantly (e.g., $\theta_A^* \cong \theta^*, \theta_R^* \cong \theta^*$; small $\Delta\theta$) before it begins to slide on a surface. A droplet that is not as mobile will tend to deform (e.g., a large θ_A^* and a small θ_R^*; large $\Delta\theta^*$) more

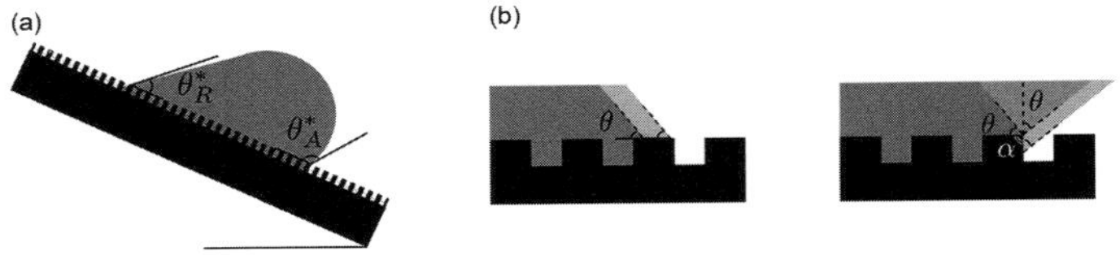

Figure 12.3 Contact angle hysteresis and pinning. (a) Contact angle hysteresis of a liquid droplet on a tilted surface. (b) Advancing contact line dynamics at a microscopic edge.

before beginning to slide. As a result, liquid repellency of a surface is typically quantified by the magnitude of the contact angle hysteresis [27] – the smaller the contact angle hysteresis, the more repellent the surface to the liquid of interest. Typical liquid-repellent surfaces have a contact angle hysteresis of <~5° to ~10°.

At the microscopic level, contact angle hysteresis is attributed to pinning of the liquid contact line at the microscale surface defects [33,34]. In the 1870s, J. W. Gibbs provided the first quantitative description of contact line pinning at a solid edge [21]. Consider a liquid contact line advancing on a surface with a contact angle θ until it pins at the edge of a surface defect (Figure 12.3b). During this stage of pinning, the local contact angle, θ_L, reaches a maximum value before the contact line is depinned from the edge with an edge angle α [35]. The local contact angle typically follows a geometrical relationship described by Gibbs as

$$\theta \leq \theta_L \leq \theta + (180° - \alpha).$$

This relationship is also known as the Gibbs inequality of pinning, and has been shown to be applicable on both rectilinear and curved surfaces [35,36].

12.2.3 Wetting on Liquid Surfaces

Up to this point, we have been considering wetting on solid surfaces. We now consider wetting on liquid surfaces. Before we characterize wetting on these surfaces, we must first define a slippery liquid-infused surface.

12.2.3.1 Slippery Liquid-Infused Porous Surfaces (SLIPS)

When a foreign droplet interacts with SLIPS, it is effectively interacting with a smooth liquid surface [37,38]. SLIPS are comprised of a roughened solid substrate with a thermodynamically stable liquid layer infused into the roughened solid, yielding a smooth liquid outer surface (Figure 12.2c). There are a few criteria that define SLIPS:

1. The infused liquid (i.e., lubricant) must wick the porous solid surface.
2. It must be energetically more favorable for the lubricant to wet the solid substrate than for the foreign fluid (i.e., the fluid to be repelled) to wet the solid substrate.
3. The lubricant and the foreign fluid must be immiscible.

The first criteria can be accomplished by roughening the surface and chemically functionalizing it as needed to achieve chemical affinity with the lubricant. The second

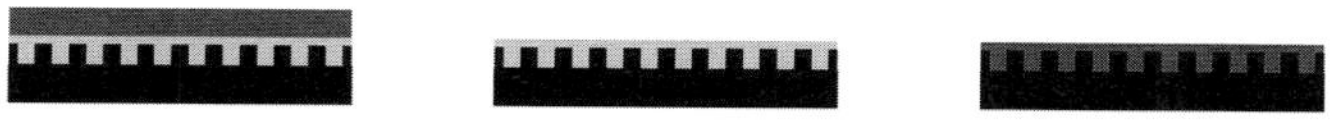

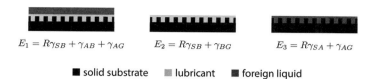

Figure 12.4 Wetting configurations on liquid-infused porous surfaces.

criteria can be achieved if the interfacial energy (E_i) of a system in which the lubricant wets the solid with (E_1) or without (E_2) a foreign liquid spreading on it is less than the interfacial energy of a system where the foreign liquid wets the surface without (E_3) the lubricant spreading on the outer surface (see Figure 12.4). That is, the following conditions must be satisfied [37]:

1. $E_3 - E_1 = R(\gamma_{BG} \cos \theta_B - \gamma_{AG} \cos \theta_A) - \gamma_{AB} > 0$;
2. $E_3 - E_2 = R(\gamma_{BG} \cos \theta_B - \gamma_{AG} \cos \theta_A) + \gamma_{AG} - \gamma_{BG} > 0$,

where R is the roughness of the solid (total area over projected area), and γ_{ij} refers to the interfacial tensions between liquid A (A), liquid B (B), and the surrounding gas (G). θ_A is the intrinsic contact angle of liquid A on the solid surface surrounded by gas (G), and θ_B is that for liquid B.

12.2.3.2 Wetting on SLIPS

When a droplet interacts with SLIPS according to the above criteria, the droplet will be interacting with a liquid. If the lubricant is thick, lubricant deformation may affect the apparent contact angle [39]. If the lubricant is thin, then we would expect to observe wetting behavior similar to that on a smooth surface [40]. That is, rather than arriving at Young's equation, we should arrive at a modified Young's equation [40],

$$\cos \theta_0 = \frac{\gamma_{AG} - \gamma_{AB}}{\gamma_{BG}},$$

where the contact angle of the foreign droplet can be predicted based on the interfacial tensions of the lubricant, foreign liquid, and surrounding gas.

12.3 Case Studies: Wetting in Nature

12.3.1 Namib Desert Beetle

The Namib Desert beetle (*Stenocara gracilipes*) lives, as the name implies, in the Namib Desert – a very dry region of southwest Africa. Like most organisms, these animals need water to survive. To meet their daily water intake in one of the driest regions of the world, these insects must harvest water from air. To do this, they travel to sand dune peaks in the early morning when dew is present. Their bodies are patterned with hydrophilic and hydrophobic regions (Figure 12.5). Researchers have found that the heterogeneous surface chemistry plays a role in increasing the beetle's

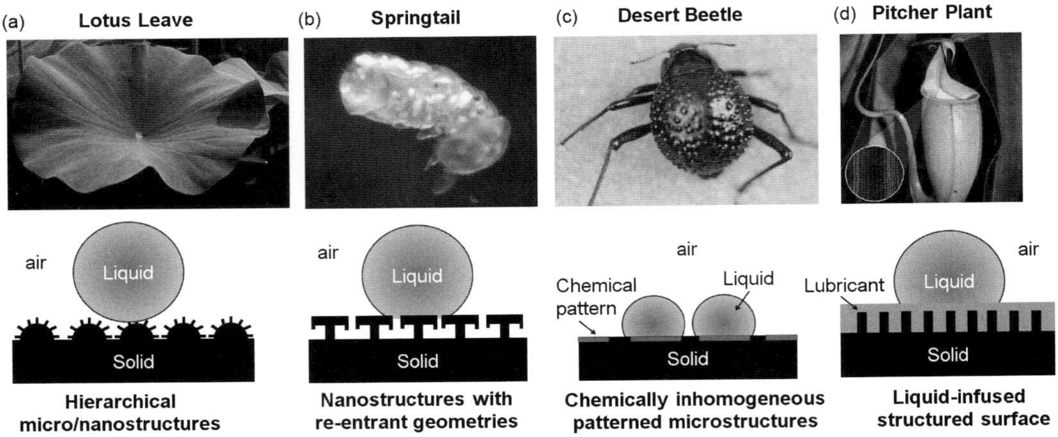

Figure 12.5 Wetting states of different biological liquid-repellent surfaces. (a) Lotus leaf. (b) Springtail. (c) Namib Desert beetle. (d) *Nepenthes* pitcher plant. (Images from panel (a, b) are reproduced under Creative Commons License from Ensikat et al. [6] and Helbig [7]. Images from (c) are reproduced from Bohn and Federle [9]. The pitcher plant image is courtesy of W. Federle and H. Bohn.)

fog-harvesting rate [10]. When the fog droplets carried by wind impact the beetle's chemically heterogeneous back, they are collected on the hydrophilic patterns. Fog droplets deposited on hydrophobic regions are either blown away by wind or migrate to the hydrophilic regions due to higher chemical affinity. The water in the hydrophilic regions accumulates until it is eventually large/massive enough to overcome pinning effects and slides down the beetle's back into its mouth.

Mimicking the strategy used by desert beetles could allow researchers to design synthetic fog-harvesting materials for accessing fresh water from fog in desert and coastal areas (Figure 12.6). Numerous methods have been developed to fabricate the heterogeneous hydrophilic–hydrophobic patterned surfaces [41–47]. In order to optimize the design of the patterned surfaces, researchers have systematically investigated a number of design parameters, such as the wettability and the dimensions of the hydrophilic patterns (shape, size, pitch) [42,47]. It has been found that the surfaces with hydrophilic patterns of the lowest contact angle were the most efficient in fog harvesting [42]. The size and pitch of the hydrophilic patterns also played an important role. The hydrophilic patterns with a diameter of 500 μm and a pitch of 1,000 μm were found to be the optimal design. The pattern shape also played a crucial role in the water collection efficiency [47]. Instead of using circular patterns, researchers designed a series of surfaces on which the patterns were star-shapes, and these star-shaped, patterned surfaces showed higher water collection efficiency than the circular ones (Figure 12.6).

12.3.2 Lotus Leaf

The lotus leaf, one of the most famous plants in the wetting community, is an aquatic plant with a water-repellent surface. As most plants rely on sunlight and photosynthesis to

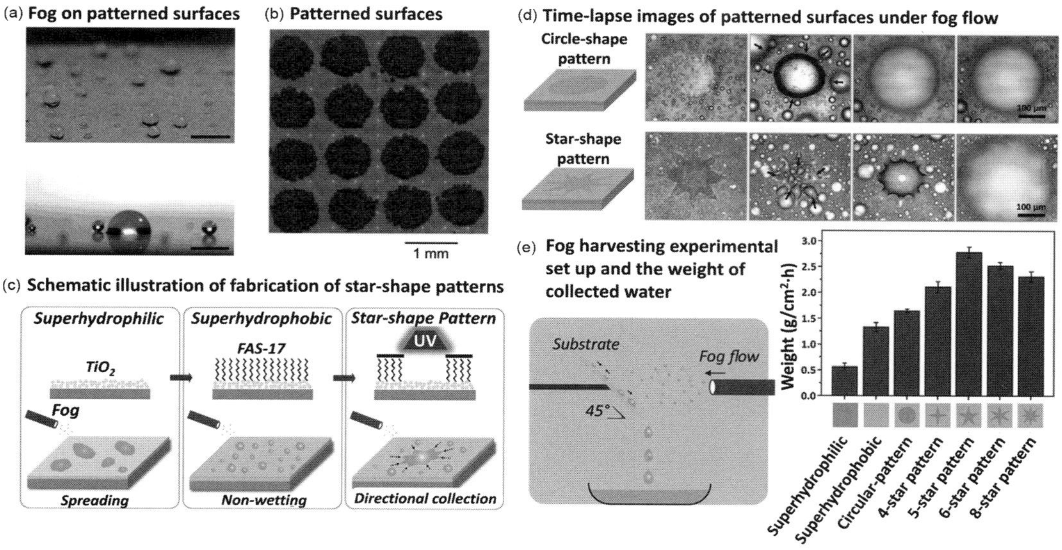

Figure 12.6 Synthetic desert beetle surfaces and their performance under fog flow. (a) Droplets that have accumulated on the patterned hydrophilic areas in the presence of fog; scale bars are 5 mm (top panel) and 750 μm (bottom panel); (b) Optical image showing a hydrophilic–hydrophobic patterned surface. (c) Schematic illustration of fabrication process of star-shaped hydrophilic patterns on hydrophobic surfaces. (d) Time-lapse optical images showing the process of fog droplets accumulating on surfaces with differently shaped patterns. (e) The experimental setup for the fog-harvesting test and the weight of collected water on surfaces with different shaped patterns. (Panel (a) reprinted from Zhai et al. [41]. Panel (b) reprinted from Garrod et al. [42]. Panels (c–e) reprinted from Bai et al. [47].)

survive, the surfaces of these plants must remain clean for optimal exposure to the sun. To maintain a clean surface, lotus leaves have an intrinsically hydrophobic, hierarchically textured micro- and nano-scale features that create a SHS (Figure 12.5). Water droplets are highly mobile on the leaf surface; as they slide off the surface, they also pick up contaminants via capillary effects, keeping the plant clean and maximizing sunlight exposure.

SHSs are able to achieve high droplet mobility when the foreign liquids are in the Cassie state. As a result, SHSs have shown remarkable self-cleaning and drag reduction abilities (Figure 12.7) [5,48–55]. When properly designed, these SHSs can be used in condensation-heat-transfer applications where tiny condensates can be rapidly removed by the surfaces through a jumping-droplet mechanism [56,57]. Researchers have also shown that SHSs can be designed for applications ranging from water–oil separation [58] to preventing ice formation under certain environmental conditions [59,60]. In some cases, SHSs have shown compatibility with biological fluids such as blood [53]. These surfaces have also been utilized as a fluid and analyte concentrator for ultra-sensitive biomolecular detection using surface enhanced Raman scattering [51].

However, droplets may transition from the Cassie state to the Wenzel state at elevated temperatures or pressures, or in the presence of contamination [61–63]. Once in the

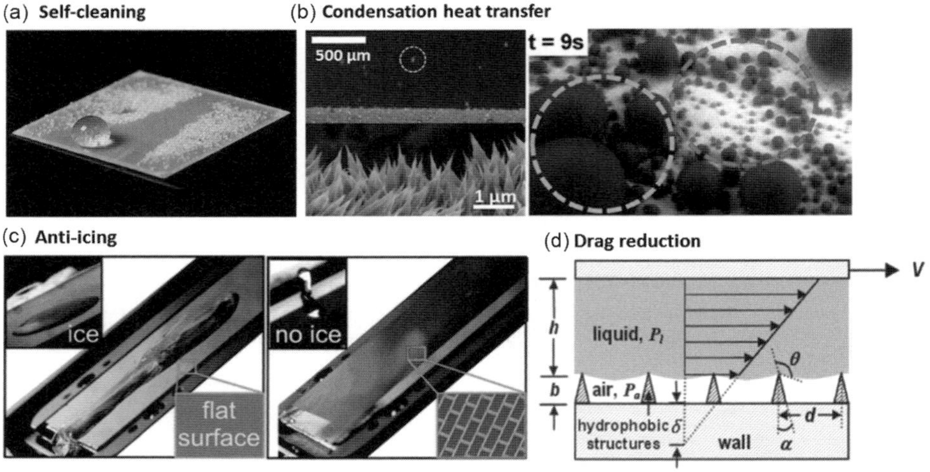

Figure 12.7 Applications of SHSs. (a) An optical image showing the self-cleaning property of a superhydrophobic surface. (b) (Left, bottom) The SEM image of a nanostructured superhydrophobic surface and (left, top) an optical image showing the jumping-droplet from this surface; (right) an environmental SEM image of condensation occurring on the nanostructured surface. (c) Ice formation on a control surface (left) and an ice-free superhydrophobic surface (right). (d) A superhydrophobic surface in Couette flow. (Panel (b) reprinted from Miljkovic et al. [57]. Panel (c) reprinted from Mishchenko et al. [60]. Panel (d) reprinted from Choi and Kim [48].)

Wenzel state, a droplet on a conventional solid-based superhydrophobic surface will likely become immobile on the surface. In certain practical applications, these conventional superhydrophobic surfaces may also become contaminated, as micro/nanoscale biological entities can adhere to the micro- to nano-scale features of the materials [63]. Furthermore, conventional superhydrophobic surfaces can lose their hydrophobicity when damaged. The durability of superhydrophobic surfaces still limits their use in practical applications; enhancing durability of these surfaces is a topic of existing research [64–66].

12.3.3 Springtail

Springtails are small, millimeter scale animals that breathe through their skin. These animals tend to reside in damp areas, where they are prone to water wetting their skin; to avoid suffocation, they must ensure clean, liquid-repellent skin [7]. To accomplish this, they have developed intricate surface structures that ensure water remains in the Cassie state and also allows for anti-fouling properties. Moreover, their structures are serif T-shaped, which allows for increased pressure stability (Figure 12.5).

The robust anti-fouling property of springtails have provided a strategy for researchers to design omniphobic synthetic materials. For a long time, the superhydrophobic surfaces mimicking lotus leaves could repel water, but were unable to repel low-surface-tension liquids. Engineered surfaces that mimic the serif T-shaped structures of springtails are capable of repelling liquids over a broad range of surface tensions (~10 mN/m $< \gamma_{LG} <$ 72.8 mN/m) [67,68]. The robust omniphobicity of these

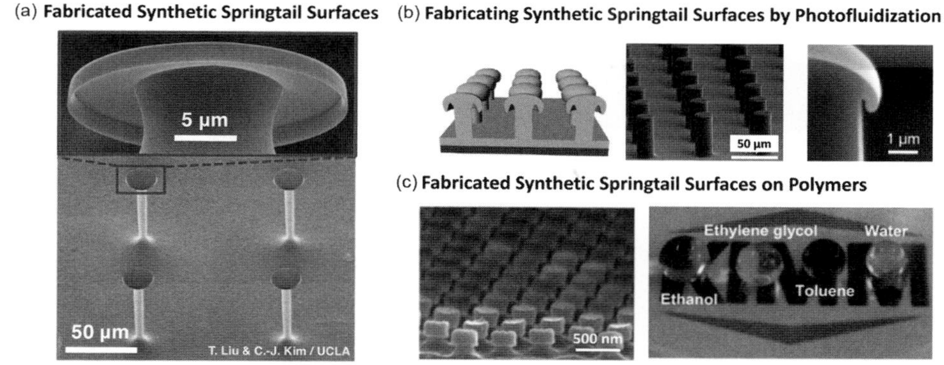

Figure 12.8 Omniphobicity on synthetic springtails surfaces. (a) Scanning electron micrographs (SEM) of doubly reentrant surfaces on a silicon wafer fabricated via reactive ion etching. (b) Fabricating synthetic springtail surfaces using localized photofluidization method. (c) (left) SEM images of synthetic springtail surfaces by nanotransfer molding and isotropic etching and (right) the optical image showing the omniphobicity and the transparency of this surface. ((a) courtesy of T. Leo Liu and C. J. Kim. (b) reprinted with permission from Choi et al. [69]. (c) reprinted with permission from Lee et al. [70].)

surfaces is attributed to the pinning effect at the edge of the T-shaped structures, also called hoodoos. In particular, the reentrant (double reentrant) structures (i.e., the serif, as opposed to sans-serif T shape) increase the number of pinning points, and thus increase the robustness of the surfaces in terms of liquid repellency. Several methods have been developed to fabricate hoodoo structures (Figure 12.8) out of various materials such as silicon and polymers [69,70].

12.3.4 Pitcher Plant

Pitcher plants are another well-known plant in the wetting community. These carnivorous plants get nutrition from ingesting insects and other organisms that may fall inside its cup-like digestive area [9]. The peristome of this plant is the rim around the pitcher cup. When dry, insects such as ants can walk on the surface without any trouble. When wet with rain or nectar, the peristome becomes slippery – insects can no longer maintain their footing and slide into the cup of the peristome. The peristome surface is textured and allows water and nectar to wick the surface, thus creating a slippery liquid interface (Figure 12.5) [9,71].

Inspired by the insect trapping mechanism of the slippery rim of the *Nepenthes* pitcher plant, synthetic slippery surfaces known as SLIPS were developed in the early 2010s [37,38]. Since then, a variety of slippery liquid-infused materials have been developed [37,72–76]. Owing to the highly mobile and smooth liquid interface, these liquid-infused surfaces outperform superhydrophobic liquid-repellent materials in terms of adhesion when in contact with various pure liquids and complex fluids; rapid and repeatable self-healing; and extreme pressure stability (up to ~676 atm) [37]. The development of the liquid-infused surfaces has opened up new scientific questions [39,77,78], as well as

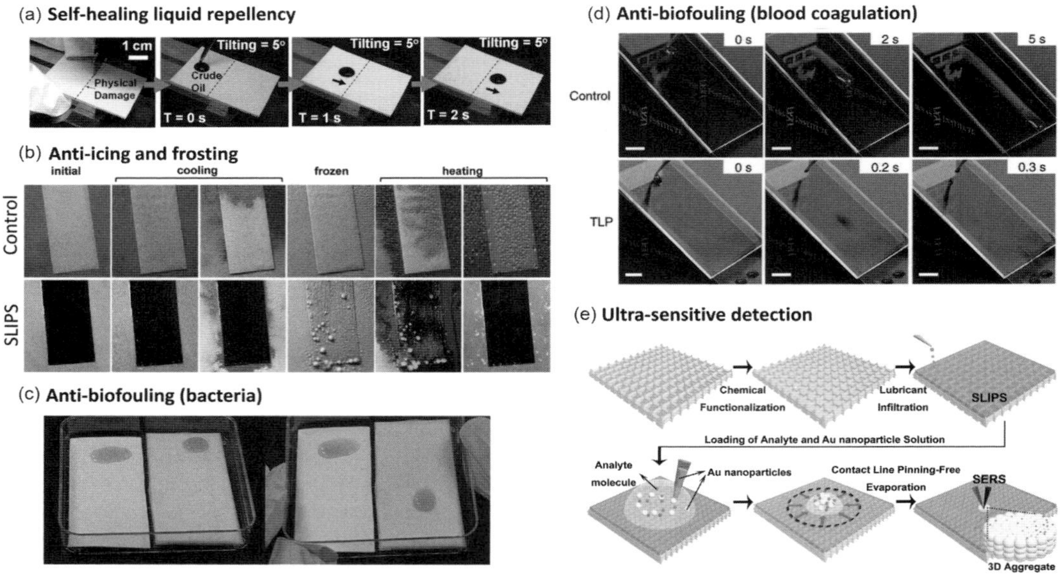

Figure 12.9 Applications of liquid-infused surfaces. (a) Self-healing liquid repellency on a SLIPS (from Wong et al. [37]). (b) Comparison of anti-icing and anti-frosting performance of (top) a bare aluminum surface and (bottom) a SLIPS (from Kim et al. [72]). (c) Comparison of anti-bacteria fouling performance on (left) a superhydrophobic surface and (right) a SLIPS (from Epstein [81]). (d) Comparison of blood repellency of a (top) glass surface and a (bottom) liquid-infused surface (from Leslie [76]). (e) Schematic showing the use of SLIPS for ultrasensitive molecular detection using surface enhanced Raman scattering (from Yang et al. [88]).

novel solutions to many challenging interfacial adhesion problems including icing and frosting [72,79,80], biofouling [74,76,81–83], fluid transport [53,84], condensation [85,86], fluid separation [87], and single molecule detection [88] (Figure 12.9).

12.3.5 Emerging Concepts: Cross-Species Bioinspired Surfaces

Many biomimetic/bioinspired studies have focused on mimicking/learning from individual biological mechanisms to create biomimetic/bioinspired materials. Recently, a number of studies have shown that one can incorporate the salient features of a number of different organisms found in nature within a single material to create cross-species materials [40,89–92] (Figure 12.10).

An example of cross-species materials is slippery rough surfaces (SRS) [89], which combine the roughness and increased surface area of lotus leaves and the locally smooth liquid interface of pitcher plants. SRS comprise a hierarchically micro- and nano-textured solid substrate that may be chemically treated to promote wetting of a liquid lubricant. The lubricant is applied such that it impregnates the nano-textures but not the microscale valleys, effectively yielding a conformally lubricated microtextured surface. In contrast to the conventional superhydrophobic surfaces, these

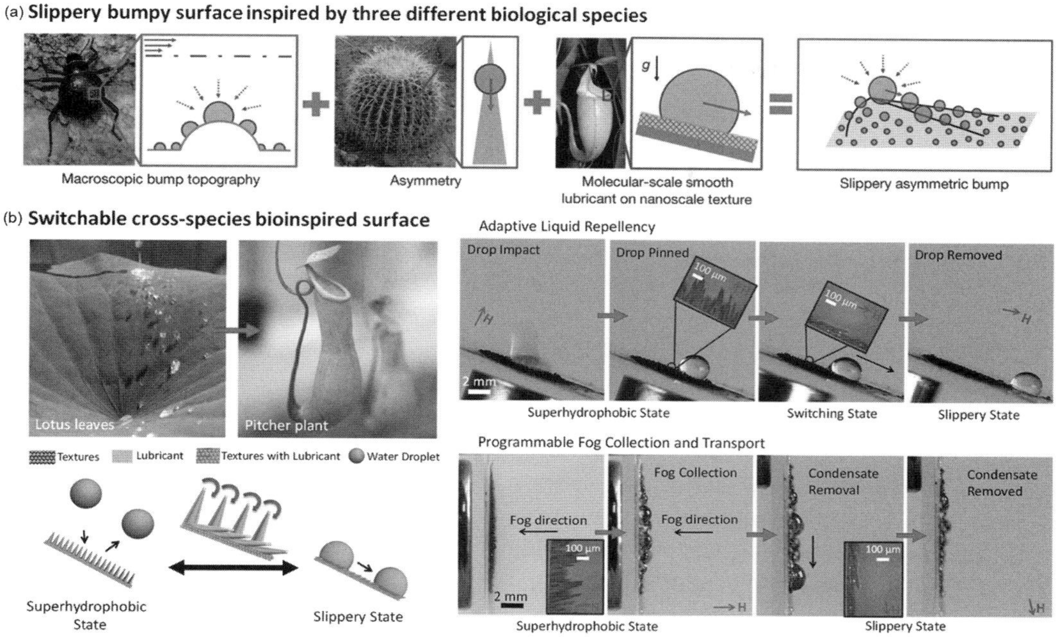

Figure 12.10 Examples of cross-species bioinspired materials. (a) A slippery bumpy surface inspired by the desert beetle, cactus, and pitcher plant (from Park et al. [90]). (b) A switchable surface inspired by the lotus leaf and pitcher plant (from Huang et al. [91]).

surfaces promote droplet mobility in both the Cassie and Wenzel state. Practically, their high surface area may be beneficial in fog-harvesting and condensation applications [40] – allowing for an increased number of nucleation sites. Moreover, droplets that nucleate and grow on textured surfaces will be in the Wenzel state, preventing superhydrophobic surfaces from being of practical use. Since Wenzel state droplets are slippery on SRS, SRS can passively overcome this pinning issue in water harvesting applications.

In addition to SRS, which combines the salient features from multiple species, smart surfaces have been developed that can transform their surface morphology to mimic liquid repellency behaviors of different natural surfaces [91,92]. For example, a switchable hydrophobic surface has been developed that combines the best of lotus leaf inspired SHS and pitcher plant inspired SLIPS (Figure 12.10). These switchable surfaces are able to switch actively from an SHS state to a SLIPS state. One way to accomplish this is to conformally lubricate flexible pillars. When the pillars are upright, the surface serves as a SHS and Cassie state droplets slide off the surface rapidly. If high pressure or temperature causes the droplet to enter the Wenzel state, the pillars can be bent such that they lay flat on the bulk of the substrate. Since these pillars are lubricated, the resulting surface when the pillars are bent is essentially a SLIPS surface. This allows pinned droplets to slide away. Once the pinned droplets are removed, the pillars can be raised back to the upright position for SHS effects.

12.4 Summary and Concluding Remarks

As we have seen here, surface chemistry and micro- to nano-scale surface features can lead to a number of fascinating macroscopic wetting properties. In particular, we saw how these small-scale properties lead to extraordinary wetting properties, such as superhydrophobicity or pressure-stable liquid repellency. Understanding the governing principles that dictate liquid wetting on biological surfaces allows us to develop generalizable design principles. These governing principles can then be harnessed to engineer materials with desired interactions with liquids. Liquid-repellent engineered materials, such as superhydrophobic surfaces and SLIPS, have relied on the design principles found in lotus leaves, pitcher plants, Namib Desert beetles, and more, and others have combined the design principles found in different species to create cross-species-inspired materials. These biologically inspired materials can lead to novel solutions to significant practical issues, such as biological fouling and icing, and in application areas such as condensation-heat transfer, water harvesting, and more.

Moving forward, there are a number of interesting fundamental and practical issues that remain in the field of wetting and liquid-repellent surfaces. Fundamentally, analytical prediction of contact angle hysteresis remains a challenge. The ability to predict the magnitude of contact angle hysteresis given the physical and chemical parameters of a surface would facilitate the design of surfaces with high liquid repellency. Practically, many engineered materials are not mechanically robust enough for real-life applications. In nature, biological surface textures can self-repair; the ability to translate this biological property to synthetic materials remains a challenge.

With an estimated > 8 million natural species [93], we have barely scratched the surface of the natural world. As a result, there remain many interesting phenomena to be discovered; translating these biological discoveries into practical applications will be an exciting endeavor for many years to come.

References

1. Young, T. (1805). An essay on the cohesion of fluids. *Philosophical Transactions of the Royal Society of London*, 95, 65–87.
2. Wenzel, R. N. (1936). Resistance of solid surfaces to wetting by water. *Industrial and Engineering Chemistry*, 28, 988–994.
3. Cassie, A. B. D., & Baxter, S. (1944). Wettability of porous surfaces. *Transactions of the Faraday Society*, 40, 546–550.
4. Cassie, A. B. D., & Baxter, S. (1945). Large contact angles of plant and animal surfaces. *Nature*, 155, 21–22.
5. Barthlott, W., & Neinhuis, C. (1997). Purity of the sacred lotus, or escape from contamination in biological surfaces. *Planta*, 202, 1–8.
6. Ensikat, H. J., Ditsche-Kuru, P., Neinhuis, C., & Barthlott, W. (2011). Superhydrophobicity in perfection: The outstanding properties of the lotus leaf. *Beilstein Journal of Nanotechnology*, 2, 152–161.

7. Helbig, R., Nickerl, J., Neinhuis, C., & Werner, C. (2011). Smart skin patterns protect springtails. *PLOS ONE*, 6, e25105.
8. Zheng, Y., Gao, X., & Jiang, L. (2007). Directional adhesion of superhydrophobic butterfly wings. *Soft Matter*, 3, 178–182.
9. Bohn, H. F., & Federle, W. (2004). Insect aquaplaning: Nepenthes pitcher plants capture prey with the peristome, a fully wettable water-lubricated anisotropic surface. *Proceedings of the National Academy of Sciences of the United States of America*, 101, 14138–14143.
10. Parker, A. R., & Lawrence, C. R. (2001). Water capture by a desert beetle. *Nature*, 414, 33–34.
11. Gao, X. F., & Jiang, L. (2004). Water-repellent legs of water striders. *Nature*, 432, 36.
12. Hu, D. L., Chan, B., & Bush, J. W. M. (2003). The hydrodynamics of water strider locomotion. *Nature*, 424, 663–666.
13. Seymour, R. S., & Hetz, S. K. (2011). The diving bell and the spider: The physical gill of *Argyroneta aquatica*. *Journal of Experimental Biology*, 214, 2175–2181.
14. Hansen, W. R., & Autumn, K. (2005). Evidence for self-cleaning in gecko setae. *Proceedings of the National Academy of Sciences of the United States of America*, 102, 385–389.
15. de Gennes, P.-G., Brochard-Wyart, F., & Quéré, D. (2004). *Capillarity and wetting phenomena: Drops, bubbles, pearls, waves.* New York: Springer.
16. Cassie, A. B. D. (1948). Contact angles. *Discussions of the Faraday Society*, 3, 11–16.
17. McHale, G. (2007). Cassie and Wenzel: Were they really so wrong? *Langmuir*, 23, 8200–8205.
18. Bico, J., Thiele, U., & Quéré, D. (2002). Wetting of textured surfaces. *Colloids and Surfaces A – Physicochemical and Engineering Aspects*, 206, 41–46.
19. Bico, J., Marzolin, C., & Quéré, D. (1999). Pearl drops. *Europhysics Letters*, 47, 220–226.
20. Marmur, A. (2009). Solid-surface characterization by wetting. *Annual Review of Materials Research*, 39, 473–489.
21. Gibbs, J. W. (1961). *The scientific papers of J. Willard Gibbs*, Ed. New Dover. New York: Dover Publications.
22. Amirfazli, A., & Neumann, A. W. (2004). Status of the three-phase line tension. *Advances in Colloid and Interface Science*, 110, 121–141.
23. Wong, T. S., & Ho, C. M. (2009). Dependence of macroscopic wetting on nanoscopic surface textures. *Langmuir*, 25, 12851–12854.
24. Zheng, Q. S., Lv, C. J., Hao, P. F., & Sheridan, J. (2010). Small is beautiful, and dry. *Science China – Physics Mechanics & Astronomy*, 53, 2245–2259.
25. Bormashenko, E. (2011). General equation describing wetting of rough surfaces. *Journal of Colloid and Interface Science*, 360, 317–319.
26. Gundersen, H., Leinaas, H. P., & Thaulow, C. (2017). Collembola cuticles and the three-phase line tension. *Beilstein Journal of Nanotechnology*, 8, 1714–1722.
27. Lafuma A., & Quéré, D. (2003). Superhydrophobic states. *Nature Materials*, 2, 457–460.
28. Quéré, D. (2008). Wetting and roughness. *Annual Review of Materials Research*, 38, 71–99.
29. Vogler, E. A. (1998). Structure and reactivity of water at biomaterial surfaces. *Advances in Colloid and Interface Science*, 74, 69–117.
30. Tian, Y., & Jiang, L. (2013). Intrinsically robust hydrophobicity. *Nature Materials*, 12, 291–292.
31. Nishino, T., Meguro, M., Nakamae, K., Matsushita, M., & Ueda, Y. (1999). The lowest surface free energy based on – CF_3 alignment. *Langmuir*, 15, 4321–4323.

32. Furmidge, C. G. (1962). Studies at phase Interfaces.I. Sliding of liquid drops on solid surfaces and a theory for spray retention. *Journal of Colloid Science*, 17, 309–324.
33. Shuttleworth, R., & Bailey, G. L. J. (1948). The spreading of a liquid over a rough solid. *Discussions of the Faraday Society*, 3, 16–22.
34. Johnson, R. E., & Dettre, R. H. (1964). Contact angle hysteresis.III. Study of an idealized heterogeneous surface. *Journal of Physical Chemistry*, 68, 1744–1750.
35. Oliver, J. F., Huh, C., & Mason, S. G. (1977). Resistance to spreading of liquids by sharp edges. *Journal of Colloid and Interface Science*, 59, 568–581.
36. Extrand, C. W., & Moon, S. I. (2008). Contact angles on spherical surfaces. *Langmuir*, 24, 9470–9473.
37. Wong, T. S., Kang, S. H., Tang, S. K. Y., et al. (2011). Bioinspired self-repairing slippery surfaces with pressure-stable omniphobicity. *Nature*, 477, 443–447.
38. Lafuma, A., & Quéré, D. (2011). Slippery pre-suffused surfaces. *Europhysics Letters*, 96, 56001.
39. Schellenberger, F., Xie, J., Encinas, N., et al. (2015). Direct observation of drops on slippery lubricant-infused surfaces. *Soft Matter*, 11, 7617–7626.
40. Dai, X. M., Sun, N., Nielsen, S. O., et al. (2018). Hydrophilic directional slippery rough surfaces for water harvesting. *Science Advances*, 4, eaaq0919.
41. Zhai, L., Berg, M. C., Cebeci, F. C., et al. (2006). Patterned superhydrophobic surfaces: Toward a synthetic mimic of the Namib Desert beetle. *Nano Letters*, 6, 1213–1217.
42. Garrod, R. P., Harris, L. G., Schofield, W. C. E., et al. (2007). Mimicking a stenocara beetle's back for microcondensation using plasmachemical patterned superhydrophobic-superhydrophilic surfaces. *Langmuir*, 23, 689–693,
43. Dorrer, C., & Rühe, J. (2008). Mimicking the stenocara beetle dewetting of drops from a patterned superhydrophobic surface. *Langmuir*, 24, 6154–6158.
44. Zhang, L., Wu, J., Hedhili, M. N., Yang, X., & Wang, P. (2015). Inkjet printing for direct micropatterning of a superhydrophobic surface: Toward biomimetic fog harvesting surfaces. *Journal of Materials Chemistry A*, 3, 2844–2852.
45. Yang, X., Song, J., Liu, J., Liu, X., & Jin, Z. (2017). A twice electrochemical-etching method to fabricate superhydrophobic-superhydrophilic patterns for biomimetic fog harvest. *Scientific Reports*, 7, 8816.
46. Kostal, E., Stroj, S., Kasemann, S., Matylitsky, V., & Domke, M. (2018). Fabrication of biomimetic fog-collecting superhydrophilic-superhydrophobic surface micropatterns using femtosecond lasers. *Langmuir*, 34, 2933–2941.
47. Bai, H., Wang, L., Ju, J., Sun, R. Z., Zheng, Y. M., & Jiang, L. (2014). Efficient water collection on integrative bioinspired surfaces with star-shaped wettability patterns. *Advanced Materials*, 26, 5025–5030.
48. Choi, C. H., & Kim, C. J. (2006). Large slip of aqueous liquid flow over a nanoengineered superhydrophobic surface. *Physical Review Letters*, 96, 066001.
49. Choi, C. H., Ulmanella, U., Kim, J., Ho, C. M., & Kim, C. J. (2006). Effective slip and friction reduction in nanograted superhydrophobic microchannels. *Physics of Fluids*, 18, 087105.
50. Lee, C., Choi, C. H., & Kim, C. J. (2008). Structured surfaces for a giant liquid slip. *Physical Review Letters*, 101, 064501.
51. Daniello, R. J., Waterhouse, N. E., & Rothstein, J. P. (2009). Drag reduction in turbulent flows over superhydrophobic surfaces. *Physics of Fluids*, 21, 085103.

52. Saranadhi, D., Chen, D. Y., Kleingartner, J. A., Srinivasan, S., Cohen, R. E., & McKinley, G. H. (2016). Sustained drag reduction in a turbulent flow using a low-temperature Leidenfrost surface. *Science Advances*, 2, e1600686.
53. Rosenberg, B. J., Van Buren, T., Fu, M. K., & Smits, A. J. (2016). Turbulent drag reduction over air- and liquid-impregnated surfaces. *Physics of Fluids*, 28, 015103.
54. Lee, C., & Kim, C. J. (2011). Underwater restoration and retention of gases on superhydrophobic surfaces for drag reduction. *Physical Review Letters*, 106, 014502.
55. Wisdom, K. M., Watson, J. A., Qu, X. P., Liu, F. J., Watson, G. S., & Chen, C. H. (2013). Self-cleaning of superhydrophobic surfaces by self-propelled jumping condensate. *Proceedings of the National Academy of Sciences of the United States of America*, 110, 7992–7997.
56. Boreyko, J. B., & Chen, C. H. (2009). Self-propelled dropwise condensate on superhydrophobic surfaces. *Physical Review Letters*, 103, 184501.
57. Miljkovic, N., Enright, R., Nam, Y., et al. (2013). Jumping-droplet-enhanced condensation on scalable superhydrophobic nanostructured surfaces. *Nano Letters*, 13, 179–187.
58. Feng, L., Zhang, Z. Y., Mai, Z. H., et al. (2004). A super-hydrophobic and super-oleophilic coating mesh film for the separation of oil and water. *Angewandte Chemie-International Edition*, 43, 2012–2014.
59. Cao, L. L., Jones, A. K., Sikka, V. K., Wu, J. Z., & Gao, D. (2009). Anti-icing superhydrophobic coatings. *Langmuir*, 25, 12444–12448.
60. Mishchenko, L., Hatton, B., Bahadur, V., Taylor, J. A., Krupenkin, T., & Aizenberg, J. (2010). Design of ice-free nanostructured surfaces based on repulsion of impacting water droplets. *ACS Nano*, 4, 7699–7707.
61. Liu, Y., Chen, X., & Xin, J. (2009). Can superhydrophobic surfaces repel hot water? *Journal of Materials Chemistry*, 19, 5602–5611.
62. Poetes, R., Holtzmann, K., Franze, K., & Steiner, U. (2010). Metastable underwater superhydrophobicity. *Physical Review Letters*, 105, 166104.
63. Hochbaum, A. I., & Aizenberg, J. (2010). Bacteria pattern spontaneously on periodic nanostructure arrays. *Nano Letters*, 10, 3717–3721.
64. Wong, T. S., Sun, T. L., Feng, L., & Aizenberg, J. (2013). Interfacial materials with special wettability. *MRS Bulletin*, 38, 366–371.
65. Deng, X., Mammen, L., Butt, H. J., & Vollmer, D. (2012). Candle soot as a template for a transparent robust superamphiphobic coating. *Science*, 335, 67–70.
66. Tian, X. L., Verho, T., & Ras, R. H. A. (2016). Moving superhydrophobic surfaces toward real-world applications. *Science*, 352, 142–143.
67. Tuteja, A., Choi, W., Ma, M. L., et al. (2007). Designing superoleophobic surfaces. *Science*, 318, 1618–1622.
68. Liu, T. Y., & Kim, C. J. (2014). Turning a surface superrepellent even to completely wetting liquids. *Science*, 346, 1096–1100.
69. Choi, J., Jo, W., Lee, S. Y., Jung, Y. S., Kim, S.-H., & Kim, H.-T. (2017). Flexible and robust superomniphobic surfaces created by localized photofluidization of azopolymer pillars. *ACS Nano*, 11, 7821–7828
70. Lee, S. E., Kim, H.-J., Lee, S.-H., & Choi, D.-G. (2013). Superamphiphobic surface by nanotransfer molding and isotropic etching. *Langmuir*, 29, 8070–8075.
71. Chen, H. W., Zhang, P. F., Zhang, L. W., et al. (2016). Continuous directional water transport on the peristome surface of Nepenthes alata. *Nature*, 532, 85–89.

72. Kim, P., Wong, T. S., Alvarenga, J. Kreder, M. J., Adorno-Martinez, W. E., & Aizenberg, J. (2012). Liquid-infused nanostructured surfaces with extreme anti-ice and anti-frost performance. *ACS Nano*, 6, 6569–6577.
73. Yao, X., Hu, Y. H., Grinthal, A., Wong, T. S., Mahadevan, L., & Aizenberg, J. (2013). Adaptive fluid-infused porous films with tunable transparency and wettability. *Nature Materials*, 12, 529–534,
74. Vogel,N., Belisle, R. A., Hatton, B., Wong, T. S., & Aizenberg, J. (2013). Transparency and damage tolerance of patternable omniphobic lubricated surfaces based on inverse colloidal monolayers. *Nature Communications*, 4, 1–10.
75. MacCallum, N., Howell, C., Kim, P., et al. (2015). Liquid-infused silicone as a biofouling-free medical material. *ACS Biomaterials Science & Engineering*, 1, 43–51.
76. Leslie, D. C., Waterhouse, A., Berthet, J. B., et al. (2014). A bioinspired omniphobic surface coating on medical devices prevents thrombosis and biofouling. *Nature Biotechnology*, 32, 1134–1140.
77. Smith, J. D., Dhiman, R., Anand, S., et al. (2013). Droplet mobility on lubricant-impregnated surfaces. *Soft Matter*, 9, 1772–1780.
78. Daniel, D., Timonen, J. V. I., Li, R. P., Velling, S. J., & Aizenberg, J. (2017). Oleoplaning droplets on lubricated surfaces. *Nature Physics*, 13, 1020–1025.
79. Irajizad, P., Hasnain, M. Farokhnia, N. Sajadi, S. M., & Ghasemi, H. (2016). Magnetic slippery extreme icephobic surfaces. *Nature Communications*, 7, 13395.
80. Golovin, K., Kobaku, S. P. R., Lee, D. H., DiLoreto, E. T. Mabry, J. M., & Tuteja, A. (2016). Designing durable icephobic surfaces. *Science Advances*, 2, e1501496.
81. Epstein, A. K., Wong, T. S., Belisle, R. A., Boggs, E. M., & Aizenberg, J. (2012). Liquid-infused structured surfaces with exceptional anti-biofouling performance. *Proceedings of the National Academy of Sciences of the United States of America*, 109, 13182–13187.
82. Amini, S., Kolle, S., Petrone, L., et al. (2017). Preventing mussel adhesion using lubricant-infused materials. *Science*, 357, 668–673.
83. Tesler, A. B., Kim, P., Kolle, S., Howell, C., Ahanotu, O., & Aizenberg, J. (2015). Extremely durable biofouling-resistant metallic surfaces based on electrodeposited nanoporous tungstite films on steel. *Nature Communications*, 6, 1–10.
84. Van Buren, T., & Smits, A. J. (2017). Substantial drag reduction in turbulent flow using liquid-infused surfaces. *Journal of Fluid Mechanics*, 827, 448–456.
85. Anand, S. Paxson, A. T., Dhiman, R., Smith, J. D., & Varanasi, K. K. (2012). Enhanced condensation on lubricant-impregnated nanotextured surfaces. *ACS Nano*, 6, 10122–10129.
86. Xiao, R., Miljkovic, N., Enright, R., & Wang, E. N. (2013). Immersion Condensation on oil-infused heterogeneous surfaces for enhanced heat transfer. *Scientific Reports*, 3, 1988.
87. Hou, X., Hu, Y. H., Grinthal, A., Khan, M., & Aizenberg, J. (2015). Liquid-based gating mechanism with tunable multiphase selectivity and antifouling behaviour. *Nature*, 519, 70–73.
88. Yang, S. K., Dai, X. M., Stogin, B. B., & Wong, T. S. (2016). Ultrasensitive surface-enhanced Raman scattering detection in common fluids. *Proceedings of the National Academy of Sciences of the United States of America*, 113, 268–273.
89. Dai, X. M., Stogin, B. B., Yang, S. K., & Wong, T. S. (2015). Slippery Wenzel State. *ACS Nano*, 9, 9260–9267.
90. Park, K.-C., Kim, P., Grinthal, A., et al. (2016). Condensation on slippery asymmetric bumps. *Nature*, 531, 78.

91. Huang, Y., Stogin, B. B., Sun, N., Wang, J., Yang, S. K., & Wong, T. S. (2017). A switchable cross-species liquid repellent surface. *Advanced Materials*, 29, 1604641.
92. Cheng, Z., Zhang, D., Lv, T., et al. (2018). Superhydrophobic shape memory polymer arrays with switchable isotropic/anisotropic wetting. *Advanced Functional Materials*, 28, 1705002.
93. Mora, C., Tittensor, D. P., Adl, S., Simpson, A. G., & Worm, B. (2011). How many species are there on Earth and in the ocean? *PLoS Biology*, 9, e1001127.

13 Biomimetic and Soft Robotics

Materials, Mechanics, and Mechanisms

Wanliang Shan and Yantao Shen

13.1 Introduction

Since the advent of the first programmable robotic arm in the early 1960s by George C. Devol, the robotics industry has seen fast growth, and nowadays robotic arms are ubiquitous in automobile assembly lines. In addition to those fixed to the ground as in the robotic arm case, autonomous mobile robots have also been designed and manufactured, and have found ample applications in many areas such as space and deep-sea exploration, thanks to synergistic progress in control, actuation, and information technology, among others. These robots are featured with high accuracy for force and position control. They are ideal for repetitive tasks that quickly bore humans. They make many fewer mistakes. Their bodies are made of hard materials, such as metals and hard plastics, while their control and actuation units use metal or semiconductors such as silicon for electronics.

Unfortunately, it is still difficult for these rigid robots to find applications where interactions with human beings or complicated environments are involved. In the case of interactions with human beings, being accurate in displacement control could be dangerous. For example, a human operator might be injured or killed if placed next to a robotic arm that is programed to do repetitive tasks of picking and releasing objects. Because of this, robotic arms are still mostly caged in factories to avoid contact with human operators. Similarly, a rigid autonomous rescue robot might get stuck in the relics of a collapsed building after a catastrophic earthquake, thus failing in its targeted tasks of searching and rescuing potential survivors.

When we look around in nature, however, we find all kinds of soft living machines: boneless caterpillars crawling and rolling off leaves; geckos with dry feet climbing up vertical walls or even beneath overhangs; human fingers with dexterous manipulation of fragile objects. Instead of using hard and stiff materials, these soft, living machines primarily use soft materials and compliant structures such as muscle, skin, and hair. While rigid materials can constrain natural human motion and cause injury, soft materials preserve natural mechanics and reduce stress concentrations during impact. It is this "softness" that enables them to fulfill tasks that are still unachievable for conventional robots.

Because of this, in recent years, there has been surging interest in biomimetic and soft materials design, as well as biomimetic and soft robotics design. Biologists, materials scientists, mechanical engineers, and roboticists have worked together to tackle these grand challenges. They first investigated the mechanisms and principles underlying the high performance and multifunctionality of soft living bodies, and then applied them to the design of high-performance multifunctional composites, and the design of biomimetic and soft robotics. In some cases, nature is still very hard to copy using existing engineering tools, while in a few other cases human kind has come up with better designs. The field of biomimetic and soft robotics is still very new, with rich research opportunities for the aforementioned multiple relevant fields. Industrial development is even lagging, which once again means ample opportunities for entrepreneurs interested in this area.

This review describes recent relevant progress made in materials engineering and solid mechanics, as well as some recent advances in biomimetic and soft robotics. Specifically, we first examine the stiffness-tuning mechanisms in nature, go through recent attempts to mimic this by designing multifunctional composites with tunable rigidity, and review existing robotic systems that utilize these rigidity-tunable composites as critical functional components, including a multifingered soft gripper and a soft gripper enabled by dynamically tunable dry adhesion. We then introduce a new morphing mechanism for biomimetic and soft robots called Rigid Elements-based Shape-Morphing Mechanism (RESMM). The mechanism is applied to produce a novel segmental muscle-mimetic design unit that efficiently mimics the earthworm's segmental muscle contractions and extensions. Meanwhile, the segmental muscle structure of the natural earthworm is reviewed to show the biomimetic features of the segmental muscle-mimetic design unit. Connected numbers of the design units with the transmission mechanism, a multisegment earthworm-like soft robot, is prototyped and its peristaltic locomotion, by controlling motors in each segment with different phase-shift angles, is analyzed and experimentally demonstrated. In addition, the relationship between phase shift and average velocity index under different numbers of segments is explored. Based on a three-segment soft robot model, the relationship between different phase-shift angles and average velocity under five contact-surface conditions is also investigated.

13.2 Stiffness Tunable Materials and Relevant Soft Robotics

In natural organisms, reversible mechanical-rigidity tuning allows soft limbs and members to support load and perform motor tasks. Organisms typically accomplish this with striated muscle tissue [1,2], hydrostatic skeletons [3], or catch-connective tissue [4,5]. These natural materials and structures are lightweight and require fractions of a second to change their elastic rigidity by an order of magnitude. These properties are necessary for an organism to perform mechanical work, support heavy loads, and control internal-stress distributions while still soft and mechanically deformable in their passive state.

Mechanical-rigidity tuning also has potential importance in bioinspired robotics, wearable technologies, and other emerging domains of engineering. Previous attempts have involved external heating of shape-memory polymers (SMP) for aerospace structures [6–8], hydraulic actuators for seismic response control [9,10], embedded heating of multilayered SMP composites [11,12], and electrostatic clamping of multi-layered beam for vibration control [13,14]. In contrast to natural organisms, these structures typically contain permanently rigid materials and/or are dependent on bulky external hardware.

In this section, recent studies in the area of soft multifunctional materials with reversibly tunable rigidity potentially for soft robotics applications are presented. Our studies are based on the onboard heating mechanism. We first fabricate composites containing low-melting-point alloys that can achieve rigidity tuning up to four orders of magnitude, activated within tens of seconds. We then explore the potential of such a design by conducting thermal analysis of a simplified multilayered composite model. Based on results from thermal analysis, we recognize the room for improvement in terms of activation time of the design, and then come up with a novel biomimetic design. This new design enables activation on the order of a few seconds. We achieve this through engineering composites containing conductive polymers with a low glass-transition temperature. The implications of these findings, and other studies contributing to artificial muscle design and soft robotics implementation, are also discussed.

13.3 Design Approaches for Soft Multifunctional Composites with Reversibly Tunable Rigidity

13.3.1 Soft Matter Multilayered Composites Embedded with Low-Melting-Point Alloys

The first design approach to soft multifunctional composites with reversibly tunable rigidity is an extension to the previous multilayered composites [11,12]. However, instead of using only one thermally active layer to tune the coupling with adjacent rigid layers, we use only soft or rigidity-tuning materials. The composite is activated with an onboard Joule heater composed of an acrylic elastomer (VHB^{TM} tape; 3M) embedded with serpentine channels filled with liquid-phase eutectic gallium indium (EGaIn) or gallium-indium-tin (Galinstan) metal alloy (Figure 13.1).

Details of the fabrication process are presented in Shan et al. [16]. We begin by producing the elastic Joule heating element using the steps shown in Figure 13.1a. A 30 W CO_2 laser engraver (VLS 3.50; Universal Laser Systems) is used to pattern the heater geometry in a sheet of VHB tape and liner film. After removing the excess tape and liner, we bond the patterned sheet to a second layer of VHB tape. Next, we deposit liquid Galinstan and remove the liner mask, leaving liquid alloy only in the exposed surface features. Last, we attach copper shim wires and seal the liquid by bonding a third layer of VHB tape. As shown in Figure 13.1b, this method allows a Galinstan heater to be embedded in a thin sheet of VHB elastomer. Since the alloy is liquid at room temperature, the heater can be stretched to several times its natural length without losing electrical functionality.

As shown in Figure 13.2, the composite contains both the stretchable Joule heating element and a thermally responsive rigidity-tuning layer. It can be represented as a

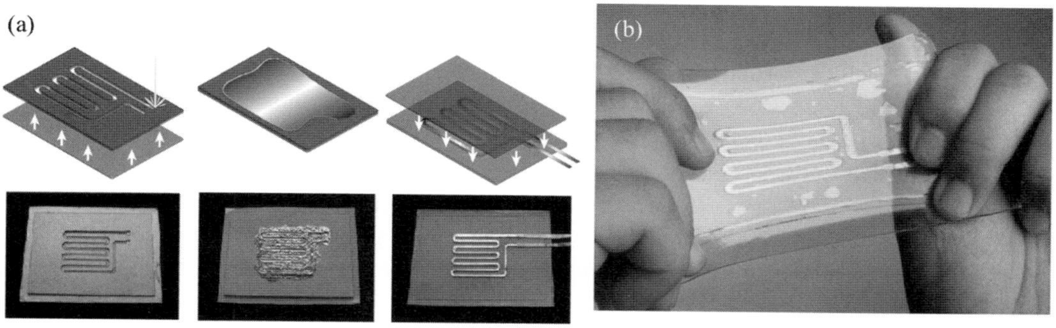

Figure 13.1 (a) Fabrication of a soft heater Galinstan heating element embedded in an acrylic VHB elastomer: (left) VHB film and liner are patterned with a laser engraver and bonded to a VHB substrate; (middle) sample is coated with Galinstan, which wets the exposed portions of the substrate; (right) the liner mask is removed and the patterned surface is sealed with a VHB film. (b) Liquid-phase Galinstan heater embedded in VHB elastomer. (Shan et al. [16].)

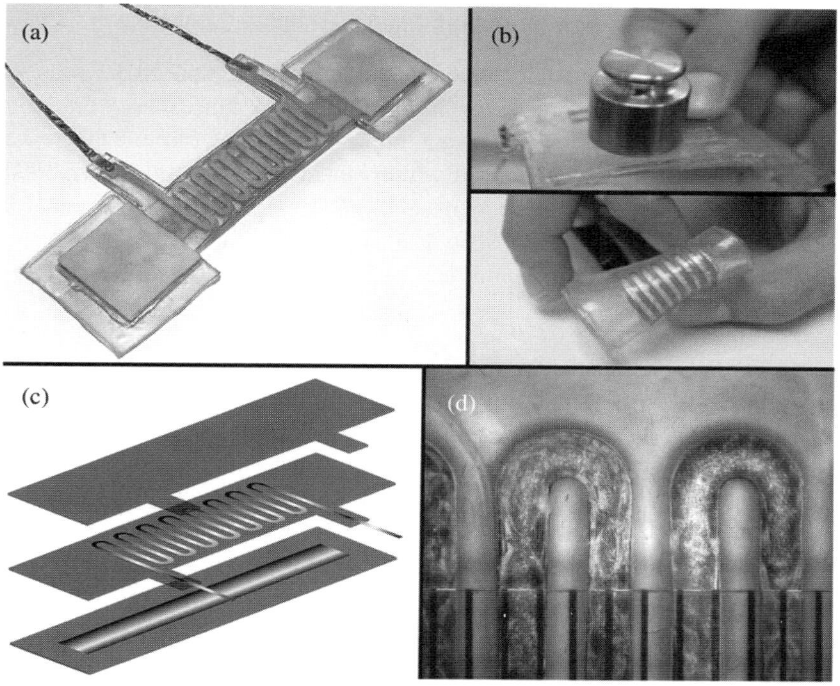

Figure 13.2 (a) Stiffness tunable composite embedded with Field's metal and a liquid Galinstan heater. (b) When electrically activated, the composite softens and easily deforms. (c) Illustration of the composite, which contains a (top) elastomer sealing layer, (middle) liquid-phase Joule heating element, and (bottom) thermally activated layer of Field's metal or SMP. (d) Close-up of the Galinstan heating element; ruler marks spaced 1 mm apart. (Shan et al. [16].)

five-layer structure, including the insulating layer between the top and bottom sealing layers of VHB acrylic elastomer. The thermal responsive layer can be fabricated using the same fabrication methods as the soft heater. Instead of a liquid metal alloy at room temperature, we use Field's metal (RotoMetals, Inc.), which is a low-melting-point alloy (melting point of 62° C). In this way, the fabrication steps described earlier can be simply carried out on a hot plate. At room temperature, the rigidity-tuning composite is rigid and can support external tensile or flexural load. When electrically activated, the Field's metal is melted by the heat generated from the onboard heating element and composite softens. As shown in Figure 13.2b, the composite bends under a small force.

Using tensile testing, we characterize the rigidity change of the composite to reversibly change between approximately 1 GPa and 100 kPa. Based on tests with several composite samples, we measured an activation time on the order of tens of seconds. If we replace the Field's metal layer with an SMP layer, the effective rigidity change is less and the activation time does not significantly change. Details of the results are presented in Shan et al. [16].

13.3.2 Thermal Analysis of Multilayered Rigidity-Tunable Composite

Although promising, the composite presented in Section 13.3.1 represents an early prototype with an ad hoc selection of materials and dimensions. Further improvements of power consumption and activation time would benefit from a theoretical model that can inform design decisions. We address this with a simplified 1D model presented in Figure 13.3 and described in Shan et al. [17]. Using this simplified model, we conduct a thermal analysis using both analytical and numerical techniques.

The governing equation for the activation process is the following:

$$\begin{aligned}
&\kappa(\chi)\frac{\partial^2\theta}{\partial\chi^2} + g(\chi,t) = \rho(\chi)C_{\mathrm{p}}(\chi)\frac{\partial\theta}{\partial t}, \quad (\chi \in [0,L]) \\
&\theta_{\mathrm{inter}^-} = \theta_{\mathrm{inter}^+}, \\
&k_{\mathrm{inter}^-}\frac{\partial\theta}{\partial\chi}\Big|_{\mathrm{inter}^-} = k_{\mathrm{inter}^+}\frac{\partial\theta}{\partial\chi}\Big|_{\mathrm{inter}^+}, \\
&\left(\frac{\partial\theta}{\partial\chi} - h_{\mathrm{air}}\theta\right)\Big|_{\mathrm{x}=0} = 0, \\
&\left(-\frac{\partial\theta}{\partial\chi} - h_{\mathrm{air}}\theta\right)\Big|_{\mathrm{x}=\mathrm{L}} = 0, \\
&\theta(\chi,0) = 0.
\end{aligned} \tag{13.1}$$

Here, $\theta(x, t)$ is the relative temperature with respect to room temperature, $k(x)$ is the thermal conductivity, $\rho(x)$ is the density, and $C_p(x)$ is the specific heat. Since the layers are of different materials, the physical properties are only piecewise continuous. Eq. (13.1) shows (i) the heat conduction equation (with nonzero generation $g(x, t)$) followed by (ii) temperature, and (iii) heat flux continuity condition on the internal

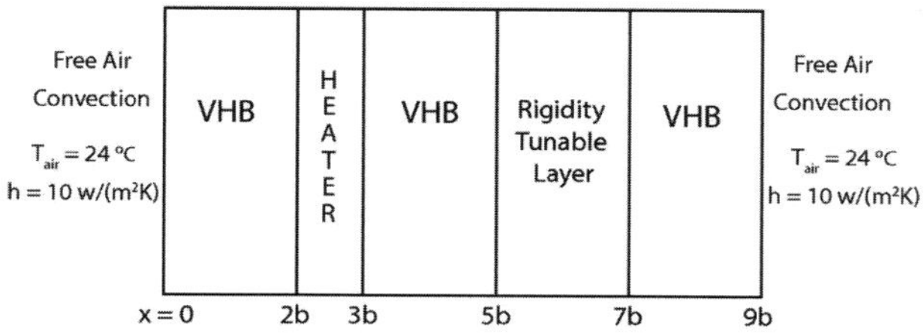

Figure 13.3 1D model of the rigidity-tunable composite for a transient thermal analysis of the activation process. (Reprinted from Shan et al. [17] with permission from Elsevier. © Elsevier 2013.)

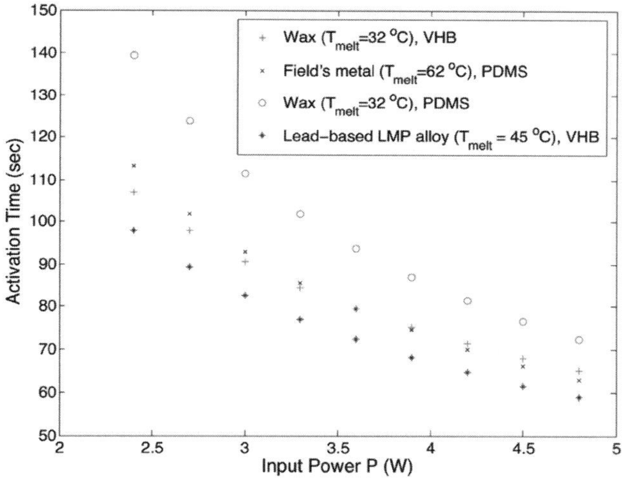

Figure 13.4 Predicted activation time of rigidity tunable composite using various thermally responsive materials and elastomers depending on power input. (Reprinted from Shan et al. [17] with permission from Elsevier. © Elsevier 2013.)

interfaces, the convective boundary conditions on the (iv) top and (v) bottom surfaces, and (vi) initial condition for the temperature field.

The analytical and numerical results agree well with each other and are also consistent with the results of tabletop experimental measurements [17]. When we consider the use of other thermally responsive materials and embedding elastomers, the modeling results show that the activation time will still be in the order of tens of seconds (Figure 13.4). The main reason for this can be attributed to the additional insulating layer between where the heat is generated and absorbed. This insulating layer is needed for containing the liquid metal alloy in the soft heater but creates a barrier to thermal diffusion. This suggests that a better design based on different mechanisms for thermal activation may be required in order to achieve more rapid rigidity tuning.

13.3.3 Soft Matter Composites Containing Conductive Elastomer with Low Glass-Transition Temperature

Recognizing the activation time limitation of the previous approach for the design of multifunctional materials with reversibly tunable rigidity, we turn to a better design addressing directly the root cause for the slow activation: the separation of where heat is generated from where heat is used for melting or glass transition. Toward this end, we combine these two processes into one component of the composite by engineering new elastomers that are conductive and exhibit a low glass-transition temperature. The elimination of liquid metal alloys avoids sealing issues, in addition to extra heat dissipation and diffusion length to cross.

Figure 13.5 shows the new rigidity-tunable composite containing a conductive propylene-based elastomer (cPBE). The cPBE is produced by combining a propylene–ethylene co-polymer with a percolating network of structured carbon black. It has a weight composition of 51/9/40 percent propylene, ethylene, and structured carbon black. Custom-ordered pellets are supplied by THEMIX Plastics, Inc (Lake Mills, WI) and pressed between steel plates at 90° C to form thin sheets. Flattened sheets of cPBE (step iii) are patterned with a CO_2 laser to form shapes with electrical

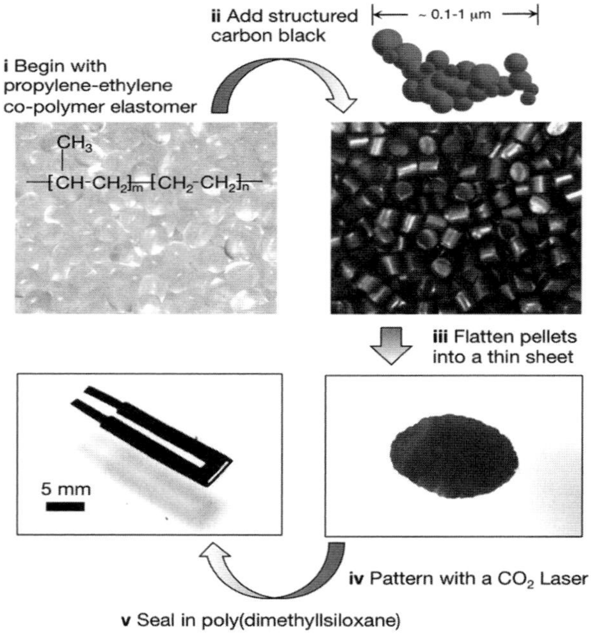

Figure 13.5 Fabrication of cPBE–PDMS composite: (i–ii) propylene–ethylene co-polymer (clear pellets) is mixed with structured carbon black to produce a conductive propylene-based elastomer (cPBE; black pellets); (iii–v). When flattened into thin sheets, the cPBE may be patterned with a CO_2 laser and embedded in PDMS. (Shan et al. [18]. © IOP Publishing. Reproduced with permission. All rights reserved.)

terminals to supply current (step iv). Finally, the cPBE strip is sealed within a polydimethylsiloxane (PDMS) matrix. With ~0.1–1 W of supplied electrical power, the effective elastic modulus of the composite reversibly changes between 1.5 and 37 MPa. This approximately corresponds to the difference in rigidity between human cartilage and skin. Thanks to its relatively low glass-transition temperature at 75° C, the activation time can be achieved within a few seconds [18].

13.3.4 Soft Robotics Enabled by Rigidity-Tunable Composites

In this subsection, we demonstrate some potential applications of these rigidity-tunable composites. As the first example, referring to Figure 13.6, three cPBE–PDMS strips are aligned on the sidewall of the finger symmetrically. When selectively activating one of the strips, the rigidity of the corresponding segment of the finger wall decreases and exhibits greater expansion. This results in the finger bending in the opposite direction. This clearly shows the potential of the integration of such materials into soft robotics and wearable devices applications.

We can keep building on this and design a multifingered soft gripper that is controlled by a single shared pneumatic control as shown in Figure 13.7 [19]. This soft gripper is composed of three individual soft fingers, each of which has structures as shown in Figure 13.6. It can pick up various objects of different shapes and weights, using different manipulation modes such as closing up, inverse bending, as well as twisting. Notably, twisting of 18° is achieved (Figure 13.7h), which has been rarely shown using soft grippers.

Another application of the rigidity-tunable composite in soft gripping is soft grippers enabled by dynamically tunable dry adhesion. Figure 13.8 shows the mechanisms of dynamically tunable dry adhesion enabled by the rigidity-tunable composites, as well as some prototype (Figure 13.8c). The underlying mechanism is that, for a composite

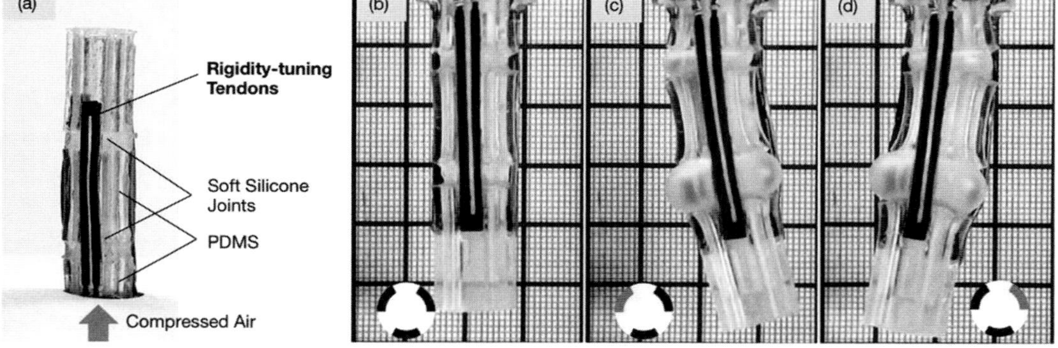

Figure 13.6 (a) Soft pneumatic robot finger integrated with cPBE "tendons" and a single air supply. (b) at rest, not inflated and not activated (c–d) the finger bends in different directions depending on which tendon element is activated to allow for stretch. (Shan et al. [18]. © IOP Publishing. Reproduced with permission. All rights reserved.)

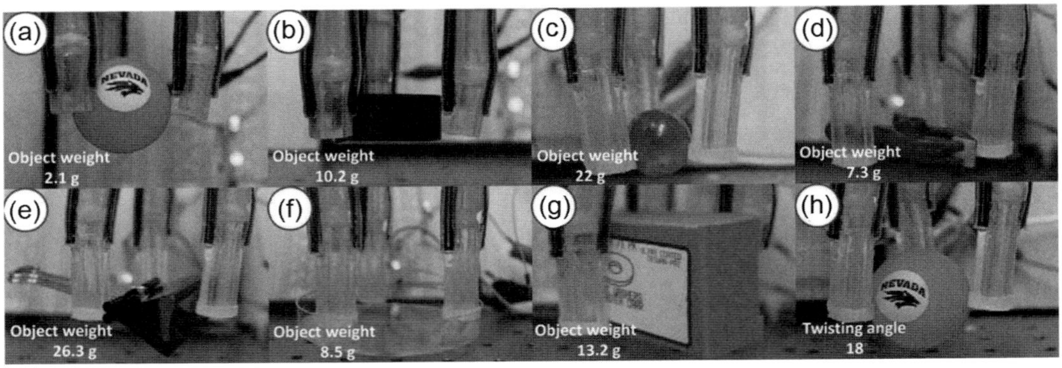

Figure 13.7 Grasping, releasing, and twisting manipulations by soft grippers. Objects manipulated: (a) a 2.1 g ping-pong ball (b) a 10.2 g tape roll; (c) a 22.0 g glass ball; (d) a 7.3 g plastic nylon spring clamp; (e) a 26.3 g paper clip; (f) an 8.5 g plastic Petri dish using the inverse bending mechanism; (g) a 13.2 g paper box; (h) Twisting of a 2.1 g ping-pong ball immediately after picking it up. (Mohammadi Nasab et al. [19]. © Mary Ann Liebert, Inc. Reproduced with permission. All rights reserved.)

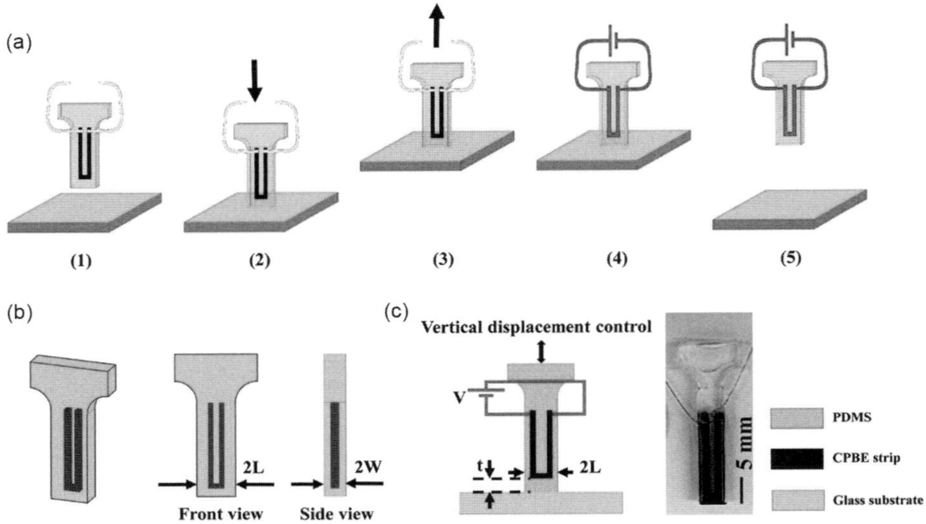

Figure 13.8 (a) Principle of operation of the tunable adhesive. The composite post can (1–3) pick up a part in the high adhesion (non-activated state), but (4) once activated the stiffness of the core is reduced and consequently the adhesion is reduced, and (5) the part is released. (b) Dimetric, front, and side views of the composite post; light gray is PDMS and dark gray is CPBE core. (c) Schematic of electrical activation and photograph of a composite post sample, consisting of PDMS embedded with a U-shaped CPBE core and wires for electrical connection. (Shan and Turner [20]. © WILEY-VCH Verlag GmbH & Co. Reproduced with permission. All rights reserved.)

core–shell soft-post structure, when the stiffness of the core is dynamically modulated so is the dry adhesion between the soft post and opposing surfaces. This is due to the fact that when the stiffness of the core is altered, the stress distribution at the interface will be altered too, which affects the interfacial crack initiation and propagation process. Using these prototypes, dry adhesion with a tuning range of 7× has been demonstrated [20].

13.3.5 Concluding Remarks and Perspectives

In this section, we present our recent efforts in the area of soft multifunctional materials with reversibly tunable rigidity. We focus on direct Joule heating by passing electrical current through an embedded conductive element. We first use low-melting-point alloys and shape-memory polymer as the thermal responsive materials. When bonded to a liquid-phase metal-alloy heater, we are able to achieve up to four orders of magnitude of reversible change in rigidity. While promising, these early prototypes require ~10–100 s for activation with ~6 W of supplied power. To reduce the activation time and power consumption, we perform a transient thermal analysis on a 1D model of the multilayered composite. Based on the results of the thermal analysis, we conclude that the inefficiency of the heat generation and diffusion fundamentally limits the performance of the rigidity-tunable composite. We then introduce an improved design composed of a conductive thermoplastic elastomer with a low glass-transition temperature. This composite exhibits a reduced activation time of ~2–5 s, which is an order of magnitude improvement compared with the first approach using low-melting-point alloys and shape-memory polymer. Still faster activation with a lower requirement on power supply is needed to make these applications adoptable for industry, which is an active research topic now.

13.4 Rigid-Elements-Based Shape-Morphing Mechanism and Enabled Biomimetic Soft Robots

The rigid-elements-based shape-morphing mechanism (RESMM) is inspired by classical foldable/deployable structures, called multijoint structures. Such structures have been made by the mechanical elements, making the structure have both stiffness and overall deformable flexibility to deform, which are more suitable for the structurally deformable robot design and its applications in complex human–robot interaction environments. Especially the multijoint foldable structures built with scissor-like elements are most widely used in engineering [21], which consist of pairs of rods connected by revolute joint with one degree-of-freedom (DOF), allowing for a compact and deployed configuration. In this foldable structure, when adding one or more elements, this system can change to two, three, or multiple dimensional system, which makes itself one of emerging application areas in general mechanical engineering, including robotics, toys manufacturing [22], and swimming pools [23]. For instance, Dai et al. introduced a group of metamorphic mechanisms that further led to several novel designs including a metamorphic robotic hand [24]. Kong et al. [25] presented a systematic approach to the design of parallel mechanisms with multiple operational modes using the structure.

In this section, enabled by the RESMM, our new class of earthworm-like soft robot is presented. The segmental muscle-mimetic design of this soft robot that relies on the RESMM is inspired by the earthworm longitudinal and circular-muscle movement segment. The segment can produce extension and contraction when it relaxes and contracts. These muscle movements are mimicked by mixing differently designed types of scissor elements together to form a controlled 3D deformable segmental muscle-mimetic structure. The driven/actuation system of the segmental muscle-mimetic structure is a combination of the actuation mechanism with the transmission mechanism in the robot system. We then built and tested the deformation of single segmental muscle-mimetic design unit, and further formed a three-segment earthworm-like soft robot using three muscle-mimetic structures [26]. Extensive locomotion simulations in different conditions were conducted in order to understand the peristaltic motion advantages of the soft robot with the new design in variety of terrains. In addition, experiments on the single muscle-mimetic structure and prototyped three-segment soft robot were performed and the results validate the effectiveness of our new class of biomimetic soft robot with the new design.

13.5 Design Approaches for Earthworm-Like Soft Robot Using Rigid-Elements-Based Shape-Morphing Mechanism

13.5.1 Inspiration from Segmental Muscle Structure of Natural Earthworm

An earthworm's body is formed of many, mostly identical, body segments. Each segment has its own discrete fluid-filled coeliac compartment. Coordinated contractions of circular and longitudinal muscles among adjacent segments generate muscular waves for locomotion. The longitudinal muscles push the worm forward and the circular muscles squeeze the worm's body inward. By working together, these different muscles move part of the worm forward. When this is happening, setae (bristles) on the underside of the worm extend or retract as segment diameter changes [27].

Figure 13.9 shows schematic diagrams of the hydraulic movement in a tubular earthworm, which demonstrates the extension and contraction condition of the one muscle segment [28,29]. A real earthworm has a volume of enclosed fluid, and contraction of circular-muscle fibers will lead to decrease in diameter and increase in pressure. Since there is no significant change in volume, the decrease in diameter will result in increase in length. Then, the decrease in length can happen because of contraction of the longitudinal-muscle fibers with re-elongating the circular-muscle fiber, giving rise to re-expand the diameter of segment. This process needs only a 25 percent deformation in diameter to generate a nearly 80 percent deformation in length [30].

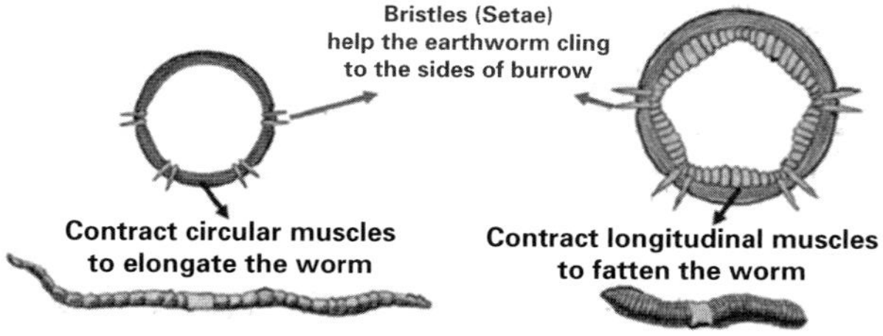

(a)

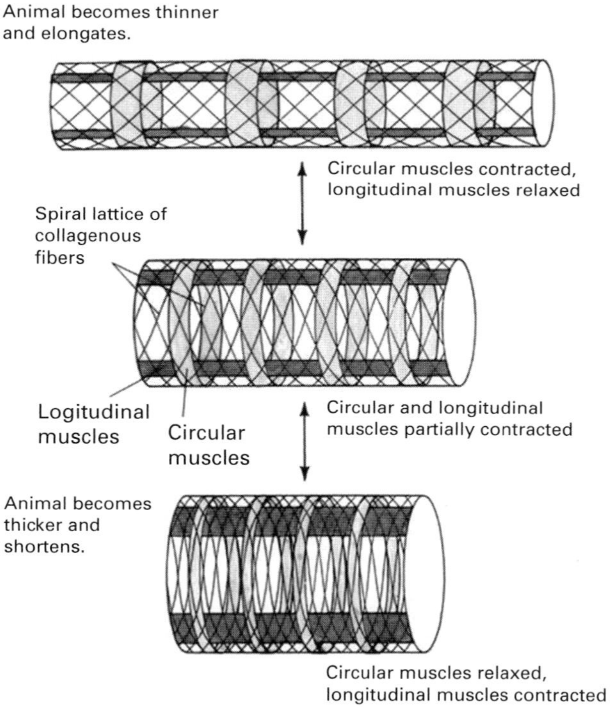

(b)

Figure 13.9 Schematic diagram of earthworm muscle deformation mechanism for locomotion: (a) natural earthworm and muscle deformation (adapted from McLaughlin [28]), and (b) simplified engineering mechanism of earthworm muscle deformation. (adapted from Lawrence et al. [29])

13.5.2 RESMM Enabled Segmental Muscle-Mimetic Structure Design

In order to use this highly effective structure of the earthworm organism, an enclosed 3D RESMM structure that is similar to the volume of enclosed fluid is designed to mimic the functions of the circular and longitudinal muscles. Figure 13.10a shows that contraction of the circular muscle causes decrease in diameter and extension in length.

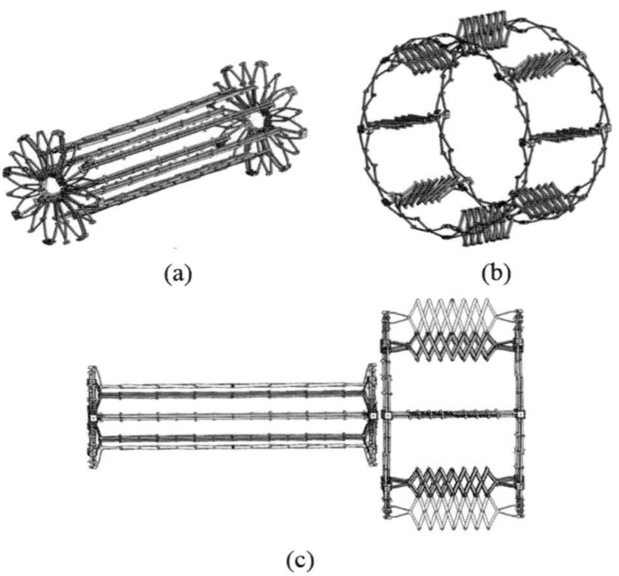

Figure 13.10 Enclosed 3D RESMM structure mimicking the circular and longitudinal muscles of the earthworm. (© 2016 IEEE. Reprinted, with permission, from Luo et al. [26].)

Figure 13.10b indicates that the contraction of longitudinal muscles leads to the increase in diameter and decrease in length. Figure 13.10c reveals the state of two adjacent segments with a certain phase. Note that, in the 3D structure, two classical scissor-element-based-morphing structures are employed. One represents the longitudinal muscle that is shown in Figure 13.11a. It demonstrates its extension and contraction states. The advantage of this linear RESMM structure lies in the consistency in direction of deformation and force transmission, which can permit effective and stable peristaltic locomotion. The second scissor elements linkage is shown in Figure 13.11b, and its off-center connection points form variable- curvature-based scissor linkage, and can represent the circular muscle.

To form an enclosed volume like earthworm organism, a special scissor adapter, called an inverse scissor, is designed to connect the basic longitudinal and circular structures shown in Figure 13.11 together. The adapter is the medium gray section in Figure 13.12.

13.5.3 Actuation and Transmission Mechanisms Design

To achieve the peristaltic locomotion of the robot, we design both actuation and transmission mechanisms. The actuation mechanism is an actuator adapter hooking up the servo with scissor elements structures, which can help perform controllable structural deformation. The transmission mechanism is to help form the multisegment robot by connecting adjacent segments. Its design ensures each segment can contract or extend in its own time sequence through independent control, even though all the

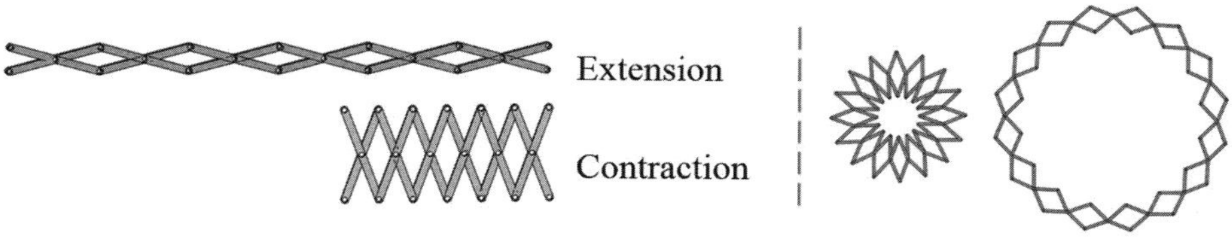

Figure 13.11 Scissor elements-based morphing structures mimicking movements of longitudinal muscle and circular muscle. (© 2016 IEEE. Reprinted, with permission, from Luo et al. [26].)

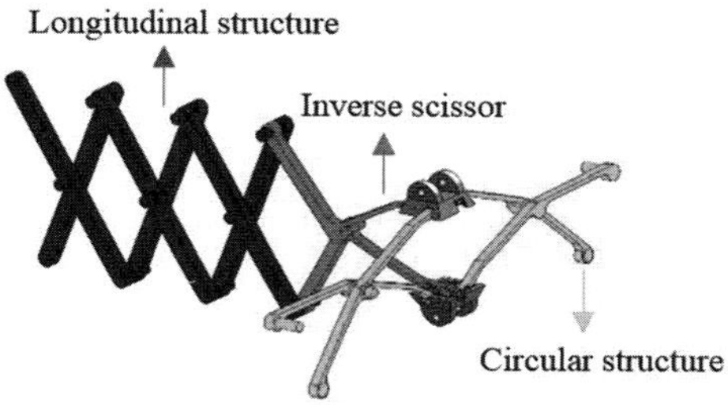

Figure 13.12 1/8 model of assembled structure consists of different types of scissor structures. (© 2016 IEEE. Reprinted, with permission, from Luo et al. [26].)

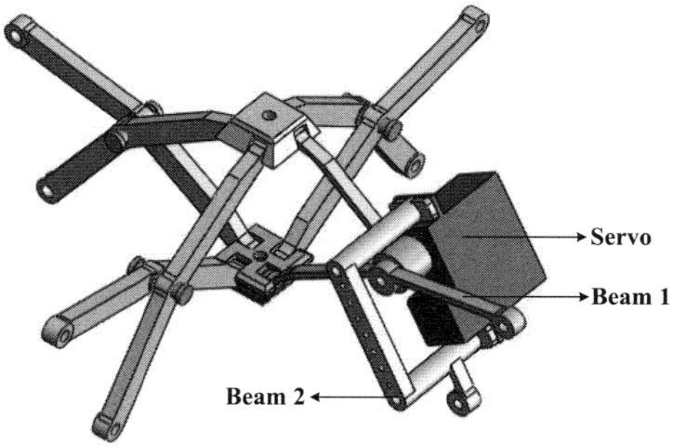

Figure 13.13 Actuation mechanism in the scissor elements structure. (© 2016 IEEE. Reprinted, with permission, from Luo et al. [26].)

segments are connected through this transmission unit. As shown in Figure 13.13, one servo motor is embedded in the scissor elements structure. Beams 1 and 2 not only act as the connecting scissors of the whole structure but also serve as the adapter of the servo motor, which makes the whole system much more compact. Multiple servo motors can be embedded to ensure smooth motion.

Figure 13.14 shows the comparison between measured and encoder output-angle results from servoing. In the figure, the darker line represents the angle recorded from the servo encoder, and the lighter line is the angle measured by the visual tracker. The

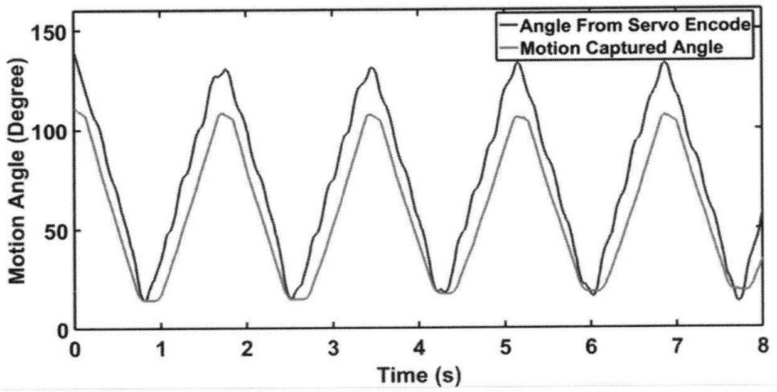

Figure 13.14 Comparison between encoder angle and measured angle during servoing. (© 2016 IEEE. Reprinted, with permission, from Luo et al. [26].)

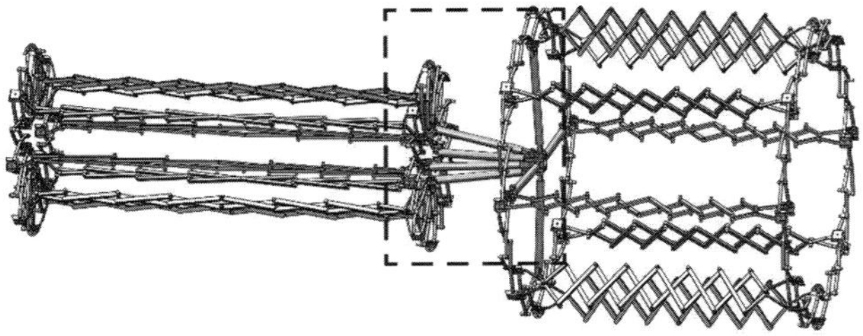

Figure 13.15 Transmission mechanism between two linked segments. (© 2016 IEEE. Reprinted, with permission, from Luo et al. [26].)

result shows two curves have good agreement, and the difference majorly comes from constrains and frictions of the scissor elements.

As shown in Figure 13.15, the transmission mechanism that is inspired by the umbrella frame structure is highlighted in the dashed rectangle. The mechanism can keep the linked adjacent segments concentric with each other as well as guaranteeing independent motion control of each segment.

13.5.4 Segment Deformation and Peristaltic Locomotion of the Soft Robot

Simulation is first conducted to test our design concept and the deformation performance of the segmental muscle-mimetic structure. The results help us to find the optimal peristaltic locomotion solutions of the earthworm-like soft robot under various conditions. The simulation is carried out in SOLIDWORKS Motion (2015 X64) powered by ADAMS.

In the first simulation, the input is servo angle, and the measured results include diameter of circular-muscle structure (DC), length of the diagonal of one scissor rhombus (DL), length of longitudinal-muscle structure (LL) and motion trajectory of leg points(A1~A7), which is shown in Figure 13.16. The side O is the fixed point during the simulation. By tracking the leg points (can serve as ineffective and/or effective

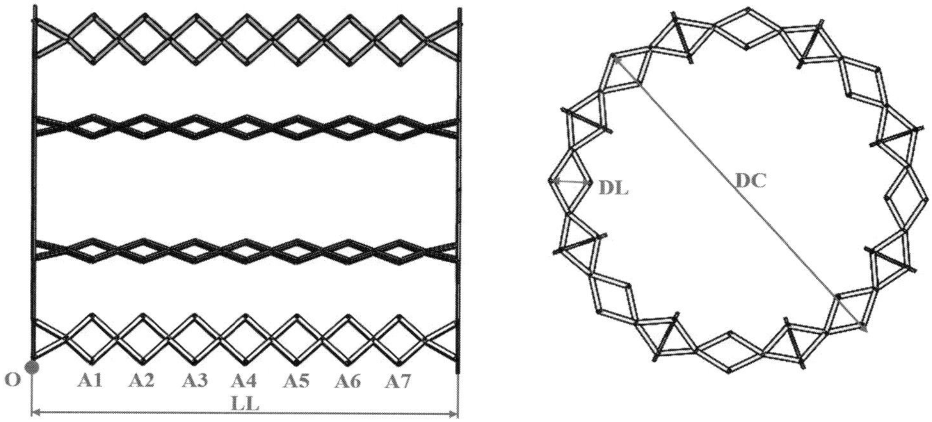

Figure 13.16 Simulation diagram of single segment. (© 2016 IEEE. Reprinted, with permission, from Luo et al. [26].)

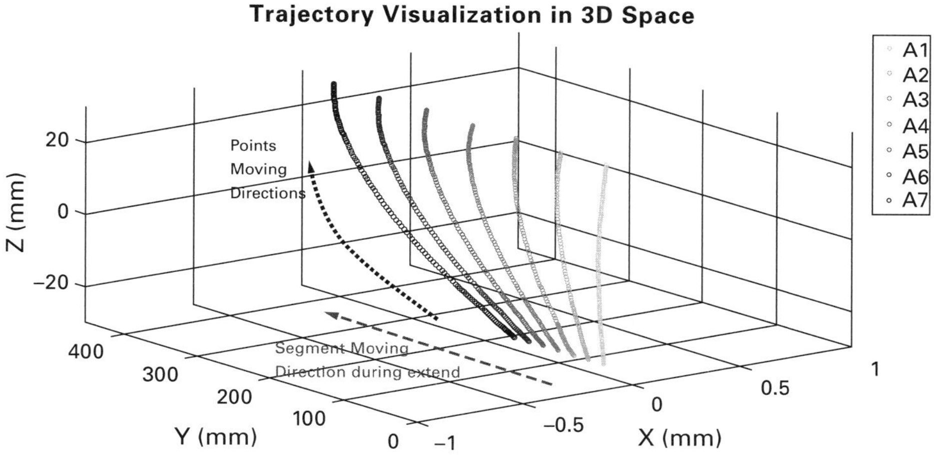

Figure 13.17 Motion trajectory of every leg point (bristle) with the side O be the fixed point. (© 2016 IEEE. Reprinted, with permission, from Luo et al. [26].)

bristles for the efficient locomotion.) from A1 to A7, Figure 13.17 shows the motion trajectory of every leg point.

Compared with the simulation results, the experimental setup, with plane view and side view of single segment, is shown in Figure 13.18. The servos work cyclically between 30 and 130 degrees, and the measured results including LL, DC, and DL are plotted in Figure 13.19. Figure 13.20 shows motion trajectory of leg points that mimic bristles (A1~A7) without the fixed point.

Figure 13.21 plots the relationship between servo angle and distance of simulated and experimental results. The lowest dot point stands for the recorded length change of the

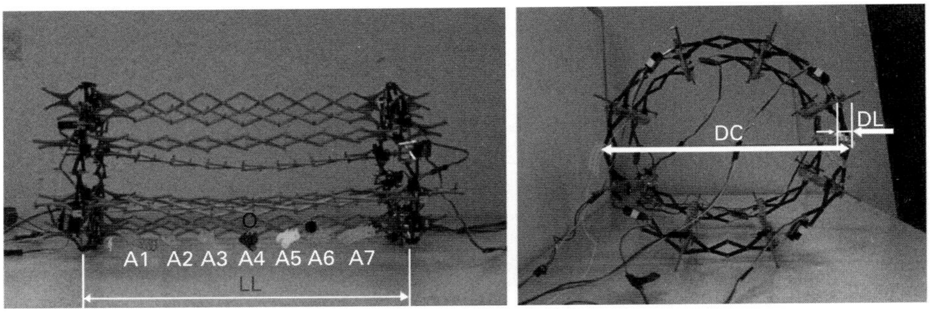

Figure 13.18 Experimental setup of single segment with (a) plane view and (b) side view. (© 2016 IEEE. Reprinted, with permission, from Luo et al. [26].)

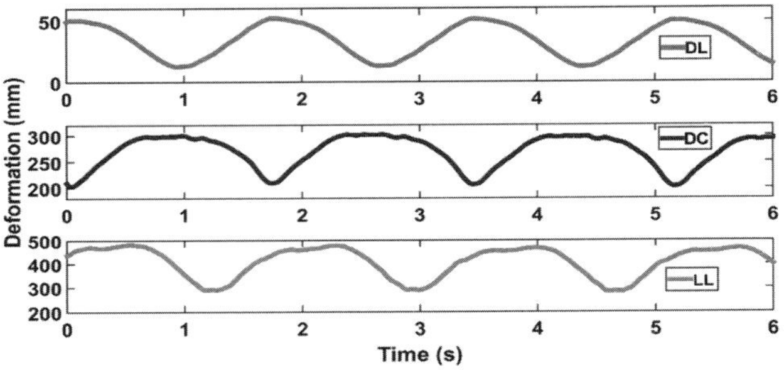

Figure 13.19 Length of the diagonal of one scissor rhombus (DL), diameter of circular-muscle structure (DC) and length of longitudinal muscle structure (LL). (© 2016 IEEE. Reprinted, with permission, from Luo et al. [26].)

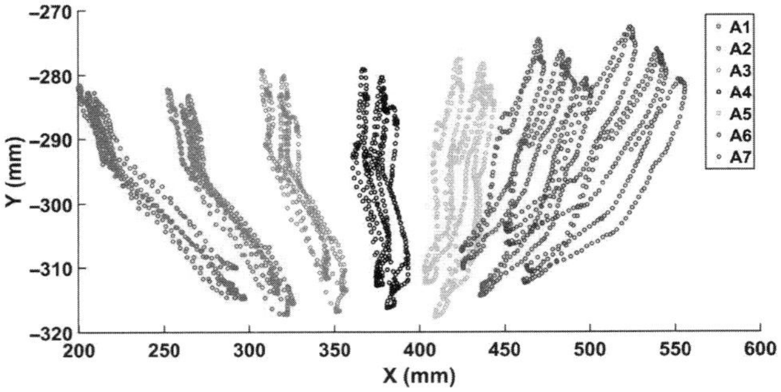

Figure 13.20 Motion trajectory of leg points (A1~A7) without the fixed point. (© 2016 IEEE. Reprinted, with permission, from Luo et al. [26].)

diagonal of one scissor rhombus (DL) in the experiment, the middle dot point is the recorded diameter deformation of circular-muscle structure (DC) and the highest dot point represents the recorded length of longitudinal-muscle structure (LL). The solid lines demonstrate the simulated data of DL, DC, and LL. By comparing both the

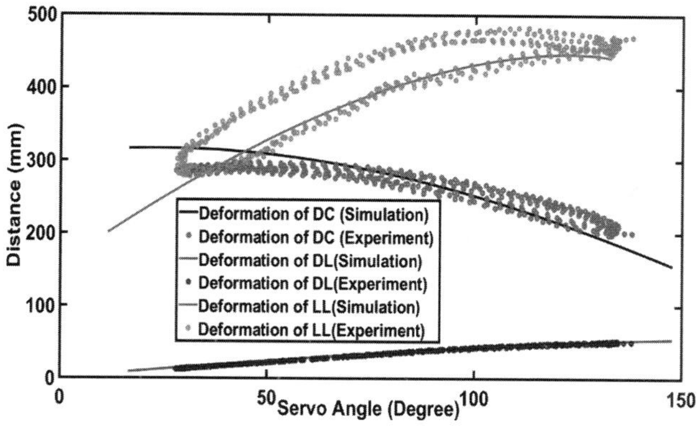

Figure 13.21 Comparison between simulation and experimental results of single segment deformation. (© 2016 IEEE. Reprinted, with permission, from Luo et al. [26].)

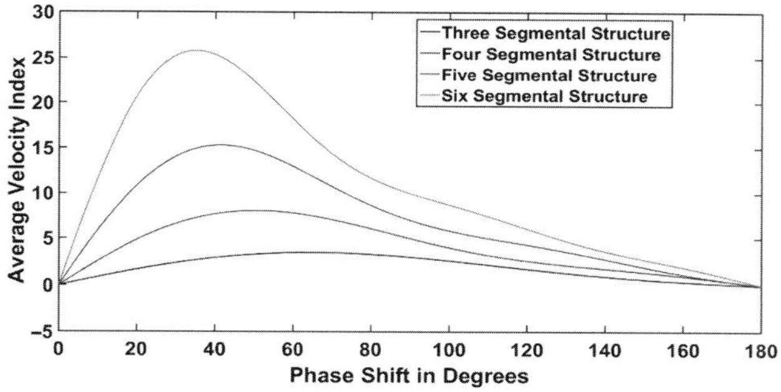

Figure 13.22 Relationship between phase shift and average velocity index. (© 2016 IEEE. Reprinted, with permission, from Luo et al. [26].)

simulated and experimental results, the prototype of single segment displays high repeatability and accuracy, which validates the effectiveness of the designed structure. Note that there is a hysteresis effect shown in the length of longitudinal-muscle structure (LL), which is mainly caused by soft or loosening factors generated by the fabrication material stiffness, inaccurate parts fabrication and assembly gaps of scissors and mechanical components.

Following the segment deformation function investigation, we then explore the peristaltic locomotion capability of the soft robot, which consists of multiple connected segmental structures. For a multisegment robot (i.e., n), the average velocity index v only in terms of the phase-shift angle $\Delta\varphi$ and number of segments can be derived as $v = \sum_{i=1}^{n}\sum_{j=1}^{n-i} j \sin(i\Delta\varphi)$. A similar principle has been mentioned in [31]. As shown in Figure 13.22, the magnitude of the average velocity has been improved significantly with increase in the number of segments, and the optimal velocity is also related with the phase-shift angle.

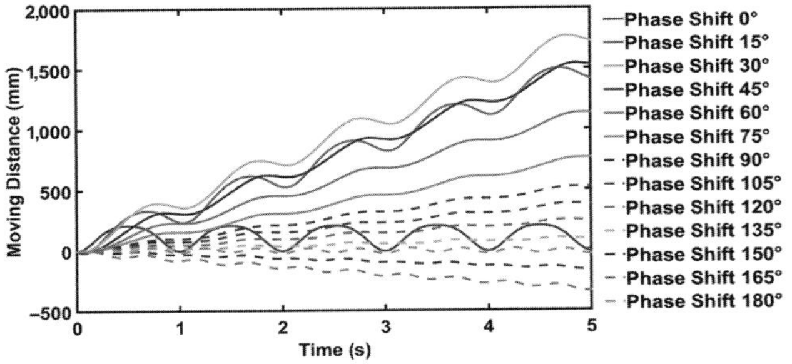

Figure 13.23 Peristaltic locomotion of the six-segment robot on Acrylic-Acrylic dry surface with different phase-shift angles. (© 2016 IEEE. Reprinted, with permission, from Luo et al. [26].)

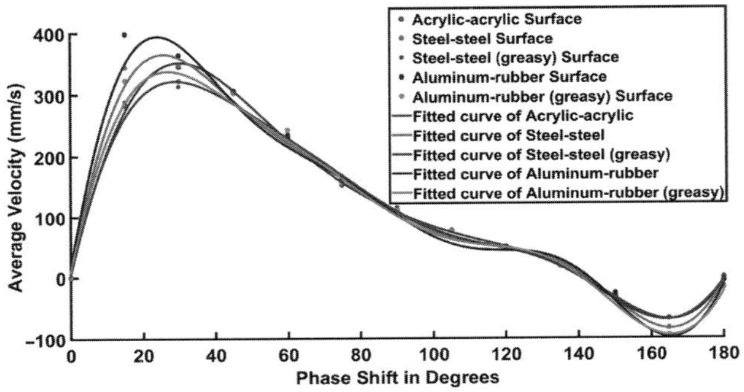

Figure 13.24 Average velocity with different phase-shift angles in five contact surfaces and their fitting curves. (© 2016 IEEE. Reprinted, with permission, from Luo et al. [26].)

Considering the real condition, friction is a necessary and inevitable factor. Therefore, locomotion simulation of six segments is carried out by considering varied friction between the robot and contacted ground surfaces. Given that the earthworm-like robot is designed symmetrically, the quarter model is used to reduce the simulation time (i.e. computing load) and improve the simulation accuracy effectively. The material of robot is defined as the Alloy 1060 and the gravity is 9.80665 m/s^2 vertical to the ground. The actuator signal is 1 Hz sine wave that drives circular-muscle structure with the diameter from 240 to 300 mm. Figure 13.23 plots the moving distance of the simulated earthworm-like soft robot during five cycles under Acrylic-Acrylic dry surface condition. It shows that when the phase-shift angle is 30 degrees, the robot achieves maximum moving distance. It can be seen that the phase-shift angle of each segment plays an important role in moving distance/speed and the backward slippage of the earthworm-like soft robot. This phenomenon is observed in four other contact-surface conditions as well. Figure 13.24 plots the relationship of different phase-shift angles and average velocity under five contact-surface conditions. The data points stand for the average velocity in thirteen different phase-shift angles. Four sine waves are used to fit

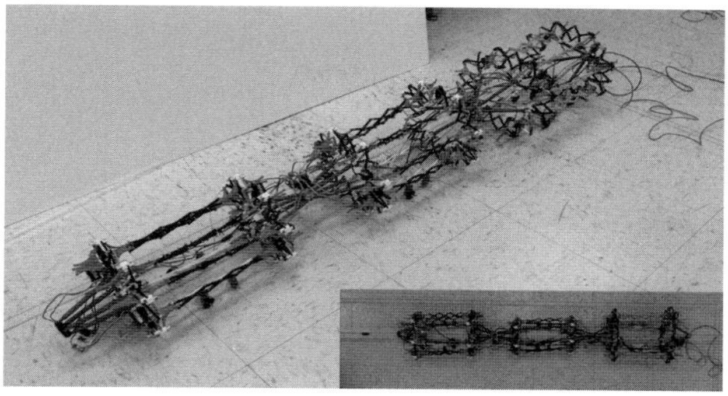

Figure 13.25 Three-segment earthworm-like robot prototype. (© 2016 IEEE. Reprinted, with permission, from Luo et al. [26].)

these simulated data in order to estimate the relationship between phase-shift angle and average velocity. All the fitting rates are around 99%. It can be seen that the contact-surface condition (like friction, stiffness, and restitution coefficient) and phase-shift angle both affect the moving speed, but phase-shift angle plays a more significant role in robot-locomotion efficiency.

As demonstrated in the above results, in each contact surface the optimal phase-shift angle of the six-segment earthworm-like robot lies around 30 degrees, which is similar to the peristaltic motion of the natural earthworm, i.e. no matter what the weight and segmental number of the earthworm are, the optimal phase-shift angle can be achieved when each segment approximately contracts and relaxes once for one wavelength period [32], which can be expressed as $\Delta\varphi_{v\max} = 180/n$. Here $\Delta\varphi_{v\max}$ is the phase-shift angle where maximum velocity occurs, and n is the number of segments.

Figure 13.25 demonstrates the prototyped earthworm-like robot with three segments, which was assembled to study and verify the peristaltic locomotion performance by testing different phase-shift angles, waveforms, and frequency. Since the actuation system shows steady features under the triangle wave compared with sin and square waves in open-loop control based on our tests, we chose the triangle wave of 1.2 Hz as the representative example to study the moving distance under three phase-shift angles, 60 degrees, 65 degrees, and 70 degrees. The testing results are shown in Figure 13.26. The solid lines plot the experimental results and the dashed lines present the simulated results. It can be seen that the three-segment earthworm-like robot prototype moves farthest within a certain time limit/fastest when the phase-shift angle is 60 degrees, which matches our simulation results and locomotion findings. As shown in Figure 13.24, the average velocity reaches 12.2 cm/s, which indicates the RESMM-based earthworm-like soft robot is faster than existing earthworm-like robots.

13.5.5 Concluding Remarks and Perspectives

Earthworms are soft, tube-shaped, segmented worms that move with waves of muscular contractions. This section presents our recently developed compliant modular

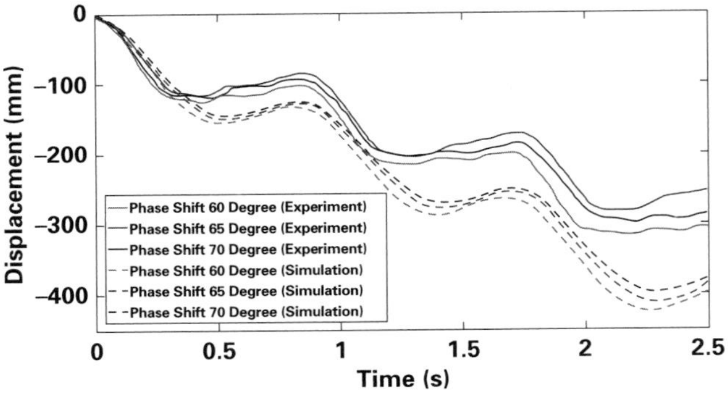

Figure 13.26 Moving displacement in simulation and experiments with 60 degrees, 65 degrees and 70 degrees phase-shift on a lab ground. (© 2016 IEEE. Reprinted, with permission, from Luo et al. [26].)

earthworm-like soft robot with the novel RESMM-based segmental muscle-mimetic design that efficiently mimics the earthworm's segmental muscle contractions. The new class of segmental muscle-mimetic design relies on the curvature of scissor-element mechanisms that can be extended and contracted smoothly through controlled servo motors. Connected numbers of the design units with the transmission mechanism, a multisegment earthworm-like soft robot can be prototyped and the peristaltic locomotion is produced by controlling motors in each segment with different phase-shift angles. In this chapter, we detail the bioinspired concept and design of the segmental muscle-mimetic unit, and demonstrate extensive simulation and experimental results on the design and the prototyped multisegment robot. All the results prove that both the design and the prototyped RESMM-based earthworm-like soft robot have excellent locomotion performance, that is, the average locomotion speed can reach a competitive value of 12.2 cm/s, which, to the best of our knowledge, indicates it is the fastest earthworm-like soft robot in the world.

References

1. Hunter I. W., & Lafontaine, S. (1992). A comparison of muscle with artificial actuators. *IEEE Solid-State Sensor and Actuator Workshop*, 5, 178–185.
2. Jung, D. W. G., Blange, T., & De Graaf, H. (1988). Elastic properties of relaxed, activated, and rigor muscle fibers measured with microsecond resolution. *Biophysics Journal*, 54, 897–908.
3. Kier, W. M. (2012). The diversity of hydrostatic skeletons. *Journal of Experimental Biology*, 215, 1247–1257.
4. Motokawa, T. (1984). Connective tissue catch in echinoderms. *Biological Reviews*, 59, 255–270.
5. Trotter, J. A., Tipper, J., Lyons-Levy, G., et al. (2000). Towards a fibrous composite with dynamically controlled stiffness: Lessons from echinoderms. *Biochemical Society Transactions*, 28, 357–362.

6. Reed, J. L., Hemmelgarn, C. D., Pelley, B. M., & Havens, E. (2005). Adaptive wing structures, smart structures and materials. In E. V. White (Ed.) *Industrial and commercial applications of smart structures technologies*, vol. 5762. *SPIE*, 132–142.
7. McKnight, G., & Henry, C. (2005). Variable stiffness materials for reconfigurable surface applications. In W. D. Armstrong (Ed.), *Smart structures and materials 2005: Active materials: Behavior and mechanics*, vol. 5761. *SPIE*, 119–126.
8. McKnight, G., Doty, R., Keefe, A. Herrera, G., & Henry, C. (2010). Segmented reinforcement variable stiffness materials for reconfigurable surfaces. *Journal of Intelligent Material Systems and Structures*, 21, 1783–1793.
9. Kobori, T., Takahashi, M., Nasu, T., Niwa, N., & Ogasawara, N. (1993). Seismic response controlled structure with active variable stiffness system. *Earthquake Engineering and Structural Dynamics*, 22, 925–941.
10. Takahashi, M., Kobori, T., Nasu, T., Niwa, N., & Kurata, N. (1998). Active response control of buildings for large earthquakes – Seismic response control system with variable structural characteristics. *Smart Materials and Structures*, 7, 522–529.
11. Gandhi, F., & Kang, S. G. (2007). Beams with controllable flexural stiffness. *Smart Materials and Structures*, 16, 1179–1184.
12. Murray, G., Gandhi, F., & Kang, S. G. (2009). Flexural stiffness control of multi-layered beams. *AIAA Journal*, 47, 757–766.
13. Bergamini, A., Christen, R., & Motavalli, M. (2007). Electrostatically tunable bending stiffness in a GFRP–CFRP composite beam. *Smart Materials and Structures*, 16, 575–582.
14. Henke, M., Sorber, J., & Gerlach, G. (2012). Multi-layer beam with variable stiffness based on electroactive polymers. In Y. Bar-Cohen (Ed.), *Electroactive polymer actuators and devices*, vol. 8340. *SPIE* 83401P.
15. Majidi, C. (2003). Soft robotics: A perspective – Current trends and prospects for the future. *Soft Robotics*, 1, 5–11.
16. Shan, W. L., Lu, T., & C. Majidi (2013). Soft-matter composites with electrically tunable elastic rigidity. *Smart Materials and Structures*, 22, 085005.
17. Shan, W. L., Lu, T., Wang, Z.H., & Majidi. C. (2013). Thermal analysis and design of a multi-layered rigidity tunable composite. *International Journal of Heat and Mass Transfer*, 66, 271–278.
18. Shan, W. L., Diller, S., Tutcuoglu, A., & Majidi, C. (2015). Rigidity-tuning conductive elastomer. *Smart Materials and Structures*, 24, 065001.
19. Mohammadi Nasab, A., Sabzehzar, A., Tatari, M., Majidi, C., & Shan, W. L. (2017). A soft gripper with rigidity tunable elastomer strips as ligaments. *Soft Robotics*, 4(4), 411–420.
20. Shan, W. L., & Turner, K. Methods for fast and reversible dry adhesion tuning between composite structures and substrates using dynamically tunable stiffness. Full patent filed in January 2018.
21. Conn, A. T., & Rossiter, J. (2012). Smart radially folding structures. *IEEE/ASME Transactions on Mechatronics*, **17**(5), 968–975.
22. Hoberman, C. (July 24, 1990). Reversibly expandable doubly curved truss structure. US Patent 4,942,700.
23. Escrig, F., Valcarcel, J. P., & Sanchez, J. (1996). Deployable cover on a swimming pool in Seville. *Bulletin of the International Association for Shell and Spatial Structures*, 37(1), 39–70.
24. Dai, J. S., & Wang, D. (2007). Geometric analysis and synthesis of the metamorphic robotic hand. *Journal of Mechanical Design*, 129(11), 1191–1197.

25. Kong, X. (2013). Type synthesis of 3-dof parallel manipulators with both a planar operation mode and a spatial translational operation mode. *Journal of Mechanisms and Robotics*, 5(4), 041015.
26. Luo, Y., Zhao, N., Shen, Y., & Kim, K. J. (2016). Scissor mechanisms enabled compliant modular earthworm-like robot: Segmental muscle-mimetic design, prototyping and locomotion performance validation. *IEEE International Conference on Robotics and Biomimetics*, 2020–2025.
27. Edwards, C. A., & Bohlen, P. J. (1996). *Biology and ecology of earthworms*, vol. 3. Springer Science & Business Media.
28. McLaughlin, M. (1994). *Nature Study*, 47, 12–15.
29. Lawrence G., Mutchmor, John A., & Dolphin Mitchell, Warren D. (1988). *Zoology*. Benjamin-Cummings Publishing Company.
30. Van Leeuwen, J. L., & William, M. K. (1997). Functional design of tentacles in squid: Linking sarcomere ultrastructure to gross morphological dynamics. *Philosophical Transactions of the Royal Society of London B: Biological Sciences*, 352(1353), 551–571.
31. Fang, H., Li, S., Wang, K.W. & Xu, J. (2015). Phase coordination and phase–velocity relationship in metameric robot locomotion. *Bioinspiration & Biomimetics,* 10(6), 066006.
32. Boxerbaum, A. S., Shaw, K.M., Chiel, H.J. & Quinn, R.D. (2012). Continuous wave peristaltic motion in a robot. *The International Journal of Robotics Research*, 31(3), 302–318.

14 Bioinspired Building Envelopes

Steven van Dessel, Mingjiang Tao, and Sergio Granados-Focil

14.1 Bioinspired Building Design

The history of architecture is laced with examples of bioinspiration, ranging from the use of decorative motifs to the implementation of functional and organizational principles found in plant and animal life. Likewise, a unique feature of our planet, one that allows life to flourish, has been named after a building type: The Greenhouse Effect. Exchanges in vocabulary from the fields of biology and construction occur quite frequently, i.e. building skin and cell wall, concrete shell and vault organelle, steel skeleton and body frame, and so on. As relative new sciences, it is no surprise that the fields of biology and earth sciences refer to things of scale and size more commonly understood, i.e. buildings. As they cope with the same environments and abide with the same physics, biology and architecture have developed similar solutions in their efforts to resist gravity's pull or provide comfort and protection. The development of buildings is one of trial and error, a slow, evolutionary process that has to date produced very different building forms. Life, as well, is quite diverse in form and has arrived at this diversity using a limited palette of building materials and sources of energy. A seminal work on the development of form in biology is a book by D'Arcy W. Thompson, *On Growth and Form*, which first appeared in 1917 and has since become a landmark for biologist and bioinspired architects alike [1]. Using mathematical reasoning and physics, Thompson sets out to illustrate nature's approach to derive shape, and illustrates how the forces at play are the same as those at play in the shaping of all matter, including buildings and bridges. An important tenet of his book concerns the importance of scale, how different physical forces work at different length-scales, and how these forces bring about vastly different results. It follows from his work that physics – the knowledge of nature – is essential to the understanding of biology and also forms a foundation for bioinspired design.

As is evident from the borrowing of vocabulary, builders have long looked at biology while using their intuition to accomplish material efficiency in structures. The formal recognition of the catenary as an ideal shape, as found in nature (i.e. egg shells and spider webs), is usually attributed to Robert Hooke who asserted, "*As hangs the flexible line, so but inverted will stand the rigid arch*" [2]. Such form-finding techniques, empirically and intuitively rooted in physics, had been used by builders long before analytical methods became available [3]. For example, Sir Christopher Wren, the

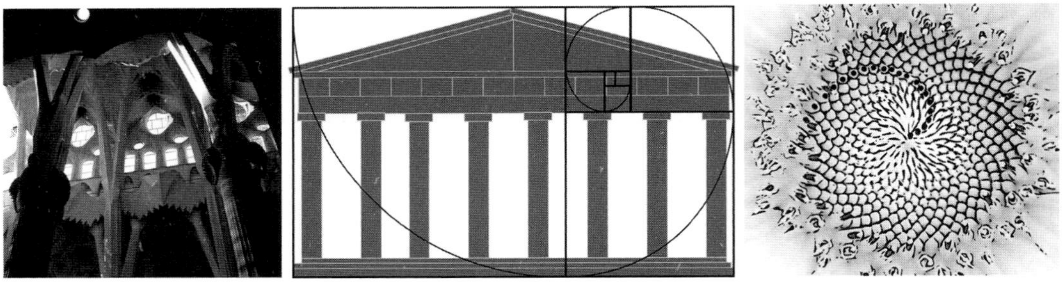

Figure 14.1 Bioinspired inverted catenary vault of the main nave of Sagrada Familia – Barcelona by Antoni Gaudi (left), Golden Ratio, and proportion of classic Greek temple (middle), and golden spiral arrangement of sunflower pellets (right).

architect and scientist, used the principle of the inverted catenary (assisted by Hooke) to design the inner load-bearing structure of London's St. Paul's Cathedral [4]. The Catalonian architect Antonio Gaudi, fascinated by nature, used inverted catenaries to shape his cathedral (see Figure 14.1) while also making reference to the shape of trees in explaining the logic of his work [5]. In a similar fashion, following the work of physicists such as Joseph Plateau [6], Frei Otto used soap bubbles to design minimal surfaces suitable for large building spans [7]. Such shape-finding traditions have continued to influence the work of many architects and engineers to this day, and as such are excellent examples of taking inspiration from biology (and physics) in building design.

The classical sense of beauty, one of ideal proportion and scale, has long been attributed to biology as well. It is here that we find many relationships between architecture, science, mathematics, and biology, namely in such concepts as the Golden Ratio, Vitruvius man, Fibonacci numbers, or the Modulor [8–12]. For example, the Golden Ratio has been used by ancient Greeks and others to proportion buildings or parts thereof. Considered to represent an aesthetic ideal, the Golden Ratio is also observable in biological phenomena, such as the spiral of the nautilus shell or the arrangement of sunflower pellets (Figure 14.1) [8]. Likewise, Vitruvius man, as drawn by Leonardo DaVinci, depicts the ideal human proportions (a person being eight heads tall) as described by Vitruvius, the architect from ancient Rome, who asserted that such human proportions are the principal source of beauty in classical Greek architecture [9,10]. In more modern times, the architect Le Corbusier used his Modular when setting scale and proportion of his buildings and paintings [12]. This logic of beauty, related to the harmony and scale of the human body, mathematical order, and patterns in nature, has long inspired the practice of architecture.

Most of the work on bioinspired architecture has followed in the footsteps of shape-finding or proportion and beauty-seeking traditions. It was not until more recently that we have started to seek inspiration in biology on matters of building energy and ecology. This chapter deals with bioinspired climate-responsive building envelopes. Building envelopes encompass the opaque, translucent, and transparent elements of façade and roof systems [13].

14.2 Bioinspired Building Envelopes

It is well known that buildings consume a considerable amount of energy and that a large fraction of this energy is needed to maintain building thermal comfort [14,15]. The energy expended on thermal comfort is largely due to heat exchange across building enclosures, which accounts for about 50% of building energy use [15–17]. There have been many efforts to reduce building energy use through optimization of building envelopes. The most common approaches include the use of thermal insulation and radiative barriers, increasing air-tightness, use of shading devices, or incorporating heat storage to dampen thermal swings. There are many examples in nature where similar strategies are used by plants and animals. For example, animals are covered in fur to stay warm, make insulated structures to provide comfort, modulate air flow to control thermal comfort, or seek shade to prevent overheating. As we occupy similar environments and are subject to the same physics, it is perhaps no surprise to see that our methods have evolved in similar fashion. Examples abound in vernacular architecture where we see many innovative solutions made possible through use of locally sourced materials. There are, however, also noteworthy differences. Fur in mammals, for example, changes its density with season and body-part to adjust its thermal performance as needed [18] and also varies among species and species size [19]. Fur also fulfills many other functions, such as shedding water and snow or providing camouflage [20,21]. The short-tailed weasel, for example, changes its brown, fluffy summer coat for a more densely packed white coat in the winter, which retains heat better, sheds snow and water more easily, and offers better camouflage in the snowy outdoors [22]. Thermal insulation in most buildings tends to have fixed thermal properties, while serving the more singular purpose of retaining heat. The inability to adapt thermal properties according to outdoor conditions often leads to reduced energy efficiency in buildings, for example, by slowing admittance of solar-heat gain on a sunny winter day. The ability to adapt to changes in the environment is perhaps one of the most inspiring attributes of biology. The concept of climate-responsive building-envelope systems has been the topic of architectural thinking for some time. Over thirty years ago, Mike Davies described a radical new proposal for the way we use glass in buildings. In an article entitled "A Wall for all Seasons," he advocated developing a polyvalent wall that protects against cold and heat while also regulating its functions automatically. Inspired by projects such as Steve Baer's thermal wall, Davies compares his wall to "*a chameleon's skin adapting itself to provide the best possible interior conditions*" [23]. Many different concepts of such bioinspired climate-responsive building envelopes have emerged recently [24].

14.3 Plant-Inspired Building-Envelope Systems

Plant leaves are cellular systems with high cell-packing-density on the sun-facing side, and reduced cell-density and increased air-permeability deeper in the leaf structure [25]. It is well known that leaves use stomata to regulate temperature through evaporative

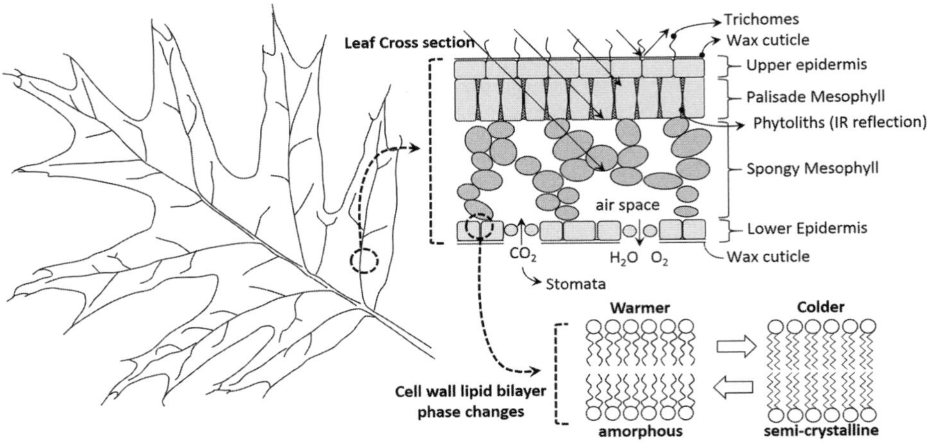

Figure 14.2 Schematic representation of plant leaf (left) and cross section showing the various leaf layers (top right), and phase changes occurring in a cell wall upon thermal cycling (bottom right).

cooling [26]. Plants also deploy many other temperature-controlling features, such as wax cuticle to affect attenuation of light [27,28], silica bodies (phytoliths) to funnel radiation [20,29,31], multi-scale surface topologies and trichomes to control solar reflection and absorption [32], hydrophobicity gradients to affect evaporative heat loss and wetting [33,34], and papillae and wax tubules to affect moisture and temperature-dependent absorption of light [35]. In addition, the cuticle layer and lipid bilayers of cell walls (and cell contents) undergo reversible phase changes, which allow them to absorb and release latent heat. Collectively, these features enable thermal control to leaf systems that can inspire building enclosure design (Figure 14.2).

In a current effort we are investigating solid–solid phase-change materials (SS-PCM) for use in plant-inspired climate-responsive building envelopes. SS-PCMs are emerging thermal-energy storage materials that avoid some of the problems associated with solid–liquid PCMs. For example, SS-PCMs can retain their solid properties when undergoing phase changes, enabling their more direct integration as functional layers of façade systems. An overview of different types of SS-PCMs and their respective properties can be found in a recent review article [36]. A promising example of an SS-PCM entails the grafting of phase-changing pendant motifs onto a polymeric backbone, where pendant groups have the ability to undergo crystalline–amorphous phase changes (Figure 14.3). Such materials undergo reversible phase changes that allow them to absorb and release latent heat to control temperature, similar to lipid bilayers in plant cells.

We investigated the use of such SS-PCMs in two plant-inspired solar-façade systems designed to respond passively to external stimuli. Such response is accomplished by the inclusion of an SS-PCM that changes from transparent to opaque to control the admittance of solar energy based on outside temperature. We investigated how the interaction between the SS-PCM and other components of the system, such as reflectors, can lead to building enclosure systems that can passively regulate building temperature.

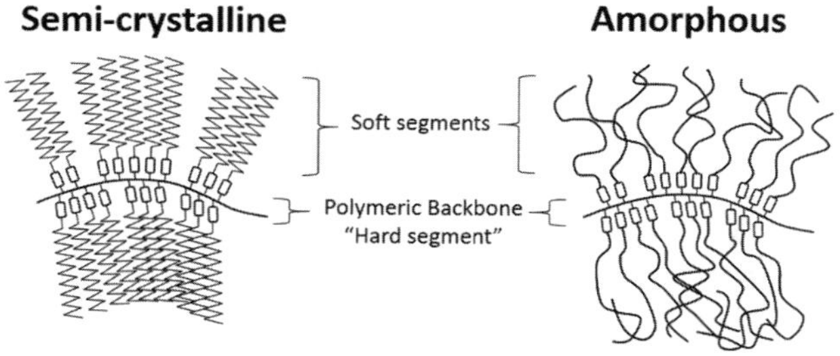

Figure 14.3 Schematic representation of polymeric solid–solid phase-change material with side-chain "soft segments" grafted onto a polymeric backbone serving as the "hard segments," The alkyl side chains transition between a semi-crystalline (opaque) and amorphous (transparent) phase upon thermal cycling. (Reprinted from Guldentops et al. [37], copyright (2018), with permission from Elsevier.)

In the first approach, similar to wax cuticle layers in plant leaves, a thin layer of SS-PCM is placed on top of a highly reflective backing layer [37]. When outdoor temperatures are low, the SS-PCM is opaque and absorbs solar energy directly into the outer surface. When outdoor temperatures exceed the phase-transition temperature, the SS-PCM turns transparent and a highly reflective backing layer becomes exposed, thus reducing solar-heat gain. This approach renders a thermochromic-like device based upon SS-PCMs, which minimizes summer solar-heat gain while enhancing winter solar-heat gain (Figure 14.4).

In the second approach we studied the use of SS-PCM foam systems [37,38] that resemble some of the cellular cross-sectional features observed in plant leaf. In this approach, a cellular SS-PCM changes transparency as a function of outdoor temperature and absorption of solar energy, as shown in Figure 14.5. When applied to the exterior of a building the SS-PCM foam slowly becomes transparent as it absorbs solar energy throughout the day. The ability to collect solar energy directly through radiation processes circumvents the low-conductivity issues associated with many PCM applications. The low thermal conductivity of the SS-PCM cellular structure, coupled with the SS-PCMs latent-heat storage capacity, allows the system to retain heat longer to keep a building at desired temperature. The goal of this strategy is to attain, at some depth into the enclosure's surface, a quasi-constant temperature that lies close to the comfort temperature. The approach thus aims to minimize undesirable heat exchange between the building's interior and exterior to render quasi-zero-energy performance at the building enclosure scale (Figure 14.5).

The feasibility of the proposed plant-inspired SS-PCM façade systems was evaluated through an exploratory numerical study, with focus on the influence of system parameters on the heat transfer during the coldest and warmest weeks of the year for a south-facing wall in a moderate climate. Finite Element Models were developed to simulate the heat transfer and quantify the influence of material properties and other features on thermal performance. The exterior reflection coefficient and the attenuation coefficient for both crystalline and amorphous phases of the SS-PCM were important variables

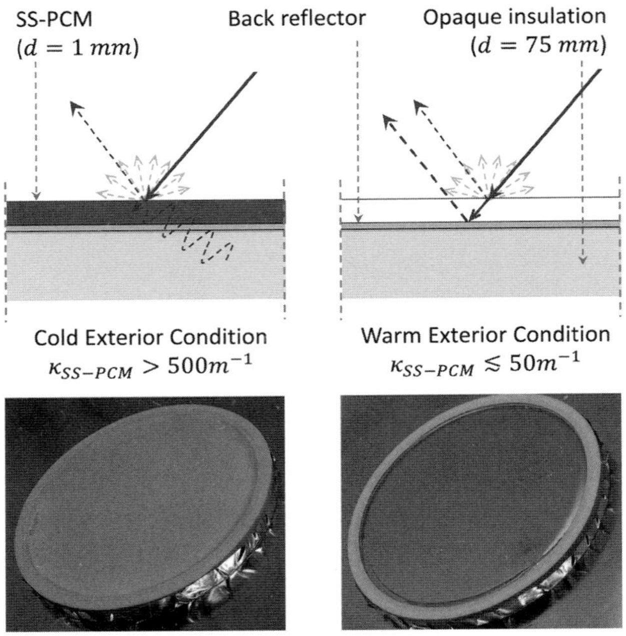

Figure 14.4 Top: Conceptual working principles of a thin-layered system with variable reflectance, based on an SS-PCM with variable transparency. Left: SS-PCM thin film is cold, crystalline, and opaque, causing considerable absorption of solar irradiance. Right: SS-PCM thin layer transitions to a transparent, amorphous state, allowing for considerable reflection through the action of the reflector behind the SS-PCM. Bottom: SS-PCM opacity change from opaque (left) to transparent (right) with exposure of the reflective film. (Reprinted from Guldentops et al. [37], copyright (2018), with permission from Elsevier.)

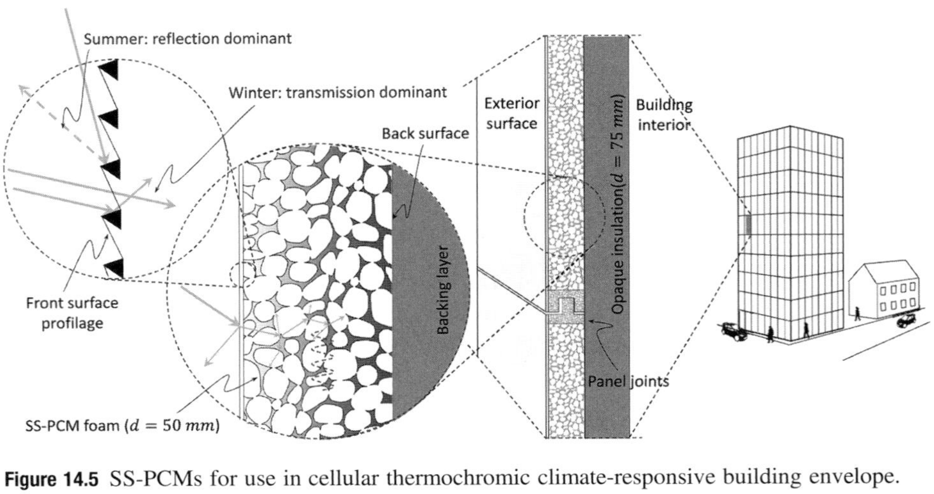

Figure 14.5 SS-PCMs for use in cellular thermochromic climate-responsive building envelope. (Reprinted from Guldentops et al. [37], copyright (2018), with permission from Elsevier.)

within the study. We assumed an SS-PCM with transition temperature of about 25°C and a latent-heat capacity of 180 kJ/kg. The performance of the two systems was compared to a reference wall consisting of similar lightweight composition and equivalent overall thermal resistance. Undesirable heat exchange across the wall (expressed in kWh/m^2 for a reference time-period) is used as the performance criteria (a value of zero represents a perfect thermal wall with zero energy demand). More details about our study can be found in our previous publication [37].

14.4 Results and Discussion

System 1 – Thin SS-PCM Layer + Reflector: Figure 14.6 shows the undesirable heat exchange for summer (left) and winter (right) for different exterior surface reflection values (0.2–0.9) and different SS-PCM attenuation coefficients ($K_{amorphous}$ from 50 to 300 m^{-1} and $K_{crystalline}$ from 500 to 5,000 m^{-1}) with a constant back-reflector surface reflection (0.95). Results indicate that the system can perform as a thermochromic-like device, as was anticipated. The temperature of the back-reflector, which becomes exposed under warm summer conditions when the SS-PCM becomes transparent, decreases from about 44–32°C as it reflects solar radiation and thus reduces heat gain. During winter the system remains opaque and absorbs solar energy as the temperature of the SS-PCM always remains below the phase-transition temperature. As was anticipated, more solar radiation is reflected by the back-reflector when a lower surface reflection is selected, and vice versa. The ability to tailor reflection of solar radiation at two different surfaces, enabled by the temperature-dependent SS-PCM attenuation coefficient, offers design flexibility to optimize the system for both winter and summer

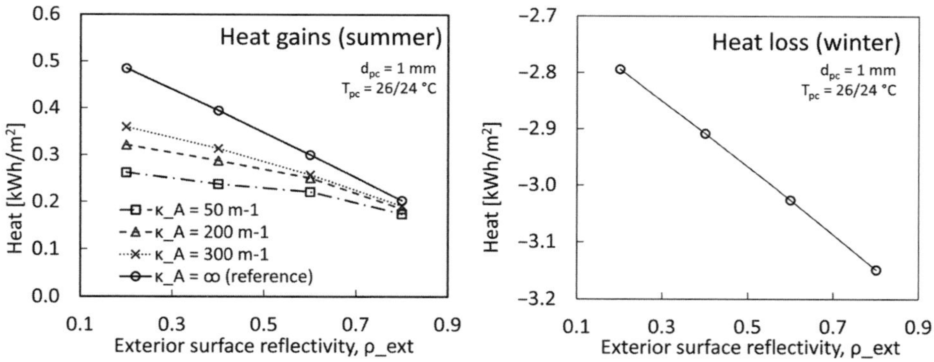

Figure 14.6 Thin SS-PCM Layer + Reflector: Total absolute heat gains and losses for the on-average warmest and coldest week (i.e. 168 h time period) of the year respectively (undesirable heat exchange). Low κA values produce a building enclosure which is less sensitive to the exterior surface reflection during summer conditions. This will allow for the selection of an exterior surface that does not reflect much solar radiation, minimizing heat losses in winter and maximizing heat gains during summer. (Reprinted from Guldentops et al. [37], copyright (2018), with permission from Elsevier.)

conditions. A low surface reflectivity is desirable, as it enables solar-heat gain during winter, while a high surface reflectivity of the back-reflector enables off-loading excessive heat during summer conditions. A low attenuation coefficient of the SS-PCM is also preferred for summer scenarios; the transition temperature of the SS-PCM can furthermore be adapted to fit specific climate types. More study details are available in [37].

The proposed system has features that resemble those found on the surface of plant leaves. The SS-PCM, for example, is similar to the cuticle in plant leaves, which consists of an insoluble matrix impregnated and covered with soluble waxes and also undergoes phase changes. The cuticle layer in plant leaves also attenuates selected wavelengths of light, similar to the SS-PCM materials we developed. Much remains to be learned about the composite thermal behavior of leaf systems and how this can further inform building envelop design. Of special note is how a combination of features found in plant leaves is synergistically tuned to cope with very different thermal environments. The thermochromic system presented here requires further research that can benefit from future discoveries of plant-inspired thermal processes. For example, our SS-PCMs have an opaque, milky appearance below their phase-transition temperature, which is not yet optimal to absorb solar radiation in winter when heat gain is desirable. Additional research can focus on the inclusion of different pigments, as is the case in plant cuticle, to affect attenuation in different seasons and solar intensities. Various surface topologies can also be investigated to tune the scattering of light and surface reflectance based on incidence angle. Such features, which can act similarly to tracheid on plant leaf, can be used to improve winter solar-heat gain while minimizing summer heat gain.

System 2 – SS-PCM foam system: Figure 14.7 depicts the undesirable heat exchange for SS-PCM systems with different foam void-ratios and attenuation coefficients for a summer and winter design week. *Winter*: Results indicate that winter heat losses decrease (relative to the reference case) when the SS-PCM foam void ratio

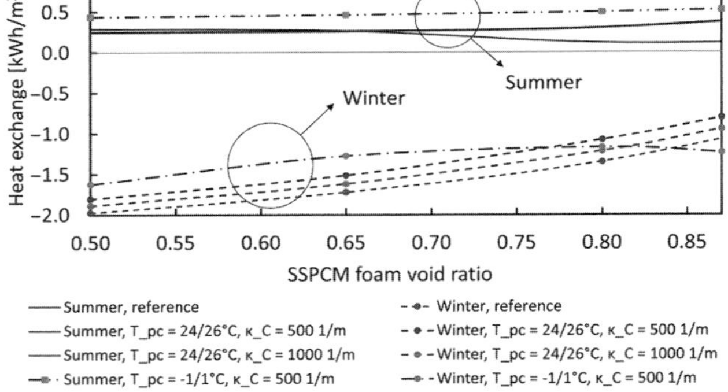

Figure 14.7 Heat loss and gains during the on-average coldest and warmest week respectively as a function of the SS-PCM foam's void ratio. (Reprinted from Guldentops et al. [37], copyright (2018), with permission from Elsevier.)

increases, and that the system with lower attenuation coefficient outperforms the system with higher attenuation coefficient (SS-PCM phase-transition temperature around 25). A lower attenuation coefficient allows solar irradiance to be absorbed deeper into the SS-PCM foam, which is more desirable in cold seasons. On the other hand, the system with high attenuation coefficient absorbs heat earlier and is able to store all available thermal energy regardless of foam void ratio. To assess the influence of the SS-PCM transition temperature we have also modeled a system with a lower phase-transition temperature (around 0°C) and lower crystalline attenuation coefficient (500 m^{-1}). Results indicate that such a system outperforms the reference case (winter) and other SS-PCMs with higher transition temperature for foam void-ratios less than 0.75. *Summer*: During summer conditions, the system with cellular SS-PCM performs marginally better compared to the reference case for a void ratio below 0.65 while undesirable heat gains in summer increase at higher foam void-ratios. This is likely due to the lower availability of latent-heat storage capacity as the amount of SS-PCM diminishes with increasing void ratio. Higher latent-heat storage capacity allows for a reduction in temperature peaks within the building enclosure. Results in Figure 14.7 show that the attenuation coefficient has little influence on the systems performance during summer (transition temperature around 25°C). In order to prevent overheating it is also important to limit the attenuation effect of the foam, thus allowing for more absorption near the exterior surface when the SS-PCM is amorphous. More study details are available in [37].

This second bioinspired system also shows features that resemble those found in plant leaves. For example, plant leaves also have a cellular composition and undergo phase transitions similar to those occurring in the proposed system. Much is still to be uncovered about how the cellular topology of plant leaves affects heat transfer, for example, how the cellular density in plant leaves, and the accompanying changes in latent-heat storage capacity and attenuation, affects the overall heat transfer for different orientations and climates. Such insights will be instrumental in broadening our ability to design bioinspired building enclosures.

14.5 Outlook

Biology offers a great source of inspiration and can offer time-tested solutions for climate-responsive building envelopes. Plant leaves, as illustrated in this chapter, already employ various useful thermal regulation processes to retain or dispose heat as needed. A more deliberate search for new thermoregulation mechanisms in biology will likely reveal many solutions that can be leveraged toward building envelopes. A high level of system adaptability seems a key in making such systems work, especially when outdoor conditions fluctuate rapidly and to large extents. New approaches are needed to accomplish material-enabled passive thermal-control systems. These new capabilities may include an ability to quickly change radiation, convection and conductive heat transfer properties of materials or surfaces, better ability to store thermal energy compactly and indefinitely for precise and on demand use, and better ability to control the flow of heat using smart material-enabled thermal-rectification strategies. Increased ability to react to semi-instantaneous weather changes will likely be instrumental in attaining thin and lightweight climate-responsive building-envelope solutions. Such approaches may lead to cost-effective building strategies that can also be more easily implemented in existing buildings. The search for solutions can focus on the following particular thermal properties: *Thermal Conductivity*: The thermal conductivity of materials lies somewhere between 10^{-2} and 10^{+3} W/mK, represented by extreme materials such as silica-aerogel and graphene at the far ends of the spectrum. While such materials are useful, adaptive materials that can quickly alter their thermal conductivity may be more useful in future smart building-envelope applications. Such capability facilitates collecting heat immediately when available outdoors, while minimizing thermal transfer when outdoor conditions are too warm or cold. Such behavior effectively causes building envelopes to behave as thermal diodes, thus possessing an ability to change the direction of heat flow as needed and in response to environmental changes. Inspiration can be found, for example, in polarity-induced heat transfer processes found in biology. *Emissivity*: The attainable emissivity of materials ranges from somewhere between 0.04 and 0.96, represented by materials such as polished metal and unpolished soda-lime glass, respectively. The ability to offload excessive heat toward the environment offers exciting new possibilities for thermal regulation in buildings. Much can be learned from biology where various photonic systems are already extensively used to affect color or radiation heat transfer [39,40]; which has already led to building-envelope applications [41]. *Heat transfer processes*: Convection and evaporative heat transfer are greatly affected by surface properties, such as surface topology and polarity. For example, hydrophobicity gradients are used by plants and animals to affect wetting and energy transfer through evaporation of water. Such cooling strategies can inspire new building-envelope systems whereby changes in surface polarity are used to alter wetting and evaporative cooling properties of surfaces, for example, using phase-change triggered hydrophobicity changes. *Heat storage*: Many approaches are currently being explored or are available for thermal-energy storage, including sensible and latent-heat storage systems [36]. New solid–solid phase-change materials are becoming available that can, in addition to their latent-heat storage functions, also possess other functional properties such as transparency change or mechanical strength. Beyond the discovery and development of new bioinspired mechanisms and materials, there is also a need to engage in a more strategic multidisciplinary research on thermoregulation processes in biological systems. This effort requires multi-physics modeling and experimentation to discover the inner workings and principles of a wide diversity of biological systems. The level of detail and intriguing hierarchical micro-structure found in biological systems are also of a scale that does not always lend itself for conventional thermal measurement techniques and present technical challenges to current manufacturing techniques. New non-intrusive experimental approaches are needed to capture the inner thermal dynamics of micrometer- and nanometer-scaled thermal systems and phenomena found in biology. In addition, we should capitalize on emerging additive manufacturing approaches to mimic the complex and intriguing structures found in biology.

References

1. Thompson, D. A. W. (1992). *On Growth and Form (Canto)*, ed. J. T. Bonner. Cambridge University Press.
2. Block, P., DeJong, M., & Ochsendorf, J. (2006). As hangs the flexible line: Equilibrium of masonry arches. *Nexus Network Journal*, 8(2), 13–24.
3. Heyman, J. (1998). *Structural analysis: A historical approach*. Cambridge University Press.
4. Pacey, A. J. (1969). Chapter 3 – Christopher Wren, 1632–1723 a2. In D. Hutchings (Ed.), *Late seventeenth century scientists*. Pergamon; pp. 72–106.
5. Md Rian, I., & Sassone, M. (2014). Tree-inspired dendriforms and fractal-like branching structures in architecture: A brief historical overview. *Frontiers of Architectural Research*, 3(3), 298–323.
6. Plateau, J. A. F. (1873). *Statique expérimentale et théorique des liquides soumis aux seules forces moléculaires*. Gauthier-Villars.
7. Otto, F., & Songel, J. M. (2010). *A conversation with Frei Otto*. Princeton Architectural Press.
8. Ripley, R. L., & Bhushan, B. (2016). Bioarchitecture: Bioinspired art and architecture – A perspective. *Philosophical Transactions of the Royal Society A: Mathematical, Physical and Engineering Sciences*, 374(2073), 1–36.
9. Murtinho, V. (2015). Leonardo's Vitruvian man drawing: A new interpretation looking at Leonardo's geometric constructions. *Nexus Network Journal*, 17(2), 507–524.
10. Pollio, V., Morgan, M. H., & Warren, H. L. (1914). *Vitruvius: The ten books on architecture*. Harvard University Press.
11. Posamentier, A. S., & Lehmann, I. (2007). *The fabulous Fibonacci numbers*. Prometheus Books.
12. Jean-Louis, C. (2014). Le Corbusier's modulor and the debate on proportion in France. *Architectural Histories*, 2(1), Art. 23.
13. Lovell, J. (2010). *Building envelopes: An integrated approach*. Princeton Architectural Press.
14. Cao, X., Dai, X., & Liu, J. (2016). Building energy-consumption status worldwide and the state-of-the-art technologies for zero-energy buildings during the past decade. *Energy and Buildings*, 128, 198–213.
15. Berry, J. (2012). *Annual energy review 2011*. U.S. Energy Information Administration, Office of Energy Statistics.
16. Hens, H. S. L. (2012). *Building physics – Heat, air and moisture: Fundamentals and engineering methods with examples and exercises*. Wiley.
17. Evans, M., Roshchanka, V., & Graham, P. (2017). An international survey of building energy codes and their implementation. *Journal of Cleaner Production*, 158, 382–389.
18. Boyles, J. G., & Bakken, G. S. (2007). Seasonal changes and wind dependence of thermal conductance in dorsal fur from two small mammal species (*Peromyscus leucopus* and *Microtus pennsylvanicus*). *Journal of Thermal Biology*, 32(7), 383–387.
19. Meyer, W., Hülmann, G., & Seger, H. (2003). REM-Atlas zur Haarkutikularstruktur mitteleuropäischer Säugetiere/SEM-Atlas on the hair cuticle structure of Central European mammals. *Mammalian Biology – Zeitschrift für Säugetierkunde*, 68(5), 328.
20. Frisch, J., Øritsland, N. A., & Krog, J. (1974). Insulation of furs in water. *Comparative Biochemistry and Physiology Part A: Physiology*, 47(2), 403–410.
21. Derocher, A. E., & Lynch, W. (2012). *Polar bears: A complete guide to their biology and behavior*. Johns Hopkins University Press.
22. Heptner, V. G. (1989). *Mammals of the Soviet Union, Volume 2 Part 2 Carnivora (Hyenas and Cats)*. Brill.

23. Mike, D. (1981). A wall for all seasons. *RIBA Journal*, 88(2), 55–57.
24. Loonen, R. C. G. M., Trčka, M., Cóstola, D., & Hensen, J. L. M. (2013). Climate adaptive building shells: State-of-the-art and future challenges. *Renewable and Sustainable Energy Reviews*, 25, 483–493.
25. Gibson, L. J. (2012). The hierarchical structure and mechanics of plant materials. *Journal of the Royal Society Interface*, 9(76), 2749–2766.
26. Matthews, J. A. (2014). Stomata. In J. A. Matthews (Ed.), *Encyclopedia of environmental change*. Sage Publishing.
27. Cutler, D. F., Botha, T., & Stevenson, D. W. (2008). *Plant anatomy: An applied approach*. Wiley-Blackwell.
28. Domínguez, E., Heredia-Guerrero, J. A., & Heredia, A. (2011). The biophysical design of plant cuticles: An overview. *New Phytologist*, 189(4), 938–949.
29. Neethirajan, S., Gordon, R., & Wang, L. (2009). Potential of silica bodies (phytoliths) for nanotechnology. *Trends in Biotechnology*, 27(8), 461–467.
30. Laue, M., Hause, G., Dietrich, D., & Wielage, B. (2006). Ultrastructure and microanalysis of silica bodies in *Dactylis glomerata* L. *Microchimica Acta*, 156(1–2), 103–107.
31. Whang, S. S., Kim, K., & Hess, W. M. (1998). Variation of silica bodies in leaf epidermal long cells within and among seventeen species of Oryza (Poaceae). *American Journal of Botany*. 85(4), 461–466.
32. Defraeye, T., Verboven, P., Tri Ho, Q., & Nicolaï, B. (2013). Convective heat and mass exchange predictions at leaf surfaces: Applications, methods and perspectives. *Computers and Electronics in Agriculture*, 96, 180–201.
33. Ensikat, H. J., Ditsche-Kuru1, P., Neinhuis, C., & Barthlott, W. (2011). Superhydrophobicity in perfection: The outstanding properties of the lotus leaf. *Beilstein Journal of Nanotechnology*, 2, 152–161.
34. Yang, H., Zhu, H., Hendrix, M. M., et al. (2013). Temperature-triggered collection and release of water from fogs by a sponge-like cotton fabric. *Advanced Materials*, 25(8), 1149–1154.
35. Schulte, A. J., Koch, K., Spaeth, M., & Barthlott, W.(2009). Biomimetic replicas: Transfer of complex architectures with different optical properties from plant surfaces onto technical materials. *Acta Biomaterialia*, 5(6), 1848–1854.
36. Fallahi, A., Guldentops, G., Tao, M., Granados-Focil, S., & van Dessel, S. (2017). Review on solid-solid phase change materials for thermal energy storage: Molecular structure and thermal properties (I). *Applied Thermal Engineering*, 127, 1427–1441.
37. Guldentops, G., Ardito, G., Tao, M., Granados-Focil, S., & van Dessel, S. (2018). A numerical study of adaptive building enclosure systems using solid–solid phase change materials with variable transparency. *Energy and Buildings*, 167, 240–252.
38. Guldentops, G., & van Dessel, S. (2017). A numerical and experimental study of a cellular passive solar façade system for building thermal control. *Solar Energy*, 149, 102–113.
39. Willot Q, S. P., Vigneron J.-P., & Aron S. (2016). Total internal reflection accounts for the bright color of the Saharan silver ant. *PLOS ONE*, 11(4), 1–14.
40. Shi, N. N., Tsai, C.-C., & Camino, F. (2015). Keeping cool: Enhanced optical reflection and radiative heat dissipation in Saharan silver ants. *Science*, 349(6245), 298–301.
41. Fan, S. (2017). Thermal photonics and energy applications. *Joule*, 1(2), 264–273.

Index